BIOLOGICAL REACTIVE INTERMEDIATES—II

Chemical Mechanisms and Biological Effects

PART B

ADVANCES IN EXPERIMENTAL MEDICINE AND BIOLOGY

BIOLOGICAL REACTIVE INTERMEDIATES—II

Chemical Mechanisms and Biological Effects

PART B

Edited by

Robert Snyder
Rutgers University
Piscataway, New Jersey

David J. Jollow
Medical University of South Carolina
Charleston, South Carolina

Dennis V. Parke
University of Surrey
Guildford, Surrey, England

C. Gordon Gibson
University of Surrey
Guildford, Surrey, England

James J. Kocsis
Thomas Jefferson University
Philadelphia, Pennsylvania

Charlotte M. Witmer
Rutgers University
Piscataway, New Jersey

Associate Editors

Beatrice N. Engelsberg
Rutgers University
Piscataway, New Jersey

George F. Kalf
Thomas Jefferson University
Philadelphia, Pennsylvania

Stephen L. Longacre
Medical University of South Carolina
Charleston, South Carolina

PLENUM PRESS • NEW YORK AND LONDON

Library of Congress Cataloging in Publication Data

Main entry under title:

Biological reactive intermediates II.

(Advances in experimental medicine and biology; v. 136)
"Proceedings of the Second International Sympsoium on Biological Reactive
Intermediates, held at the University of Surrey, July 14–17, 1980" – T.p. verso.
Includes bibliographical references and indexes.
1. Toxicology – Congresses. 2. Biotransformation (Metabolism) – Congresses.
3. Xenobiotic metabolism – Congresses. I. Snyder, Robert, 1935–
II. International Symposium on Biological Reactive Intermediates (2nd: 1980:
University of Surrey) III. Title: Reactive intermediates II. IV. Series.
RA1191.B55   1981   615'.704   81-12003
ISBN 0-306-40802-3 (2-vol set)   AACR2

Second half of the proceedings of the Second International Symposium on
Biological Reactive Intermediates, held at the University of Surrey, July 14-17, 1980

© 1982 Plenum Press, New York
A Division of Plenum Publishing Corporation
233 Spring Street, New York, N.Y. 10013

Printed in the United States of America

METABOLISM AND TOXICITY OF AFLATOXINS

D.P.H. Hsieh and J.J. Wong

Department of Environmental Toxicology
University of California, Davis
Davis, California 95616

INTRODUCTION

Aflatoxins are a family of compounds related to four toxic metabolites produced by the mold Aspergillus flavus. Since their discovery in England some twenty years ago, aflatoxins have become a world wide food safety concern due to the potent toxicity, carcinogenicity, and widespread occurrence of certain members in food commodities (Stoloff, 1977). To date, over a dozen aflatoxins have been described in the literature and the structures of the more common members are shown in Fig. 1. Of these, aflatoxin B_1, B_2, G_1 and G_2 are products of the fungi A. flavus and A. parasiticus which frequently invade and propagate in corn, peanuts, cottonseed, and other commodities; M_1, M_2, GM_1 and GM_2 are metabolites of the first four produced by animals exposed to the mycotoxins.

TOXICITY OF AFLATOXINS

Studies on aflatoxins have largely been focused on aflatoxin B_1 (AFB_1) because it is the most abundant and the most toxic member. It is highly hepatotoxic, hepatocarcinogenic, and teratogenic to test animals and highly mutagenic to microbial assay systems. Strong evidence exists to indicate that AFB_1 requires metabolic activation for its mutagenic and carcinogenic effects. For example, it requires mixed function oxygenase (MFO) activation to elicit its potent mutagenicity or genetic toxicity to the bacterium Salmonella typhimurium used in the Ames mutagen assay (Ames et al., 1973). Also, its covalent binding to nucleic acids occurs only in vivo or in vitro in the presence of a MFO activation system (Lin et al., 1977). The marked difference in the

847

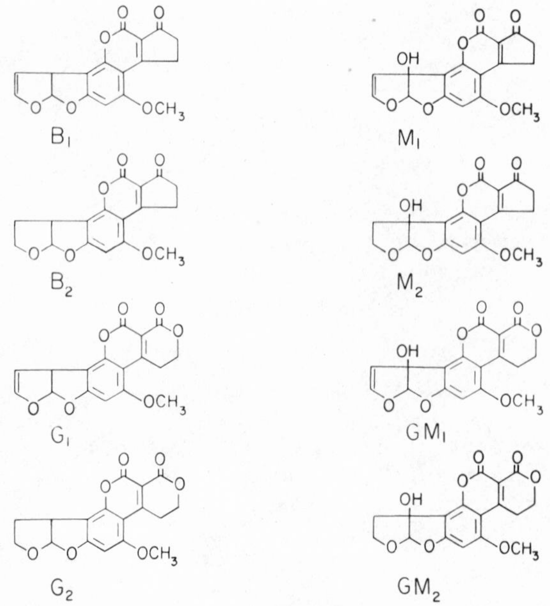

Fig. 1.   The structures of some aflatoxins

susceptibility to AFB$_1$ toxicity and carcinogenicity among animal
species reflects, at least in part, its varying metabolic fates
in the different species (Patterson, 1973).   In terms of carcinogenic effects, the rat and rainbow trout are extremely sensitive
while the mouse and rhesus monkey are relatively resistant.

IN VITRO METABOLISM OF AFLATOXIN B$_1$

Based on the apparent association between metabolism and
toxicity of AFB$_1$, attempts have been made to quantitate the
relative susceptibility to AFB$_1$ as a function of respective in
vitro metabolic profiles of AFB$_1$ (Patterson, 1973; Edwards et
al., 1975; Hsieh et al., 1977).   Certain in vitro metabolic parameters of AFB$_1$ for various animal species were determined and
compared with animal susceptibilities to aflatoxin toxicity.   The
same metabolic parameters were then determined for human tissue
preparations and used to estimate the unknown human susceptibility
based upon the correlations established for the test animals.

In vitro metabolism studies using liver enzyme preparations
have revealed that there are at least five types of metabolic
reactions characteristic of AFB$_1$: reduction, hydroxylation,
hydration, O-demethylation, and epoxidation leading to the
respective formation of aflatoxicol (AFL), aflatoxin M$_1$ (AFM$_1$)
and aflatoxin Q$_1$ (AFQ$_1$), aflatoxin B$_{2a}$ (AFB$_{2a}$), aflatoxin P$_1$
(AFP$_1$), and the aflatoxin B$_1$-2,3-oxide (AFB$_1$-2,3-oxide) (Fig. 2)

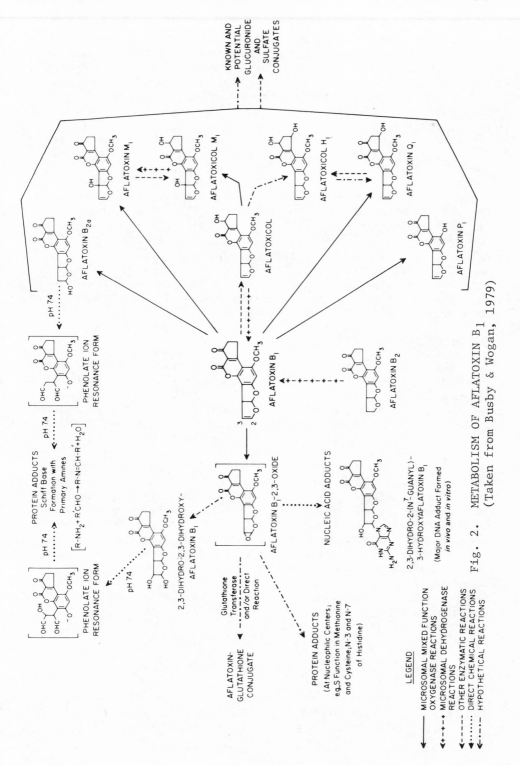

Fig. 2.   METABOLISM OF AFLATOXIN B₁
(Taken from Busby & Wogan, 1979)

(Campbell and Hayes, 1976). Except for the epoxide, all the metabolites have been isolated and characterized. They all contain hydroxyl-groups which are amenable to conjugation with glucuronic acid and sulfate, resulting in detoxification. Evaluation of these metabolites in the Ames mutagen assay showed that they are less potent than the parent compound, AFB_1, and that they all require metabolic activation for activity (Wong and Hsieh, 1976; Wei et al., 1978). However, for reasons yet to be elucidated, the reductive activity of an animal to convert AFB_1 to AFL, by and large, seems to correlate with its sensitivity to the acute toxicity of AFB_1 as indicated by the LD_{50} values (Edwards et al., 1975; Hsieh et al., 1977). Also the sensitivity of an animal to the carcinogenic effect of AFB_1 seems to correlate with its in vitro activity to convert AFB_1 to AFL as relative to the activity to form AFQ_1 or to the activity to form water soluble metabolites (Hsieh et al., 1977). Thus, the percentage of AFL formation from AFB_1 in vitro and the ratio of the percentage of AFL to that of AFQ_1 or to that of water-soluble metabolites appear to be metabolic parameters useful in predicting the animal's relative sensitivity to the acute and the carcinogenic effects of AFB_1, respectively.

THE AFLATOXIN B_1-2,3-OXIDE

The AFB_1-2,3-oxide has been proposed by various investigators as the active form or ultimate carcinogen of AFB_1 (Swenson et al., 1974; 1975). The metabolic fates of the epoxide are summarized in Fig. 2. Once formed, the epoxide can either exert electrophilic attack on the target nucleophiles such as in nucleic acids and proteins to cause biochemical lesions; or be conjugated by reduced glutathione (GSH) (Degen and Neuman, 1978) and be detoxified (Wong and Hsieh, 1980; Lotlikar et al., 1980); or be hydrolyzed to form the 2,3-dihydrodiol of AFB_1; or be reduced to the hemiacetal form of AFB_1, AFB_{2a}. The dihydrodiol and the hemiacetal have been proposed as another class of possible active forms and will be discussed in the following section. The major adduct formed between AFB_1 and DNA or RNA has been identified as the 2,3-dihydro-2-(N^7-guanyl)-3-hydroxy-AFB_1 (Essigman et al., 1977; Lin et al., 1977). The AFB_1-GSH conjugate and the dihydrodiol have also been isolated and characterized (Degen and Neuman, 1978; Swenson et al., 1974; Lin et al., 1977; Neal and Colley, 1979). There is circumstantial evidence to indicate that the covalent binding of the epoxide of AFB_1 with nucleophiles in DNA, RNA, and proteins is the main mode of toxic action of AFB_1.

THE DIHYDRODIOL AND HEMIACETAL OF AFLATOXIN B_1

The dihydrodiol and AFB_{2a} have also been proposed as two of the active forms of AFB_1 (Patterson and Roberts, 1970; Gurtoo and Campbell, 1974; Ashoor and Chu, 1975; Neal and Colley, 1979).

The key reactions of these two metabolites are the formation of
the hypothetical dialdehyde phenolate intermediate (Fig. 2),
which would presumably react with the primary amines in proteins
to form Schiff base as a biochemical lesion. This mechanism has
received support from experiments which showed reactions between
the hemiacetal and amino acids or proteins (Gurtoo and Campbell,
1974; Gurtoo and Dahms, 1974; Neal and Colley, 1979) and from the
isolation and characterization of the adducts (Ashoor and Chu,
1975). However, the isolation of free dihydrodiol and AFB_{2a} from
protein-containing reaction mixtures (Gurtoo and Campbell, 1974;
Neal and Colley, 1979) and from the urine of certain exposed
animals (Dann et al., 1972) seem to indicate that these metabolites
are not very reactive and hence their role in AFB_1 toxicity needs
further substantiation.

AFLATOXIN$_1$-2,3-OXIDE FORMATION AND ANIMAL SUSCEPTIBILITY

Even though the AFB_1-2,3-oxide has not been isolated, its
formation and toxic action can be measured indirectly by the Ames
mutagen assay. Using mutagenesis in the test bacterium Salmonella
typhimurium TA98 as the model biochemical lesion, differing AFB_1
mutagenic responses elicited by post-mitochondrial liver fractions
prepared from animals of different susceptibilities to AFB_1
carcinogenicity indicate that net epoxide formation appears to be
another metabolic parameter associated with animal susceptibility
(Hsieh et al., 1977). Data presented in Table 1 demonstrate that
decreasing sensitivity to the carcinogenic effect of AFB_1 in
vivo is reflected by a decreased ability of the appropriate
hepatic preparation to generate a mutagenic species in vitro.
Structure-activity evaluations using mutagenicity and DNA-binding
activity as model biological lesions (Wong and Hsieh, 1976; Coles
et al., 1977; Gurtoo et al., 1978) support the involvement of the
2,3-vinyl-ether double bond in toxicity, with the epoxide as the
active species, and do not support the involvement of AFB_{2a} in
DNA-based lesions. Based on the comparative Ames test on AFB_1 as
shown in Table 1, the relatively low metabolic activity of the S-
9 system prepared from human tissues would indicate that humans
may be less sensitive than the mouse and monkey.

QUESTIONS ABOUT MIGRATION OF AFB-2,3-OXIDE TO TARGET SITES

Despite the isolation of aflatoxin-macromolecule adducts,
mutagenicity to microbial systems, and DNA-binding data, there
are still unresolved questions concerning the epoxide and its
role in the toxic action of AFB_1. One question concerns how the
extremely reactive epoxide survives the migration from its proposed
site of synthesis within the endoplasmic reticulum, through the
cytosol, and then across the nuclear membrane to the target site
on DNA. Recently, it has been reported (Vaught et al., 1977;
Guengerich, 1979) that rat liver nuclei were capable of metabolizing

TABLE 1. Aflatoxin B$_1$ Induced In Vitro Salmonella Mutagenicity and In Vivo Animal Carcinogenicity

SPECIES (#animals)	MUTAGENIC ACTIVITY rev./μg AFB$_1$/mg prot.	AFB$_1$ EXPOSURE level	time (yrs)	HEPATOMA INCIDENCE %	REFERENCE
DUCK (3)	8493 (6508–10489)	35ppb	1.2	72	Carnaghan (1965)
RAT (9)	2005 (1262–2643)	15ppb	1.3–1.5	100	Wogan and Newberne (1967)
		100ppb	0.8–1.2	86	Newberne and Rogers (1972)
		100ppb	2.0	48	
MONKEY (5)	814 (119–2056)	360ppb	3.0	0	Cuthbertson et al. (1967)
		1800ppb	3.0	0	
		99–842mg	4–6.0	7	Adamson et al. (1976)
MOUSE (9)	491 (420–605)	1000ppb	1.3	15	Newberne (1965)
		1000ppb	1.7	0	Wogan (1973)
		100μg/day	1.0	0	Louria et al. (1974)
HUMAN (5)	286 (100–755)	?	?	?	

AFB_1 to both less toxic hydroxylation metabolites and reactive specie(s) capable of covalent DNA interaction. Electron microscopy has revealed that in the Ames mutagen assay, microsomes found in the hepatic postmitochondrial or S-9 fraction, required for AFB_1 bioactivation, adhere to the bacterial cells (Glatt and Oesch, 1977). This close proximity of the activation and target site offers an explanation for the survival of the epoxide in vitro during its migration to the target site and may simulate activation at the nuclear membrane and subsequent DNA binding in vivo.

We have recently performed the Ames assay using tri-layer agar plates instead of the standard bi-layer plates to demonstrate the inhibitory effect of GSH on AFB_1 mutagenesis (Wong and Hsieh, 1980). In the standard bi-layer technique, the S-9 system, tester bacteria, AFB_1, and GSH are mixed in the top agar. Since the microsomes adhere to the bacterial surface, the epoxide formed there would directly react with DNA, escaping from detoxification by GSH and the cytosolic GSH-transferase enzyme. Previous investigators (Garner and Wright, 1973; Schoenhard et al., 1976), based on the apparent inability of GSH to prevent nucleic acid interaction with the active form of AFB_1, have questioned the GSH involvement in AFB_1 detoxification. In our tri-layer technique, the liver S-9 system and the bacterial cells were distributed into separate layers to grossly simulate the intracellular architecture of the activation, detoxification, and target sites. The results of the tri-layer Ames assay on AFB_1 (Table 2) indicate that, compared to the standard Ames assay, the tri-layer technique has considerably lower sensitivity, possibly reflecting the increased distance between the activation and target sites. However, AFB_1 is still definitely mutagenic, indicating that some molecules of the activated form of AFB_1 can indeed survive the migration to the target site. When GSH was added to the top layer of the tri-layer plates, the mutagenicity of AFB_1 was reduced (Table 3), indicating that GSH is involved in AFB_1 detoxification. This further supports the role of the epoxide in AFB_1 mutagenicity.

QUESTIONS REGARDING THE EFFECTS OF PHENOBARBITAL PRETREATMENT IN VIVO AND IN VITRO

Another unresolved question is related to an apparent discrepancy in the effects of phenobarbital (PB) pretreatment on AFB_1 toxicity in vivo and epoxide formation in vitro. The drug PB, which enhances hepatic transformation activities, exerts a protective effect against AFB_1-induced toxicity, carcinogenicity, and DNA-binding in vivo (Garner, 1975; Mgbodile et al., 1975; McLean and Marshall, 1971; Swenson et al., 1977). In contrast, the hepatic enzyme preparations of so pretreated animals are more effective in producing the DNA-bound species of AFB_1 (Gurtoo and Bejba, 1974; Gurtoo and Dave, 1975; Neal and Godoy, 1976; Vaught

TABLE 2. Effects of Ames Assay Modifications of AFB_1-Mutagenesis

Assay Modification	Liver S-10 a) (5.6 mg/plate)	Revertants b)
Bi-layer (standard)	I	1709
Bi-layer (standard)	N	921
Tri-layer	I	313
Tri-layer	N	237
Quad-layer	I	137
Quad-layer	N	109

a) I = Induced; N = Noninduced.

b) Revertants/μg AFB_1/mg S-10 protein are means ± standard deviation obtained from 4 plates.

TABLE 3. Influence of Glutathione of AFB_1- Mutagenesis: Trilayer Modification

No. of Replicate Experiment	Glutathione (mM/plate)	Liver S-10 (5.6 mg/plate) a)	Revertants
16	0	I	399 ± 85
1	5	–	314
7	10	–	169 ± 34
2	25	–	108 ± 11
9	0	N	241 ± 18
2	10	N	158 ± 25

a) I = Induced; N = Noninduced.

b) Revertant/μg AFB_1/mg protein are means ± standard deviation of calculated values derived from the number of replicate experiments.

et al., 1977), DNA-linked bacteriacidal activity (Garner et al., 1971; Garner and Wright, 1973), mutagenicity (Ames et al., 1973, 1975), and inhibition of RNA polymerase (Neal and Godoy, 1976).

The discrepancy between the protective effect of PB in vivo and the enhanced activation of AFB1 in vitro could also be explained by the destruction of the cellular architecture of the activation and detoxification enzymes, during the preparation of these enzyme systems. PB is known to induce not only the MFO which activates AFB_1, but also the epoxide hydrolase (Bresnick et al., 1977), GSH-transferase (Hales and Neims, 1977; Van Canfort et al., 1979) and UDP-glucuronyl-transferase (Bock, 1977) which are probably all involved in the decomposition and detoxification of the AFB_1-epoxide. Under in vivo conditions after PB treatment, presumably all of these induced enzymes function in concert to more effectively protect the cell from AFB_1 toxicity. Whereas under in vitro conditions, the detoxification enzymes such as UDP-glucuronyl-transferase may be buried in the lipid interior of the endoplasmic reticulum during the formation of the microsomal particles (Bock, 1977). Not discounting the influences of extrahepatic metabolism and in vivo toxicokinetics, the artificial arrangement of cellular metabolic structures in vitro may account for disparity between the in vitro and in vivo fates of AFB_1.

METABOLISM AND TOXICITY IN PRIMARY CELL CULTURES

In order to maintain the intracellular architecture and the simplicity of an in vitro system, we have also used primary hepatocyte cultures of the rat and mouse to demonstrate the relationship between metabolism and aflatoxin toxicity. ^{14}C-labelled AFB_1 was incubated in the cultured hepatocytes from adult male rats and mice, under conditions where the cytochrome P-450 was maintained at the in vivo levels (Decad et al., 1977). The total covalent binding was used to indicate lesion or toxicity. After a 10 hour incubation, the total AFB_1 metabolized in the rat culture was 57% and in the mouse culture, 90%, indicating that the mouse was more active in metabolism. The percentage of the dose covalently bound to cellular macromolecules was approximately 16% in the rat culture, compared to 0.41% in the mouse culture, indicating that the mouse has more effective detoxification activities. When the hepatocyte cultures were treated with 10^{-3}M of cyclohexene oxide (Table 4), an inhibitor of epoxide hydrolase the binding was increased by 22% in the rat culture and by 61% in the mouse culture, indicating that hydrolysis of the epoxide is a detoxification pathway and that the mouse has considerably greater capacity of this pathway than the rat (Decad et al., 1979). When the cultures were treated with 6 x 10^{-4}M of diethyl maleate (Table 5), which is known to reduce the cellular level of GSH, the binding of AFB_1 to macromolecules was increased by 12% in the rat culture and by a dramatic 800% in the mouse culture,

TABLE 4

EFFECT OF CYCLOHEXENE OXIDE ON BINDING OF
[^{14}C]AFLATOXIN B$_1$ IN CULTURED MOUSE AND RAT
HEPATOCYTES

	Mouse	Rat
dpm bound	1,656 ($\pm$206)	36,869 ($\pm$426)
Percentage dose bound	0.41	16.3
Cultures treated with 10^{-3} M cyclohexene oxide		
dpm bound	2,669 ($\pm$174)	44,873 ($\pm$888)
percentage dose bound	0.66	19.9
Percentage increase in bound dpm	61.2	21.7

(Decad et al., 1979)

TABLE 5

EFFECT OF DIETHYL MALEATE ON BINDING OF [^{14}C]-
AFLATOXIN IN CULTURED MOUSE AND RAT HEPA-
TOCYTES

	Mouse	Rat
Control cultures		
dpm bound	639 ($\pm$153)	36,869 ($\pm$426)
percentage dose bound	0.34	16.3
Cultures treated with 6.2 $\times$ 10^{-4} M diethyl maleate		
dpm bound	5,784 ($\pm$269)	41,117 ($\pm$1,391)
percentage dose bound	3.1	18.2
Percentage increase in bound dpm	805	11.5

(Decad et al., 1979)

indicating that GSH is indeed playing a major role in the detoxification of AFB_1 in the mouse (Decad et al., 1979).

CONCLUSION

The carcinogenic potency of AFB_1 in an animal appears to be determined by competing pathways of activation and detoxification in the cells, involving the MFO in the endoplasmic reticulum and nuclear membrane and the reductase, epoxide hydrolase, GSH-transferase, and other enzymes in the cytoplasm. These enzymes appear to be localized intracellularly in a specific manner so that no single in vitro metabolism system is able to simulate the in vivo conditions to produce a simple set of metabolic parameters for the prediction of aflatoxin toxicity. However, by using a battery of in vitro assays to evaluate the production and degradation of the reactive aflatoxin intermediates, it may be possible to estimate the AFB_1 toxicity in any animal species, including humans.

ACKNOWLEDGEMENT

    Our research of this subject matter was supported in part by
U.S. Public Health Service Grants ES00612 and ES00814, NIEHS
Training Grant in Environmental Toxicology ES07059-01, National
Cancer Institute Grant 1-ROI-CA27246-01, and by Western Regional
Research Project W-122.

REFERENCES

Adamson, R.H., Correa, P., Sieber, S.M., McIntire, K.R., and
    Dalgard, D.W., 1976, Carcinogenicity of aflatoxin B_1 in
    rhesus monkeys: Two additional cases of primary liver
    cancer. J. Natn. Cancer. Inst. 50:549.

Ames, B.N., McCann, J., and Yamasaki, E., 1975, Methods for
    detecting carcinogens and mutagens with the Salmonella/
    mammalian-microsome mutagenicity test. Mutation Res.
    31:347.

Ames, B.N., Durston, W.E., Yamasaki, E., and Lee, F.D., 1973,
    Carcinogens are mutagens: A simple test system combining
    liver homogenates for activation and bacterial for detection.
    Proc. Natl. Acad. Sci. USA 70:2281.

Ashoor, S.H., and Chu, F.S., 1975, Interaction of aflatoxin B_{2a}
    with amino acids and proteins. Biochem. Pharmacol. 24:1799.

Bock, K.W., 1977, Dual role of glucuronyl-and sulfo-transferases
    converting xenobiotics into reactive or biologically inactive
    and easily excretable compounds. Arch. Toxicol. 39:77.

Bresnick, E., Mukhtar, H., Stoming, T.A., Dansette, P.M., and
    Jerina, D.M., 1977, Effect of phenobarbital and 3-methyl-
    cholanthrene administration on epoxide hydrase levels in
    microsomes. Biochem. Pharmacol. 26:891.

Busby, W.R., Jr., and Wogan, G.N., 1979, Mycotoxins and alimentary
    mycotoxicosis. In: "Foodborne Infections and Intoxications,"
    H. Riemann and F.L. Bryan, eds. Academic Press, New York,
    pp. 519-610.

Campbell, T.C., and Hayes, J.R., 1976, The role of aflatoxin
    metabolism in its toxic lesion. Toxicol. and Appl. Pharmacol.
    35:199.

Carnaghan, R.B., 1965, Hepatic Tumors in ducks fed a low level of
    toxic ground nut meal. Nature (Lond.) 208:308.

Coles, B.F., Smith, J.R.L., and Garner, R.C., 1977, Mutagenicity of 3a,8a-dihydrofuro [2,3-6] benzofuran, a model of aflatoxin B_1 for Salmonella typhimurium TA 100. Biochem. Biophys. Res. Comm. 76:888.

Cuthbertson, W.F., Larsen, A.C., and Piatt, D.A.H., 1967, Effect of groundnut meal containing aflatoxins on cynomolgus monkeys. Brit. J. Nutr. 21:893.

Dahms, R., and Gurtoo, H.L., 1976, Metabolism of aflatoxin B_1 to aflatoxin Q_1, M_1 and P_1 by mouse and rat. Res. Comm. in Chem. Path. and Pharmacol. 15:11.

Dann, R.E., Mitscher, L.A., and Couri, D., 1972, The in vivo metabolism of ^{14}C-labeled aflatoxin B_1, B_2, G_1 in rats. Res. Commun. Chem. Pathol. Pharmacol. 3:667.

Decad, G.M., Hsieh, D.P.H., and Byard, J.L., 1977, Maintenance of cytochrome P450 and metabolism of aflatoxin B_1 in primary hepatocyte cultures. Biochem. Biophys. Res. Commun. 78:279.

Decad, G.M., Dougherty, K.K., Hsieh, D.P.H., and Byard, J.L., 1979, Metabolism of aflatoxin B_1 in cultured mouse hepato-cytes: comparison with rat and effects of cyclohexene oxide and diethyl maleate. Toxicol. and Appl. Pharmacol. 50:429.

Degen, G.N. and Neumann, H, 1978, The major metabolite of afla-toxin B_1 in the rat is a glutathione conjugate. Chem. Biol. Interactions   22:239-255.

Edwards, G.S., Rintel, T.D., and Parker, C.M., 1975, Aflatoxicol as a possible predictor for species sensitivity to aflatoxin B_1. Proc. Am. Assoc. Cancer Res. 16:133.

Essigman, J.M., Croy, R. G., Nadzan, A.M., Busby, W.F., Reinhold, V.N., Büchi, G., and Wogan, G.N., 1977, Structural identification of major DNA adduct formed by aflatoxin B_1 in vitro. Proc. Natl. Acad. Sci. U.S. 74:1870.

Garner, R.C., 1975, Reduction in binding of ^{14}C-aflatoxin B_1 to rat liver macromolecules by phenobarbitone pre-treatment. Biochem. Pharmacol. 24:1553.

Garner, R.C., Miller, E.C., Miller, J.A., Garner, J.V., and Hanson, R.S., 1971, The formation of a factor lethal for S. typhimurium TA 1530 and 1531 on incubation of aflatoxin B_1 with rat liver microsomes. Biochem. Biophys. Res. Comm. 45:774.

Garner, R.C., and Wright, C.M., 1973, Induction of Mutations in DNA-repair deficient bacterial by liver microsomal metabolite of aflatoxin B_1. Brit. J. Cancer. 28:544.

Glatt, H., and Oesch, F., 1977, Inactivation of electrophillic metabolites by glutathione S-transferases and limitation of the systems due to subcellular localization. Arch. Toxicol. 39:87.

Guengerich, F.P., 1979, Similarity of nuclear and microsomal cytochromes P-450 in the in vitro activation of aflatoxin B_1. Biochem. Pharmacol. 28:2883.

Gurtoo, H.L. and Bejba, N., 1974, Hepatic microsomal mixed function oxygenase: enzyme multiplicity for the metabolism of carcinogens to DNA-binding metabolites. Biochem. Biophys. Res. Comm. 61:735.

Gurtoo, H.L., and Campbell, T.C., 1974, Metabolism of aflatoxin B_1 and metabolism-dependent and independent binding of afla-toxin B_1 to rat hepatic microsomes. Mol. Pharmacol. 10:776.

Gurtoo, H.L., and Dahms, R., 1974, On the nature of the binding of aflatoxin B_{2a} to rat hepatic microsomes. Res. Comm. Chem. Path. and Pharmacol. 9:107.

Gurtoo, H.L., Dahms, R.P., and Paigen, B., 1978, Metabolic activa-tion of aflatoxins related to their mutagenicity. Biochem. Biophys. Res. Comm. 81:965.

Gurtoo, H.L., and Dave, C.V., 1975, In vitro metabolic conversion of aflatoxins and benzo(a)pyrene to nucleic acid-binding metabolites. Cancer Res. 35:382.

Hales, B.F., and Neims, A.H., 1977, Induction of rat hepatic glutathione S-transferase by phenobarbital and 3-methylcho-lanthrene. Biochem. Pharmacol. 26:555.

Hsieh, D.P.H., Wong, Z.A., Wong, J.J., Michas, C., and Ruebner, B.H., 1977, Comparative metabolism of aflatoxin in: "Mycotoxins in Human and Animal Health," J.V. Rodricks, C.W. Hesseltine, M.A. Mehlamn, eds. Pathotox Publishers, Inc. Park Forest Smith, IL.

Lin, J.K., Miller, J.A., and Miller, E.C., 1977, 2,3-dihydro-2-(N^7-guanyl)-(guanyl-7)-3-hydroxy-aflatoxin B_1, a major acid hydrolysis product of aflatoxin B_1-DNA or-ribosomal RNA adducts formed in hepatic microsome-mediated reactions and in rat liver in vivo. Cancer Res. 37:4430.

Lotlikar, P.D., Insetta, S.M., Lyons, P.R., and Jhee, E., 1980,
    Inhibition of Microsome-Mediated Binding of Aflatoxin B_1 to
    DNA by Glutathione S-Transferase. Cancer Lett. 9:143.

Louria, D.B., Finkel, G., Smith, J.K., and Buse, M., 1974,
    Aflatoxin induced tumors in mice, Sobonraudia 12:371.

McLean, A., and Marshall, A., 1971, Reduced carcinogenic effects
    of aflatoxin in rats given phenobarbitone. Br. J. Exp.
    Path. 52:322.

Mgbodile, M., Holscher, M., and Neal, R., 1975, A possible pro-
    tective role of reduced glutathione in aflatoxin B_1 toxicity:
    effect of pretreatment of rats with phenobarbital and
    3-methylcholanthrene on aflatoxin toxicity. Toxicol. Appl.
    Pharmacol. 34:128.

Neal, G.E., and Colley, P.J., 1979, The formation of 2,3-dihydro-
    2,3-dihydroxy aflatoxin B_1 by the metabolism of aflatoxin B_1
    in vitro by rat liver microsomes. FEBS Let. 101:382.

Neal, G.E., and Godoy, H.M., 1976, The effect of pre-treatment
    with phenobarbitone on the activation of aflatoxin B_1 by
    rat liver. Chem. Biol. Interactions 14:279.

Patterson, D.S.P., 1973, Metabolism as a factor in determining
    the toxic action of the aflatoxins in different animal
    species. Food Cosmet. Toxicol. 11:287.

Patterson, D.S.P., and Roberts, B.A., 1970, The formation of
    aflatoxin B_{2a} and G_{2a} and their degradation products during
    the in vitro detoxification of aflatoxin by livers of certain
    avian and mammalian species. Fd. Cosmet. Toxicol. 8:527.

Roebuck, B.D., and Wogan, G.N., 1977, Species comparison of
    in vitro metabolism of aflatoxin B_1. Cancer Res. 37:1649.

Schabort, J.C., and Steyn, M., 1972, Aflatoxin B_1 and phenobarbital
    inducible aflatoxin Δ^2-hydration by rat liver microsomes.
    Biochem. Pharmacol. 21:2931.

Schoenhard, G.L., Lee, D.J., Howell, S.F., Pawlowski, N.E.,
    Libbey, L.M., and Sinnhuber, R.O., 1976, Aflatoxin B_1
    Metabolism to aflatoxicol and derivatives lethal to Bacillus
    subtilis GS4 by rainbow (Salmo gairdneri) trout. Cancer Res.
    36:2040.

Stoloff, L., 1977, Aflatoxins-an overview in: "Mycotoxin in Human and Animal Health," J.V. Rodricks, C.W. Hesseltine, M.A. Mehlman, eds. Pathotox. Publisher, Inc. Park Forest South 116.

Swenson, D.H., Miller, J.A., Miller, E.C., 1973, 2,3-dihydro-2,3-dihydroxyaflatoxin B_1: An Acid hydrolysis product of an RNA-aflatoxin B_1 adduct formed by hamster and rat liver microsomes in vitro. Biochem. and Biophys. Res. Comm. 53:1260.

Swenson, D.H., Miller, J.A., and Miller, E.C., 1974, Aflatoxin B_1-2,3,-oxide: Evidence for its formation in rat liver in vivo and by human liver microsomes in vitro. Biochem. and Biophys. Res. Comm. 60:1036.

Swenson, D.H., Miller, J.A., and Miller, E.L., 1975, The reactivity and carcinogenicity of aflatoxin B_1-2,3-dichloride, a model for the putative 2,3-oxide metabolite of aflatoxin B_1. Cancer Res. 35:3811.

Swenson, D.H., Lin, J.K., Miller, E.C., and Miller, J.A., 1977, Aflatoxin B_1-2,3-oxide as a probable intermediate in the covalent binding of aflatoxin B_1 and B_2 to rat liver DNA and ribosomal RNA in vivo. Cancer Res. 37:172.

Vaught, J.B., Klohs, W., and Gurtoo, H.L., 1977, In vitro metabolism of aflatoxin B_1 by rat liver nuclei. Life Sciences 21:1497.

Van Cantfort, J., Manil, L., Gielen, J.E., Glatt, H.R., and Oesch, F., 1979, A new assay for glutathione S-transferase using [^{3}H-]-benzo(a)pyrene 4,5-oxide as substrate. Biochem. Pharmacol. 28:455.

Wei, C.I., Decad, G.M., Wong, Z.A., Byard, J.L., and Hsieh, D.P.H., 1978, Characterization and mutagenicity of water-soluble conjugates of aflatoxin B_1 meeting abstract #126. Paper presented 17th Annual Meeting of the Society of Toxicology, San Francisco, CA.

Wogan, G.N., 1973, Aflatoxin carcinogenesis. In: "Methods in Cancer Research," H. Bush, ed., Academic Press, New York, London.

Wogan, G.N., and Newberne, P.M., 1967, Dose response characteristics of aflatoxin B_1 carcinogenesis in the rat. Cancer Res. 27:2370.

Wong, J.J., and Hsieh, D.P.H., 1976, Mutagenicity of aflatoxins
    related to their metabolism and carcinogenic potential.
    Proc. Natl. Acad. Sci. USA. 73:2241.

Wong, J.J., and Hsieh, D.P.H., 1980, Inhibition of Aflatoxin B_1
    Mutagenesis by Conjugative Metabolism. Paper presented at
    the Nineteenth Annual Meeting of the Society of Toxicology.
    Washington, D.C. March 9-13.

TOXICITY MEDIATED BY REACTIVE METABOLITES OF FURANS

Michael R. Boyd

Molecular Toxicology Section, Clinical Pharmacology
Branch, Division of Cancer Treatment, National
Cancer Institute, Bethesda, Maryland 20205

ABSTRACT

Furan derivatives occur widely in the environment, and
several of these compounds cause necrosis of target cells
within certain organs, including the liver, the kidneys, and
the lungs. The tissue specificity may vary from compound to
compound. For individual compounds, the specificity may be
greatly influenced by the species, sex, and age of the test
animal and by the prior exposure of the animal to inducers of
drug metabolism. Studies in vitro and in vivo indicate that
cytochrome P-450 enzymes in the target tissues mediate the
formation of highly reactive, electrophilic furan metabolites
that bind covalently to tissue macromolecules. Epoxides are
suspected, but not proven, to be the proximate or ultimate
toxic metabolites of furans. One study suggested that epoxide
hydratase might influence the covalent binding of a furan
derivative in vitro, but similar investigations with other
furans have been negative. Glutathione (GSH) can inhibit the
covalent binding of reactive furan metabolites in vitro, presumably by forming less reactive, water-soluble conjugates with
the activated furans. GSH-furan conjugate formation can occur
nonenzymatically, and a study with 4-ipomeanol indicated that
cytosolic enzyme preparations did not enhance the amounts of
conjugates produced. It is likely that GSH provides a major
mechanism for detoxification of some furans in vivo.

INTRODUCTION

Chemicals containing the furan nucleus occur widely in the
environment. As indicated in Table I, they may be found as natural

865

TABLE I

SOME SOURCES OF OCCURRENCE OF FURANS IN THE ENVIRONMENT

Source	Examples
Natural products	ipomeamarone, ngaione, 4-ipomeanol, perilla ketone
Intermediates and/or solvents in chemical and industrial processes	furfural, furan
Constituents of food and medicinal preparations	ipomeamarone, 4-ipomeanol, perilla ketone, furosemide
By-products of photolytic and/or thermal degradation of plant and animal material	3-methylfuran, furan, furfural

TABLE II

SOME EXAMPLES OF STRUCTURES OF TOXIC FURANS SHOWING
DIVERSE NATURE OF SUBSTITUENT GROUPS

Compound	Structure
Ngaione	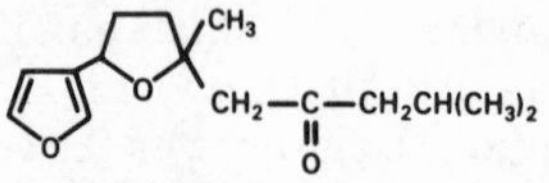
Furosemide	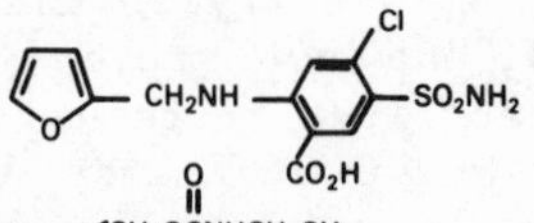
3-(N-Ethylcarbamoylhydroxymethyl) furan	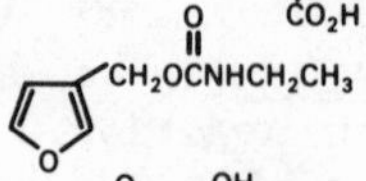
4-Ipomeanol	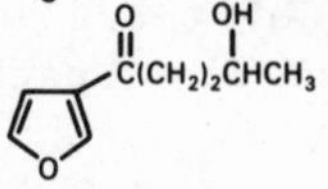
3-Methylfuran	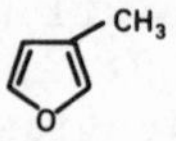
Furan	

products, as intermediates or solvents in chemical and industrial processes, as constituents of food and medicinal preparations, and as by-products of photolytic and/or thermal degradation of plant and animal materials. Some examples of structures of toxic furans are given in Table II, illustrating the diverse nature of substituent groups.

A number of toxicological and pharmacological properties of furans have been described, but only relatively recently has detailed attention been given to the biochemical mechanisms involved in toxicities of furan compounds. In particular, there has been considerable interest in the role of metabolic activation and the involvement of chemically reactive metabolites in furan-induced acute toxicity both to the liver and to extrahepatic tissues. There also is increasing interest in the potential carcinogenicity of simple furan derivatives as well as more structurally complex compounds (e.g., aflatoxin) which contain furan-related structural features. Detailed reviews of the roles of metabolism in the toxicity of simple furans are already available (see Boyd, 1976, 1980a-c; and Boyd et al, 1979a, 1980a). Therefore, the scope of the present paper is restricted to a brief, general overview of some of the major findings. Hopefully, this will provide the groundwork for future discussions. I will include relevant findings from several laboratories that have contributed substantially to the area. In this overview, I plan to include considerations of a) the nature of the tissue lesions caused acutely by the furans, b) the types of toxification and detoxification metabolic pathways that may be involved in their dispostion, c) some characteristics of their ultimate toxic metabolites, and d) some other features of particular interest or importance, including the effects of species, age, sex, and prior treatment with inducers, on the hepatic and extrahepatic target organ metabolism and toxicity, and, finally, e) the significance of cell-specific metabolic activation in extrahepatic target tissues. I have not included the nitrofurans in this discussion; some of these compounds have interesting toxic effects, but a possible role for metabolism is more difficult to analyze for these agents, since they conceivably can be activated by nitroreduction instead of (or in addition to) oxidation of the furan ring.

METABOLIC ACTIVATION OF FURANS IN VITRO

Studies with 4-ipomeanol and 3-methylfuran by Boyd and co-workers (1974a,b; 1975; 1978a,b), with furosemide by Mitchell and colleagues (1974, 1976), with 2-furamide by McMurtry and Mitchell (1977), and with 2-(N-ethylcarbamoyl-hydroxymethyl)furan [CMF] by Guengerich (1977a) all indicated that the in vitro covalent binding of these toxic furans was mediated by microsomal cytochrome P-450 enzymes present in target tissues for toxicity by the respective compounds (Table III). The involvement of cytochromes P-450 in the

TABLE III

STUDIES CHARACTERIZING THE METABOLIC

ACTIVATION OF FURANS IN VITRO

Compounds	References
4-ipomeanol, 3-methylfuran	Boyd et al, 1974; Boyd, 1976, 1978
furosemide, 2-furamide	Mitchell et al, 1974, 1976; McMurtry and Mitchell, 1977
2-(N-ethylcarbamoyl-hydroxymethyl)furan	Guengerich, 1977

covalent binding pathway was inferred by its dependency upon NADPH
and oxygen, and by its inhibition by carbon monoxide and other drug
metabolism inhibitors including piperonyl butoxide, SKF-525A, and
cobaltous chloride.   Additionally, Guengerich (1977a,b) demonstrated that CMF and 4-ipomeanol could be activated in vitro in highly
purified, reconstituted cytochrome P-450 systems. In vivo pretreatments with drug metabolism inducers have shown varying effects on
the in vitro hepatic microsomal covalent binding of the different
furans.   Thus, phenobarbital increased the covalent binding activity
for furosemide, but not for 2-furamide, in mouse liver microsomes
(Mitchell et al, 1976; McMurtry and Mitchell, 1977).   On the other
hand, 3-methylcholanthrene enhanced the binding of 2-furamide (Mc-
Murtry and Mitchell, 1977) and also of 4-ipomeanol (Boyd and Dutcher, 1978) in mouse liver microsomes.   In rats, phenobarbital, but
not 3-methylcholanthrene or pregnenolone-16α-carbonitrile, increased the hepatic microsomal covalent binding of CMF, whereas both
phenobarbital and 3-methylcholanthrene increased the covalent binding of 4-ipomeanol (Boyd et al, 1978a).   No significant enhancements
of the in vitro covalent binding of the furans, however, have been
demonstrated with any of these pretreatments in extrahepatic tissue
microsomes.

STUDIES ON THE NATURE OF REACTIVE FURAN METABOLITES AND
POTENTIAL PATHWAYS FOR THEIR DETOXIFICATION

There has been considerable speculation from several laboratories as to the possible involvement of epoxides in the metabolic

activation of furans. The likelihood seems especially attractive,
given the precedents implicating a 2,3-epoxide intermediate in the
metabolic activation of the dihydrofuran moiety of aflatoxin-B1
(Swenson et al, 1973). However, definitive proof for the formation
of epoxides during metabolism of simple furans has not been forthcoming. Nevertheless, several studies (Table IV) have demonstrated
the furan moiety as the essential structural locus for metabolic
activation of toxic furan derivatives, and circumstantially these
investigations still point to furan epoxides as likely reactive
intermediates. For example, radiolabeled analogues of 4-ipomeanol
in which the furan moiety was replaced by methyl- or phenyl-groups
did not bind covalently to a substantial extent in vitro or in vivo,
nor were they toxic, even at doses many-fold higher than the furan
analogue. Likewise, covalent binding studies by Wirth et al (1975)
and by Mitchell et al (1976) with furosemide, and by Guengerich
(1977a) with CMF, using compounds containing radiolabeled atoms at
various structural positions, similarly indicated the primary involvement of the furan group, and not the side-chain substituents,
in the covalent binding reactions. Wirth et al (1976) also demonstrated that the tetrahydrofuran analogue of furosemide did not bind
covalently in vitro, indicating the requirement for the unsaturated
furan ring in the binding reaction.

TABLE IV

STUDIES ON THE NATURE OF
THE REACTIVE FURAN METABOLITES

Compound	References
4-ipomeanol	Boyd 1976, 1980
furosemide	Wirth et al, 1975; Mitchell et al, 1976
2-(N-ethylcarbamoyl-hydroxymethyl)furan	Guengerich, 1977

There is evidence that the reactive metabolites of some furan
derivatives, such as 4-ipomeanol (Boyd, 1977) and aflatoxin-B1
(Neal et al, 1979), are extremely reactive, binding predominantly
with nucleophilic cellular constituents near their site of formation. On the other hand, there also is evidence (Guengerich, 1977a;
Neal et al, 1979) that alkylating metabolites of other furans, such
as CMF, are somewhat stable. Indeed, Guengerich (1977a) estimated
the half-life of "activated" CMF to be as much as 30 minutes in
aqueous media in the absence of tissue nucleophiles.

Some possible detoxification pathways for reactive furan metabolites are illustrated in Table V. Studies with furosemide
(Mitchell et al, 1976), 4-ipomeanol (Boyd et al, 1978a) and CMF
(Guengerich, 1977a) all have shown that the nucleophilic compound,
reduced glutathione (GSH), inhibits the covalent binding of these
furans in vitro, presumably by reacting with highly reactive, electrophilic furan metabolites to form less reactive conjugates. In
the case of 4-ipomeanol, we have found that at least two different
GSH conjugates are formed, in approximately equal ratios, in aerobic
incubation mixtures containing rat lung or liver microsomes, NADPH,
4-ipomeanol, and GSH (Buckpitt and Boyd, 1980a). Moreover, the
formation of GSH/4-ipomeanol conjugates was not enhanced by cytosol
preparations, suggesting that the cytosolic GSH-transferases have
little or no importance in the detoxification of 4-ipomeanol by GSH.
In vivo evidence, too, has been presented that GSH plays a major
in vivo role in the detoxification of lung-toxic 4-ipomeanol metabolites in the rat (Boyd et al, 1979b) and with hepatotoxic 4-ipo-
meanol metabolites in the bird (Buckpitt and Boyd, 1980b). On the
other hand, the possibility was suggested by Mitchell et al (1974)
that GSH might not be a major protective factor in mice in vivo with
a hepatotoxic metabolite of another type of furan compound, furosemide, because toxic doses of this agent did not deplete hepatic GSH.

TABLE V

STUDIES OF POTENTIAL DETOXIFICATION
PATHWAYS FOR REACTIVE FURAN METABOLITES:

1. GSH; GSH-transferase
2. epoxide hydratase
3. nonenzymatic degradation and/or rearrangements

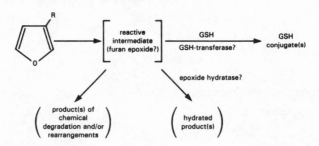

Only one study has suggested a modulatory role for epoxide hydratase in the covalent binding of a furan compound. Wirth and
co-workers (1976) reported that the covalent binding of furosemide
to microsomal protein was significantly enhanced in the presence of
an epoxide hydratase inhibitor, 1,2-epoxy-3,3,3-trichloropropane.
In contrast, Boyd et al (1978a) found that epoxide hydratase inhibitors did not alter the in vitro covalent binding of 4-ipomeanol,
and Guengerich (1977a) likewise obtained a similar negative result

with CMF. It is possible that furan-epoxides of the latter compounds, if indeed they are formed, are poor substrates for epoxide
hydrase, or they may be too reactive to reach the active site of
the enzyme. A dihydrodiol metabolite has not yet been definitively
identified as a metabolite of a simple furan in vitro or in vivo.

An epoxide intermediate produced initially during the metabolism of a furan compound conceivably could rearrange spontaneously
to form such products as dihydrofurans and dialdehydes, or they
might otherwise undergo nonenzymatic degradation to form more stable
products (Wirth et al, 1976; Boyd, 1976). Identification of such
products might thereby implicate an epoxide precursor. However,
the definitive identification of such key products has not yet been
forthcoming.

EXAMPLES OF FURAN-INDUCED TARGET ORGAN TOXICITY IN WHICH
REACTIVE METABOLITES ARE LIKELY TO BE INVOLVED

Factors Affecting Target Organ Toxicity

Three prominent target tissues for toxic furans are the liver,
the kidneys, and the lungs. Among the furans that have been adequately studied, there is considerable variability in their relative
tissue selectivity. Some sources of such variations are summarized
in Table VI. The variations may be related to differences in the
chemical nature and/or biological fate of the various compounds
themselves or, for individual compounds, variability in response
may result from host-related factors, such as species, age, sex,
and prior exposure to other chemicals. The following examples
illustrate many of these features.

Hepatotoxicity

Table VII indicates some examples of furan-induced liver toxicity in which reactive metabolites are likely to be involved. The
first studies that suggested a role for metabolism in the acute
toxicity of a furan compound were described by Seawright and his
colleagues (Seawright, 1968; Seawright and Hrdlicka, 1972). These
workers found that pretreatments with phenobarbital or SKF-525A
enhanced or decreased, respectively, the hepatic necrosis caused in
mice or rats by the naturally occurring sesquiterpene oil, ngaione.
Likewise, the hepatotoxicity of dehydrongaione could be prevented by
SKF-525A (Allen and Seawright, 1973). Studies in other laboratories
with several other compounds, including furosemide (Mitchell et al,
1976), 2-furamide (McMurtry and Mitchell, 1977), and 4-ipomeanol
(Buckpitt and Boyd, 1979), have similarly shown that the hepatic
necrosis caused in the particular test species studied could be modified by pretreatments with drug metabolism inducers or inhibitors.
The latter studies additionally showed that the toxic compounds were

TABLE VI

SOME IMPORTANT FACTORS DETERMINING
TARGET ORGAN SELECTIVITY OF TOXIC FURANS

I. Nature of specific test compound:

   a) distribution to target tissues

   b) metabolic toxification/detoxification
       pathways to which it is susceptible

   c) nature and stability of reactive
       intermediate

II. Particular test animal used:

   a) species

   b) age

   c) sex

   d) prior treatment with inducers

TABLE VII

EXAMPLES OF FURAN-INDUCED LIVER TOXICITY IN
WHICH REACTIVE METABOLITES ARE LIKELY TO BE INVOLVED

Compound	Test Species	Reference
ngaione	mouse, rats	Seawright & Hrdlicka, 1972
dehydrongaione	sheep	Allen & Seawright, 1973
furosemide	mouse	Mitchell et al, 1976
2-furamide	mouse	McMurtry & Mitchell, 1977
4-ipomeanol	bird	Buckpitt & Boyd, 1979

bound covalently to the target tissue in vivo and that the pretreat-
ment-induced changes in toxicity were closely paralleled by similar
alterations in the amounts of material covalently bound. These
results clearly implicated reactive metabolites in these toxicities.

Inducer pretreatments can strikingly modify the type of hepatic
lesion caused by furans or they can shift to the liver the organ
specificity of certain toxic furans that normally damage only extrahepatic tissues in a particular host. To exemplify, phenobarbital
pretreatment changed the pattern of ngaione-induced hepatic necrosis
in mice from a midzonal to a periportal distribution (Lee et al,
1979), and in rats pretreated with 3-methylcholanthrene, 4-ipomeanol
caused severe hepatic necrosis (4-ipomeanol normally causes only
lung damage in rats [Boyd and Burka, 1978]).

Renal Toxicity

The role of metabolism in renal toxicity has been studied in
some detail for two furans, 2-furamide (McMurtry and Mitchell, 1977)
and 4-ipomeanol (Boyd and Dutcher, 1980). The results of investigations with both of these compounds were consistent with the view
that renal damage resulted from the metabolic activation of the
toxic furan in situ within the kidney (Table VIII). Autoradiograph-
ic studies with 4-ipomeanol have shown that large amounts of 4-ipo-
meanol metabolite are bound covalently in the proximal renal tub-

TABLE VIII

EXAMPLES OF FURAN-INDUCED KIDNEY TOXICITY IN
WHICH REACTIVE METABOLITES ARE LIKELY TO BE INVOLVED

Compound	Test Species	Reference
2-furamide	mouse	McMurtry & Mitchell, 1978
4-ipomeanol	mouse	Boyd & Dutcher, 1978

ules, the major locus of kidney damage by this furan. Moreover, the
kidney lesion occurs only in adult male mice, and these are the only
animals that show extraordinarily high levels of in vivo renal alkylation by 4-ipomeanol (Jones et al, 1979). Female mice, immature
male mice, and other species tested (rats, rabbits, guinea pigs,
hamsters) do not show the renal lesion, and only relatively small or
insignificant levels of covalent binding of 4-ipomeanol occur in the
kidneys (Longo and Boyd, 1980). In vitro investigations indicate

that the age-related development of susceptibility to renal damage
by 4-ipomeanol in male mice is due to the development of renal enzyme activity required for the metabolic activation of 4-ipomeanol.
These studies with 4-ipomeanol emphasize the importance of species,
age, and sex of the test animals as potential determinants of target
organ alkylation and toxicity by compounds such as the furans, which
require metabolic activation to produce toxicity.

Pulmonary Toxicity

One of the most consistent, and certainly one of the most
striking, toxic effects of many furans is their ability to profound-
ly damage the lungs of animals given the compounds by any of several
different routes of administration (Boyd, 1980a). Table IX lists
some examples of lung-toxic furans. The two furan carbamates shown
were synthesized and evaluated as possible models for naturally
occurring toxic furans such as 4-ipomeanol. Methylfuran was investigated because of its apparent prominence as an atmospheric contaminant present in certain photochemical smogs. 4-Ipomeanol, however,
is the most extensively studied lung-toxic furan. It has been of
continuing interest because of its extraordinary potency and specificity, compared to other furans, for alkylation and damage of the
lungs. It has served as an important tool for the elucidation of
the potential role of in situ metabolic activation in drug-induced
lung injury (Boyd, 1976, 1980a, 1980b) and for the localization of
an extremely active cytochrome P-450 mixed-function oxidase system
in a specific family of lung cells, the Clara cells (Boyd, 1977).

TABLE IX

EXAMPLES OF FURAN-INDUCED LUNG TOXICITY IN WHICH
REACTIVE METABOLITES ARE LIKELY TO BE INVOLVED

Compound	Test Species	Reference
3-(N-ethylcarbamoyl-hydroxymethyl)furan	mouse, rat	Seawright and Mattocks, 1973
2-(N-ethylcarbamoyl-hydroxymethyl)furan	rat	Guengerich, 1977
3-methylfuran	mouse	Boyd et al, 1978
4-ipomeanol	rat, mouse, hamster, rabbit, guinea pig	Boyd, 1972, 1974, 1976

These studies have already been reviewed in detail elsewhere, and I should only like to emphasize here a few of the features that seem especially relevant to this discussion.

After i.p. administration of toxic doses of 4-ipomeanol, large amounts of 4-ipomeanol metabolites are covalently bound preferentially in the lungs of most species tested, including rats, female and immature male mice, guinea pigs, hamsters, and rabbits, and the lungs are the primary target site for toxicity of the compound in these species (Boyd et al, 1979a; Dutcher and Boyd, 1979). But, two interesting species differences in this specificity are noteworthy. As I have already discussed, the kidneys, in addition to the lungs, are a major target site for the metabolism, covalent binding, and toxicity of 4-ipomeanol in adult male mice. Birds, on the other hand, are the only species that we have investigated which are not susceptible to pulmonary damage by 4-ipomeanol (Buckpitt and Boyd, 1979). Avian lungs contain little or none of the mixed-function oxidase activity necessary for metabolically activating 4-ipomeanol, and very little 4-ipomeanol is bound to bird lungs _in vivo_. It is of interest that bird lungs do not contain cells morphologically or functionally analogous to the Clara cells present in the airways of other species. Bird liver microsomes, on the other hand, are extremely active in mediating the covalent binding of 4-ipomeanol, and, as I have indicated before, 4-ipomeanol is a potent hepatotoxin in avians.

Aside from the exception of avian species, when one examines in detail the lungs of other species injected with 4-ipomeanol, some striking features are present. As shown by autoradiography, the covalently bound 4-ipomeanol metabolite is found predominantly in the Clara cells (i.e., the nonciliated bronchiolar cells) present in the small airways. 4-Ipomeanol is also actively metabolized and covalently bound _in vitro_ in lung slice preparations, and the cellular specificity is entirely comparable to that seen _in vivo_ (Longo and Boyd, 1979). The covalent binding of the reactive 4-ipomeanol metabolite to the Clara cells _in vivo_ is followed by the selective necrosis of this cell population. Both the alkylation and the necrosis of the Clara cells by 4-ipomeanol can be prevented by pretreatment of the animals with piperonyl butoxide, an inhibitor of the metabolism of 4-ipomeanol.

It should be emphasized that it is only the covalently bound 4-ipomeanol metabolite that accumulates in the lungs; the parent 4-ipomeanol itself is not preferentially concentrated in the tissue. Moreover, when the pulmonary levels of glutathione are drastically decreased by pretreatments with diethylmaleate, the total amounts of bound 4-ipomeanol metabolite and the potency of the toxin are increased, but the Clara cell specificity is not altered (conjugation with GSH is a major mode of detoxification of the toxic metabolite

olite of 4-ipomeanol). These and related studies indicated further
that the specificity of 4-ipomeanol for Clara cells is due to the
in situ metabolic activation of the compound preferentially in
these target cells. These investigations also led directly to the
conclusion that the Clara cells are a major site of cytochrome P-450
monooxygenase activity in lungs, a view which has numerous implications concerning the Clara cells as potential targets for other
toxic or carcinogenic chemicals that require metabolic activation
(Boyd, 1977). The toxicological significance of the monooxygenase
system in Clara cells is being tested and supported further in our
continuing investigations with other structurally diverse chemicals,
such as carbon tetrachloride (Boyd et al, 1980), other halocarbons
and hydrocarbons, and carcinogenic nitroso compounds, which affect
the Clara cells (Boyd, 1980b). Recently, additional evidence for
the Clara cells as a site of cytochrome P-450 enzymes in lung was
provided by an elegant immunohistochemical study by Dr. Serabjit-
Singh and co-workers (1980) in Dr. Philpot's laboratory, using an
antibody prepared against a purified cytochrome P-450 (Wolfe et al,
1979) isolated from rabbit lungs.

A method has recently been developed in Dr. Fout's laboratory
(Devereux et al, 1979; Devereux and Fouts, 1980) for the isolation
of homogeneous preparations of Clara cells and type 2 pneumocytes
from rabbit lungs. We have been fortunate in being able to under-
take some collaborative studies with Drs. Fouts, Philpot, Bend and
colleagues on the metabolism of 4-ipomeanol in reconstituted systems
containing two different forms of rabbit pulmonary cytochrome P-450,
and in isolated rabbit lung cells, and in isolated perfused rabbit
lungs. Our preliminary results are very exciting and we anticipate
being able to report these investigations in detail in the near
future. We believe that these new studies will integrate with the
previous in vitro and in vivo investigations on 4-ipomeanol to provide an unprecedented depth of understanding of an extrahepatic
toxicity involving a biological reactive intermediate.

REFERENCES

Allen, J.G. and Seawright, A.A.: The effect of prior treatment with
    phenobarbitone, dicophane (DDT) and β-diethylaminoethylphenyl-
    propylacetate (SKF-525A) on experimental intoxication of sheep
    with the plant Myoporum deserti. Cunn. Res. Vet. Sci. 15: 167-
    179, 1973.
Boyd, M.R.: Isolation and characterization of 4-ipomeanol, a lung-
    toxic furanoterpenoid produced by sweet potatoes (Ipomoea bata-
    tas). J. Agric. Food Chem. 20: 428-430, 1972.
Boyd, M.R., Burka, L.T., Neal, R.A., Wilson, B.J., and Holscher,
    M.M.: Role of covalent binding in the toxic mechanism of the
    lung-edemagenic agent, 4-ipomeanol. Fed. Proc. 33: 234, 1974a.

Boyd, M.R., Burka, L.T., Harris, T.M., and Wilson, B.J.: Lung-toxic furanoterpenoids produced by sweet potatoes (Ipomoea batatas) following microbial infection. Biochim. Biophys. Acta 337: 184-195, 1974b.

Boyd, M.R., Burka, L.T., and Wilson, B.J.: Distribution, excretion and binding of radioactivity in the rat after intraperitoneal administration of lung-toxic furan, 4-ipomeanol-C^{14}. Toxicol. Appl. Pharmacol. 32: 147-157, 1975.

Boyd, M.R.: Role of metabolic activation in the pathogenesis of chemically induced pulmonary disease: Mechanism of action of the lung-toxic furan, 4-ipomeanol. Environ. Hlth. Perspectives 16: 127-138, 1976.

Boyd, M.R.: Evidence for the Clara cell as a site of cytochrome P-450-dependent mixed-function oxidase activity in lung. Nature 269: 713-715, 1977.

Boyd, M.R. and Burka, L.T.: In vivo studies on the relationship between target organ alkylation and the pulmonary toxicity of a chemically reactive metabolite of 4-ipomeanol. J. Pharmacol. Exp. Therap. 207: 687-697, 1978.

Boyd, M.R. and Dutcher, J.S.: Renal toxicity due to reactive metabolites formed in situ in the kidney: Investigations with 4-ipomeanol in the mouse. J. Pharmacol. Exp. Therap., in press, 1980.

Boyd, M.R., Burka, L.T., Wilson, B.J., and Sasame, H.A.: In vitro studies on the metabolic activation of the pulmonary toxin, 4-ipomeanol, by rat lung and liver microsomes. J. Pharmacol. Exp. Therap. 207: 677-686, 1978a.

Boyd, M.R., Statham, C.N., Franklin, R.B., and Mitchell, J.R.: Pulmonary bronchiolar alkylation and necrosis by 3-methylfuran, a potential atmospheric contaminant derived from natural sources. Nature 272: 270-271, 1978b.

Boyd, M.R., Dutcher, J.S., Buckpitt, A.R., Jones, R.B., and Statham, C.N.: Role of metabolic activation in extrahepatic target organ alkylation and cytotoxicity by 4-ipomeanol, a furan derivative from moldy sweet potatoes: Possible implications for carcinogenesis. In "Naturally-Occurring Carcinogens-Mutagens and Modulators of Carcinogenesis," E.C. Miller, J.A. Miller, O. Hirono, T. Sugimura, and S. Takayama (Eds.), University Park Press, Baltimore, pp. 35-56, 1979a.

Boyd, M., Statham, C., Stiko, A., Mitchell, J., and Jones, R.: Possible protective role of glutathione in pulmonary toxicity by 4-ipomeanol. Toxicol. Appl. Pharmacol. 48: A66, 1979b.

Boyd, M.R.: Biochemical mechanisms in pulmonary toxicities by furan derivatives. In "Reviews in Biochemical Toxicology," E. Hodgson, J. Bend, and R. Philpot (Eds.), Elsevier/North Holland, New York, pp. 71-102, 1980a.

Boyd, M.R.: Biochemical mechanisms in chemical-induced lung injury: Roles of metabolic activation. CRC Crit. Rev. Toxicol. 7: 103-176, 1980b.

Boyd, M.R.: Effects of inducers and inhibitors on extrahepatic drug metabolism and toxicity. In "Environmental Chemicals, Enzyme Function and Human Disease" (Ciba Fdn. Symposium No. 76), Excerpta Medica, Amsterdam, in press, 1980c.

Boyd, M.R., Buckpitt, A.R., Jones, R.B., Statham, C.N., and Longo, N.S.: Metabolic activation of toxins in extrahepatic target organs and target cells. In "The Scientific Basis of Toxicology Assessment", H. Witschi (Ed.), Elsevier/North Holland, New York, pp. 141-152, 1980a.

Boyd, M.R., Statham, C.N., and Longo, N.S.: The pulmonary Clara cell as a target for toxic chemicals requiring metabolic activation: Studies with carbon tetrachloride. J. Pharmacol. Exp. Ther. 212: 109-115, 1980b.

Buckpitt, A.R. and Boyd, M.R.: Xenobiotic metabolism in birds, species lacking pulmonary Clara cells. The Pharmacologist 20: 181, 1978.

Buckpitt, A.R. and Boyd, M.R.: The in vitro formation of glutathione conjugates with the microsomally activated pulmonary bronchiolar alkylating agent and cytotoxin, 4-ipomeanol. J. Pharmacol. Exp. Ther., in press, 1980a.

Buckpitt, A.R. and Boyd, M.R.: Protective role of glutathione in the alkylation and hepatotoxicity by 4-ipomeanol in the bird. Toxicol. Appl. Pharmacol., in press, 1980b.

Devereux, T.R., Hook, G.E.R., and Fouts, J.R.: Foreign compound metabolism by isolated cells from rabbit lung. Drug Metab. Dispos. 7: 70-75, 1979.

Devereux, T.R. and Fouts, J.R.: Isolation and identification of Clara cells from rabbit lung. In Vitro, in press, 1980.

Dutcher, J.S. and Boyd, M.R.: Species and strain differences in target organ alkylation and toxicity by 4-ipomeanol: Predictive value of covalent binding in studies of target organ toxicities by reactive metabolites. Biochem. Pharmacol. 28: 3367-3372, 1979.

Guengerich, F.P.: Studies on the activation of a model furan compound--toxicity and covalent binding of 2-(N-ethylcarbamoylhydroxymethyl)furan. Biochem. Pharmacol. 26: 1909-1915, 1977a.

Guengerich, F.P.: Preparation and properties of highly purified cytochrome P-450 and NADPH-cytochrome P-450 reductase from pulmonary microsomes of untreated rabbits. Mol. Pharmacol. 13: 911-923, 1977b.

Jones, R., Allegra, C., and Boyd, M.: Predictive value of covalent binding in target organ toxicity: Age-related differences in target organ alkylation and toxicity by 4-ipomeanol. Toxicol. Appl. Pharmacol. 48: A129, 1979.

Lee, J.S., Hrdlicka, J., and Seawright, A.A.: The effect of size and timing of a pretreatment dose of phenobarbitone on the liver lesion caused by ngaione in the mouse. J. Path. 127: 121-127, 1979.

Longo, N. and Boyd, M.: In vitro metabolic activation of the pulmonary toxin, 4-ipomeanol, by lung slices and isolated whole lungs. Toxicol. Appl. Pharmacol. 48: A130, 1979.

Longo, N.S. and Boyd, M.R.: Sex differences in renal alkylation and
toxicity by 4-ipomeanol in the mouse. Toxicol. Appl. Pharmacol.,
in press, 1980.

McMurtry, R.J. and Mitchell, J.R.: Renal and hepatic necrosis after
metabolic activation of 2-substituted furans and thiophenes, including furosemide and cephaloridine. Toxicol. Appl. Pharmacol.
42: 285-300, 1977.

Mitchell, J.R., Potter, W.Z., Hinson, J.A., and Jollow, D.J.: Hepatic Hepatic necrosis caused by furosemide. Nature (London) 251:
508-510, 1974.

Mitchell, J.R., Nelson, W.L., Potter, W.Z., Sasame, H.A., and Jollow
D.J.: Metabolic activation of furosemide to a chemically reactive, heptatotoxic metabolite. J. Pharmacol. Exp. Ther. 199: 41-
52, 1976.

Neal, G.E., Mattocks, A.R., and Judah, D.J.: The microsomal activation of aflatoxin B1 and 2-(N-ethylcarbamoylhydroxymethyl)-furan
in vitro using a novel diffusion apparatus. Biochim. Biophys.
Acta 585: 134-142, 1979.

Seawright, A.A.: Patterns of liver lesions caused by ngaione in the
rat. Aust. Vet. J. 44: 426-430, 1968.

Seawright, A.A. and Hrdlicka, J.: The effect of prior dosing with
phenobarbitone and β-diethylaminoethyldiphenylpropyl acetate
(SKF-525A) on the toxicity and liver lesion caused by ngaione in
the mouse. Br. J. Exp. Path. 53: 242-252, 1972.

Seawright, A.A. and Mattocks, A.R.: The toxicity of two synthetic
3-substituted furan carbamates. Experientia 29: 1197-1200, 1973.

Serabjit-Singh, C.J., Wolf, C.R., Philpot, R.M., and Plopper, C.G.:
Cytochrome P-450: Localization in rabbit lung. Science 207: 1469-
1470, 1980.

Swenson, D.H., Miller, J.A., and Miller, E.C.: 2,3-Dihydro-2,3-
dihydroxyaflatoxin B1: An acid hydrolysis product of an RNA-
aflatoxin B1 adduct formed by hamster and rat liver microsomes
in vitro. Biochem. Biophys. Res. Commun. 53: 1260-1267, 1973.

Wirth, P.J., Bettis, C.J., and Nelson, W.L.: Microsomal metabolism
of furosemide. Evidence for the nature of the reactive intermediate involved in covalent binding. Mol. Pharmacol. 12: 759-768,
1976.

Wolf, C.R., Smith, B.R., Ball, L.M., Serabjit-Singh, C., Bend, J.R.,
and Philpot, R.M.: The rabbit pulmonary oxygenase system. Catalytic differences between two purified forms of cytochrome P-450
in the metabolism of benzo(a)pyrene. J. Biol. Chem. 254: 3658-
3663, 1979.

THE RELATIONSHIP BETWEEN THE METABOLISM AND TOXICITY OF BENZODIOXOLE COMPOUNDS

James W. Bridges and Timothy R. Fennell

Robens Institute of Industrial and Environmental Health
and Safety
University of Surrey
Guildford, Surrey GU2 5XH, U.K.

INTRODUCTION

Man is continually exposed to benzodioxole compounds which are found as common secondary metabolites of many herbs and spices (e.g. parsley, carrot, mint, dill, cinammon, ginger, black pepper, nutmeg, mace, star anise oil, sesame oil and camphor oil) and are extensively used as insecticide synergists. Most of the benzodioxole compounds which have been examined have a low acute toxicity. The oral LD_{50} values for safrole, isosafrole and dihydrosafrole in the rat are in the range 1 - 2.5 g/kg body weight (Jenner et al., 1964). Few studies have been made of the chronic toxicity of benzodioxole compounds. It has been demonstrated that in rodents safrole, dihydrosafrole and isosafrole all cause hepatotoxicity (Taylor et al., 1964) and that safrole and isosafrole are weak hepatocarcinogens while dihydrosafrole produces oesophageal tumours (Homburger et al., 1961, 1962; Long et al., 1961, 1963; IARC 1976). In man the only known toxic effect of benzodioxole compounds is contact dermatitis which has been observed with piperonyl butoxide, sesamin and sesamolin (Haley, 1978; Neering et al., 1975).

Safrole and isosafrole have been extensively investigated in bacterial and mammalian mutation assays. Although there have been one or two reports that they are mutagenic, in the recent international blind trial (De Serres and Ashby, in press) the majority of laboratories failed to identify safrole as a mutagen. This raises the interesting question of whether these compounds provoke their carcinogenic effect in rodents predominantly through initiating somatic mutations or via an epigenetic mechanism. The early effects of safrole and isosafrole on liver are compatible with the concept

of tumour initiation through an exagerated work hypertrophy involving
the endoplasmic reticulum in particular (Crampton et al., 1977).
The changes in liver biochemistry and morphology caused by safrole
appear to involve several phases (see Fig. 1):

   i)   safrole is rapidly taken up into the tissues reaching high
   concentrations in the liver, pancreas and body fat. It causes
   non-competitive inhibition of a number of cytochrome P-450
   dependent drug metabolising enzyme activities, inhibition of
   synthesis at least in vitro, covalent binding of safrole
   related material to cellular constituents and depletion of
   cellular glutathione (Bridges, in press)

   ii) proliferation of the smooth endoplasmic reticulum,
   induction of various cytochrome P-450 dependent activities

   iii) loss of ribosomes from the endoplasmic reticulum, a fall in
   cytochrome P-450 dependent activities, proliferation of the
   endoplasmic reticulum

   iv) development of preneoplastic nodules

   v) appearance of well differentiated hepatic tumours.

The purpose of the present paper is to discuss the possible
contribution of safrole metabolites to the above processes.

   The metabolism of benzodioxole compounds in vivo in mammals,
involves in many cases opening of the dioxole ring with the production
of CO_2 from the methylene carbon atom (Casida et al., 1966). During
the metabolism of benzodioxole compounds in vitro by purified
cytochrome P-450, hepatic microsomes or freshly isolated hepatocytes,
the production of a highly stable metabolite complex with cytochrome
P-450 can be demonstrated spectrally with the appearance of peaks at
427 and 455 nm in the reduced state and 438 nm in the oxidised state
(see Fig. 2). The formation of this metabolite complex is inhibited
by cytochrome P-450 inhibitors such as metyrapone and SKF 525A, and it
results in the non-competitive inhibition of many cytochrome P-450
mediated reactions (Hodgson and Philpot, 1974). NADPH and oxygen are
required for the production of the metabolite complex (Franklin, 1971)
but cumene hydroperoxide may substitute for these (Elcombe et al.,
1975b; Kulkarni and Hodgson, 1978) with the production of a single
peak at 438 being observed. The formation of the 427/455 spectrum
results in a decrease in the amount of spectrally detectable cytochrome
P-450 and indicates the formation of a stable complex between a
benzodioxole metabolite or metabolites and cytochrome P-450. The
formation of a 427 nm peak requires the presence of a dioxole ring,
whereas the double Soret spectrum with peaks at 427 and 455 nm are
only produced by benzodioxole compounds (Hodgson et al., 1973; Dahl
and Hodgson, 1979).

The ratio of the 427/455 nm peaks is influenced by chemical structure, concentration and pH. Compounds with large lipophilic side chains e.g. piperonyl butoxide, tend to favour the formation of a 455 nm peak. Low 427/455 ratios tend to occur at high pH values while high 427/455 ratios occur at low pH. When isosafrole and NADPH are added to hepatic microsomes obtained from male rats pretreated with 3-methylcholanthrene, a spectrum consisting predominantly of a 455 nm peak is formed.

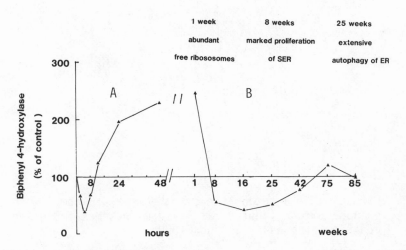

Fig. 1.   Inhibition and induction of hepatic microsomal biphenyl 4-hydroxylase following treatment of rats with isosafrole by injection (A, 150 mg/kg i.p) or safrole in the diet (B, 0.25% w/w). Modified from Elcombe et al. (1977) and Crampton et al., (1977).

On reduction of hepatic microsomes obtained from rats pre-treated with isosafrole or safrole, two peaks similar to those observed in vitro are produced, with maxima at 427 nm and 455 nm. The nature of this peak at 427 nm has yet to be properly defined, one obfuscating factor is that cytochrome b_5 absorbs at this wavelength. Addition of NADPH and benzodioxole compounds to these microsomes results in an intensification of the spectrum. Despite the similarity of their spectral appearance, there are major differences in the properties of the metabolite-cytochrome P-450 complexes generated in vivo and those generated in microsomes from

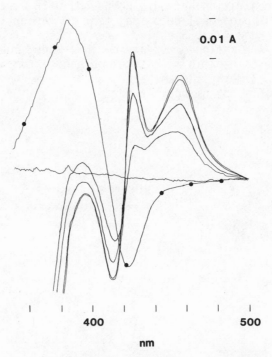

Fig. 2.   Formation of a metabolite cytochrome P-450 complex in
          phenobarbitone induced rat liver microsomes following
          addition of isosafrole and NADPH.

rats pretreated with phenobarbitone or 3-methylcholanthrene. On
removal of non-utilised NADPH enzymically, each metabolite cytochrome
P-450 ligand gives rise to an oxidised absorbance maximum at 438 nm,
and when microsomes from phenobarbitone or 3-methylcholanthrene
pretreated rats are used, this peak disappears with time.
However, the metabolite complex generated in vivo is stable in the
oxidised state, and, on addition of a suitable ligand for cytochrome
P-450, undergoes 'displacement' to yield catalytically active
cytochrome P-450 (Elcombe et al., 1975a; Dickins et al., 1979).
The progressive loss of the 438 nm peak with time can be correlated
with an increase in the binding spectrum, indicating the displacement
of the metabolite from cytochrome P-450 followed by the binding of
displacing substance with this free cytochrome P-450. Although

displacement occurs readily in oxidised microsomes, it will not occur in the reduced state. The difference in properties of the metabolite-cytochrome P-450 complex isolated from animals pretreated with safrole can be attributed to the fact that safrole and a number of other benzodioxole compounds are able to induce the synthesis of a particular form of cytochrome P-450, and it is this novel form of cytochrome P-450 which forms the displaceable metabolite-cytochrome complex.

The stoichiometry of displacement indicates that one molecule of benzodioxole related material is lost for each molecule of cytochrome P-450 released (Elcombe et al., 1976), but the nature of the displaced metabolite has yet to be ascertained. The displacement occurs equally well in the absence of oxygen and is linear with temperature between 4° and 40°. The isosafrole metabolite can be displaced by a wide range of compounds other than benzodioxoles including fatty acids, aliphatic hydrocarbons and carbamates, 2-alkylbenzimidazoles and some steroids. The steric requirements for optimal displacement are rather precise (see Fig. 3) with optimal displacement occuring with an aliphatic chain of ten carbon atoms or its equivalent (Dickins et al., 1979). Following reduction of microsomes during displacement of the isosafrole metabolite complex, it can be shown that the 455 nm peak decreases virtually to zero, but there is little change in the 427 nm peak.

The differences between the 427 and 455 nm peaks could be due to: i) different metabolites bound to the same species of cytochrome P-450, ii) a single metabolite on the same cytochrome P-450 species but in different protonation states; iii) a single metabolite on the same cytochrome P-450 species but bound to two different binding sites; iv) a single metabolite bound to different species of cytochrome P-450. These will be discussed in more detail later.

THE NATURE OF THE METABOLITE RESPONSIBLE FOR COMPLEX FORMATION

The evidence for ligand interaction of a metabolite with the haem of cytochrome P-450 is derived from studies on photo-dissociation of the complex, ligand competition and the similarity of absorbance maxima to those of other well documented ligands and has been reviewed by Franklin (1977). The wide variety of benzo-dioxole compounds capable of forming metabolite complexes shows that metabolism of the methylene carbon atom results in the formation of a metabolite complex. Indeed metabolism of this carbon atom of certain benzodioxole compounds is known to result in vivo in the production of carbon dioxide (Casida et al., 1966). There is evidence that the metabolism of benzodioxole compounds in vitro can result either in the formation of a metabolite cytochrome P-450 complex as is the case with safrole, or the production of carbon monoxide from the methylene carbon atom with compounds such as

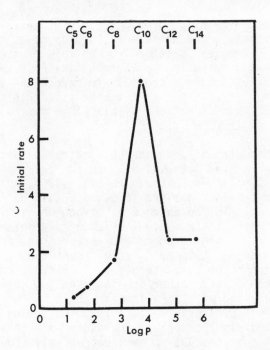

Fig. 3.   Effect of chain length and Log P (octanol/water) on the
          rate of displacement of the isosafrole metabolite cytochrome
          P-450 complex, by a series of aliphatic fatty acids.

4,5-dichloromethylenedioxybenzene (Yu et al., 1980). Whether all
these processes are related by a common precursor has yet to be
determined.

     The replacement of the hydrogen atoms of the methylene group
with oxygen or methyl groups abolished the synergistic activity of
benzodioxoles, while substitution with deuterium decreased
synergistic activity (Weiden and Moorefield, 1965). Hennessy (1965)
proposed that abstraction of a hydride ion could lead to the
formation of an electrophilic benzodioxolium ion capable of inhibiting
metabolism. Hansch (1968), using a structure activity relationship
to equate synergist activity with steric, homolytic and hydrophobic
constants suggested that abstraction of a hydrogen atom would lead
to production of a free radical capable of inhibiting insecticide
metabolism. A free radical mechanism is unlikely since radical
trapping agents e.g. dithiothreitol have no effect on the formation
of the complex. It has been suggested that benzodioxole compounds
could form a carbanion by loss of a proton, in a manner analogous to
the formation of a carbanion by fluorene (Ullrich and Schnabel,
1973 a,b). However, the methylene carbon atom is not acidic and
the complex is not discharged by either CO_2 or oxygen, which would
be expected by analogy with fluorene. By analogy with work on

polyhalogenated methanes, Ullrich et al., (1975) later proposed that
the metabolite could be a carbene, formed by hydroxylation and

Carbene   Carbanion

elimination of water. This proposal is mechanistically appealing
and is compatible with the properties of the metabolite-cytochrome
P-450 complex. It would account for the spectral similarities
between the cytochrome P-450 complex produced by benzodioxole
compounds and that formed by cyanide and alkyl isocyanides.
Furthermore, Mansuy et al. (1979) have demonstrated that the model
synthetic complex of iron (II) (tetraphenylporphyrin) (1,3-benzo-
dioxole-2-carbene) complex has similar spectral characteristics to
the benzodioxole metabolite-cytochrome P-450 complex. If one
accepts that a carbene metabolite is responsible for formation of
the stable benzodioxole metabolite cytochrome P-450 complex, how
does this help to explain the nature of the interactions involved
in the 427 and 455 nm peaks?

We envisage that in the oxidised state a carbene metabolite
interacts with the haem of cytochrome P-450 by a single σ bond and
that a second lipophilic interaction occurs at the Type I binding
site to form a bidentate ligand (Dickins et al., 1979). From their
work on dioxolanes and other ligands capable of forming double
Soret spectra, Dahl and Hodgson (1979) identified ligands forming
double Soret peaks as lipophilic and strong π- acceptor ligands and
proposed at 455 nm peak is formed by essentially a ligand interaction
with the haem, while the 427 peak is caused by a strong lipophilic
interaction which results in distortion of the 5th ligand of
cytochrome P-450. From displacement studies it seems that the isosafrole metabolite cytochrome P-450 complex formed in vivo must be
stabilised in the oxidised state by a lipophilic interaction
(Dickins et al., 1979). Since displacement of this particular
complex results in a decrease predominantly of the 455 nm peak, and
purified cytochrome P-450 from isosafrole treated rat forms mainly
a 455 nm complex (Ryan et al., 1980), it seems probable that the
455 nm peak in this case at least is stabilised by a lipophilic
interaction which does not result in distortion of the metal -
mercaptide bond of the fifth ligand. However only indirect
measurement of the lipophilic interactions of benzodioxole compounds
in complex formation has so far been possible using displacement.
The effect of benzodioxole substitution on the 427/455 ratio of
complexes formed with purified preparations of cytochrome P-450
should yield valuable information about the nature of lipophilic
stabilization of these complexes in the reduced state.

INDUCTION OF A NOVEL FORM OF CYTOCHROME P-450

Dickins et al. (1978) have demonstrated that isosafrole induces the formation in rats of a novel form of cytochrome P-450 which is electrophoretically dissimilar to the forms of cytochrome P-450 induced by phenobarbitone and methylcholanthrene. These findings have been supported by Sharma et al. (1979) and Ryan et al. (1980). Isosafrole also induces the synthesis of cytochrome P-450 in both C57BL/10 and DBA/2 mice (Fennell et al., 1980).

Using a series of benzodioxole compounds Fennell and Bridges (1979; Bridges and Fennell, 1980) have shown that in rats, the presence of the dioxole ring is necessary for the induction of the novel form of cytochrome P-450. Furthermore, there appears to be a good correlation between the ability of benzodioxole compounds when administered in vivo to form a stable, isolatable metabolite cytochrome P-450 complex, and their ability to induce the synthesis of the novel form of cytochrome P-450 measured by SDS polyacrylamide gel electrophoresis (see Table 1). This suggests that the formation of a metabolite complex may be a requisite for induction of this particular form of cytochrome P-450. Induction is favoured by the presence of a lipophilic side chain and the absence of alternative sites of preferential metabolism.

One may postulate that the initial events leading to the induction of the novel form of cytochrome P-450 are as follows: the benzodioxole compound binds to one or more species of cytochrome P-450 including the novel form which may be present in small amounts in normal liver and whose primary function may be to metabolise endogenous lipids. The benzodioxole is then metabolised to form a ligand complex leading to inhibition of cytochrome P-450 mediated metabolism. As a consequence of the interference in metabolism of endogenous lipids, accumulation of substrates could provide the trigger for the selective induction of the novel form of cytochrome P-450. Although this hypothesis fits the experimental observations, much work remains to be done to ascertain its validity.

In addition to inducing the novel form of cytochrome P-450, benzodioxole compounds induce other forms of cytochrome P-450 in the liver to a lesser extent (Dickins et al., 1978; Ryan et al., 1980) and also induce in many extra hepatic organs (Lake et al., 1973). Despite the inducing potency of the benzodioxole compounds, rather a low increase in the oxidative metabolism of foreign compounds is observed because the induced forms of cytochrome P-450 react with the inducing benzodioxole compound to form a metabolite complex, resulting in a loss of functionality. This phenomenon of a continually frustrated enzyme induction could be the initiating event for the excessive hypertrophy and loss of normal function of the endoplasmic reticulum. This would of course only occur at the rather high dose levels required to inactivate the majority of the of the cytochrome P-450 molecules.

Table 1. Induction of cytochrome P-450 and formation of a metabolite cytochrome P-450 complex _in vivo_ with various benzodioxole compounds.

		$\Delta A_{455-490}/$ mg protein	Induction of novel haemoprotein
Hexenyl benzodioxole	$R_1 = CH=CH-C_4H_9$	0.048	++
Octenyl benzodioxole	$R_1 = CH=CH-C_5H_{13}$	0.028	++
1'-hydroxy safrole	$R_1 = CHOH\ CH=CH_2$	0.005	+
Oxopentenyl benzodioxole	$R_1 = CH=CH-CO-C_2H_5$	0.011	+
Oxohexenyl benzodioxole	$R_1 = CH=CH-CO-C_3H_7$	0.021	++
Oxooctenyl benzodioxole	$R_1 = CH=CH-CO-C_5H_{11}$	0.030	++
Myristicin	$R_1 = CH_2-CH=CH_2$ $R_2 = OMe$	0.014	++
Apiol	$R_1 = CH-CH=CH_2$ $R_2 = R_3 = OMe$	0.002	-

OTHER METABOLIC INTERACTION WITH CELLULAR CONSTITUENTS

In addition to its effect on cytochrome P-450, safrole when added to isolated hepatocytes is an effective inhibitor of protein synthesis and causes depletion of glutathione. In the presence of NADPH it is very effective at degranulating _in vitro_ preparations of rough endoplasmic reticulum. Both _in vivo_ and _in vitro_ safrole and isosafrole are converted to covalently bound metabolites which are found associated with all the major cell organelles in liver. In rats which received ^{14}C isosafrole (0.925 nmoles/kg body weight) the extent of covalent binding was 1.4 nmoles/mg of microsomal protein after 24 h. Covalent binding can also be demonstrated

in <u>vitro</u> both in hepatic microsomes fortified with NADPH and in
<u>freshly</u> isolated hepatocytes. The nature of the active metabolite(s)
involved in covalent binding and their role in safrole and isosafrole
toxicity is uncertain. A number of metabolites of safrole involving
modification of the allyl side chain have been described (Oswald
et al., 1969; Borchert et al., 1973 a,b ; Stillwell et al., 1974;
Benedetti et al., 1977). Formation of 1'-hydroxysafrole leading to
the 1'-keto derivative and subsequent condensation with endogenous
amines is a likely route for covalent binding, as is epoxide
formation across the double bond of the allyl group. Metabolism
at several sites in the same molecule may also occur. Isosafrole
probably undergoes similar metabolic transformations.

REFERENCES

Benedetti, M.S., Malnoe, A., and Broillet, A.L., 1977, Absorption,
    metabolism and excretion of safrole in the rat and man,
    <u>Toxicology</u>, 7:69.
Borchert, P., Miller, J.A.,Miller, E.C., and Shires, T.K., 1973a,
    1'-Hydroxysafrole, a proximate carcinogenic metabolite of
    safrole in the rat and mouse, <u>Cancer Res.</u>, 33:590.
Borchert, P., Wislocki, P.G., Miller, J.A., and Miller, E.C.,
    1973 b, The metabolism of the naturally occurring hepato-
    carcinogen safrole to 1'-hydroxysafrole and the electrophilic
    reactivity of 1'-acetoxysafrole, <u>Cancer Res.</u>, 33:575.
Bridges, J.W., In press, The use of isolated hepatocytes in
    assessing chemically induced toxicity, in: "Toxicity Testing
    of Pharmaceutical and Other Products," J.W. Gorrod, ed.,
    Taylor and Francis, London.
Bridges, J.W. and Fennell, T.R., 1980, Characterisation of the form
    of P-450 induced by methylenedioxy compounds in rats and mice,
    in: "Microsomes, Drug Oxidations, and Chemical Carcinogenesis,
    Vol. II," Coon, M.J., Conney, A.H., Estabrook, R.W., Gelboin,
    H.V., Gillette, J.R., and O'Brien, P.J., Academic Press, New
    York.
Casida, J.E., Engel, T.T., Esaac, E.G., Kamienski, F.X., and
    Kuwatsuka, S., 1966, Methylene-^{14}C-dioxyphenyl compounds:
    Metabolism in relation to their synergistic action, <u>Science</u>,
    153:1130.
Crampton, R.F., Gray, T.J.B., Grasso, P., and Parke, D.V., 1977,
    Long-term studies on chemically induced liver enlargement
    in the rat II. Transient induction of microsomal enzymes
    leading to liver damage and nodular hyperplasia produced by
    safrole and Ponceau MX, <u>Toxicology</u>, 7:307.
Dahl, A.R., and Hodgson, E., 1979, The interaction of aliphatic
    anologs of methylenedioxyphenyl compounds with cytochromes
    P-450 and P-420, <u>Chem.-Biol. Interact.</u>, 27:163.
De Serres, F.J., and Ashby, J., In press, "Short Term Test for
    Carcinogens," Elsevier North-Holland, New York.

Dickins, M., Bridges, J.W., Elcombe, C.R., and Netter, K.J., 1978,
    A novel haemoprotein induced by isosafrole pretreatment in the
    rat, Biochem.Biophys,Res.Commun., 80:89.
Dickins, M., Elcombe, C.R., Moloney, S.J., Netter, K.J., and Bridges,
    J.W., 1979, Further studies on the dissociation of the isosafrole
    metabolite-cytochrome P-450 complex, Biochem.Pharmacol., 28:231.
Elcombe, C.R., Bridges, J.W., Gray, T.J.B., Nimmo-Smith, R.H.,
    and Netter, K.J., 1975 a, Studies on the interaction of safrole
    with rat hepatic microsomes, Biochem.Pharmacol., 24:1427.
Elcombe, C.R., Bridges, J.W., Nimmo-Smith, R.H., and Werringloer,
    J., 1975 b, Cumene hydroperoxide-mediated formation of inhibited
    complexes of methylenedioxyphenyl compounds with cytochrome
    P-450, Biochem.Soc.Trans. 3:967.
Elcombe, C.R., Bridges, J.W., and Nimmo-Smith, R.H., 1976, Substrate-
    elicited dissociation of a complex of cytochrome P-450 with a
    methylenedioxyphenyl metabolite, Biochem.Biophys.Res.Commun.,
    71:915.
Elcombe, C.R., Dickins, M., Sweatman, B.C., and Bridges, J.W.,
    1977, Substrate-elicited dissociation of the isosafrole
    metabolite cytochrome P-450 complex and the consequential
    reactivation of monooxygenation, in: "3rd Intl. Symp. on
    Microsomes and Drug Oxidations," Ullrich, V., Roots, I.,
    Hildebrandt, A.G., Estabrook, R.W., and Conney, A.H., eds.,
    Pergamon Press, Oxford.
Fennell, T.R., and Bridges, J.W., 1979, Structure-activity
    relationship for 'safrole-type' cytochrome P-450 induction,
    Biochem.Soc.Trans., 7:1104.
Fennell, T.R., Sweatman, B.C., and Bridges, J.W., 1980, The
    induction of hepatic cytochrome P-450 in C57BL/10 and DBA/2
    mice by isosafrole and piperonyl butoxide. A comparative study
    with other inducing agents, Chem.-Biol.Interact., 31:189.
Franklin, M.R., 1971, The enzymic formation of a methylenedioxyphenyl
    derivative exhibiting an isocyanide-like spectrum with
    reduced cytochrome P-450 in hepatic microsomes, Xenobiotica,
    1:581.
Franklin, M.R., 1977, Inhibition of mixed-function oxidations by
    substrates forming reduced cytochrome P-450 metabolic-
    intermediate complexes, Pharmac.Ther.A., 2:227.
Haley, T.J., 1978, Piperonyl butoxide, α[2-(2-butoxyethoxy) ethoxy]-
    4,5-methylenedioxy-2-propyltoluene: A review of the literature,
    Ecotoxicol.Environ.Saf., 2:9.
Hansch, C., 1968, The use of homolytic, steric, and hydrophobic
    constants in a structure-activity study of 1,3-benzodioxole
    synergists, J.Med.Chem., 11:920.
Hennessy, D.J., 1965, Carbamate insecticides hydride transferring
    ability of methylenedioxybenzenes as a basis of synergistic
    activity, J.Agric.Food.Chem., 13:218.
Hodgson, E., and Philpot, R.M., 1974, Interaction of methylenedioxy-
    phenyl [1,3-benzodioxole] compounds with enzymes and their
    effects on mammals, Drug Met.Rev., 3:231.

Hodgson, E., Philpot, R.M., Baker, R.C., and Mailman, R.B., 1973,
    Effect of synergists on drug metabolism, Drug Metab.Disp.,
    1:391.
Homburger, F., Kelley, T.J., Friedler, G., and Russfield, A.B.,
    1961, Toxic and possible carcinogenic effects of 4-allyl-1,2-
    methylenedioxybenzene (safrole) in rats on deficient diets,
    Med.Exp., 4:1.
Homburger, F., Kelley, T., Baker, T.R., and Russfield, A.B.,
    1962, Sex effect on hepatic pathology from deficient diet
    and safrole in rats, Arch. Pathol., 73:118.
IARC Monographs on the Evaluation of Carcinogenic Risk of Chemicals
    to Man, Vol. 10, 1976, International Agency for Research in
    Cancer, Lyon.
Jenner, P.M., Hagan, E.C., Taylor, J.M., Cook, E.L., and
    Fitzhugh, O.G., 1964, Food flavourings and compounds of
    related structure. I. Acute oral toxicity, Food Cosmet. Toxicol.,
    2:327.
Kulkarni, A.P., and Hodgson, E., 1977, Cumene hydroperoxide-
    generated spectral interactions of piperonyl butoxide and
    other synergists with microsomes from mammals and insects,
    Pest. Biochem. Physiol., 9:75.
Lake, B.G., Hopkins, R., Chakraborty, J., Bridges, J.W., and Parke,
    D.V.W., 1973, The influence of some hepatic enzyme inducers
    and inhibitors on extrahepatic drug metabolism, Drug Metab.
    Disp., 1:342.
Long, E.L., Hansen, W.H., and Nelson, A.A., 1961, Liver tumours
    produced in rats by feeding safrole, Fed. Proc., 20:287.
Long, E.L., Nelson, A.A., Fitzhugh, O.G., and Hansen, W.H., 1963,
    Liver tumours produced in rats by feeding safrole, Arch.
    Pathol., 75:595.
Mansuy, D., Battioni, J.-P., Chottard, J.-C., and Ullrich, V.,
    1979, Preparation of a porphyrin-iron-carbene model for
    the cytochrome P-450 complexes obtained upon metabolic
    oxidation of the insecticide synergists of the 1,3-benzodioxole
    series, J.Am.Chem.Soc., 101:3971.
Neering, H., Vitanyi, B.E.J., Malten, K.E., Van Ketel, W.G., and
    Von Dijk, E., 1975, Allergens in sesame oil contact dermatitis,
    Acta Der.-Venereol., 55:31.
Oswald, E.O., Fishbein, L., and Corbett, B.J., 1969, Metabolism
    of naturally occurring propenylbenzene derivatives. 1.
    Chromatographic separation of ninhydrin-positive materials
    of rat urine, J. Chromatog., 45:437.
Ryan, D.E., Thomas, P.E., and Levin, W., 1980, Hepatic microsomal
    cytochrome P-450 from rats treated with isosafrole, purification
    and characterisation of four enzymic forms, J. Biol. Chem.,
    255:7941.
Sharma, R.N., Cameron, R.G., Farber, E., Griffin, M.J., Joly, J.-G.,
    and Murray, R.K., 1979, Multiplicity of induction patterns of
    rat liver microsomal mono-oxygenases and other polypeptides
    produced by administration of various xenobiotics,

Biochem. J., 182:317.
Stillwell, W.G., Carman, M.J., Bell, L., and Horning, M.G., 1974, The metabolism of safrole and 2',3'-epoxysafrole in the rat and guinea pig, Drug Metab. Disp., 2:489.
Taylor, J.M., Jenner, P.M., and Jones, W.I., 1964, A comparison of the toxicity of some allyl, propenyl and propyl compounds in the rat, Toxicol.Appl. Pharmacol., 6:378.
Ullrich, V., and Schnabel, K.H., 1973 a, Formation and binding of carbanions by cytochrome P-450 of liver microsomes, Drug Metab. Disp., 1:176.
Ullrich, V., and Schnabel, K.H., 1973 b, Formation of and ligand binding of fluorenyl carbanion by hepatic cytochrome P-450, Arch. Biochem. Biophys., 159:240.
Ullrich, V., Nastianczyk, W., and Ruf, H.H., 1975, Ligand reactions of cytochrome P-450, Biochem. Soc. Trans., 3:803.
Weiden, M.H.J., and Moorefield, H.H., 1965, Synergism and species specificity of carbamate insecticides, J. Agric. Food Chem., 13:200.
Yu, L.-S., Wilkinson, C. F., and Anders, M.W., 1980, Generation of carbon monoxide during the microsomal metabolism of methylenedioxyphenyl compounds, Biochem. Pharmacol., 29:1113.

DISCUSSION

In response to Hsieh's presentation, Vainio wondered whether the
considerable interindividual variation in drug metabolism found in
a human population would affect their rank ordering among other
species. Hsieh responded that he had also observed intraspecies
variation in rats but that predicting toxicity within an individual
species is difficult. Oesch suggested that the species differences
observed may be due, in part, to differences in hepatic epoxide
hydrolase activity. While the microsomal epoxide hydrolase activity
is greater in rat than in the mouse, the cytosolic activity is greater
in the mouse than in the rat. If the species differences demonstrated are due to epoxide hydrolase activity, then the enzyme in
the cytosol is responsible for the difference. Gillette asked Hsieh
whether the amount of aflatoxin metabolized in the Ames test was
measured, since if only a fraction (50%) is metabolized, the side
reactions which may not affect the mutagenesis *in vitro* may pro-
foundly affect adduct formation *in vivo*. Hsieh responded that while
the amount of aflatoxin metabolized had not been measured, the Ames
test data were obtained over a low dose range of aflatoxin, suggesting that most of the compound was probably converted to metabolites.
Hsieh, answering a question from Remmer, stated that binding of
aflatoxin was not studied in cultures of hepatocytes derived from
phenobarbital-treated animals. Sugimura asked whether S9 from trout
had ever been used and was told by Hsieh that no data were obtained
with S9 from trout because of technical difficulties in maintaining
the reaction at the optimal temperature for trout, which, he stated,
is 15°C.

Following Boyd's presentation on furan toxicity, Cottrell asked·
whether furans affected the production of lung surfactant. Boyd
said that while some people think that Clara cells make the surfactant believed to be essential for a normal airway and normal lung
surface properties, furans did not seem to affect its production, at
least early in furan toxicity. Menzel then asked Boyd whether he
had any data to support his interesting proposal that the origin of
bronchogenic tumors was the Clara cell, and if so, what compound was
used to evoke the tumors? Boyd stated that he and others are coming
to the conclusion that while polycyclic hydrocarbon-type compounds

895

do not produce the pathological changes in Clara cells, nitrosamines do. Nitrosamines have also produced lung tumors, and such tumor tissue still activates nitrosamine compounds. However, lung tumor production by these compounds is much more complicated than just a simple cytochrome P-450-mediated activation of these compounds. Boyland questioned whether any simple furans had been shown to be carcinogenic and Boyd referred the question to Elizabeth Miller. She replied that simple furans had been tested in the A/Jax strain of mouse which has a high propensity for lung adenomas. Ipomeanol, ipomeadiol and ipomearone were each tested with 10 doses of 25% of the LD_{50}. None of these increased the incidence of lung adenomas. Ahmed asked whether the furan analogs bind to nucleic acids and was told that they bind to both DNA and protein, whereupon Mitchell commented that furosemide (another furan) also binds to DNA and RNA *in vivo*.

The first question posed to Bridges following his talk on the mechanism of activation of benzoxazoles came from Brooks. Brooks commented that the dioxole ring itself apparently forms a strong complex with cytochrome P-450 which leads to a difference spectrum with peaks at 427 and 455 nm and then asked Bridges whether there is any evidence that dioxole is a monooxygenase inhibitor. Bridges responded that Professor Hodgson had shown that if the benzene ring is not present, peaks at 427 and 458 nm are seen, so the assumption is made that the peak at 427 nm relates to the oxygenated ring but that benzene itself is not necessary. Liebman asked if any information was available on the spectrum of metabolic reactions catalyzed by the novel form of cytochrome P-450 induced by safrole. Bridges replied that it is difficult to study that because a carbene bound to cytochrome P-450 has to be displaced before you can realize the full metabolic potential of P-450. To displace the carbene, one has to use a lipophilic compound; this will, however, in turn bind to P-450. Therefore, a full measure of the metabolic competence of P-450 can only be determined if lipophilic substrates which are also displacing agents, are used. For example, one can show with biphenyl, which will displace the carbene, that an increased 4-hydroxylation of biphenyl occurs. Para-nitroanisole also displaces the carbene and shows an increased metabolism. Jerina asked whether the iso-safrole-catechol listed as a metabolite had been isolated and identified since he had tried for six months to make that compound without any success. Bridges replied that the compound had not actually been isolated. Oesch made the point that catechol was shown on one of the slides as a detoxification product and wondered whether it represented a major detoxification product or the major product. Since a lot of catechol is produced, Bridges presumed that it represents the major detoxification product.

METABOLIC ACTIVATION OF 2-ACETYLAMINOFLUORENE

Snorri S. Thorgeirsson, Peter J. Wirth, Norma Staiano
Carole L. Smith

Biochemical Pharmacology Section, Laboratory of Chemical
Pharmacology, National Cancer Institute, National
Institutes of Health, Bethesda, Maryland 20205

INTRODUCTION

Since the demonstration by Yoshida in 1933 (Yoshida, 1933)
of the formation of liver tumors in rats following administration
of 0-amino-azotoluene (2,2'-dimethyl-4-aminoazobenzene), aromatic
amines and amides have been among the most studied chemical
carcinogens. Studies with 2-acetylaminofluorene (AAF) and its
derivatives have provided the most extensive data on the mechanism
by which aromatic amides cause neoplasia (Weisburger and Weisburger
1973, Miller 1970). Our laboratory has for some time been studying the mechanism of metabolic activation of N-acetylarylamines,
in particular that of AAF, and how the metabolic processing of
these compounds influences their capacity to cause genotoxic
damage (Thorgeirsson and Nebert 1977, Thorgeirsson et al., 1977;
Schut et al., 1978; Sakai et al., 1978; Wirth et al., 1980 a,b).
In this paper we present a summary of our data on the metabolic
and mutagenic activation of AAF.

MATERIALS AND METHODS

Animals

Male Sprague Dawley rats (200-300 g) were obtained from the
National Institutes of Health Animal Supply and were used for
purification of both N-OH-AAF N-O acyltransferase and N-OH-AAF
sulfotransferase as well as for obtaining the subcellular liver
fractions employed in the mutagenesis assay. Prior to sacrifice
the animals were housed in plastic cages on standard hardwood
bedding and allowed water and food (Purina Lab Chow) ad libitum.

Chemicals

        AAF and 2-aminofluorene (AF) were obtained from Eastman
Organic Chemicals Co. (Rochester, NY). N-Hydroxy-2-acetylamino-
fluorene (N-OH-AAF) was a generous gift from Dr. Elizabeth Weisburger, National Cancer Institute. N-OH-AAF (9-^{14}C) (32 mCi/mmol)
was obtained from ICN Pharmaceuticals Inc., Irvine, California.
Labelled N-OH-AAF was purified by thin layer chromatography
(chloroform:methanol, 97:3 v/v) to more than 99.9%.

        3'-Phosphoadenosine-5'-phosphosulfate (PAPS) was obtained
from PL Biochemicals, Milwaukee, Wisconsin. Adenosine 3':5'
diphosphate (PAP) p-nitrophenyl sulfate, yeast tRNA and L-ascorbic
acid were purchased from Sigma Chemical Company, St. Louis,
Missouri and paraoxon (diethyl p-nitrophenyl phosphate) was
obtained from Aldrich Chemical Company. Salmonella tester strains
TA 98 and TA 100 were generous gifts from Dr. Bruce N. Ames,
University of California (Berkeley). All other chemicals were
the best reagent grade and were obtained commercially.

Mutagenesis Assay

        Mutagenesis assay was performed according to Ames et al.
(1975). To 2.2 ml of molten top agar containing 17 μmol MgCl$_2$,
0.125 μmol biotin, 0.125 μmol histidine, 33 μmol KCl, and 100 μmol
phosphate buffer pH 7.4 at 45° were added 0.1 ml of the bacterial
tester strain TA 98 (2 x 10^8 bacteria), 0.1 ml of a solution
containing the mutagen dissolved in 0.1 ml of DMSO and 0.2 ml of
the cytosolic fractions or purified enzyme fractions. Prior to
addition cytosolic fractions or purified enzyme fractions were
diluted with phosphate buffer to the desired protein concentrations
and filtered through a 0.45 mm Swinnex filter unit (Millipore).
The concentrations of protein in the filtrates were then determined after filtration to estimate losses during this process. In
assays where paraoxon was used, paraoxon was added in 0.1 ml of
DMSO to the mutagenesis mixture. In experiments where PAPS was
used, PAPS was added in 0.1 ml of phosphate buffered saline pH
7.4. The colonies on each plate (histidine independent revertants)
were scored on a Count-all (Model 600) colony counter (Fisher
Scientific Co., Pittsburg, Pa.) after a 48 hr incubation at 37°C.
The toxicity of the test compounds to the bacteria was tested
by determining the number of colonies formed in histidine enriched
(4.5 mM) agar after the bacteria had been exposed to varying
concentrations of the test compounds for 30 min at 37°C and
diluted to approximately 10^4/ml before plating. The bacterial
mutagenesis and DNA damage in the Salmonella/hepatocyte system was
assayed as previously described (Staiano et al., 1980).

Enzyme Assays

N- and C-hydroxylations of AAF by purified cytochrome P-450
forms were measured as previously described (Thorgeirsson and
Nelson 1976; Johnson et al., 1980). The activity of N-hydroxy-2-
acetylaminofluorene (N-OH-AAF) N-O acyltransferase was measured
essentially as described by King (1974). The activity of N-OH-AAF
sulfotransferase was determined spectrophotometrically as previously described (Mulder et al., 1977). This assay system measures
the capacity of the various enzyme fractions to generate the
active sulfate conjugate of N-OH-AAF and reflects the net effect
of PAPS formation and the transfer of the sulfate from PAPS to N-
OH-AAF. p-Nitrophenyl sulfate was used as the sulfate donor for
the in situ synthesis of PAPS from PAP. The incubations (3 ml)
were carried out at room temperature in 1 cm light-path length
cuvettes. The incubation mixtures contained (final concentrations):
0.5 mM N-OH-AAF, 10 mM p-nitrophenyl sulfate, 20 μM PAP, 150 mM
KCl and 50 mM sodium phosphate (pH 7.4). The incubations were
initiated by the addition of either cytosolic protein or purified
enzyme fractions (1 mg/ml). The rates of sulfation were calculated
from the rates of release of p-nitrophenol from p nitrophenyl
sulfate and were corrected for changes in absorbance (405 nm) in
the absence of N-OH-AAF. Enzyme activity was expressed as nmol
p-nitrophenol released per mg protein per 5 min. Assays and
controls were done in duplicate.

An alternative assay for the determination of sulfotransferase
activity was also used in which the PAPS dependent binding of N-
OH-AAF to tRNA was measured (King 1974). Incubation mixtures
contained in a final volume of one milliliter: 100 μmol Tris
buffer pH 7.4, 2 mg yeast tRNA, 100 μg (0.4 μmol) N-OH-AAF [9-
^{14}C] and 0.65 μmol PAPS. Reactions were initiated by the addition
of either cytosolic protein or partially purified enzyme fractions
and incubations were performed at 37°C for 20 min. The tRNA was
isolated and processed as described for the N-O acyltransferase
assay (King 1974).

Purification of Enzyme Activities

Sulfotransferase was purified as described by Wu and Straub
(1976) through the hydroxyapatite chromatography stage. Sulfotransferase activities of the various fractions were monitored
spectrometrically as described earlier in this paper (Mulder et
al., 1977).

Acyltransferase was purified using the same procedure as
reported by King (1974) except that the final chromatography step
(Sephadex G-100) was performed under air.

Cytochrome P-450 forms from rabbit liver were purified as previously described (Johnson et al., 1980).

RESULTS AND DISCUSSION

The major metabolic pathways for AAF are shown in Fig. 1. The metabolic activation of AAF proceeds via a two step process, namely, a cytochrome P-450 dependent N-hydroxylation with subsequent ester-formation and/or deacetylation of the hydroxamic acid. Since AAF is extensively hydroxylated on both the amide nitrogen and on the various carbon atoms on the fluorene ring by the cytochrome P-450 dependent monooxygenases, and since these metabolic pathways represent metabolic activation and detoxification respectively for AAF, we have examined the specificity of these pathways with four purified forms of cytochrome P-450 from rabbit liver (Johnson et al., 1980). The data is shown in Table 1.

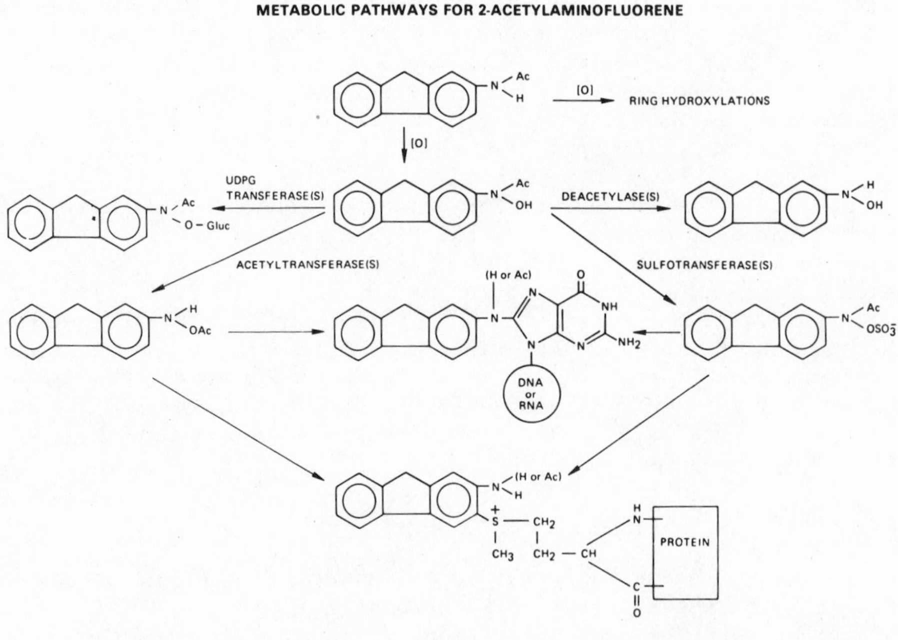

METABOLIC PATHWAYS FOR 2-ACETYLAMINOFLUORENE

Figure 1   Major metabolic pathways for AAF

Forms 3 and 6, which are the major constitutive and polycyclic hydrocarbon induced forms in adult and neonatal rabbit liver, respectively, catalyze almost exclusively the formation of 7-OH-AAF. Form 4, the major form that is induced by polycyclic hydrocarbons in the adult rabbit liver, is apparently the only form that catalyses the N-hydroxylation of AAF. Form 4 in addi-

Table 1. Metabolism of 2-Acetylaminofluorene by Reconstituted
Forms of Cytochrome P-450

Metabolite[a] Formation	Form 2	Form 3	Form 4	Form 6
7-OH-AAF	0	0.94 ± 0.07	0.58 ± 0.13	1.13 ± 0.15
5-OH-AAF	0	0	0.13 ± 0.03	0
3-OH-AAF	0	0	0.04 ± 0.01	0
1-OH-AAF	0	0	0.14 ± 0.02	0
N-OH-AAF	0	0	2.09 ± 0.16	0

[a]Rates are expressed as mol product formed/min/mol cytochrome
P-450 for the mean S.D. of experiments performed in triplicate
with two different preparations of each cytochrome. Zero
indicates a value less than 0.020 mol/min/mol. (Data from
Johnson et al. 1980).

tion to catalyzing the N-hydroxylation of AAF, also catalyzes the
formation of 7-OH-AAF, 5-OH-AAF, 3-OH-AAF and 1-OH-AAF. However,
seventy percent of the catalytic activity of form 4 is involved
with the N-hydroxylation (Table 1). Form 2, which is the major
form induced by phenobarbital in rabbit liver, does not exhibit
any measurable oxidative metabolism of AAF. In the case of the
four cytochromes described here, forms 3 and 6 catalyze almost
exclusively the detoxification pathway, whereas form 4 catalyzes
predominantly the activation of AAF. Form 2 does not appear to
participate to a significant degree in the metabolism of AAF.
Since the occurrence of each cytochrome is dependent on many
factors and the relative role of each cytochrome must be integrated
with other processes occurring during metabolism and carcinogenesis, it is difficult to predict with accuracy the impact of
these metabolic differences on a multi-step process such as carcinogenesis. However, it seems clear that the proportion of the AAF
dose that is activated (i.e., N-hydroxylated) in the first step of
the metabolic activation process, depends on the relative composition of the different cytochrome P-450 forms in the organism.

Previous studies from our laboratory on the in vitro metabolism and mutagenic activation of AAF and N-OH-AAF by microsomal
liver and kidney fractions from rat and mice have shown that the
rate limiting step in the mutagenic activation of AAF is N-
hydroxylation (Felton et al., 1976; Thorgeirsson et al., 1977)
with subsequent deacetylation of N-OH-AAF to N-OH-AF via the
membrane bound N-OH-AAF deactylase (Schut et al., 1978). Further-

more the mutagenicity of N-OH-AAF mediated by liver and kidney
microsomal preparations from rats and mice can be completely
inhibited by paraoxon (diethyl p-nitrophenyl phosphate) a known
carboxyesterase/amide inhibitor (Schut et al., 1978; Sakai et
al., 1978). Paraoxon, when added at concentrations as low as
10^{-7}M, markedly inhibits the mutagenicity of N-OH-AAF mediated by
liver of kidney microsomes from mice and rats (Schut et al.,
1978) and at a concentration of 10^{-4} M, a concentration that is
neither mutagenic nor toxic to the bacteria, the mutagenicity of
N-OH-AAF is completely inhibited (Fig. 2). Concomitant with
this inhibition of the mutagenicity of N-OH-AAF by paraoxon the
in vitro deacetylation of both AAF and N-OH-AAF mediated by liver
and kidney microsomes from rats and mice is also inhibited by
paraoxon (Fig. 3). As shown in Fig. 3 paraoxon at 10^{-4}M completely inhibits the deacetylation of both AAF and N-OH-AAF by
liver and kidney microsomes from DBA/2N mice.

    N-OH-AAF, in addition to being activated to a mutagen in the
Salmonella test system by the microsomal fraction, is also activated to a mutagen by rat liver post microsomal fractions (Stout et
al., 1976; Sakai et al., 1978; Mulder et al., 1977, Weeks, et
al., 1978, Andrews et al., 1978). In the cytosolic fraction the
mutagenic activation of N-OH-AAF is thought to be mediated by
either or both, N-O acyltransferase or sulfotransferase both of
which are also thought to be involved in the in vivo metabolic
activation of N-OH-AAF to its ultimate carcinogenic species in
the rat (DeBaun et al., 1970; King 1974). While the sulfate
ester of N-OH-AAF has been proposed to be the ultimate carcinogenic species in the rat liver (DeBaun et al., 1970), addition of
sulfation cofactors (PAPS) to the standard Salmonella test system
resulted in a marked decrease in the mutation frequency of N-OH-
AAF mediated by rat liver cytosol (Schut et al., 1978; Sakai et
al., 1978; Mulder et al., 1977; Andrews et al., 1979). In an
attempt to establish the relative importance of N-O acyltransferase and sulfotransferase in the mutagenic activation of N-OH-
AAF in the Salmonella test system we have partially purified both
N-O acyltransferase and sulfotransferase from male Sprague Dawley
rat liver and utilized the purified enzyme fractions in studies
of both the metabolic and mutagenic activation of N-OH-AAF in the
Salmonella test system.

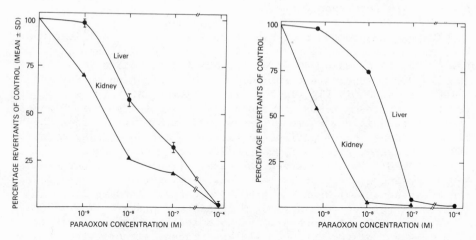

Figure 2   Effect of paraoxon on N–OH–AAF mutagenicity mediated
by liver and kidney microsomes from MC-treated C57BL/6N
mice (A) and PB-treated rats (B).   (Data from Schut et
al. 1978).

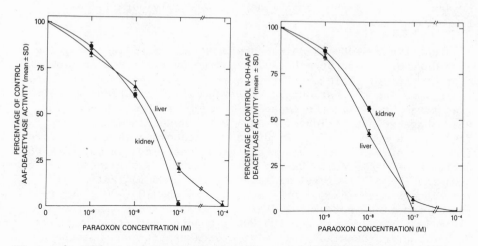

Figure 3   Effect of paraoxon on microsomal AAF deacetylase (A)
and N–OH–AAF deacetylase (B) of liver (Δ) and kidney (o)
from untreated DBA/2N mice (Data from Schut et al.
1978).

Mutagenicity of N-OH-AAF mediated by partially purified rat liver
N-OH-AAF N-O acyltransferase

    Using the procedure of King (1974) with minor modifications
N-OH-AAF N-O acyltransferase was purified from adult Sprague
Dawley rat liver using ammonium sulfate fractionation (45-65%) of
the 105,000 g supernatant followed by gel filtration on Sephadex
G-100. As shown in Fig. 4 partially purified rat liver N-O
acyltransferase markedly increased the mutagenicity of N-OH-AAF
when utilized in the Salmonella test system similar to that which
is observed using unfractionated rat liver cytosol (Sakai et al.,
1978; Andrews et al., 1978) the addition of either ascorbate (2
mM) or NADPH(1 mM) to mutagenesis assay mixtures using purified
N-O acyltransferase results in a four to five-fold increase in
the mutagenicity of N-OH-AAF (6000-7000 revertants vs 1500
revertants for controls at 5 μg N-OH-AAF per plate). The addition
of paraoxon (10^{-6}M), however, has no effect on the mutagenicity
of N-OH-AAF mediated by N-O acyltransferase. Although the
addition of PAPS resulted in a 50% decrease in the mutagenicity
of N-OH-AAF mediated by rat liver cytosol (Schut et al., 1978;
Wirth and Thorgeirsson, 1980a) the addition of PAPS has no effect
on the mutagenicity of N-OH-AAF mediated by N-O acyltransferase
(Fig.4).

Mutagenicity of N-OH-AAF mediated by partially purified N-OH-AAF
sulfotransferase

    Rat liver N-OH-AAF sulfotransferase was purified using the
procedure of Wu and Straub (1976) and an overall purification of
approximately 55 fold was obtained after hydroxyapatite chromatography. Using the various partially purified fractions of sulfotransferase activity the capacity of each fraction to catalyze
the mutagenic activation of N-OH-AAF was determined and the
results are summarized in Figure 5. At each successive purification step of rat liver sulfotransferase activity the capacity to
activate N-OH-AAF to a mutagen is decreased. The mutagenicity of
N-OH-AAF (5 μg per plate) mediated by the ammonium sulfate
fraction was only 67% of that mediated by the 105,000 g supernatant fraction whereas the DEAE cellulose fraction was three
times less active than the 105,000 g supernatant fraction. The
final fraction obtained following hydroxyapatite chromatography
and the fraction with the greatest sulfotransferase activity
(assayed colorimetrically as described in Materials and Methods)
was completely inactive in its capacity to activate N-OH-AAF to a
mutagen in the Salmonella test system (Fig. 5). Addition of
either ascorbate, NADPH, or PAPS has no effect (no revertants
observed) on the mutagenicity of N-OH-AAF mediated by the hydroxyapatite fraction of rat liver sulfotransferase.

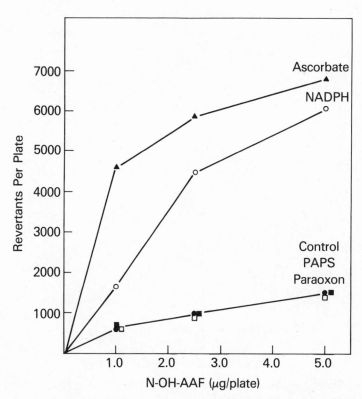

Figure 4   Effect of various agents on the in vitro mutagenicity
of N-OH-AAF in Salmonella TA 98 mediated by partially
purified rat liver N-O acyltransferase.

The mutagenesis assay was performed as described
under Material and Methods. Points are the mean
number of revertants per plate observed in two
experiments and have been corrected for spontaneous (background) revertants observed with N-OH-AAF
in the absence of N-O acyltransferase. Protein concentration of N-O acyltransferase was 0.3 mg per
plate in all experiments. □ , experiments with PAPS
(100 μg); ■ , paraoxon (10^6 M); ●, control; o, NADPH
(1 mM); and ▲, ascorbate (2 mM). (Data from Wirth
and Thorgeirsson 1980).

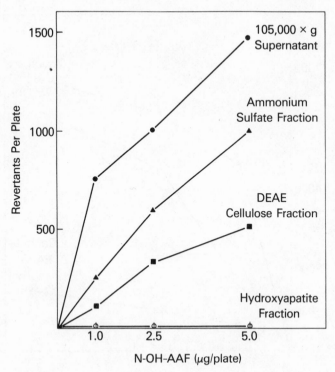

Figure 5   In vitro mutagenicity of N–OH–AAF in Salmonella TA 98
            mediated by partially purified fractions of rat liver
            sulfotransferase.

            The mutagenesis assay was performed as described under
            Materials and Methods. Points are the mean number of
            revertants per plate observed in two experiments and
            have been corrected for spontaneous (background)
            revertants observed with N–OH–AAF in the absence of
            purified sulfotransferase fractions. Purification of
            various fractions are described under Materials and
            Methods. Protein concentration of each fraction was
            1.0 mg per plate. ●, experiments utilizing 105,000 g
            supernatant; ▲, fraction obtained after ammonium sulfate
            precipitation;■ , fraction obtained after DEAE cellulose
            chromatography; and o, following hydroxyapatite chroma-
            tography. (Date from Wirth and Thorgeirsson (1980).

Covalent binding of N-OH-AAF to tRNA mediated by partially
purified N-O acyltransferase

       In addition to activating N-OH-AAF to a mutagen in the
Salmonella test system rat liver cytosol also activates N-OH-AAF
to an electrophilic species capable of reacting covalently with
various cellular macromolecules such as proteins, DNA, RNA, etc.
Incubation of partially purified N-O acyltransferase with N-OH-
AAF and yeast tRNA results in an approximately 15 fold increase
in the extent of covalent binding of (9-^{14}C) N-OH-AAF to tRNA
(Fig. 6) as compared to that catalyzed by rat liver cytosol.   In

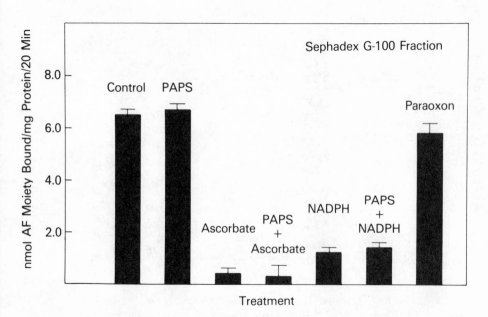

Figure 6   Effect of various agents on the in vitro covalent
           binding of N-OH-AAF to yeast tRNA mediated by
           partially purified rat liver N-O acyltransferase.

           The incubation mixture, conditions of acyltransferase
           assay, and isolation of tRNA-AF bound adducts were
           performed as described under Materials and Methods.
           Concentration of various agents: PAPS (40 µg per ml),
           ascorbate (2mM); NADPH (1 mM), paraoxon (10^{-6} M);
           PAPS (40 µg per ml in combination with either ascor-
           bate (2mM) or NADPH (1 mM). Values represent the
           mean ± standard derivation of three or more exper-
           iments. (Data from Wirth and Thorgeirsson 1980).

contrast to that observed using unfractionated rat liver cytosol
in which the addition of PAPS results in a marked increase in the
extent of covalent binding of N-OH-AAF to tRNA, the addition of
PAPS has no effect on the covalent binding of N-OH-AAF to tRNA
mediated by partially purified N-O acyltransferase (Figs. 6 and 7).
Both ascorbate and NADPH markedly inhibit N-O acyltransferase
mediated covalent binding of N-OH-AAF to tRNA. The addition of
PAPS in combination with either ascorbate or NADPH has no effect
on the inhibition of binding of N-OH-AAF by either ascorbate or
NADPH. Paraoxon (10^{-6} M) has no significant effect on the covalent binding of N-OH-AAF to tRNA.

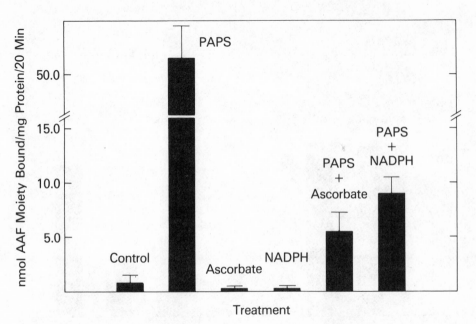

Figure 7   Effect of various agents on the in vitro covalent
            binding of N-OH-AAF to yeast tRNA mediated by
            partially purified rat liver sulfotransferase.

            The incubation mixture, conditions of sulfotrans-
            ferase assay (PAPS dependent covalent binding), and
            isolation of tRNA-AAF bound adducts were performed
            as described under Materials and Methods. Sulfo-
            transferase activity obtained following hydroxy-
            apatite chromatography was utilized in all exper-
            iments. Concentrations of various agents: PAPS
            (40 µg per ml), ascorbate (2 mM); NADPH (1 mM),
            PAPS (40 µg per ml) in combination with either
            ascorbate (2 mM) or NADPH (1 mM). (Data from Wirth
            and Thorgeirsson 1980).

In vitro covalent binding of N-OH-AAF to tRNA mediated by partially
purified rat liver N-OH-AAF sulfotransferase

In the absence of PAPS the capacity of partially purified
sulfotransferase (hydroxyapatite fraction) to catalyze the covalent binding of N-OH-AAF to tRNA (Fig. 7) is not significantly
different than that of 105,000 g supernatant fraction. However
following the addition of PAPS there is a 60 fold increase in
extent of the covalent binding of N-OH-AAF to tRNA. Similar to
that observed with both the rat liver 105,000 g supernatant
fraction and the partially purified N-O acyltransferase fraction
both ascorbate (2 mM) and NADPH (1 mM) cause a marked inhibition
in the extent of covalent binding of N-OH-AAF to tRNA mediated by
rat liver sulfotransferase. The addition of PAPS in combination
with either ascorbate or NADPH results in a marked increase (50-
70 fold) in the extent of covalent binding of N-OH-AAF although
the total amount bound was only 10-20% of that observed with PAPS
alone.

Although the mechanism of N-OH-AAF mutagenic activation by
the rat liver 105,000 g supernatant fraction is not fully understood, existing evidence shows the essential involvement of N-O
acyltransferase. Sulfotransferase, however, is not involved in
the mutagenic activation of N-OH-AAF in the Salmonella test
system. (Wirth and Thorgeirsson 1980a). Evidence indicates that
deacetylation of N-OH-AAF via N-O acyltransferase and possibly
free radical formation similar to the mechanism proposed for the
mutagenic activation of N-OH-AAF by rat liver microsomal and
nuclear fractions is responsible for the mutagenic activation of
N-OH-AAF (Sakai et al., 1978). Rat liver N-OH-AAF N-O acyltransferase is a sulfhydryl dependent enzyme that catalyzes the
intramolecular acetyl transfer of N-OH-AAF to form N-acetoxy-
aminofluorene (King 1974). Weeks et al. (1978) have shown that
the mechanism of metabolite activation of N-OH-AAF to an intermediate capable of forming adducts with nucleic acids could be
dissociated from that which induced mutations in Salmonella TA
1538. The addition of varying concentrations of either guanine
monophosphate (GMP) or AF to incubations contining N-OH-AAF,
tRNA, and N-O acyltransferase resulted in a 70% inhibition of
adduct formation while similar additions to the Salmonella test
system had no effect on the mutagenicity of N-OH-AAF mediated by
N-O acyltransferase (Weeks et al., 1978). Therefore if N-acetoxy-
aminofluorene were the mutagenic intermediate formed from N-OH-
AAF then the addition of either AF or GMP would be expected to
decrease the mutagenicity of N-OH-AAF. The role of free radical
formation from N-OH-AAF in the in vitro mutagenic activation as
well as the covalent binding to nucleic acids and proteins either
directly or indirectly by forming, e.g., 2-nitrosofluorene and N-
acetoxy-AAF from nitroxyl free radical is unclear (Bartsch et

al., 1972; Floyd and Soong, 1977).   2-Nitrosofluorene is a potent
direct acting frameshift mutagen in Salmonella typhimurium,
whereas N-acetoxy-2-acetylaminofluorene is only weakly mutagenic.
The mutagenicity of N-acetoxy-2-acetylaminofluorene is increased
several-fold by the addition of either rat liver S-9 or cytosolic
fractions (Stout et al., 1976) or the inclusion of mouse liver
nuclei in the Salmonella test system (Sakai et al., 1978).   Since
ascorbate markedly increases the mutagenicity of N-OH-AAF mediated by either microsomal preparations or purified N-O acyltransferase a possible explanation may be that ascorbate (or NADPH) is
reducing any nitroxyl free radicals formed from N-OH-AAF back to
N-OH-AAF which is then deacetylated to form the highly mutagenic
N-OH-AF via either the membrane bound deacetylase or the soluble
N-O acyltransferase.   In the hydrogen peroxide peroxidase model
system of Walker and Floyd (1979) ascorbate inhibited N-OH-AAF
oxidation and is preferentially oxidized presumably by reducing
any nitroxyl free radicals formed back to N-OH-AAF.

        The role of free radical formation from N-OH-AAF in the in
vitro mutagenic activation as well as covalent binding to tRNA is
not clear.   Both ascorbate and NADPH cause a marked increase in
the mutagenicity of N-OH-AAF while at the same time cause a
marked decrease in the covalent binding of N-OH-AAF to yeast
tRNA.   Since both ascorbate and NADPH reacts via 1 electron
reductions with free radicals the possibility exists that either
ascorbate or NADPH may reduce any free radicals derived from N-
OH-AAF back to the parent hydroxamic acid which could then serve
as a substrate for deacetylation via N-O acyltransferase.   The
net effect would be an increase in the formation of the highly
mutagenic N-OH-AF.   Ascorbate in itself has no effect on the
mutagenicity of N-OH-AF.   Furthermore, ascorbate causes an
increase in the release of acetate from N-OH-AAF when N-OH-AAF
and ascorbate are incubated in the presence of rat liver 105,000
g supernatant fractions (Andrews et al., 1978).   Further mutagenic activation of the initially formed N-OH-AF to either a free
radical or a nitrenium ion within the bacteria similar to that
which we have proposed for the mutagenic activation of N-OH-AAF
via the membrane bound deacetylase is also possible (Sakai et
al., 1978).

        Although sulfotransferase is not involved in the mutagenic
activation of N-OH-AAF in the Salmonella test system, the sulfate
ester of N-OH-AAF is directly mutagenic in the Bacillus subtilis
system (Maher et al., 1968).   The susceptibility of rats and mice
to hepatic carcinogenesis by N-OH-AAF correlates well with liver
sulfotransferase activity and carcinogenesis can be enhanced by
the administration of sodium sulfate (Weisburger et al., 1972).
The sulfotransferase contributes significantly to the in vitro
covalent binding of N-OH-AAF to cytosolic proteins (Stout et al.,
1976) and yeast tRNA (Fig. 8) but it does not contribute signifi-

cantly to the in vitro covalent binding of N–OH–AAF to nuclear
DNA, RNA or proteins (Sakai et al., 1978). Since presumably the
process of N–OH–AAF induced carcinogenesis in the liver involves
both initiation and promotion we would like to propose that the
role of N–OH–AAF sulfotransferase in N–OH–AAF induced liver
carcinogenesis is that of a promoter (Fig. 9, Thorgeirsson et
al., 1980b). N–OH–AAF is metabolized via sulfotransferase to the
highly reactive and cytotoxic sulfate ester which causes cell
death and induces subsequent cellular proliferation (Thorgeirsson
et al., 1976; Irving, 1975). This cellular proliferation may in
turn promote the process of carcinogenesis of "initiated" cells
which have been transformed by N–OH–AF or subsequent metabolites
of N–OH–AF.

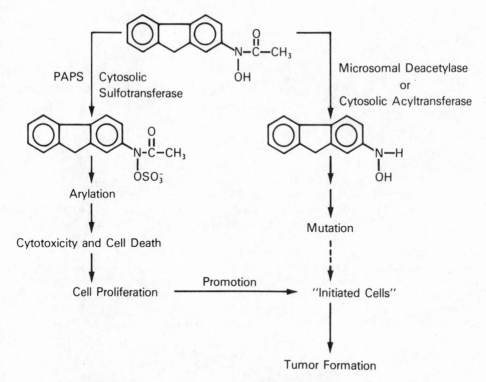

Figure 8   Proposed role of sulfotransferase in N–OH–AAF induced
           carcinogenesis

Recently we have been utilizing isolated hepatocytes together
with the Salmonella tester strains as a useful system to assess
the formation and excretion of mutagenic metabolites from a whole
cell system. The isolated hepatocyte possesses both the activating (P-450, deacetylase, acyltransferase) and detoxifying enzymes
(P-450, glucuronyltransferase) and such a system might be more
informative than the use of subcellular fractions in that the role
of competing metabolic pathways can simultaneously be studied.

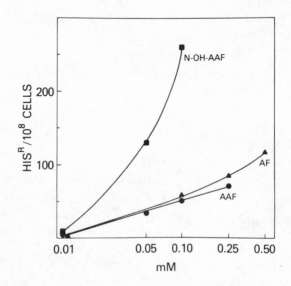

Figure 9   Mutagenicity of AAF, AF, and N-OH-AAF in the
            Salmonella/hepatocyte system. The data points
            (●——● AAF, ▲——▲ AF, and ■——■ N-OH-AAF were deter-
            mined after 1 hr. incubation and are means of three
            experiments. (Data from Staiano et al. 1980)

Fig. 9 shows the mutation frequency of histidine independent
revertants per 10^6 hepatocytes after one hour incubation with
increasing concentrations of AAF, AF, and N-OH-AAF. N-OH-AAF was
5-6 times more mutagenic than either AAF or AF. The mutagenic
effect of AF was slightly greater than that of AAF (Fig. 9) possibly due to greater formation of the hydroxylamine from AF than
from AAF (Thorgeirsson et al., 1980a).

In the isolated hepatocyte system, the mutagenic species derived from the promutagen and/or procarcinogen must escape the detoxification processess of the hepatocyte in order to mutate the bacteria. It is likely, however, that these reactive metabolites would also cause toxicity, including genotoxicity, in the hepatocyte. In addition the more reactive metabolites might not escape the cell at all and may exert their toxic effects solely on the hepatocyte. In addition to determining the mutation frequency of various promutagens in the Salmonella/hepatocyte system this same system can also be utilized to determine the genotoxic effects of these agents to the hepatocyte, as measured by a DNA alkaline elution technique (Kohn et al., 1976; Kohn et al., 1979; Kohn 1979). The DNA alkaline elution technique has proven to be a powerful and sensitive method for the measurement of a variety of DNA lesions produced in mammalian cells by various types of DNA damaging agents. Briefly the technique measures the rate at which single strands of DNA, released in alkali, are able to pass through filters. The rate of passage is dependent upon DNA strand length and the presence of either DNA-DNA interstrand crosslinks or DNA-protein crosslinks. The rate of elution of the DNA through the filters gives sensitive and convenient measurements of chemically induced DNA strand breaks and crosslinks. A more detailed discussion of the methodology and mechanisms has appeared elsewhere (Kohn 1979). In Fig. 10 are shown the alkaline elution curves for DNA damage in rat hepatocytes after one hour incubation with AAF (0.10 and 0.25 mM), AF (0.10 and 0.50 mM), and N-OH-AAF (0.01 and 0.10 mM). The DNA damage induced by these compounds shows the same pattern as is observed for their mutagenicity, namely, N-OH-AAF induces the most DNA damage whereas AAF and AF cause similar DNA damage (Fig. 9,10). Fig. 9 and 10 also clearly illustrate the differences in sensitivity between the bacterial mutagenicity and DNA damage by N-OH-AAF. Since the metabolic activation of N-OH-AAF may proceed via several enzyme systems (Thorgeirsson et al., 1980a,b) it seems likely that more than one of the reactive metabolites formed could damage the hepatocyte DNA. However, the mutagenic activation of N-OH-AAF, in the classical Salmonella system in which either liver microsomal or S-9 fractions are used, has been shown to proceed almost exclusively via deacetylation. The difference in the sensitivity between the hepatocyte DNA damage and the bacterial mutagenicity of N-OH-AAF may therefore reflect selectivity in the bacterial mutagenicity of this compound. Whether the deacetylation of N-OH-AAF is an important activation process, as is the case for the bacterial mutagenicity, for damaging the hepatocyte DNA remains to be established.

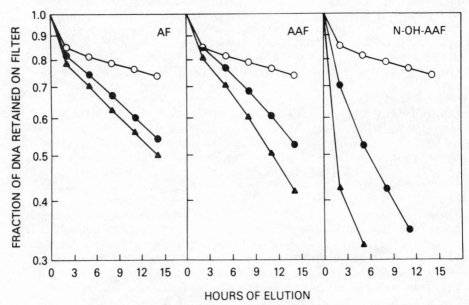

Figure 10   Kinetics of DNA alkaline elution in hepatocytes
            after 1 hr incubation with AF, AAF, and N-OH-AAF
            in the Salmonella/hepatocyte system The data
            illustrate representative experiments, o——o,
            control; •——• and ▲——▲, are respectively
            0.10 and 0.50 mM for AF; 0.10 and 0.24 mM for
            AAF and 0.01 and 0.10 mM for N-OH-AAF. (Data
            from Staiano et al. 1980)

        The alkaline elution technique allows for quantitation of
chemically induced DNA damage to be compared with radiation
induced DNA damage. This is shown in Fig. 11 in which hepatocytes were exposed to increasing dose of X-rays and the alkaline
elution measured. Since the DNA break frequency for X-rays is
2.7×10^{-12} breaks per dalton per rad (Kohn et al., 1976), the
approximate number of breaks for AF (0.5 mM), AAF (0.25 mM) and
N-OH-AAF (0.01 mM) is 1.7×10^{-10}, 2.4×10^{-10} and 7.8×10^{-10}
per dalton, respectively.

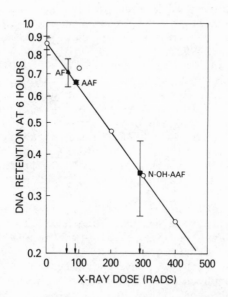

Figure 11   Comparison of DNA damage after X-ray and AF, AAF
            and N-OH-AAF treatment of hepatocytes. The data is
            expressed as mean ± S.D. of three or four experiments.
            (Data from Staiano et al. 1980)

        At present we do not know whether the breaks are due to a
direct action of the compounds on DNA, or result from repair
processes (nuclease action) of the hepatocyte. However, the
Salmonella/hepatocyte system can also be used to study the DNA
repair in the hepatocytes and thereby complement both the bacterial mutagenesis and DNA elution data when genotoxicity of
chemicals are evaluated.

REFERENCES

Ames, B. N., McCann, J., and Yamaski, E., 1975, Methods for de-
    tecting carcinogens and mutagens with Salmonella/mammalian
    microsome mutagenicity test, Mutat. Res., 31:347.
Andrews, L. S., Hinson, J. A., and Gillette, J. R., 1978, Studies
    on the mutagenicity of N-hydroxy-2-acetylaminofluorene in
    the Ames-Salmonella mutagenicity test system, Biochem.
    Pharmacol., 27:2399.
Andrews, L. S., Fysh, J. M., Hinson, J. A., and Gillette, J. R.,
    1979, Ascorbic acid inhibits covalent binding of enzyma-
    tically generated 2-acetylaminofluorene-N-sulfate to DNA
    under conditions in which it increases mutagenesis in
    Salmonella TA 1538, Life Sci., 24:59.
Bartsch, H., and Hecker, E., 1971, On the metabolic activation
    of the carcinogen N-hydroxy-2-aminofluorene. III Oxidation
    with horseradish peroxidase to yield 2-nitrosofluorene and
    N-acetoxy-N-2-acetylaminofluorene, Biochim. Biophys. Acta.,
    237:567.
DeBaun, J. R., Miller, E. C., and Miller, J. A., 1970, N-Hydroxy-
    2-acetylaminofluorene sulfotransferase: Its probable role
    in carcinogenesis and in protein-(methion-S-yl) binding in
    rat liver, Cancer Res., 30:577.
Felton, J. S., Nebert, D. W., and Thorgeirsson, S. S., 1976,
    Genetic differences in 2-acetylaminofluorene mutagenicity
    in vitro associated with mouse hepatic aryl hydrocarbon
    hydroxylase induced by polycyclic aromatic compounds.
    Mol. Pharmacol., 12:255.
Floyd, R. A., and Soong, L. M., 1977, Obligatory free radical
    intermediate in the oxidative activation of the carcinogen
    N-hydroxy-N-acetyl-2-aminofluorene, Biochim. Biophys. Acta.,
    498:244.
Floyd, R. A., Soong, L. M., and Culver, P. L., 1977, Horseradish
    peroxidase/hydrogen peroxide catalyzed oxidation of the car-
    cinogen N-hydroxy-N-acetyl-2-aminofluorene as effected by
    cyanide and ascorbate, Cancer Res., 36:1510.
Irving, C. C., 1975, Comparative toxicity of N-hydroxy-2-acetyl-
    aminofluorene in several strains of rats, Cancer Res.,
    35:2959.
Johnson, E. F., Levitt, D. S., Muller-Eberhard, U., and
    Thorgeirsson, S. S., 1980, Evidence that multiple forms of
    cytochrome P-450 catalyze divergent pathways of 2-acetyl-
    aminofluorene metabolism, Cancer Res., (in press).
King, C. M., 1974, Mechanism of reaction, tissue distribution,
    and inhibition of arylhydroxamic acid acyltransferase,
    Cancer Res., 34;1503.

Kohn, K. W., 1979, DNA as a target in cancer chemotherapy: Measurement of macromolecular DNA damage produced in mammalian cells by anticancer agents and carcinogens, in: "Methods of Cancer Research," Vol. 16, Busch, H. and DeVita, V. eds., Academic Press, New York.

Kohn, K. W., Erickson, L. C., Ewig, R. A. G., and Friedman, C. A., 1976, Fractionation of DNA from mammalian cells by alkaline elution, Biochemistry, 15:4629.

Kohn, K. W., Ewig, R. A. G., Erickson, L. C., and Zwelling, L. A., 1979, Measurements of strand breaks and crosslinks in DNA by alkaline elution, in: "DNA Repair: Laboratory Manual of Research Techniques," Friedberg and P. Hanawalt, eds, Marcel Dekker, New York.

Maher, V. M., Miller, E. C., Miller, J. A., and Szybalski, W., 1968, Mutations and decreases in density of transforming DNA produced by derivatives of the carcinogens 2-acetylaminofluorene and N-methyl-4-aminoazobenzene, Mol. Pharmacol., 4:411.

Miller, J. A., 1970, Carcinogenesis by chemicals: An overview, G. H. A. Clowes memorial lecture, Cancer Res., 30:559.

Mulder, G. J., Hinson, J. A., Nelson, W. L., and Thorgeirsson, S. S., 1977, Role of sulfotransferase from rat liver in the mutagenicity of N-hydroxy-2-acetylaminofluroene in Salmonella typhimurium, Biochem. Pharmacol., 26:1356.

Sakai, S., Reinhold, C. E., Wirth, P. J., and Thorgeirsson, S. S., 1978, Mechanism of in vitro mutagenic activation and covalent binding of N-hydroxy-2-acetylaminofluorene in isolated liver cell nuclei from rat and mouse, Cancer Res., 38:2058.

Schut, H. A. J., Wirth, P. J., and Thorgeirsson, S. S., 1978, Mutagenic activation of N-hydroxy-2-acetylaminofluorene in the Salmonella test system: The role of deacetylation by liver and kidney fractions from mouse and rat, Mol. Pharmacol., 14:682.

Staiano, N., Erickson, L. C., and Thorgeirsson, S. S., 1980, Bacterial mutagenesis and host cell DNA damage by chemical carcinogens in the Salmonella/hepatocyte system, Biochem. Biophys. Res. Comm., 94:837.

Stout, D. L., Baptist, J. N., Matney, T. S., and Shaw, D. R., 1976, N-Hydroxy-2-aminofluorene: The principle mutagen produced from N-hydroxy-2-acetylaminofluorene by a mammlian supernatant enzyme preparation, Cancer Lett., 1:269.

Thorgeirsson, S. S., and Nelson, W. L., 1976, Separation and quantitative determination of 2-acetylaminofluorene and its hydroxylated metabolites by high pressure liquid chromatography, Anal. Biochem., 75:133.

Thorgeirsson, S. S., and Nebert, D. W., 1977, The Ah locus and the metabolism of chemical carcinogens and other foreign compounds, Advan. Cancer Res., 25:149.

Thorgeirsson, S. S., Jollow, D. J., Sasame, H. A., Green, I.,
    and Mitchell, J. R., 1973, The role of cytochrome P-450 in
    N-hydroxylation of 2-acetylaminofluorene, Mol. Pharmacol.,
    9:398.
Thorgeirsson, S. S., Mitchell, J. R., Sasame, H. A., and Potter,
    W. Z., 1976, Biochemical changes after hepatic injury by
    allyl alcohol and N-hydroxy-2-acetylaminofluorene, Chem.
    Biol. Interact., 15:139.
Thorgeirsson, S. S., Wirth, P. J., Nelson, W. L., and Lambert,
    G. H., 1977, Genetic regulation of metabolism and mutageni-
    city of 2-acetylaminofluorene and related compounds in mice,
    in: "Origin of Human Cancer," J. D. Watson and H. Hiatt, eds.,
    Cold Spring Laboratory, New York.
Thorgeirsson, S. S., Schut, H. A. J., Wirth, P. J., and Dybing, E.,
    1980a, Mutagenicity and carcinogenicity of aromatic amines,
    in "Molecular Basis of Environmental Toxicity," R. S.
    Bhatnagar, ed., Ann Arbor Science, Ann Arbor, Michigan.
Thorgeirsson, S. S., Schut, H. A. J., Staiano, N., Wirth, P. J.,
    and Everson, R. B., 1980b, Mutagenicity of N-substituted
    aryl compounds in microbial systems, J. Natl. Cancer Inst.
    Monographs, (in press).
Walker, R. N., and Floyd, R. A., 1979, Free radical activation of
    N-hydroxy-2-acetylaminofluorene by methemoglobin and hydro-
    gen peroxide, Cancer Biochem. Biophys., 4:87.
Weeks, C. E., Allaben, W. T., Louis, S. C., Lazear, E. J., and
    King, C. M., 1978, Role of arylhydroxamie acid acyltrans-
    ferase in the mutagenicity of N-hydroxy-N-2-fluorenylace-
    tamide in Salmonella typhimurium, Cancer Res., 38:613.
Weisburger, J. H. and Weisburger, E. K., 1973, Biochemical formation
    and pharmacological, toxicological, and pathological pro-
    perites of hydroxylamines and hydroxamic acids, Pharmacol.
    Rev., 25:1.
Weisburger, J. H., Yamamoto, R. S., Williams, G. M., Grantham,
    P. H., Matsushima, T. and Weisburger, E. K., 1972, On the
    sulfate ester of N-hydroxy-2-fluorenylacetamide as a key
    ultimate hepatocarcinogen in the rat, Cancer Res., 32:491.
Wirth, P. J., and Thorgeirsson, S. S., 1980a, Mechanism of N-
    hydroxy-2-acetylaminofluorene mutagenicity in the Salmonella
    test system: Role of N-O acyltransferase and sulfotransfer-
    ase from rat liver, Mol. Pharmacol., (in press).
Wirth, P. J., Dybing, E., von Bahr, C., and Thorgeirsson, S. S.,
    1980b, Mechanism of N-hydroxyacetylarylamine mutagenicity in
    the Salmonella test system: Metabolic activation of N-hydroxy-
    phenacetin by liver and kidney fractions from the rat, mouse,
    hamster, and man, Mol. Pharmacol., 18:117.
Wu, S-C, G., and Straub, K. D., 1976, Purification and characteri-
    zation of N-hydroxy-2-acetylaminofluorene sulfotransferase
    from rat liver, J. Biol. Chem., 251:6529.

Yoshida, T., 1933, Uber die Serien Weise Verfolgung der Veranderungen der Leber bei der Experimentellen Hepatomerzeugung durch O-Amidoazotoluol, Trans. Soc. Pathol. Japan, 23:636.

MICROSOMAL METABOLISM OF 2-ACETYLAMINOFLUORENE, ROLE IN MUTAGENICITY AND CARCINOGENICITY

M.Roberfroid, C.Razzouk, E.Agazzi-Léonard,
M.Batardy-Grégoire, F.Poncelet and M.Mercier
Laboratory of Biotoxicology
Université Catholique de Louvain
UCL 73.69 B-1200 Brussels, Belgium

INTRODUCTION

The microsomal metabolism of 2-acetylaminofluorene (2-AAF) includes N-and ring-hydroxylation (Lotlikar et al, 1976), N-deacetylation (Irving, 1966) and, at least in some cases, N → C transoxygenation (Gutmann and Erikson, 1969). N-hydroxylation is recognized as the initial critical step in the metabolic activation leading to mutagenicity and carcinogenicity. Hydroxylation at various positions of the fluorene ring and N → C transoxygenation play an essential role in the detoxification processes. N-deacetylation of the N-hydroxy-metabolite is most probably involved in the activation of 2-AAF to a mutagenic intermediate in the Ames test in vitro (Schut et al, 1978). Pretreatment of the animal with 3-methylcholanthrene (3-MC) largely induces the N-hydroxylase activity in various species including rat, hamster and mouse (Lotlikar et al, 1976). It also increases the N → C transoxygenase activity at least in the rat. (Gutmann and Erikson, 1969). Its effects on the various C-hydroxylases activity are both selective and specific (Lotlikar et al, 1976).

The endoplasmic reticulum-mediated metabolism of 2-AAF thus includes both the reactions of its primary activation and some of the main steps of its detoxification process. By carefully analyzing the biochemical properties of the various microsomal enzymes which catalyze those reactions we hypothesized that it could be possible to better understand the biological effects of 2-AAF.

Such a research requires highly specific and very sensitive methods for quantitatively assaying the various metabolites. We have previously published such a method for measuring both N-hydroxy-2-AAF and N-hydroxy-2-aminofluorene (Razzouk et al, 1977 and Razzouk et al, 1978).
A manuscript is in preparation which will report on a methodology for measuring the ring-hydroxy metabolites (M.Batardy-Grégoire, manuscript in preparation, 1980).

MATERIALS AND METHODS

All details concerning the chemicals, the animals and their pretreatment, the apparatus and the methodologies, the enzymes assay and the conditions of their incubation have been reported in previous publications (Razzouk et al, 1977 and Razzouk et al, 1980 d) including one communication in those proceedings (Razzouk et al.).

RESULTS AND DISCUSSION

Cytochrome P-448 dependent N-hydroxylation is the first, most probably, rate limiting step in the microsomal metabolic activation of 2-AAF to its mutagenic form (Felton et al, 1976).
It plays also a critical role in the mechanism of carcinogenicity by that arylamide (Cramer et al, 1960). It is well known that marked species differences exist in both the enzyme (microsomes)-mediated mutagenicity and the carcinogenicity of 2-AAF. In term of carcinogenicity the rat appears to be the most susceptible species followed by the hamster and finally the mouse (Miller et al, 1960 and Miller et al, 1964). When 2-aminofluorene (2-AF) is used, as a carcinogen, the situation is different : the mouse is the most susceptible species followed by the hamster and finally the rat for which that chemical appears to be a relatively weak carcinogen (Miller et al, 1964). In term of microsomes-mediated mutagenicity in the Ames test in vitro, the hamster liver enzymes are the most potent in activating 2-AAF, followed by the liver enzymes from mouse and finally the rat liver microsomes. Even though those species are resistant to the carcinogenic effect of 2-AAF, both Guinea-pig and monkey liver microsomes are very active in mediating the mutagenicity of 2-AAF in vitro (McGregor, 1975). The first question we asked was thus : does the biochemical analysis of the microsomal N-hydroxylation of 2-AAF and 2-AF explain those biological effects and their species differences ? Table 1 correlates the kinetic parameters of the liver microsomal N-hydroxylase and the susceptibility to the carcinogenic effect of 2-AAF and 2-AF for rat, hamster and mouse.

Table 1. Relationships between species specificity to the liver
carcinogenic effects of 2-AAf and 2-AF and the bio chemical properties of liver microsomal N-hydroxylase.

Animal species	Chemical used	Susceptibility to carcinogenic effect	Kinetic parameters of liver microsomal N-hydroxylase	
			$KM \times 10^{-6}M$	V. max. P. moles $min^{-1}. mg^{-1}.$
Rat	2-AAF	● ● ●	0.53	13.2
	2-AF	*	4.40	41.0
Hamster	2-AAF	● ●	0.93	139.6
	2-AF	* *	2.90	210.0
Mouse	2-AAF	●	1.00	225
	2-AF	* * *	1.60	413.0

With 2-AAF as a carcinogen, there is no correlation between
the N-hydroxylase activity (V_{max}) and the susceptibility to the
carcinogenic effect. The rat is the most susceptible species but
it showed the lowest activity; the mouse is the least susceptible
species but it has the highest enzyme activity. The only correlation which clearly shows up concerns the enzyme affinity. The rat
liver microsomal N-hydroxylase shows the highest affinity for
2-AAf (lowest K_M) whereas the affinity of the mouse liver enzyme
is the lowest (highest K_M). Species differences in enzyme affinity seem thus to correlate with the species susceptibility to the
carcinogenic effect of 2-AAF.

With 2-AF as a carcinogen the rat is the less susceptible species when compared to hamster and mouse. The rat is also the species which shows the lowest liver enzyme affinity for that substrate.
The mouse which is the most susceptible species, also has the
highest liver enzyme affinity. In that case however increase in
enzyme affinity parallels increase in enzyme activity (V_{max}).
Even if it is evident that the affinity of the enzyme which catalyzes the first step in the metabolic activation of 2-AAF or 2-AF
is not the only parameter which controls the species susceptibility to their carcinogenic effect, those results demonstrate that
such a parameter needs further attention.

In terms of mutagenicity in the Ames test in vitro, the differences in the activity of the various microsomal N-hydroxylases
correlated well with the differences in their activating capacities.

With 2-AAF as a promutagen, hamster and mouse preparations are
equally active but rat liver enzymes are much less effective.
Differences in enzyme affinity does not apparently play any role.
This is most probably due to the fact that in vitro mutagenicity
tests are performed at saturating concentrations of cofactors and
substrates.

The resistance of both Guinea-pig and monkey to the carcinogenic effect of 2-AAF has been explained by the inability of those
species to N-hydroxylate that substrate (Irving, 1964). Such an
hypothesis fails however to explain the high capacity of their
liver subcellular preparations to activate 2-AAF to a mutagenic
species in the Ames test in vitro. By applying our methodology we
have clearly demonstrated that both Guinea-pig and monkey liver
microsomes do in fact N-hydroxylate 2-AAF and 2-AF (Razzouk et al,
1980 b and Razzouk et al, 1980 c). Relative to rat, hamster and
mouse, the Guinea-pig is even the most active species in
N-hydroxylating the arylamine. The resistance of Guinea-pig and
monkey to the carcinogenic effect of 2-AAF and 2-AF appears rather
to be due to a high capacity of their endoplasmic reticulum-bound
enzymes to rapidly metabolize the N-hydroxy derivative to a yet
unknown product. Very preliminary results seem to indicate that
Guinea-pig liver microsomes transform N-hydroxy-2-AAF to 7-hydro-
xy 2-AAF. Such a reaction is not catalyzed by rat liver microsomes. Those data, which need further investigations, would be in
agreement with the in vivo findings of high urinary excretion of
7-hydroxy-2-AAF upon feeding Guinea-pig with 2-AAF or N-hydroxy-
2-AAF (Miller et al, 1964). The high capacity of Guinea-pig liver
microsomes to activate 2-AAF to a mutagen in the Ames test in
vitro could then be explained by a difference in the balance
between activation and inactivation, difference due to incubation
conditions which could favor the activation process. The high
capacity of those microsomes to deacetylate (Irving, 1966)
N-hydroxy-2-AAF and perhaps also 2-AAF could be another pathway
which could prevent the transformation to 7-hydroxy 2-AAF.

Among the various position-isomers of acetylaminofluorene,
the isomer in position 2 is the only one which is strongly carcinogenic and mutagenic. This specificity is most probably related
to the ability of the cytochrome P-448-dependent enzymes to N-hydroxylate only 2-AAF.

By determining the kinetic parameters of liver microsomal N-hydroxylase from animals previously pretreated with various position
isomers of acetylaminofluorene, we have shown that, among those
isomers, 2-AAF and 2-AF are the only compounds which modify the
enzyme (Fig.1).

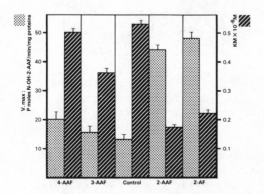

Figure 1. Kinetic parameters of rat liver microsomal N-hydroxy-
          lase. Effects of pretreatments by 2-AAF and its isomers.

Pretreatments of rat with 2-AAF (ip. 10 mg/kg) indeed not
only induce the enzyme activity (V_{max}) by a factor of 3-4 but they
also significantly reduce the K_M by a factor of 2.5-3. Pretreatments with either 4-AAF or 3-AAF neither induce the enzyme activity nor do they modify the enzyme properties. Such effects of
2-AAF on both the V_{max} and the K_M of rat liver microsomal N-hydroxylase are dose-dependent. The effect on the K_M lasts for at least
3 days after one dose of 2-AAF. Upon chronic feeding rat with
2-AAF (0.03 % in diet) this effect is maximum after 12 weeks
(C.Razzouk manuscript submitted for publication 1980 e). As shown
in figure 2, 2-AAF produces the same effects in both hamster and
mouse. In this last species which has the highest basal level of
microsomal N-hydroxylase activity, pretreatment with 2-AAF decreases the enzyme activity. This could be another fact which could
explain the relatively low susceptibility of that species to the
carcinogenic effect of 2-AAF.

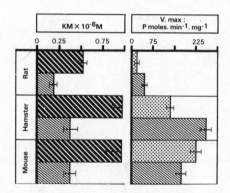

Figure 2. Effects of acute i.p. pretreatment by 2-AAF (10 mg/kg)
        on the kinetic parameters of rat, hamster and mouse
        liver microsomal N-hydroxylase. (Top bars : control,
        bottom bars = 2-AAF treated animals).

   Another difference which needs to be mentioned, refers to
the effects of chronic feeding of 2-AAF and 2-AF to rat on the urinary excretion of their various primary metabolites (Miller et al,
1964).   Chronically feeding rat with 2-AAF increases the excretion
of N-hydroxy-2-AAF and 7-hydroxy-2-AAF without significantly changing the excretion of 3- and/or 5-hydroxy-2-AAF.   If chronically
feeding rat with 2-AF similarly acts on the excretion of both N-
and 7-hydroxy-2-AAF, it also drastically increases the excretion
of the 3- and 5-hydroxy metabolites.   Since we have shown previously (Razzouk et al, 1980 d) that 3- and 5-hydroxy-2-AAF are potent
competitive inhibitors of microsomal N-hydroxylase, that specific
effect of 2-AF could be part of the explanation for its weak carcinogenic effect : chronic feeding increases the production of
metabolites which inhibit the N-hydroxylation, an essential prerequisite for the carcinogenicity.

   Such an indirect effect could also be part of the explanation
for the inhibitory effect of 3-MC on the carcinogenicity of 2-AAF.
It is indeed well-known that simultaneous feeding with 3-MC and
2-AAF protects the rat (Miller et al, 1964) but not the hamster
(Enomoto et al, 1968) against the carcinogenicity of that chemical.
That inhibitory effect has been explained by the inducibility of
ring hydroxylases by 3-MC. We have already reported that this is
only part of the explanation. If it is true that 3-MC or other
polycyclic aromatic hydrocarbons induce the ring hydroxylases, they
also largely induce the N-hydroxylase particularly in the rat liver
(30 fold). We thus suggest two main explanations for that inhibitory effect of 3-MC. The first one derives from the evidences we
have published that, in vitro, 3-MC acts as a very potent competitive inhibitor of 2-AAF N-hydroxylation (Razzouk et al, 1980 a
and Razzouk et al, 1980 d).

The second one derives from the well known effect of 3-MC on the ring hydroxylations of 2-AAF. Chronic feeding with 3-MC (0.05 % in diet), specifically in the rat, induces the urinary excretion of 3- and 5-hydroxy metabolites whereas it does not significantly modifies that excretion in hamster or mouse (Table 2).

Table 2. Effect of chronic 3-MC feeding (0.05 % in diet) on the in vivo microsomal metabolism of 2-AAF as measured by the urinary excretion of the various metabolites.

Metabolites	Animal species		
	Rat	Hamster	Mouse
N-OH	~	↗	↘
C_7-OH	~	~	~
$C_3 + C_5$OH	↗	~	~

Upon feeding of rat with 2-AAF + 3-MC, both compounds reach the liver endoplasmic reticulum at the same time and a competitive inhibition reduces its N-hydroxylating capacity. Upon chronic feeding of rat with 2-AAF + 3-MC, this last compound induces the 3- and 5-hydroxylases and their products, the 3- and 5-hydroxy metabolites, competitively inhibit the N-hydroxylase, thus reducing the primary activation step. In the hamster, if 3-MC acts also as a competitive inhibitor of the N-hydroxylase, it does not induce the C-hydroxylases. Since the hamster liver N-hydroxylase activity is far higher than that of the rat liver the net result is a production of enough N-hydroxy-metabolite to initiate cancer. That indirect effect of 3-MC on the rat liver N-hydroxylase is evidenced by the increase in K_M which was reported previously (Razzouk et al, 1980a and Razzouk et al, 1980 d). The microsomal metabolism of 2-AAF and 2-AF includes both activation and detoxification pathways. There are marked species differences in the biochemical properties of the enzymes catalyzing that metabolism. Various pretreatments of the animals either by the carcinogen itself or by other chemicals (3-MC) markedly induce or modify those enzymes.

An understanding of the mechanism of action of those compounds
require a complete analysis of the kinetic and biochemical properties of their metabolizing enzymes in order to precisely determine
the balance between activation and detoxification pathways and the
interactions between them.

REFERENCES

Cramer, J.W., Miller, J.A. and Miller, E.C., 1960, N-hydroxyla-
        tion, a new metabolic reaction observed in the rat with
        the carcinogen 2-acetylaminofluorene,
        J.Biol.Chem., 235:885-888.
Enomoto, M., Miyake, M. and Sata, K., 1968, Carcinogenicity in
        the hamster of simultaneously administered 2-acetylami-
        dofluorene and 3-methylcholanthrene,
        Gann, 59:177-186.
Felton, J.S., Nebert, D.W. and Thorgeirsson, S.S., 1976, Genetic
        differences in acetylaminofluorene mutagenicity in vitro
        associated with mouse hepatic aryl hydrocarbon hydroxy-
        lase activity induced by polycyclic aromatic compounds,
        Mol.Pharmacol., 12:225-233.
Gutmann, H.R. and Erikson, R.R., 1969, The convefsion of the
        carcinogen N-hydroxy-2-fluorenylacetamide to 0-amidophe-
        nols by rat liver in vitro.
        J.Biol.Chem., 244:1729-1740.
Irving, C.C., 1964, Enzymatic N-hydroxylation of the carcinogen
        2-acetylaminofluorene-9C^{14} in vitro.
        J.Biol.Chem., 239:1589-1596.
Irving, C.C., 1966, Enzymatic deacetylation of N-hydroxy-2-ace-
        tylaminofluorene by liver microsomes.
        Cancer Res., 26:1390-1396.
Lotlikar, P.D.', Enomoto, M., Miller, J.A., Miller, E.C., 1976,
        Species variations in the N- and ring-hydroxylation of
        2-acetylaminofluorene and effects of 3-methylcholanthre-
        ne pretreatments.
        Proc.Soc.Exp.Biol.Med., 125:341-346.
McGregor, 1975, The relationship of 2-acetamidofluorene mutage-
        nicity in plate tests with its in vivo liver cell
        component distribution and its carcinogenic potential,
        Mutat.Res., 30:305-306.
Miller, J.A., Cramer, J.W. and Miller, E.C., 1960, The N- and
        ring-hydroxylation of 2-acetylaminofluorene during carci-
        nogenesis in the rat.
        Cancer Res., 20:950-962.

Miller, E.C., Miller, J.A. and Enomoto, M., 1964, The comparative carcinogenicity of 2-acetylaminofluorene and its N-hydroxy metabolite in mice, hamster and Guinea-pig, Cancer Res., 24:2018-2031.

Razzouk, C., Agazzi-Léonard, E., Batardy-Grégoire, M., Mercier, M., Poncelet, F. and Roberfroid, M., 1980 a, Competitive inhibitory effect of microsomal N-hydroxylase. A possible explanation for the in vivo inhibition of 2-acetylaminofluorene carcinogenicity by 3-methylcholanthrene, Toxicol.Lett., 5:61-67.

Razzouk, C., Evrard, E., Lhoest, G., Roberfroid, M. and Mercier, M., 1978, Isothermal gas chromatography by wall-coated glass capillary columns, electron-capture detection and a solid injector. Application to the assay of 2-fluorenylacetamide N-hydroxylade activity in a rat-liver microsomal system, J.Chromatogr., 161:103-109.

Razzouk, C., Lhoest, G., Roberfroid, M. and Mercier, M., 1977, Subnanogram estimation of the proximate carcinogen N-hydroxy-2-fluorenylacetamide by gas-liquide chromatography, Anal.Biochem., 83:194-203.

Razzouk, C., Mercier, M. and Roberfroid, M., 1980 b, Characterization of the Guinea-pig liver microsomal 2-fluorenylamine and N-2-fluorenylacetamide N-hydroxylase, Cancer Lett., 9:123-131.

Razzouk, C., Mercier, M and Roberfroid, M., 1980 c, Biochemical basis for the resistance of Guinea-pig and monkey to the carcinogenic effects of arylamine and arylamide, Xenobiotica in press.

Razzouk, C., Mercier, M. and Roberfroid, M., 1980 d, Induction, activation and inhibition of hamster and rat liver microsomal arylamide and arylamine-N-hydroxylase, Cancer Res., in press.

Razzouk, C., Mercier, M. and Roberfroid, M., 1980 e, Induction and modification of rat liver microsomal arylamide N-hydroxylase by various pretreatments, Mol.Pharmacol., submitted.

Schut, H.A.J., Wirth, P.J. and Thorgeirsson, S.S., 1978, Mutagenic activation of N-hydroxy-2-acetylaminofluorene in the Salmonella test systems : The role of deacetylation by liver and kidney fractions from mouse and rat, Mol.Pharmacol., 14:682-692.

FORMATION OF CHEMICALLY REACTIVE METABOLITES OF

PHENACETIN AND ACETAMINOPHEN

J.R. Gillette, S.D. Nelson,[1] G.J. Mulder,[2] D.J. Jollow,[3]
J.R. Mitchell,[4] L.R. Pohl and J.A. Hinson

Laboratory of Chemical Pharmacology, Natl. Heart, Lung,
and Blood Institute, National Institutes of Health
Bethesda, MD 20205 USA

Studies during the past several years have revealed that phenacetin can be converted to chemically reactive metabolites by at least three general pathways (Fig.1). The initial steps of these pathways appear to be the formation of (1) phenacetin-3,4-epoxide, (2) N-hydroxyphenacetin and (3) acetaminophen (paracetamol).

---

[1] Present address: Department of Medicinal Chemistry
University of Washington
School of Pharmacy
Seattle, Washington 98195

[2] Present address: Department of Pharmacology
State Univ. of Groningen
Bloemsingel 1
Groningen, The Netherlands

[3] Present address: Department of Pharmacology
Medical University of South Carolina
Charleston, South Carolina 29401

[4] Present address: Department of Internal Medicine
Baylor College of Medicine
1200 Moursund Ave.
Houston, Texas 77030

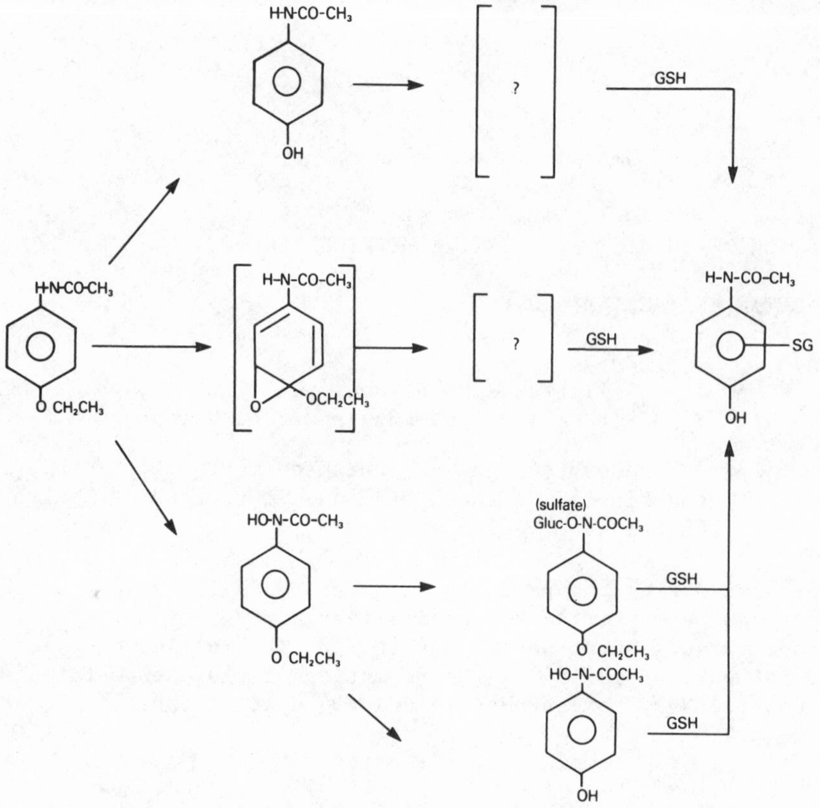

Fig. 1. Pathways for formation of glutathione conjugates from
phenacetin

In this paper, we shall review briefly the evidence for these
pathways and their possible relative importance in the formation
of toxic metabolites of phenacetin and acetaminophen.

Phenacetin-3,4-epoxide pathways. It was discovered several
years ago that incubation of phenacetin with hamster liver microsomes, glutathione and a NADPH generating system resulted in the
formation of glutathionyl acetaminophen but not glutathionyl phenacetin (Hinson et al., 1977). It was also discovered that when
the reaction was carried out in the presence of $^{18}O_2$, exactly 50%
of the phenolic oxygen in the conjugate originated from atmospheric
oxygen. Moreover, studies with [^{18}O-ethoxy] phenacetin revealed
that exactly 50% of the ethereal oxygen of phenacetin was lost
during the formation of the glutathionyl acetaminophen (Hinson et
al., 1979a). The results of these studies thus suggested to us
that phenacetin was converted to phenacetin 3,4-epoxide, which

apparently does not react with glutathione. Instead it appears to
lose the ethyl group to form an intermediate (Intermediate 1),
which may be acetaminophen-3,4-epoxide. The intermediate either
before or after it reacts with glutathione is converted to a symmetrical intermediate in which two hydroxyl groups are attached
to the 4-position of the ring. One of the hydroxyl groups is then
randomly lost during dehydration to form the phenol. At least three
different sequences, however, can be envisioned to occur after the
loss of the ethyl group: 1) Glutathione reacts with Intermediate 1
to form the symmetrical intermediate followed by dehydration. 2)
Intermediate 1 rearranges to the symmetrical intermediate followed
by reaction with glutathione and dehydration. 3) Intermediate 1
rearranges to the symmetrical intermediate which undergoes dehydration to form an intermediate that reacts with glutathione. From
studies of the properties of the intermediates formed during the decomposition of phenacetin-NO-glucuronide and N-hydroxyacetaminophen
(see below), it seems unlikely that the intermediate which reacts
with GSH is either an 4-hydroxy-4-ethoxy derivative or N-acetylimidoquinone.

Studies with $[^{14}C$-ring] phenacetin revealed that hamster liver
microsomes convert phenacetin to a metabolite that becomes covalently
bound to protein (Fig.2) (Nelson et al., 1981). Nevertheless, the
the finding that radiolabel of $[^{14}C$-acetyl] phenacetin was not covalently bound indicated that the bound moiety was neither phenacetin
nor acetaminophen. Addition of various concentrations of glutathione
to the hamster liver microsomal system resulted in a decrease in the
covalent binding of the radiolabel of $[^{14}C$-ring] phenacetin and an
increase in the radiolabels of both $[^{14}C$-acetyl] phenacetin and $[^{14}C$-
ring] phenacetin trapped as a glutathione conjugate. At [GSH] above
1.0 mM the covalent binding of the ring did not appear to be decreased
further. In accord with the finding that glutathionyl acetaminophen
is formed, a plot of $([^{14}C$-ring] conjugate$)^{-1}$ vs $[GSH]^{-1}$ was similar
to a plot of $([^{14}C$-acetyl)] conjugate$)^{-1}$ vs $[GSH]^{-1}$ (Fig.2A and B).
The plots revealed that the calculated GSH concentrations required
for 50% of the maximum amount of GSH conjugate formed, $K_{[N]}$, was
about 1.03 mM with $[^{14}C$-ring] phenacetin and 1.86 mM with $[^{14}C$-acetyl]
phenacetin. The discrepancy between these concentrations is probably
statistically insignificant; a plot (not shown) of $[^{14}C$-acetyl] conjugate vs $[^{14}C$-ring] conjugate formed in the presence of various [GSH]
had a correlation coefficient of 0.996 and a slope of 0.95. By
contrast, a plot of $([^{14}C$-ring] covalent binding$)^{-1}$ vs [GSH] was not
linear (Fig.2D), but on neglecting the data with the highest [GSH]
we found that the [GSH] required to inhibit covalent binding of
the ring label was considerably smaller (about 0.14 mM). Accordingly,
the effects of GSH on the amount of radiolabel trapped as both
covalent binding and the conjugate are biphasic; as the [GSH] is
increased the sum of the trapped radiolabel is decreased and then
increased. Moreover, a plot of the ratio of covalently bound

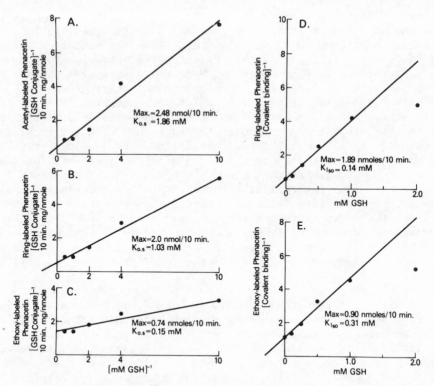

Fig. 2. Plots of the (glutathione conjugates)$^{-1}$ vs [GSH]$^{-1}$
and (covalent binding)$^{-1}$ vs [GSH]. Possible interpre-
tations of the parameters of such plots of steady-state
first order systems are given by Gillette (1980).

radiolabel to the conjugated radiolabel, that is CB/Conjugate, vs
[GSH]$^{-1}$ was nonlinear (Fig.3A). As pointed out by Gillette
(1980), this plot should be linear if GSH reacted with only one
intermediate or with intermediates that are in equilibrium. Thus,
the data suggest that GSH affects covalent binding of the radiolabel
and the formation of the conjugate by reacting with two intermediates;
one results in the formation of the GSH conjugate and the other
leads to covalently bound material. It was still possible, however,
that one of the intermediates was derived from the other. To determine the plausibility of this view one of us (Gillette) derived
equations for a linear plot of sequential reactions under steady-

state conditions as follows:

Model A

$$\frac{III}{}$$

$$I \xrightarrow{k_{12}} II \xrightarrow{k_{25}} V \xrightarrow{k_{56}} VI \begin{bmatrix} \text{Covalently} \\ \text{Bound} \end{bmatrix}$$

with k_{23} (II → III), $k_{24}[N]$ (II → IV), $k_{57}[N]$ (V → VII)

IV — Conjugate   VII

Equation (1)

$$\frac{IV}{[N]VI} = \frac{k_{24}}{k_{25}}\left[1 + \frac{k_{57}[N]}{k_{56}}\right]$$

Thus a plot of the amount of glutathionyl acetaminophen divided by [GSH] and the amount of covalently bound radiolabel vs [GSH] should be linear.

By contrast, if GSH reacted with two intermediates that were formed completely independently of each other, the plot would be curved.

Model B

$$\begin{bmatrix} \text{Covalently} \\ \text{Bound} \end{bmatrix} VI \longleftarrow V \longleftarrow I \longrightarrow II \longrightarrow III$$

with $k_{57}[N]$ (V → VII) and $k_{24}[N]$ (II → IV)

VII   IV — Conjugate

Equation (2)

$$\frac{IV}{[N]VI} = \frac{k_{12}k_{24}\left(k_{56} + k_{57}[N]\right)}{k_{15}k_{55}\left(k_{23} + k_{24}[N]\right)}$$

The finding that the plot of (Conjugate/[GSH] Binding) vs [GSH]
was linear (r = 0.97; Fig.3B) thus suggests that the intermediate
which reacts with GSH to form glutathionyl acetaminophen is a precursor of the other intermediate.

Equation 1 does not contain k_{23} and thus the parameters of the
plot would be independent of any side reaction of the first
intermediate. A comparison of the y intercept of this plot (1.14;
Fig.3B) with the x intercept of the double reciprocal plot [i.e. x =
$-k_{24}/(k_{23} + k_{25})$ = -0.97; Fig.2B) suggests that about 15% of II is
converted to III (Model A). Whether the 15% represents a side pathway or experimental error in the data remains to be determined,
but the simulation of a model lacking the side pathway closely
resembles the plot of the actual data (Fig.3C).

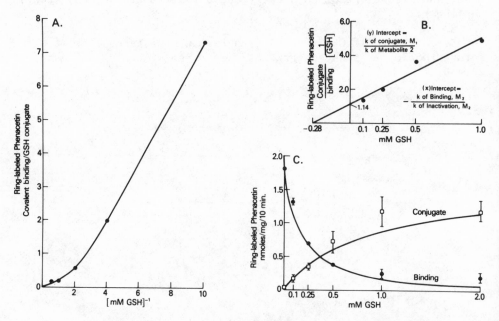

Fig. 3.   The effects of [GSH] on the formation of ring labeled
          glutathione conjugates and ring labeled covalently bound
          metabolites formed from (^{14}C-ring) phenacetin.

In the absence of GSH, the acetyl group of phenacetin is released as acetamide (Nelson et al., 1981). Ascorbic acid inhibits the formation of acetamide in these systems. But ascorbic acid
also inhibits the covalent binding of the ring-label of phenacetin
and, in the presence of GSH, the formation of glutathionyl acetaminophen. It is not possible, therefore, to conclude at this time
whether the ascorbic acid acts solely on the intermediate that reacts with GSH to form glutathionyl acetaminophen or also on the
intermediate derived from it. Amidases in hamster liver microsomes also catalyze the hydrolysis of phenacetin to acetic acid and
phenetidine, but this reaction does not appear to lead to the
chemically reactive metabolites discussed here.

Studies on the metabolism of [^{14}C-ethoxy] phenacetin have clari-
fied the fate of the ethoxy group in these reactions. In the absence
of GSH the radiolabel of [^{14}C-ethoxy] phenacetin becomes covalently
bound to protein in amounts equivalent to about one-half the covalent binding of the label of [^{14}C-ring] phenacetin (Cf.Fig.2D and
2E) (Nelson et al.(1981). Various concentrations of GSH inhibit
the covalent binding of the ^{14}C-ethoxy moiety and lead to the
formation of radiolabeled GSH conjugates. The major conjugate
is S-ethyl glutathione (Nelson et al., 1981). Moreover, studies
with [d_5-ethyl] phenacetin in the presence of N-acetyl cysteine
revealed that all 5 hydrogens of the ethyl group are retained in
the S-ethyl-acetyl cysteine (Nelson et al., 1979). Thus the
conjugate cannot be mediated by acetaldehyde. A plot of (S-ethyl-
glutathione)$^{-1}$ vs [GSH]$^{-1}$ revealed that the [GSH] required to
form one-half the maximum amount of S-ethyl glutathione, $K_{[N]}$,
was about 0.15 mM (Fig.2C). Similarly a plot of (covalently bound
label)$^{-1}$ vs [GSH] revealed that the [GSH] required to inhibit
covalent binding by 50% was about 0.31 mM (Fig.2E). The similarity
of these results suggested that the S-ethyl glutathione and the
covalently bound metabolites may be derived from the same intermediate. In accord with this view a plot of the ratio, covalent
binding/S-ethyl glutathione, vs [GSH]$^{-1}$ was linear (r > .99),
(Fig.4), but the finding that the y intercept is greater than zero
(0.15) suggests that about 13% of the covalent binding may occur
through a completely different mechanism. This is in accord with
the conclusion of Nelson et al.(1981) that GSH in concentrations
greater than 1.0 mM has little further effect on the covalent binding
of the ethoxy label. Inspection of Figs. 2B and C reveals that
the maximum amount of S-ethyl glutathione is only about one-half
of the amount of glutathionyl acetaminophen formed at infinite
concentrations of glutathione. The fate of the other half of the
ethoxy-label remains obscure but clearly it is lost either by a
completely different mechanism or by an intermediate formed
prior to the intermediate that reacts with GSH to form S-ethyl
glutathione. Ascorbic acid also inhibits the covalent binding of
the ethoxy metabolite, but the mechanism remains obscure.

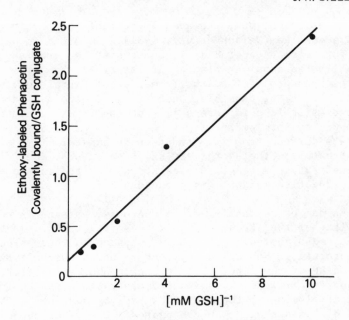

Fig. 4.   Effects of [GSH] on the formation of ethoxy labeled
          glutathione conjugate and ethoxy labeled covalently
          bound metabolites formed from (^{14}C-ethoxy) phenacetin.

        Much has been learned concerning the chemical and kinetic
properties of the chemically reactive metabolites formed directly
from phenacetin, but a discussion of the identity of the intermediates would be largely speculation, until we learn the fate of
the rest of the ethoxy group and whether the covalently bound ethoxy
group is still attached to the ring of acetaminophen. We also need
to identify the products formed in the presence of ascorbic acid
and low glutathione concentrations.

        N-Hydroxyphenacetin. Phenacetin is converted to this metabo-
lite by cytochrome P-450 enzymes in hamster liver microsomes
(Hinson and Mitchell, 1976). Although the metabolite is chemically
rather stable and does not rapidly combine covalently with the
protein it may be converted to chemically reactive metabolites by
several intracellular enzyme systems (Mulder et al., 1977,1978).

        Liver microsomes, treated with Triton X-100, rapidly convert
N-hydroxyphenacetin to phenacetin-NO-glucuronide, but there is a
delay in the covalent binding of the metabolite to protein indicat-

ing that the glucuronide is rather stable. Indeed, the glucuronide
has been isolated and found to have a half-life of 8.7 hr in Tris
buffer, pH 7.4 (Mulder et al., 1978).

Studies of the decomposition products of phenacetin-NO-glucuronide formed in Tris buffer (Mulder et al., 1978) revealed that it
was converted directly to phenacetin, 2-glucuronyl phenacetin and
a chemically reactive intermediate. The chemically reactive intermediate could be covalently bound to protein, reduced to acetaminophen by ascorbic acid or GSH, converted to a GSH conjugate and to a
minor extent hydrolyzed to acetamide (Fig.5). Studies with [^{14}C-
acetyl]-phenacetin-NO-glucuronide and [^{14}C-ethoxy]-phenacetin-NO-
glucuronide revealed that the acetyl label but not the ethoxy
label is covalently bound to protein. The chemically reactive
metabolites formed from phenacetin-NO-glucuronide are thus different
from those formed directly from phenacetin.

In the presence of potassium phosphate buffer, however, phenacetin-NO-glucuronide also decomposes to phenacetin-3-phosphate (Hinson
et al., 1979). Increasing the concentration of inorganic phosphate
results in no changes in the formation of either phenacetin or
phenacetin-2-glucuronide but decreases in the formation of acetamide,
acetaminophen, glutathionyl acetaminophen and covalent binding to
serum albumin. Thus, in the presence of phosphate, phenacetin-NO-
glucuronide must decompose to an intermediate that either reacts
with the phosphate ion or decomposes to a second intermediate
which leads to the formation of acetamide, acetaminophen, glutathionyl acetaminophen and the covalently bound metabolite. Using
a graphical method (Residual/(Uninhibited-Residual) vs [P_i]$^{-1}$)
developed by Gillette (1980), however, Hinson et al. (1979) calculated that the formation of acetamide would have been completely
blocked at an infinite concentration of phosphate, but that only
about 2/3 of the formation of acetaminophen, glutathionyl acetaminophen and covalently bound metabolite would have been blocked.
Similar conclusions may be drawn from plots of acetaminophen/phe-
nacetin-3-phosphate vs [P_i]$^{-1}$ (unpublished results). In the
presence of protein, however, the formation of acetaminophen would
have been completely blocked at an infinite concentration of P_i.
There are two mechanisms that can account for these findings (Fig.5):

Mechanism I; Phenacetin-NO-glucuronide decomposes to a single
intermediate that either decomposes to acetamide, acetaminophen,
glutathionyl acetaminophen, and covalently bound metabolite or
reacts with phosphate to form a phosphate intermediate. The phosphate intermediate, in turn, either is converted to phenacetin-3-
phosphate or decomposes to a secondary intermediate that leads to
the formation of acetaminophen, glutathionyl acetaminophen and
covalently bound metabolite but not to acetamide. It seems likely
that GSH reacts solely with the secondary intermediate because

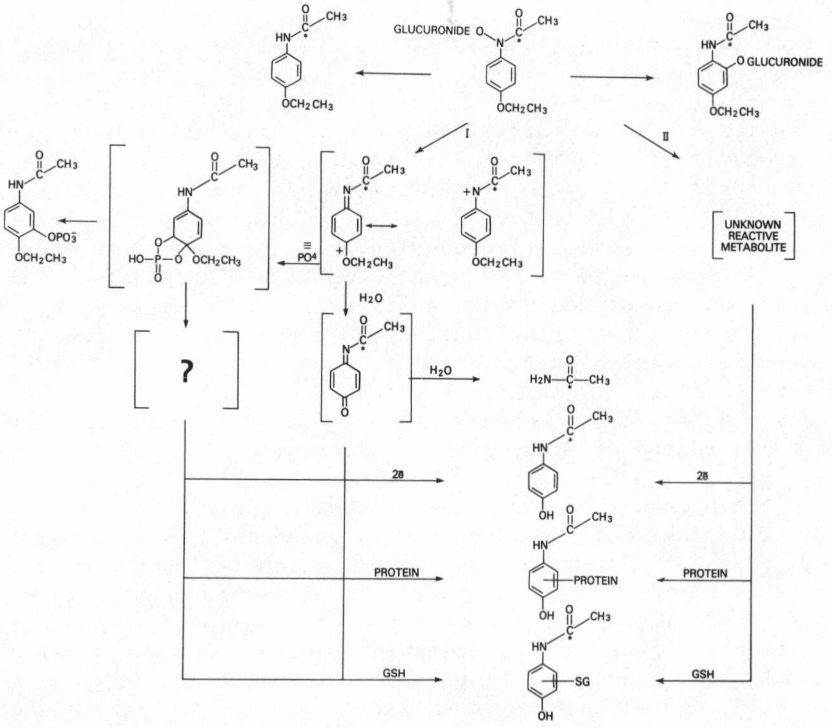

Fig. 5.   Decomposition products of phenacetin-NO-glucuronide in
          Tris and phosphate buffers.

glutathionyl phenacetin was not found as a decomposition product.

    Mechanism II; Phenacetin-NO-glucuronide decomposes to two inter-
mediates that lead to the formation of acetaminophen, glutathionyl
acetaminophen and covalently bound metabolite. But only one of
the intermediates reacts with phosphate or leads to the formation
of acetamide (and presumably quinone). Unfortunately we cannot differentiate between mechanisms I and II from the available data.

    Since the ^{14}C-ethoxy label was not covalently bound to serum
albumin (Mulder et al., 1978), the protein must react with an
intermediate that lacks the ethoxy group. The mechanism by which
the ethoxy group is lost, however, is rather unusual. After incubation of phenacetin-NO-glucuronide and glutathione in the presence
of $H_2{}^{18}O$, the glutathione conjugate was isolated and treated
with Raney nickel, which reductively cleaves glutathione from the

conjugate. Mass spectrometry revealed not only that the conjugate
was glutathionyl acetaminophen, but it also revealed that the
ethoxy oxygen of phenacetin was completely replaced by H_2O
oxygen (Hinson et al., 1979). These findings thus support the
view that phenacetin-NO-glucuronide decomposes to an intermediate
with which hydroxyl ions react in the para position to form N-acetyl-
imido-4-hydroxy-4-ethoxy-cyclohexyldiene. This secondary intermediate
would then decompose to ethanol and N-acetylimidoquinone, which in
turn would react with glutathione or protein. Thus, unlike the metabolite formed directly from phenacetin by hamster liver microsomes,
significant amounts of the acetyl label of N-acetylimidoquinone
are covalently bound to protein, and the phenolic oxygen of the
glutathionyl acetaminophen comes from water.

The soluble fraction of liver contains sulfo transferases
which in the presence of a 3'-phosphoadenosine-5'-phosphosulfate
generating system convert N-hydroxyphenacetin to phenacetin-NO-
sulfate (Mulder et al., 1977, 1978). The metabolite decomposes
rapidly to a chemically reactive metabolite; indeed there is no
delay in the covalent binding of the metabolite to protein as there
is during the generation of phenacetin-NO-glucuronide. However,
the chemically reactive metabolites derived from the sulfate ester
are similar to those generated from phenacetin-NO-glucuronide in
that the reactive metabolites react with protein to form covalently bound metabolites that contain the N-acetyl group but not the
ethoxy group of phenacetin. Moreover, the reactive metabolite is
rapidly reduced to acetaminophen by ascorbic acid.

A cytochrome P-450 enzyme in liver microsomes catalyzes the
oxidative O-dealkylation of N-hydroxyphenacetin to form N-hydroxy-
acetaminophen (Hinson et. al., 1979b). Recently, this metabolite
was synthesized (Healey et al., 1978; Gemborys et al., 1978) and
its properties studied. Its stability in buffer is markedly dependent on the pH value. At pH values up to 9, the $t_{1/2}$ decreases with increasing pH values. Indeed the data of Gembroys
et al., (1980) obtained at 37° indicates that the $t_{1/2}$ fits the
equation: $\log (t_{1/2}) = 8.86 - 1.03$ pH within a pH range 3-9.
Above pH 9, the $t_{1/2}$ increases with increasing pH. At pH 7.6,
the $t_{1/2}$ ranged from 6.5 - 9.4 min. Regardless of the pH range,
however, about equivalent amounts of acetaminophen and p-nitrosophenol are formed. Although various mechanisms for the formation
of p-nitrosophenol have been suggested none of them is completely
satisfactory. Healey and Calder (1979) have suggested that N-
hydroxyacetaminophen undergoes a base catalyzed dehydration to form
N-acetylimidoquinone, which reacts with another molecule of N-hydroxyacetaminophen to form acetaminophen and a nitrone. The nitrone
then rearranges N-acetoxy-p-benzoquinoneimine, which undergoes hydrolysis. Although this mechanism would account for the finding of
equivalent amounts of acetaminophen and nitrosobenzene, it does not

account for the findings of Gemborys et al. (1980) that 20 mM
concentrations of ascorbic acid, cysteine, cysteamine and N-acetylcysteine not only prevent the formation of nitrosobenzene but also
increase the half-life of N-hydroxyacetaminophen about 10-fold.
As Gemborys et al. (1980) rightly pointed out, if these substances
acted solely by reducing the nitrone to N-hydroxylacetaminophen the
half-life of N-hydroxyacetaminophen would be increased 2-fold at
most. Gemborys et al. (1980) have suggested that the discrepancy
could be explained by an equilibrium between N-hydroxyacetaminophen and N-acetylimidoquinone, but their reasoning is fallacious.
They apparently failed to recognize that if ascorbic acid and the
thio compounds acted solely by reducing the nitrone to N-hydroxy-
acetaminophen, the effects of these substances on the half-life of
acetaminophen would be independent of whether N-hydroxyacetaminophen
and the N-acetylimidoquinone were in equilibrium or the conversion
of N-hydroxyacetaminophen to N-acetylimidoquinone was irreversible.
Thus, the maximum increase in half-life caused by ascorbic acid or
the thio compounds would still be only 2-fold. Moreover, if N-
hydroxyacetaminophen and the N-acetylimidoquinone were in equilibrium, any reaction between the N-acetylimidoquinone and ascorbic
acid or the thio compounds (whether it results in the formation of
acetaminophen or acetaminophen conjugates) should decrease the half-
life of N-hydroxyacetaminophen, not increase it.

It should be evident that any mechanism that one might wish
to propose must have one of two features: 1) the ascorbic acid
and thio compounds slow the dehydration of N-hydroxyacetaminophen
to N-acetylimidoquinone (which seems unlikely) or 2) in the absence
of ascorbic acid or the thio compounds, N-acetylimidoquinone in
some way promotes its own formation from N-hydroxyacetaminophen.
In the latter kind of mechanism, the half-life of N-hydroxyacetaminophen in the presence of the thio compounds would represent the
true rate of dehydration of N-hydroxyacetaminophen to N-acetylimidoquinone.

In this context it is noteworthy that Hinson, Pohl and
Gillette (unpublished results) found that the half-life of N-
hydroxyacetaminophen in the presence of GSH (51 min) was increased
rather than decreased by the addition of hamster liver microsomes
and a NADPH generating system (61 min). Thus, liver microsomes do
not appear to contain an enzyme that catalyzes the dehydration of
N-hydroxyacetaminophen.

Although N-acetylimidoquinone is presumably formed from both
phenacetin-NO-glucuronide and N-hydroxyacetaminophen, the mechanisms
are quite different. After N-hydroxyacetaminophen was allowed to
decompose in $H_2^{18}O$ in the presence of GSH the glutathionyl acetaminophen was isolated and treated with Raney nickel. Mass spectroscopy of the resulting acetaminophen revealed that the phenolic

oxygen was not exchanged with H_2O oxygen as it was when phenacetin-NO-glucuronide was allowed to decompose. These findings
are thus in accord with the mechanism proposed by Healey et al.
(1978) and Gembroys et al. (1978) that the initial event in the
decomposition of N-hydroxyacetaminophen is the ionization of the
phenolic hydrogen followed by a rearrangement to form an hydroxyl
ion and N-acetylimidoquinone. Although it is likely that N-acetylimidoquinone also undergoes hydrolysis to form acetamide and quinone,
its rate of reaction with protein is considerably faster than the
rate of hydrolysis and thus most of it becomes covalently bound to
protein. Thus, the covalently bound metabolite contains nearly
equivalent amounts of the acetyl and ring labels.

Acetaminophen. The oxidative dealkylation of phenacetin to form
acetaldehyde and acetaminophen was among the first reactions demonstrated to be catalyzed by a mixed-function oxidase in liver microsomes (Axelrod, 1958). In accord with the view that the reaction
occurs through the formation of an unstable hemiacetal that decomposes to the products (Brodie et al., 1958), $[\alpha,\alpha-^2H_2$-ethoxy]-phe-
nacetin is dealkylated less rapidly than phenacetin, which suggests
breakage of the C-D bond is a rate-limiting step (Nelson et al.,
1978). Moreover, when $[p-^{18}O]$-phenacetin was incubated with liver
microsomes, virtually all of the ^{18}O in phenacetin was retained
in acetaminophen (Hinson et al., 1979a).

It is now well established that acetaminophen may be converted
by liver microsomes to chemically reactive metabolites that become
covalently bound to protein. Unlike the pattern of covalently bound
metabolites of phenacetin, however, the N-acetyl group and the ring
of acetaminophen are covalently bound in almost equivalent amounts
(Potter et al., 1973; Hinson et al., 1981b). Thus very little
acetamide is formed during the activation of acetaminophen (Hinson
et al., 1981b).

GSH inhibits the covalent binding of acetaminophen metabolites
(Potter et al., 1973) presumably by forming glutathionyl acetaminophen. As the concentration of GSH is increased, however, the total
amount of metabolite trapped as glutathionyl acetaminophen and covalently bound metabolite increases (Borner et al., 1976; Rollins
and Buckpitt, 1979). Moreover, GSH increases the disappearance of
acetaminophen (Rollins and Buckpitt, 1979), indicating that the
increase in trapped metabolite was not due to incomplete recovery
of covalently bound material. Instead the findings suggested
that GSH may stimulate the formation of the reactive metabolite
(Rollins and Buckpitt, 1979) or that GSH prevents the reduction of
the reactive metabolite by trapping it as glutathionyl acetaminophen.
If GSH stimulates the formation of the reactive metabolite, however,
GSH should increase the rate of acetaminophen dependent NADPH
oxidation by liver microsomes. But if GSH decreased the rate of

conversion of the metabolite back to acetaminophen by trapping the intermediate, GSH should either not alter the acetaminophen dependent NADPH oxidation or decrease it depending on the ultimate source of the electrons. The finding that GSH decreases the acetaminophen dependent NADPH oxidation by an amount equivalent to the increase in trapped metabolites (Table 1), therefore, suggests that the reactive metabolite is easily reduced by NADPH either enzymatically or nonenzymatically (Hinson et al., 1981b). In accord with the view that the reactive metabolite is easily reduced, ascorbic acid markedly inhibits the covalent binding of the metabolite (Hinson et al., 1981a).

Thus, the chemically reactive metabolite appeared to be short lived and easily reducible. At first we suggested that formation of the chemically reactive metabolite was formed by way of N-hydroxyacetaminophen (Jollow et al., 1973; Hinson et al., 1977; Buckpitt et al., 1977). This hypothesis was initially based primarily on the finding of Calder et al. (1974) that N-acetylimidoquinone, a known electrophile, could be formed from N-hydroxyphenacetin under acidic conditions. Thus, it seemed possible that liver microsomes might convert acetaminophen to N-hydroxyacetaminophen which in turn undergoes spontaneous dehydration to N-acetylimidoquinone. Subsequent studies tended to support this hypothesis. For example, treatments of animals that increased the activity of the liver microsomal

Table 1

NADPH oxidation by hamster liver microsomes and
trapped metabolites of acetaminophen*

	Trapped metabolites** (nmoles/6 mg/10 min)	NADPH oxidation***
+ Glutathione (1 mM)	55.5	131
– Glutathione	15.2	170
Difference	40.3	–39

*Data taken from Hinson et al. (1981b).
**Covalently bound metabolites plus glutathione-acetaminophen.
***Increase in NADPH oxidation caused by acetaminophen.

enzymes that catalyzed the N-hydroxylation of p-chloroacetanilide (Hinson and Mitchell, 1976) also increased the covalent binding of acetaminophen by the liver microsomes. Moreover, studies with $H_2^{18}O$ and $[^{18}O]$-acetaminophen indicated that the phenolic oxygen of acetaminophen and N-hydroxyacetaminophen were retained in the formation of glutathionyl acetaminophen (Hinson et al., 1979a).

It is now clear, however, that the principal chemically reactive metabolite of acetaminophen is not formed by way of N-hydroxyacetaminophen. Recent studies have shown that enzymes in hamster liver microsomes catalyze the formation of N-hydroxyacetaminophen from N-hydroxyphenanetin but not from acetaminophen (Fig.6). Nevertheless considerably more covalently bound metabolites occur with acetaminophen than with N-hydroxyphenacetin (Hinson et al., 1979b). Moreover, after incubation of ^3H-acetaminophen and unlabeled N-hydroxyacetaminophen with liver microsomes from mice, no radiolabel was detected in the isolated N-hydroxyacetaminophen fraction (Nelson et al., 1980). It might be suggested that liver microsomes contain an enzyme that catalyzes the dehydration of N-hydroxyacetaminophen to N-acetylimidoquinone. But addition of liver microsomes does not decrease the half-life of N-hydroxyacetaminophen (Hinson, unpublished results). It thus seems more likely that the chemically reactive metabolite is formed directly by cytochrome P-450. In this context, it may be relevant that the activated N-hydroxylase-substrate complex may be visualized as [acetaminophen-N-O-Fe (Cytochrome P-450)]. If the O-Fe bond in the transitory complex is broken the product would be the N-hydoxy derivative, but if the N-O bond is broken, N-acetylimidoquinone would be formed (Fig.7). In the case of acetaminophen, the N-O-bond would preferentially break because of the resonance stabilizing influence of the phenolic group on the incipient nitrenium ion. Rapid ionization of the phenolic group of this intermediate would lead to formation of N-acetylimidoquinone. This mechanism is especially attractive because it can explain why p-chloroacetanilide and phenacetin are N-hydroxylated apparently by the same hepatic microsomal cytochrome P-450 that converts acetaminophen to its reactive metabolite (Hinson et al., 1975; 1976; Hinson and Mitchell, 1976). Since these compounds have no ionizable phenolic group, their nitrenium ions would not be stabilized by resonance to the same degree as would occur with acetaminophen. Consequently their transitory complexes would decompose by cleavage at the O-Fe bond to form the N-OH metabolite. The mechanism would also be consistent with our view that the chemically reactive metabolite of acetaminophen is N-acetylimidoquinone.

It has been suggested from time to time that acetaminophen may be converted to a chemically reactive metabolite by way of an epoxide. The finding that $^{18}O_2$ is not incorporated into glutathionyl acetaminophen by hamster liver microsomes incubated with acetaminophen, however, precludes the possibility that acetaminophen-3,4-epoxide

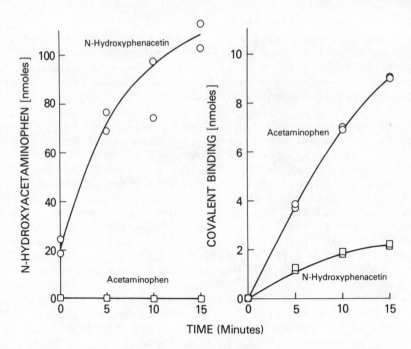

Fig. 6.   Formation of N-hydroxyacetaminophen from N-hydroxy-
          phenacetin and acetaminophen by hamster liver micro-
          somes. Taken from Hinson et al. (1979b).

is a major metabolite of acetaminophen. Nevertheless, the finding
of 3-hydroxyacetaminophen as a metabolite of acetaminophen both in
vivo (as 3-methoxy acetaminophen) (Klutch et al., 1978) and in vitro
(Hinson et al., 1981a; Forte et al., 1981) raises the possibility
that acetaminophen may be converted to acetaminophen-2,3-epoxide.
In accord with this view, studies with $^{18}O_2$ revealed that atmospheric
oxygen served as the source of the 3-hydroxyl group (Hinson et
al., 1981a). Thus, 3-hydroxyacetaminophen cannot be formed by
hydration of N-acetylimidoquinone. Strangely, however, the rate

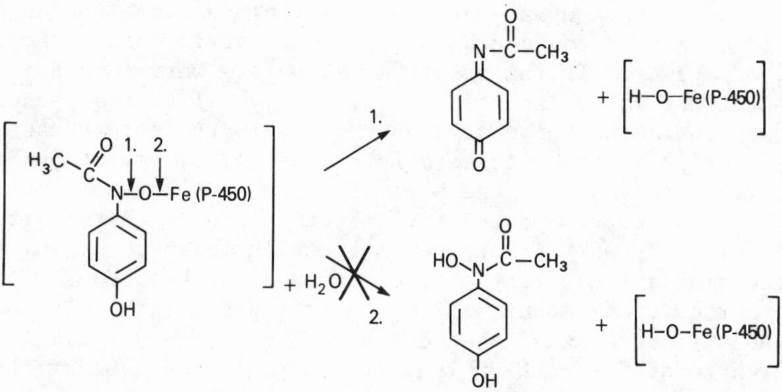

Fig. 7. Possible mechanism for formation of N-acetylimidoquinone directly from acetaminophen.

of formation of 3-hydroxyacetaminophen is not inhibited by GSH, ascorbic acid or epoxide hydrolase, which suggests either that acetaminophen-2,3-epoxide is rearranged almost instantaneously to the catechol derivative or that the catechol derivative is formed by a direct insertion of the oxygen (Hinson et al., 1981a). In either case, there is no evidence that appreciable covalent binding of acetaminophen occurs by way of an epoxide.

The formation of a catechol raised the possibility that it might be oxidized by superoxide generated by liver microsomes to form a chemically reactive metabolite. When ^{3}H-acetaminophen was incubated with hamster liver microsomes, the amount of covalently bound metabolite was not significantly changed by the addition of superoxide dismutase or by a combination of superoxide dismutase and catechol-O-methyl transferase (Hinson et al., 1981a). Thus, this pathway contributes little to the total covalent binding of acetaminophen metabolites.

Relative importance of the pathways of phenacetin metabolism
in vivo. Because of the plethora of possible pathways by which
phenacetin is converted to chemically reactive metabolites, it
seemed important to evaluate their relative importance in vivo.
To this end advantage was taken of the fact that the source of the
phenolic oxygen in the acetaminophen mercapturic acid differed
with the pathway. If the reactive metabolite were formed in
vivo mainly by way of acetaminophen, nearly all of the phenolic
oxygen of the mercapturic acid should originate from phenacetin.
If the reactive metabolite were formed mainly by way of phenacetin-
3,4-epoxide, only 50% of the phenolic oxygen of the mercapturate
should originate from phenacetin. If the reactive metabolite were
formed mainly by way of N-hydroxyphenacetin followed by sulfation
(or glucuronidation), very little, if any, of the phenolic oxygen
of the mercapturate should originate from phenacetin. By administration of $[^{18}O$-ethoxy] phenacetin to hamsters and isolation of
the mercapturate from the urine, it was found that 90% of the ^{18}O
was retained in the mercapturates (Hinson et al., 1979). Thus at
least 80% of the mercapturate was formed by way of acetaminophen.
On the other hand, direct measurements of the relative amounts of
the ethoxy, ring and acetyl labels that are covalently bound to
liver proteins of animals treated with cyloheximide would suggest
that as much as 40-50% of the covalent binding of the ring of phenacetin could occur through the formation of phenacetin-3,4-epoxide
(Nelson et al., 1981a). However, as the authors point out cycloheximide may not have completely inhibited protein synthesis and thus
may not have blocked the incorporation of nonessential amino acids
derived from the two-carbon pools in the body or cycloheximide may
have altered the relative activities of the cytochrome P-450 enzymes.
Thus, the latter estimates may be incorrect.

It is nevertheless evident from this discussion that even rather
simple molecules can be converted to chemically reactive metabolites
by several different pathways. But the relative importance of the
pathways in accounting for various toxicities will depend on several
interrelated factors. Moreover, some precursors may affect the
relative importance of the various pathways. For example, Smith and
Jollow (1978) have shown that large concentrations of phenacetin
inhibit the formation of the chemically reactive metabolite by
hamster hepatocytes. Thus at high doses of phenacetin the pathways
mediated by phenacetin-3,4-epoxide may be initially rather important,
whereas later the acetaminophen pathway may become predominant.

References

Axelrod, J. (1956). The enzymatic cleavage of aromatic ethers.
Biochem. J. 63: 634.
Borner, H., Hinson, J.A., Mitchell, J.R. and Gillette, J.R.
(1976). The enzymatic conjugation of glutathione with the
reactive metabolite of acetaminophen. Pharmacologist 18: 114.
Brodie, B.B., Gillette, J.R. and La Du, B.N. (1958). Enzymatic
metabolism of drugs and other foreign compounds. Ann. Rev.
Biochem. 27: 427.
Buckpitt, A.R., Rollins, D.E., Nelson, S.D., Franklin, R.B. and
Mitchell, J.R. (1977). Quantitative determination of the
glutathione cysteine and N-acetyl cysteine conjugates of acetaminophen by high pressure liquid chromaatography. Analytical
Biochemistry 83: 168.
Calder, I.C., Creek, M.J. and Williams, P.J. (1974). N-Hydroxy-
phenacetin as a precursor of 3-substituted 4-hydroxy-acetamide
metabolites of phenacetin. Chem. Biol. Interact. 8: 87.
Forte, A.J., Wilson, R.J., McMurtry, R. and Nelson, S.D. (1981).
Formation and toxicity of catechol metabolism of acetaminophen.
(in preparation).
Gemborys, M.W., Mudge, G.H. and Gribble, G.W. (1979). Mechanism
of decomposition of N-hydroxyacetaminophen, a postulated toxic
metabolite of acetaminophen. J. Med. Chem. 23: 304.
Gillette, J.R. (1980). Kinetics of decomposition of chemically
unstable metabolites in the presence of nucleophiles: Derivation
of equations used in graphical analyses. Pharmacology 20: 64.
Healey, K., Calder, I.C., Yong, A.C., Crowe, C.A., Funder, C.C.,
Ham, K.N., and Tange, J.D. (1978). Liver and kidney damage induced
by N-hydroxyparacetamol. Xenobiotica 8: 403.
Healey, K. and Calder, I.C. (1979). The synthesis and reaction
of N-hydroxyparacetamol (N,4'-dihydroxyacetanilide). Aust. J.
Chem. 32: 1307.
Hinson, J.A., Mitchell, J.R. and Jollow, D.J. (1975). Microsomal
N-hydroxylation of p-chloroacetanilide. Mole. Pharmacol. 11: 462.
Hinson, J.A. and Mitchell, J.R. (1976). N-Hydroxylation of
phenacetin by hamster liver microsomes. Drug Metab. Disposit.
4: 430.
Hinson, J.A., Mitchell, J.R. and Jollow, D.J. (1976). N-Hydroxylation of p-chloroacetanilide in hamsters. Biochem. Pharmacol.
25: 599.
Hinson, J.A., Pohl, L.R. and Gillette, J.R. (1976). N-Hydroxy-
acetaminophen: A microsomal metabolite of N-hydroxyphenacetin but
apparently not of acetaminophen. Life Sciences 24: 2133.

Hinson, J.A., Nelson, S.D. and Mitchell, J.R. (1977). Studies on the microsomal formation of arylating metabolites of acetaminophen and phenacetin. Mole. Pharmacol. 13: 625.

Hinson, J.A., Nelson, S.D. and Gillette, J. (1979a). Metabolism of (p-^{18}O)-phenacetin: the mechanism of activation of phenacetin to reactive metabolites in hamsters. Mole. Pharmacol. 15: 419.

Hinson, J.A., Pohl, L.R., Monks, T.J., Gillette, J.R. and Guengerich, F.P. (1981a). 3-Hydroxyacetaminophen: A microsomal metabolite of acetaminophen. Evidence against an epoxide as the chemically reactive metabolite of acetaminophen. Drug. Metabolism and Disp. in press.

Hinson, J.A., Sasame, H.A. and Gillette, J.R. (1981b). Some properties of the chemically reactive metabolite formed from acetaminophen by hamster liver microsomes (submitted for publication)

Jollow, D.J., Mitchell, J.R., Potter, W.Z., Davis, D.C., Gillette, J.R. and Brodie, B.B. (1973). Acetaminophen-induced hepatic necrosis. II. Role of covalent binding in vivo. J. Pharmacol. Exp. Ther. 187: 195.

Klutch, A., Levin, W., Chang, R.L., Vane, F. and Conney, A.H. (1976). Formation of a thiomethyl metabolite of phenacetin and acetaminophen in dogs and humans. Clin. Pharmacol. Therap. 24: 287.

Mulder, G.J., Hinson, J.A. and Gillette, J.R. (1977). Generation of reactive metabolites of N-hydroxy-phenacetin by glucuronidation and sulfation. Biochem. Pharmacol. 26: 189.

Mulder, G.J., Hinson, J.A. and Gillette, J.R. (1978). Conversion of the N-O-glucuronide and N-O-sulfate conjugate of N-hydroxy phenacetin to reactive intermediates. Biochem. Pharmacol. 27: 1641.

Nelson, S.D., Vaishav, Y., Mitchell, J.R., Gillette, J.R. and Hinson, J.A. (1979). The use of ^{3}H and ^{18}O to examine arylating and alkylating pathways of phenacetin metabolism in stable isotopes. Proceedings of the Third International Conference (R.E. Klein and P.D. Klein eds.) Academic Press, New York, p.385.

Nelson, S.D., Garland, W.A., Mitchell, J.R., Vaishnav, Y., Statham, C.N. and Buckpitt, A.R. (1978). Deuterium isotope effects on the metabolism and toxicity of phenacetin in hamsters. Drug Metab. Disp. 6: 363.

Nelson, S.D., Forte, A.J. and Dahlin, D.C. (1980). Lack of evidence for N-hydroxyacetaminophen as a reactive metabolite of acetaminophen in vitro. Biochem. Pharmacol. 29: 1617.

Nelson, S.D., Forte, A.J., Vaishnav, Y., Mitchell, J.R., Gillette, J.R. and Hinson, J.A. (1981a). The formation of arylating and alkylating metabolites of phenacetin in hamsters and hamster liver microsomes. Mole. Pharmacol., in press.

Rollins, D.E. and Buckpitt, A.R. (1979). Liver cytosol catalyzed conjugation of reduced glutathione with a reactive metabolite of acetaminophen. Toxicology and Applied Pharmacology 47: 331.

Smith, C.L. and Jollow, D.J. (1978). Acetaminophen hepatotoxicity: Predictive ability of isolated hamster hepatocytes for drug-drug interactions. Federation Proc. 37: 644.

REACTION OF MICROSOMAL AND CYTOSOLIC ENZYMES WITH

N-ARYLACETOHYDROXAMIC ACIDS IN THE PRESENCE OF OXYGEN

Werner Lenk and Ulrich Scharmer

Pharmakologisches Institut der Universität
München, Nussbaumstrasse 26, D-8000 München 2
Fed. Rep. Germany

INTRODUCTION

Thanks to the work of Drs. Elizabeth and James
Miller it is known, that certain polycyclic N-arylacetamides, such as 2-acetylaminofluorene (= 2-AAF) or 4-
acetylaminobiphenyl (= 4-AAB), are carcinogenic in most
animal species, because they are N-hydroxylated in vivo
(1-5), and that the corresponding N-hydroxy derivatives,
the N-arylacetohydroxamic acids, are more toxic than
the parent compounds (1-3). N-Hydroxylation of 2-AAF
and 4-AAB was also demonstrated in vitro (6-9). In contrast, the ferrihemoglobin-forming capacity of N-hydro-
xy-2-acetylaminofluorene (= N-OH-2AAF) and N-hydroxy-4-
acetylaminobiphenyl (= N-OH-4AAB) was found to be lower
than that of certain monocyclic analogs (10).
Monocyclic N-arylacetamides were considered non-toxic
until 1971, because N-hydroxylation in vivo or in vitro
could not be demonstrated at all or was found to proceed at such a slow rate that concentrations of the
corresponding N-hydroxy derivatives large enough to
produce toxic effects were not formed (11-16).
In 1971 Nery (17) put forward the hypothesis that N-
hydroxyphenacetin might be the metabolite responsible
for the toxic effect in humans of an abuse of phenace-
tin-containing analgesics (18), and Jollow et al. (19)
proposed that N-hydroxy-acetaminophen might be the metabolite responsible for the toxic effects of large doses of acetaminophen in laboratory animals and humans.
Own experiments on the toxicity of N-hydroxy-4-chloro-
acetanilide (= N-OH-4ClAA) and N-hydroxyphenacetin

(= NHP) in the rat have shown, that these N-hydroxy
acetanilides exerted low chronic, but high acute toxicity in rats due to their high ferrihemoglobin-forming
capacity (10).
Since such differences in the acute and chronic toxicity between poly- and monocyclic N-arylacetohydroxamic
acids may reflect differences in the metabolic behaviour of the compounds themselves or their derivatives,
experiments on the metabolic fate of N-OH-2AAF, N-OH-
4AAB, N-OH-4ClAA, and NHP and their derivatives have
been carried out, of which the first part is reported
here.

REACTION OF N-ARYLACETOHYDROXAMIC ACIDS WITH MICROSOMAL
AND CYTOSOLIC N-ARYLAMIDASES FROM RABBIT OR RAT LIVER
IN THE PRESENCE OR ABSENCE OF OXYGEN AND NADPH

1. ANALYSIS OF THE REACTION BY SPECTROPHOTOMETRY

On addition of N-OH-2AAF, N-OH-4AAB, N-OH-4ClAA or
NHP to suspensions of rabbit or rat liver microsomes
from phenobarbital-treated animals, at first a type
I binding spectrum is observed, followed by a gain
of absorbance at 455 nm and between 400 and 310 nm,
and a loss of absorbance at 415 nm, indicative for
the formation of a product-adduct of cytochrome p-
450 at the expense of its low spin form. But whereas the increase at 455 nm leveled off after approximately 30 min, indication for a limited binding capacity, the increase between 400 and 310 nm continued for a longer time, indication for a mere accumulation of one or several metabolites. If the N-
arylacetohydroxamic acid was added to microsomal
suspensions reduced with NADPH, a gain of absorbance
at 455 nm was accompanied by a loss of absorbance
at 424 nm and a gain at 407 nm, indication for a
dependence of the product-adduct formation on the
redox state of cytochrome b$_5$. This became even more
obvious, when the microsomes were reduced with NADH
prior to substrate addition. Within the first 36 min
a gain of absorbance at 455 nm was accompanied by a
loss of absorbance at 418 nm and 380 nm, indicative
for the formation of the product-adduct at the expense of the low spin form of cytochrome p-450. But
then, after depletion of NADH, a further increase
of absorbance at 455 nm occurred, accompanied by a

loss of absorbance at 424 nm and a gain at 406 nm,
indication for the increased formation of the product-adduct on re-oxidation of cytochrome b_5. These
optical changes are illustrated in Figure 1 A-C.
In order to analyse the kinetics of the Soret peak
formation, the optical changes were pursued with the
dual wave length mode of an Aminco DW-2a. The results show that:

i) the formation of the 455 nm peak occurred with a
   distinct lag phase, indication for an autocatalytic reaction, by which the relevant ligand of cytochrome P-450 is formed,
ii) after 30 min the increase of absorbance at 455
    nm leveled off, an observation for which the
    true explanation is not known,
iii) the extent of the product-adduct formation dif-
     fered among the four substrates, the order be-
     ing the same for rabbit and rat liver micro-
     somes:

$$N-OH-4ClAA > N-OH-2AAF \geq N-OH-4AAB > NHP$$

This is evidence, that the rate of utilization differs among the substrates, indication for marked
differences in substrate affinity of the microsomal
enzymes. Whereas large differences between the two
monocyclic substrates are apparent, the differences
between the two polycyclic substrates are less distinct. These results are illustrated in Figure 2.
With the iron atom of cytochrome P-450 in the ferric
state, no such spectral changes were observed with
the corresponding N-arylacetamides, arylamides, nitroarenes, or bis-azoxyarenes, however, with nitrosoarenes and arylhydroxylamines the corresponding
product-adducts were formed immediately without detectable lag-phase, since arylhydroxylamines are
instantaneously converted into nitrosoarenes in the
presence of oxygen. This is indication, that the
actual ligand either is the nitrosoarene itself or
structurally closely related to it. This is shown
in Figure 3.
In the presence of NADPH, however, nitroarenes as
well as arylamines produced a transient increase of
absorbance at 455 nm, indication that a small concentration of arylhydroxylamine was formed either
by reduction of the nitroarene or by N-hydroxylation of the arylamine.

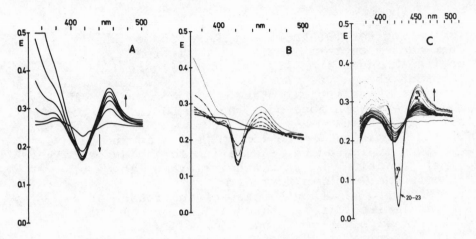

Fig. 1. Spectral changes of microsomal suspensions
         caused by N-hydroxy-4-chloroacetanilide.
         To liver microsomal suspensions from PB-treated
         rabbits (1 mg/ml) in 0.05 M Tris-buffer pH 7.5
         were added:
         Part A. N-Hydroxy-4-chloroacetanilide to give
                 a final concentration of 10^{-3}M. Repeti-
                 tive scanning: 0.5 nm/sec, rate:
                 1 cycle /5 min.
         Part B. NADPH to R- and S-cell to give a final
                 concentration of 10^{-4}M; then N-hydroxy-
                 4-chloroacetanilide was added to the
                 S-cell to give a final concentration
                 of 6 mM. Repetitive scanning: 0.5 nm/
                 sec, 5,10,15,25, and 30 min after the
                 addition of S.
         Part C. NADH to R- and S-cell to give a final
                 concentration of 10^{-4}M. Then N-hydroxy-
                 4-chloroacetanilide was added to the
                 S-cell to give a final concentration
                 of 10^{-3}M. Repetitive scanning: 0.5 nm/
                 sec, rate: 1 cycle/2 min.

When the kinetics of the 455 nm peak formation were
studied in the presence of NADPH, different results
were obtained. With rat liver microsomes, the absorbance at 455 nm more rapidly increased, no lag-phase

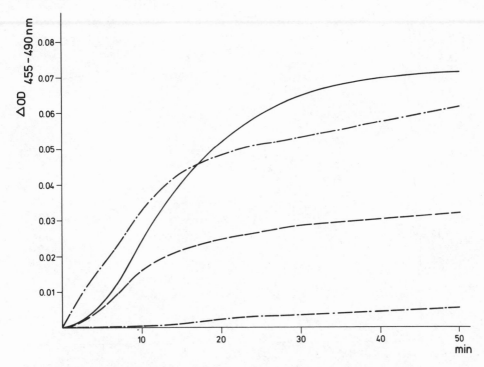

Fig. 2. Kinetics of the product-adduct formation.
Dual wave length scannings of liver microsomal
suspensions from PB-treated rats (1 mg/ml) in
0.05 M Tris-buffer pH 7.5, to which were suc-
cessively added to give final concentrations
of 10^{-3}M:
- - · N-hydroxy-2-acetylaminofluorene
- - - - N-hydroxy-4-acetylaminobiphenyl
———— N-hydroxy-4-chloroacetanilide
- - ·--N-hydroxyphenacetin

being detectable, and suddenly dropped after 3-4 min,
when NADPH was completely oxidized. Thereafter a
slow increase followed comparable to that seen without NADPH. Also with rabbit liver microsomes, the
increase of absorbance at 455 nm occurred rapidly,
but the lag-phase was still detectable. After 10-12
min, when NADPH was completely oxidized, the absorbance at 455 nm decreased, but later on increased
as in the experiments without NADPH. These results

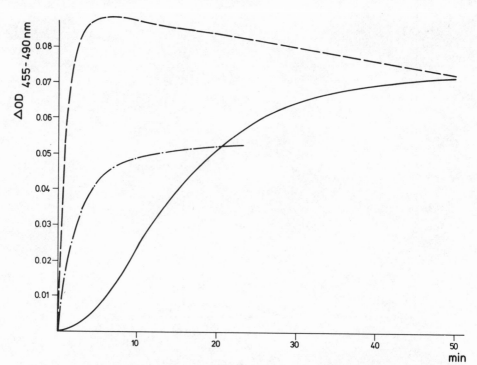

Fig. 3. Kinetics of the product-adduct formation.
Dual wave length scannings of liver microsomal
suspensions from PB-treated rats (1 mg/ml) in
0.05 M Tris-buffer pH 7.5, to which were suc-
cessively added to give final concentrations
of 10^{-3}M:
——— N-hydroxy-4-chloroacetanilide
- - - - N-hydroxy-4-chloroaniline
- · - · 4-nitrosochlorobenzene

indicate,that the product-adduct is more rapidly
formed during the catalytic cycle of cytochrome P-
450. When N-OH-4ClAA was added in the presence of
10^{-6}M paraoxon, a type I binding spectrum was the
only spectral change to be observed. This proves
that N-deacetylation of the N-arylacetohydroxamic
acid is a prerequisite for the formation of the ligand of cytochrome P-450 and for the formation of
the metabolite(s) with an absorption maximum between
400 and 310 nm.
On addition of N-OH-4ClAA to a suspension of rabbit

liver microsomes in an atmosphere of helium, at
first a type I binding spectrum was observed, followed by a gain of absorbance at 424 nm and a loss
of absorbance at 407 nm, indicative for a reduction
of cytochrome b_5 (redox potential: 0 to + 0.013 V,
(20)) by the accumulating arylhydroxylamine (redox
potential: -0.087 to -0.242, depending on the aryl-
substituent, (21)), but not of cytochrome P-450 (redox potential: -0.34 to -0.400 V, (22)). On admis-
sion of oxygen, the 455 nm peak appeared. This proves that oxygen is the second prerequisite for the
formation of the relevant ligand of cytochrome P-
450 and for the metabolite(s) with an absorption
maximum between 400 and 350 nm.

2. ANALYSIS OF THE REACTION BY CHROMATOGRAPHIC RESOLU-
   TION OF THE METABOLITES ACCUMULATED DURING A 30 MIN
   INCUBATION IN THE PRESENCE OF OXYGEN AND IN THE AB-
   SENCE AND PRESENCE OF NADPH

   After incubation of 10^{-3}M N-OH-2AAF, N-OH-4AAB,
   N-OH-4ClAA, and NHP with suspensions of rabbit or
   rat liver microsomes or with the 105 000 x g supernatant, both in the absence or presence of NADPH,
   the major metabolites were identified and determined. Inclusion of 1.2 mM NADPH was found to cause
   only quantitative, but not qualitative changes of
   the metabolite pattern.
   1) <u>Metabolites of N-OH-2AAF</u>: 2-nitrofluorene, cis-
      and trans-bis-azoxyfluorene, 2-aminofluorene,
      2-acetylaminofluorene.
      Rabbit liver microsomes (cytosolic fraction*) N-
      deacetylated 6.9 (13.8) nmol/min/mg protein in
      the absence or presence of NADPH; rat liver micro-
      somes (cytosolic fraction) N-deacetylated 4.5 nmol/
      min/mg protein in the absence and 4.2 (2.4) nmol
      in the presence of NADPH.
   2) <u>Metabolites of N-OH-4AAB</u>: cis- and trans-bis-azo-
      xybiphenyl, 4-aminobiphenyl, 4-acetylaminobiphenyl.
      Rabbit liver microsomes (cytosolic fraction)
      N-deacetlyated 6.6 (11.4) nmol/min/mg protein in
      the absence or presence of NADPH; rat liver mi-
      crosomes (cytosolic fraction) N-deacetylated 6.5
      nmol/min/mg protein in the absence and 5.6 (1.6)
      nmol/min/mg protein in the presence of NADPH.
   3. <u>Metabolites of N-OH-4ClAA</u>: 4-nitrochlorobenzene,
      4-nitrosochlorobenzene, bis-azoxychlorobenzene,
      4-chloroaniline, 4-chloroacetanilide.
      <u>Rabbit liver microsomes (cytosolic fraction)</u>
 * Cytosolic fraction refers to 105,000 x g supernatant.

N-deacetylated 10.5 nmol/min/mg protein in the absence and 12.9 nmol (11.2) nmol/min/mg protein in
the presence of NADPH; rat liver microsomes (cytosolic fraction) N-deacetylated 6.5 (2.2) nmol/min/
mg protein in the absence and presence of NADPH.

4) <u>Metabolites of NHP</u>: 4-nitrophenetole, 4-nitroso-
phenetole, 4-phenetidine, phenacetin.
Rabbit liver microsomes (cytosolic fraction) N-deacetylated 4.9 nmol/min/mg protein in the absence
and 5.7 (3.7) nmol/min/mg protein in the presence
of NADPH; rat liver microsomes (cytosolic fraction) N-deacetylated 3.9 nmoles/min/mg protein in
the absence and 5.8 (4.3) nmol/min/mg protein in
the presence of NADPH.

These results demonstrate a certain order of the rate
of N-deacetylation by rabbit and rat liver microsomes
in the absence of NADPH:

rabbit: N-OH-4ClAA > N-OH-2AAF ≥ N-OH-4AAB > NHP

rat: N-OH-4ClAA = N-OH-4AAB > N-OH-2AAF > NHP

Since the rate of N-deacetylation varied considerably
among the substrates, significant differences could
not be proved in each case due to a limited number of
experiments. This may be the reason that the results
of the chromatographic analysis with rabbit liver
microsomes are consistent with the spectroscopic findings, but not with rat liver microsomes.
These results exceed the results of the spectroscopic
experiments in so far, as the transiently detectable
2-nitrosofluorene and 4-nitrosobiphenyl were quantitatively, but 4-nitrosochlorobenzene was only partly
converted to the corresponding bis-azoxyarenes, whereas 4-nitrosophenetole accumulated. This is certainly true only for the concentration applied (10^{-3}M).
On increasing the concentration of the arylhydroxylamines, first 4-nitrosochlorobenzene and at last 4-
nitrosophenetole were converted to the bis-azoxyarenes, indication for marked differences in the reactivity of the nitrosoarenes. It was found, however,
that bis-azoxyarenes are not formed solely by condensation of nitrosoarenes with an excess of arylhydroxylamine, but also from the nitroso dimer by
stepwise liberation of oxygen, the first step leading to bis-azoxyarenes and the second to azoarenes,
which were occasionally observed in trace amounts.
The explanation being, that in nitrosoarenes with
maximal resonance interaction the tendency to undergo

bimolecular reactions, such as condensation with
nucleophiles or dimerisation is weak as compared to
analogs with reduced resonance interaction.
The results are summarized in Scheme 1.

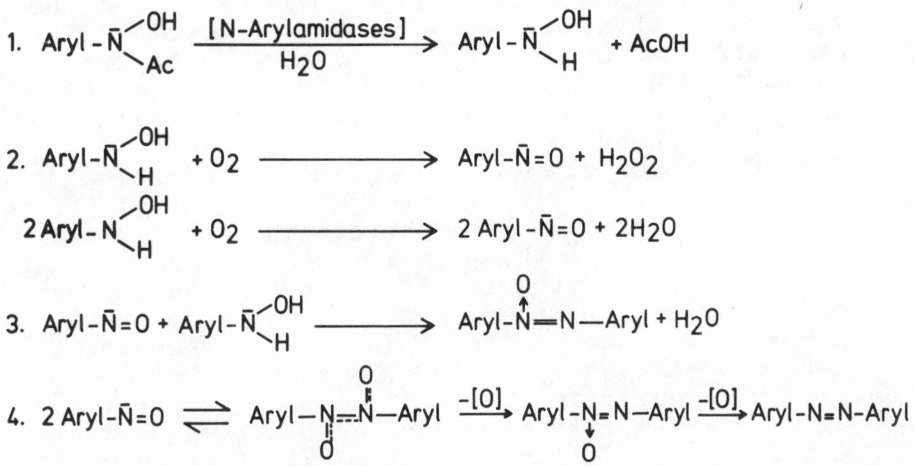

1. $\text{Aryl} - \bar{N} \genfrac{}{}{0pt}{}{\nearrow \text{OH}}{\searrow \text{Ac}}$ $\xrightarrow[\text{H}_2\text{O}]{\text{[N-Arylamidases]}}$ $\text{Aryl} - \bar{N} \genfrac{}{}{0pt}{}{\nearrow \text{OH}}{\searrow \text{H}}$ + AcOH

2. $\text{Aryl} - \bar{N} \genfrac{}{}{0pt}{}{\nearrow \text{OH}}{\searrow \text{H}}$ + O_2 $\longrightarrow$ $\text{Aryl} - \bar{N} = O$ + H_2O_2

   $2\,\text{Aryl} - N \genfrac{}{}{0pt}{}{\nearrow \text{OH}}{\searrow \text{H}}$ + O_2 $\longrightarrow$ $2\,\text{Aryl} - \bar{N} = O$ + $2H_2O$

3. $\text{Aryl} - \bar{N} = O$ + $\text{Aryl} - \bar{N} \genfrac{}{}{0pt}{}{\nearrow \text{OH}}{\searrow \text{H}}$ $\longrightarrow$ $\text{Aryl} - \overset{O}{\overset{\uparrow}{N}} = N - \text{Aryl}$ + H_2O

4. $2\,\text{Aryl} - \bar{N} = O$ $\rightleftharpoons$ $\text{Aryl} - \overset{O}{\underset{O}{\overset{\parallel}{\underset{\parallel}{N}}}} = N - \text{Aryl}$ $\xrightarrow{-[O]}$ $\text{Aryl} - \underset{O}{\overset{}{N}} = N - \text{Aryl}$ $\xrightarrow{-[O]}$ $\text{Aryl} - N = N - \text{Aryl}$

Scheme 1. The reaction of N-arylacetohydroxamic acids
          with microsomal or cytosolic N-arylamidases
          in the presence of oxygen. Depicted are re-
          actions of N-arylacetohydroxamic acids,
          arylhydroxylamines, and nitrosoarenes, which
          have been observed with aryl= -2-fluorenyl,
          -4-biphenyl, -4-chlorophenyl, and -4-ethoxy-
          phenyl.

3. THE REACTION OF ARYLHYDROXYLAMINES WITH OXYGEN

   Since it was the aim of this study to detect differences in the reactivity between polycyclic carcinogenic and monocyclic non-carcinogenic N-arylacetohydroxamic acids and their derivatives, we have also
   studied the reaction of the following arylhydroxylamines with oxygen in the absence and presence of
   microsomes: N-hydroxy-2-aminofluorene (=N-OH-2AF),
   N-hydroxy-4-aminobiphenyl (= N-OH-4AB), N-hydroxy-
   4-chloroaniline (= N-OH-4ClA), and N-hydroxy-4-phe-
   netidine (= N-OH-4Ph).
   Methanol solutions of the individual compounds were
   mixed with 0.066 M phosphate buffer pH 7.4 to give
   final concentrations of 0.05 to 0.07 mM. Time-

dependent optical changes in the cuvette were recorded by using i) split-beam spectrophotometry and ii) dual wavelength spectrophotometry.

The results of the wavelength scannings show, that N-OH-2AF was converted into bis-azoxyfluorene, N-OH-4AB into bis-azoxybiphenyl, N-OH-4ClA into 4-nitrosochlorobenzene, and N-OH-4Ph into 4-nitrosophenetole, indication for clear-cut differences in the reactivity between polycyclic and monocyclic nitrosoarenes, in as much as the polycyclic compounds reacted with an excess of arylhydroxylamine to give the corresponding bis-azoxyarenes, whereas the monocyclic compounds accumulated.

The results of the dual wave length measurements are shown in Figure 4. It is obvious from the graph, that the kinetics of the conversion are complicated by the fact, that the primary oxidation product (nitrosoarene) affected the reaction rate and the formal order of the reaction by accelerating (N-OH-2AF, N-OH-4Ph) or decreasing it (N-OH-4AB, N-OH-4ClA).

Rate constants and the formal order of the conversion were determined to be: N-OH-2AF: $2.0 \times 10^{-3} M^{0.15} sec^{-1}$ (app. formal order of 0.85), N-OH-4AB: $8.2 \times 10^{-4} sec^{-1}$ (app. formal order of 1), N-OH-4ClA: $7.2 \times 10^{-4} sec^{-1}$ (app. formal order of 1), and N-OH-4Ph: $1.5 \times 10^{-3} M^{0.3} sec^{-1}$ (app. formal order of 0.7). From these experiments the following order of oxidation rate is apparent:

N-OH-2AF $>$ N-OH-4Ph $>$ N-OH-4AB $>$ N-OH-4ClA

These results indicate that differences exist in the rate of oxidation, but that they are not clear-cut as to permit a differentiation between the poly- and monocyclic arylhydroxylamines.

It is not surprising that N-OH-4Ph was more rapidly oxidized than N-OH-4AB, and N-OH-4AB more rapidly than N-OH-4ClA, because the electron density at the nitrogen atom (for which the pK_a-value is a good measure) determines the rate of oxidation, and the pK_a-values decrease in this order. As yet unexplained, however, is the unexpected high oxidation rate of N-OH-2AF in relation to its pK_a-value.

When the utilization of oxygen was measured during the conversion of N-OH-4ClA and N-OH-4Ph in the absence of microsomes, a molar ratio of oxygen to arylhydroxylamine utilized of 1 : 3.9 and 1 : 2 was found, respectively. A molar ratio of 1 : 2 is expected, if the primarily formed nitrosoarene does

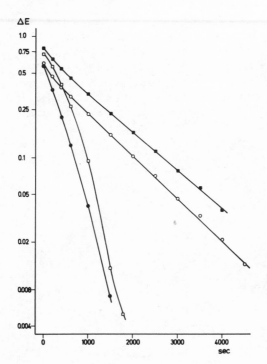

Fig. 4   Analysis of the kinetics and order of the
oxidation of arylhydroxylamines. Plots of
ΔE ($E_{max}-E_t$) versus time in a semi-logarith-
mic scale. E-values were obtained from dual
wave length scannings of arylhydroxylamines
in 0.066 M phosphate buffer pH 7.4, concentration: 0.05 to 0.07 mM. Symbols represent:
●-●-● N-hydroxy-2-aminofluorene (app. $t_{1/2}$ =
260 sec)
o-o-o N-hydroxy-4-aminobiphenyl (app. $t_{1/2}$ =
895 sec)
■-■-■ N-hydroxy-4-chloroaniline (app. $t_{1/2}$ =
950 sec)
□-□-□ N-hydroxy-4-phenetidine   (app. $t_{1/2}$ =
210 sec)

not react further, and a ratio of 1 : 4, if the nitrosoarene reacts further with an excess of arylhydroxylamine. The corresponding values for the initial velocity of oxygen uptake were 6.2 and 10.4 nmol O_2/min,
respectively, see Figure 5. In the presence of microsomal protein, however, a molar ratio of 1 : 2.4 and
1 : 2.9, respectively, was found, and the initial

velocity of oxygen uptake dropped to 4.7 and 6.3 nmol/
min, respectively, indication for the effect of par-
titioning of the lipophilic arylhydroxylamines between the microsomal lipids and the aqueous phase. Due
to the uptake of arylhydroxylamines by microsomal
lipids, the actual concentration was decreased, thereby diminishing the probability for bimolecular reactions to occur.

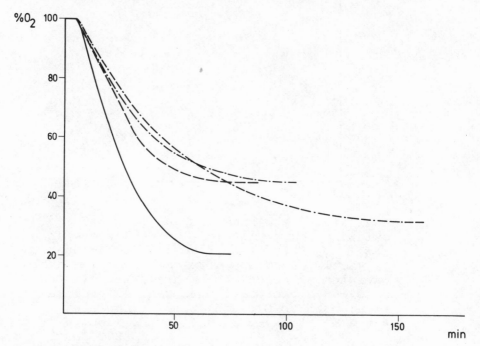

Fig. 5 Measurement of the utilization of oxygen
during the oxidation of N-hydroxy-4-chloro-
aniline (= N-OH-4ClA) and N-hydroxy-4-phene-
tidine (= N-OH-4Ph).
—·—·— N-OH-4ClA in 0.066 M phosphate buffer
pH 7.4, concentration: 0.4 mM.
——·—— N-OH-4ClA in 0.066 M phosphate buffer
pH 7.4, concentration: 0.4 mM in the
presence of microsomal protein (2 mg/ml)
———— N-OH-4Ph in 0.066 M phosphate buffer
pH 7.4, concentration: 0.4 mM.
————— N-OH-4Ph in 0.066 M phosphate buffer
pH 7.4, concentration: 0.4 mM in the
presence of microsomal protein (2 mg/ml)

CONCLUSIONS

1. Due to differences in substrate affinity, microsomal
   and cytosolic N-arylamidases N-deacetylated N-OH-
   2AAF, N-OH-4AAB, N-OH-4ClAA, and NHP with different
   rate, yet the differences are not clear-cut as to
   permit a differentiation between the carcinogenic
   polycyclic and the non-carcinogenic monocyclic compounds.
2. Due to different electron densities at the nitrogen
   atom, differences in the rate of oxidation of the
   corresponding arylhydroxylamines were observed, yet
   these differences are not clear-cut as to permit a
   differentiation between the poly- and monocyclic
   compounds.
3. Due to differences in resonance interaction between
   the aryl substituent and the nitrogen atom, differences in the reactivity between poly- and monocyclic
   nitrosoarenes were observed, which are clear-cut. But
   whether these clear-cut differences can explain differences in the toxicity of the corresponding N-arylacetohydroxamic acids, will further experiments show.
4. The prominent reactivity of N-OH-2AF and of 2-nitro-
   sofluorene has born the working hypothesis, that the
   primary lesion, by which tumor growth is initiated,
   may be caused by this arylhydroxylamine or (and) its
   nitrosoarene rather than by the esters of this N-
   arylacetohydroxamic acid as is postulated in the
   Scheme of Chemical Carcinogenesis accepted today,
   which is shown in Scheme 2.
5. This working hypothesis is supported by: i) results
   of toxicological studies, which prove the carcinogenic and mutagenic properties of N-OH-2AF (26-29) and
   2-nitrosofluorene (28, 30-33), of N-OH-4AB (29, 34)
   and 4-nitrosobiphenyl (28), ii) results of toxicological studies, which prove that only those synthetic
   N-hydroxy derivatives of polycyclic N-aryl-N-acyl-
   amines or N-aryl-N-phenylamines are carcinogenic,
   which are N-deacetylated or N-dearylated in vivo to
   yield the arylhydroxylamine (30, 31, 35), and iii)
   reports on the formation of nitroxide radicals with
   microsomal lipids (36) or membranes (37) or on the
   incorporation of "reactive metabolites" in microsomes, which occurred to a larger extent and was more
   persistent with a metabolite of 2-aminofluorene than
   with a metabolite of aniline (38).

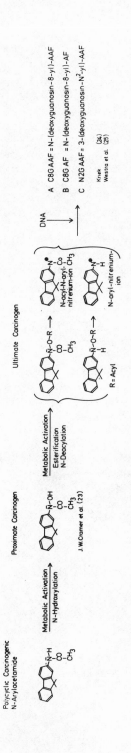

Scheme 2. Biochemical mechanisms involved in chemical carcinogenesis. R may represent different acyl residues, such as the glucuronyl, sulfonic, or acetyl residue. The primary lesion is seen in the formation of adducts with guanine.

REFERENCES

1. E. C. Miller, J. A. Miller, and H. A. Hartmann,
   N-Hydroxy-2-acetylaminofluorene: A metabolite of
   2-acetylaminofluorene with increased carcinogenic
   activity in the rat, Cancer Res. 21: 815 (1961).
2. J. A. Miller, M. Enomoto, and E. C. Miller, The carcinogenicity of small amounts of N-hydroxy-2-ace-
   tylaminofluorene and its cupric chelate in the
   rat, Cancer Res. 22: 1381 (1962).
3. J. A. Miller, C. S. Wyatt, E. C. Miller, and H. A.
   Hartmann, The N-hydroxylation of 4-acetylamino-
   biphenyl by the rat and dog and the strong carcinogenicity of N-hydroxy-4-acetylaminobiphenyl
   in the rat, Cancer Res. 21: 1465 (1961).
4. C. C. Irving, N-Hydroxylation of 2-acetylaminofluo-
   rene in the rabbit, Cancer Res. 22: 867 (1962).
5. W. Fries, M. Kiese, and W. Lenk, Oxidation of polycyclic N-arylacetamides to glycolamides and hydroxamic acids in rabbits,Xenobiotica 3: 525
   (1973).
6. P. D. Lotlikar, M. Enomoto, J. A. Miller, and E. C.
   Miller, Species variations in the N- and ring-
   hydroxylation of 2-acetylaminofluorene and effects of 3-methylcholanthrene pretreatment,
   Proc. Soc. Exptl. Biol. Med. 125: 341 (1967).
7. C. C. Irving, N-Hydroxylation of the carcinogen
   2-acetylaminofluorene by rabbit liver microsomes,
   Biochim. Biophys. Acta 65: 564 (1962).
8. K. Benkert, W. Fries, M. Kiese, and W. Lenk, N-(9-
   Hydroxy-9H-fluoren-2yl)-acetamide and N-(9-oxo-
   9H-fluoren-2yl)-acetamide: metabolites of N-(9H-
   fluoren-2yl)-acetamide, Biochem. Pharmacol. 24:
   1375 (1975).
9. J. Booth and E. Boyland, The Biochemistry of Aromatic Amines 10. Enzymic N-hydroxylation of arylamines and conversion of arylhydroxylamines into
   o-aminophenols, Biochem. J. 91: 362 (1964).
10. H. Hertle, T. Fischbach, M. Kiese, W. Lenk, and
    P. Meister, Toxicity of certain polycyclic and
    monocyclic N-arylacetohydroxamic acids in rats,
    in preparation.
11. M. Kiese and W. Lenk, Metabolites of 4-chloroaniline
    and chloroacetanilides produced by rabbits and
    pigs, Biochem. Pharmacol. 20: 379 (1971).
12. J. A. Hinson, J. R. Mitchell, and D. J. Jollow, N-
    Hydroxylation of p-chloroacetanilide in hamsters,
    Biochem. Pharmacol. 25: 599 (1976).

13. J. A. Hinson, J. R. Mitchell, and D. J. Jollow,
    Microsomal N-hydroxylation of p-chloroacet-
    anilide, Mol. Pharmacol. 11: 462 (1975).
14. W. Lenk, unpublished
15. T. Fischbach and W. Lenk, in preparation.
16. J. A. Hinson and J. R. Mitchell, N-Hydroxylation
    of phenacetin by hamster liver microsomes,
    Drug Metab. Disp. 4: 430 (1976).
17. R. Nery, The possible role of N-hydroxylation in
    the biological effects of phenacetin, Xeno-
    biotica 1: 27 (1971).
18. G. Carro-Ciampi, Phenacetin abuse: A Review,Toxi-
    cology 10: 311 (1978).
19. D. J. Jollow, J. R. Mitchell, W. Z. Potter, D. C.
    Davies, J. R. Gillette, and B. B. Brodie,
    Acetaminophen-induced hepatic necrosis. III.
    Cytochrome P-450-mediated covalent binding
    in vitro, J. Pharmacol. Exp. Ther. 187: 203
    (1973).
20. T. Iyanagi, Redox properties of microsomal reduced
    nicotinamide adenine dinucleotide-cytochrome
    b5 reductase and cytochrome b5, Biochemistry
    16: 2725 (1977).
21. L. Chuang, I. Fried, and P. J. Elving, Voltametric
    behaviour of the system nitrosobenzene-phenyl-
    hydroxylamine at the graphite electrode, Ana-
    lyt. Chem. 36: 2426 (1964).
22. M. R. Waterman and H. S. Mason, Redox properties of
    liver cytochrome P-450, Archs. Biochem. Bio-
    phys. 150: 57 (1972).
23. J. W. Cramer, J. A. Miller, and E. C. Miller,
    N-Hydroxylation: A new metabolic reaction ob-
    served in the rat with the carcinogen 2-ace-
    tylaminofluorene, J. biol. Chem. 235: 885
    (1960).
24. E. Kriek, Difference in binding of 2-acetylamino-
    fluorene to rat liver deoxyribonucleic acid
    and ribosomal ribonucleic acid in vivo, Bio-
    chim. Biophys. Acta 161: 273 (1968).
25. J. G. Westra, E. Kriek, and H. Hillenhausen, Iden-
    tification of the persistently bound form of
    the carcinogen N-acetyl-2-aminofluorene to
    rat liver DNA in vivo, Chem.-Biol. Interac-
    tions 15: 149 (1976).
26. E. C. Miller, J. A. Miller, and M. Enomoto, The
    comparative carcinogenicities of 2-acetylamino-
    fluorene and its N-hydroxy metabolite in mice,
    hamsters, and guinea pigs, Cancer Res. 24:
    2018 (1964).

27. C. C. Irving and R. Wiseman, jr., Studies on the carcinogenicity of the glucuronides of N-hydroxy-2-acetylaminofluorene and N-2-fluorenylhydroxylamine in the rat, Cancer Res. 31: 1645 (1971).

28. B. N. Ames, F. D. Lee, and W. E. Durston, An improved bacterial test system for the detection and classification of mutagens and carcinogens, Proc. Nat. Acad. Sci. USA 70: 782 (1973).

29. R. E. McMahon, J. C. Cline, and C. Z. Thompson, Assay of 855 test chemicals in ten tester strains using a new modification of the Ames Test for bacterial mutagens, Cancer Res. 39: 682 (1979).

30. H. R. Gutmann, D. S. Leaf, Y. Yost, R. E. Rydell, and C. C. Chen, Structure-activity relationships of N-acylarylhydroxylamines in the rat, Cancer Res. 30: 1485 (1970).

31. D. Malejka-Giganti, H. R. Gutmann, R. E. Rydell, and Y. Yost, Activation of the carcinogen, N-hydroxy-2-fluorenylbenzenesulfonamide, by desulfonation to N-2-fluorenylhydroxylamine in vivo, Cancer Res. 31: 778 (1971).

32. E. C. Miller, D. McKechnie, M. M. Poirier, and J. A. Miller, Inhibition of amino acid incorporation in vitro by metabolites of 2-acetylaminofluorene and by certain nitroso compounds, Proc. Soc. Exptl. Biol. Med. N.Y. 120: 538 (1965).

33. E. Hecker, M. Traut, and M. Hopp, The carcinogenic activity of 3-amino-estra-1,3,5(10)-triene and 2-nitrosofluorene, Z. Krebsforsch. 71: 81 (1968).

34. J. W. Gorrod, R. L. Carter, and F. J. C. Roe, Induction of hepatomas by 4-aminobiphenyl and three of its hydroxylated derivatives administered to newborn mice, J. Natl. Cancer Inst. 41: 403 (1968).

35. H. R. Gutmann, S. B. Galitski, and W. A. Foley, The conversion of non-carcinogenic aromatic amides to carcinogenic arylhydroxamic acids by synthetic N-hydroxylation, Cancer Res. 27: 1443 (1967).

36. R. A. Floyd, L. M. Soong, M. A. Stuart, and D. L. Reigh, Free Radicals and Carcinogenesis: Some properties of the nitroxyl free radicals produced by covalent binding of 2-nitrosofluorene to unsaturated lipids of membranes, Archs. Biochem. Biophys. 185: 450 (1978).

37. A. Stier, I. Reitz, and E. Sackmann, Radical accumulation in liver microsomal membranes during biotransformation of aromatic amines and nitro compounds, <u>Naunyn-Schmiedeberg's Arch. Pharmacol.</u> 274: 189 (1972).
38. T. Hultin, Reactions of 2-aminofluorene-9-C^{14} with rat liver proteins, <u>Acta Unio Intern. contra Cancrum (Louvain)</u> 16: 1115 (1960).

DISCUSSION

Elizabeth Miller, commenting on the scheme proposed by Thorgeirsson for the metabolic activation of 2-acetyl-2-aminofluorene,
stated that the acyl transferase may be quite important in tumor
promotion and should be studied further. The main point raised by
Miller was that when activation is carried out by acyl transferase,
presumably all the DNA adducts must bear an acetyl group; however,
when one gives N-OH acetylaminofluorene, only 30% of the adducts in
rat liver bear an acetyl group. On this basis Miller felt that the
system described by Thorgeirsson is not sufficiently analogous to
make a total extrapolation and say that sulfotransferase does not
have an effect in initiation; more data are needed. Oesch asked
whether it was possible that sulfotransferase played no role. Since
the data show a good correlation of sulfotransferase activity with
hepatocarcinogenicity, Oesch raised the question whether a correlation with the N-O acetyl transferase was not also possible. Thorgeirsson pointed out that since sulfotransferase seems to be present
in all target organs for AAF-induced cancer in the rat, one can
probably associate this enzyme with carcinogenicity. Elizabeth
Miller described the study carried out by the Weisbergers who blocked
the carcinogenicity of N-OH acetylaminofluorene by depletion of
tissue sulfate through hydroxyacetanilide administration. If sulfate
was then provided, AAF again became carcinogenic, indicating that
sulfate did play a role. If the sulfate is indeed a promoting
factor, it may be more powerful than an initiating event since small
changes in sulfate concentrations may affect the outcome to a greater
extent than an initiating event. Thorgeirsson then responded negatively to a question by Kato as to whether there was any information
available on the rate of direct covalent binding of N-OH acetylaminofluorene to DNA *in vitro*.

In response to the talk by Roberfroid, McIntosh asked whether
the antagonism seen between chronic treatment with 3-methylcholan-
threne (3-MC) and the 2-acetylaminofluorene (AAF)-induced hepato-
carcinogenesis was also found in female rats. There appears to be
a form of cytochrome P-450 present in adult male rats that is not
present in adult female rats and this male specific form of P-450
is decreased by 3-MC treatment. Consequently, a decrease in this

cytochrome P-450 might provide an explanation for the antagonism;
but this hypothesis would then predict that there would be no antagonism in female rats. McIntosh asked whether such an antagonism
between 3-MC and AAF did occur in female rats. Elizabeth Miller
commented that the original studies demonstrating this antagonism
were done by Richardson who was trying to induce murine tumors in
female rats. Snyder noted that the fact that N-OH-AAF compounds
can produce cancer in guinea pigs seems to conflict with Roberfroid's
observation that such compounds are rapidly destroyed by guinea pigs.
Greim also made the comment that Roberfroid's finding that guinea
pig liver has a high N-hydroxylation activity is in contrast to
previous opinion. Roberfroid had proposed that the resistance of
the guinea pig to AAF carcinogenesis is the result of rapid inactivation of the N-hydroxy product, causing Greim to wonder whether
N-OH acetylaminofluorene undergoes reduction or conjugation with
glutathione. Gorrod offered the observation that his results indicate that aromatic amines are N-hydroxylated by LM2 whereas Thorgeirsson had indicated earlier that they are N-hydroxylated by LM4.
Gorrod wondered if Roberfroid had any data on the comparative
enzymology of the two N-oxidative processes. Roberfroid had no
information on this.

Following the talk by Gillette on reactive metabolites of
phenacetin and acetaminophen, Lenk commented on the interesting finding that the ethyl group of phenacetin was integrated into glutathione
which suggests that further oxidation of the ethyl group is not a
possible alternative pathway. Lenk felt that the ethyl group is in
fact oxidized because during the incubation of phenacetin with
microsomes from phenobarbital-treated animals one can detect the
oxidized ethyl derivative in amounts that are comparable with acetaminophen. Gillette stated that he did detect this derivative,
whereupon Lenk commented that such a pathway was not listed in
Gillette's summary of the metabolism. Gillette pointed out that the
conversion of phenacetin to acetaminophen is an oxidative dealkylation which releases CO_2. Lenk cited work in his laboratory which
indicated that up to 10% of an administered dose of phenacetin to
rabbits pre-induced with phenobarbital was excreted as para-acetyl-
amino phenoxyacetic acid, indicating that this reaction may be a
major pathway. Gillette pointed out that in his studies he was
interested in trapping the SH conjugates of glutathione as mercapturic acids and therefore did not follow up on any side pathways
not involved with GSH conjugation. Gillette then informed Boyland
that the phosphate ester described in the talk is an unstable
compound.

In commenting on the talk by Lenk, Gorrod reinforced what
Elizabeth Miller had stated earlier, that despite the elegant experiments carried out by Dr. Lenk, we are still left with the problem
of the acetyl group still being intact in the DNA adducts in the
sensitive tissue.

METABOLIC ACTIVATION OF HYDRAZINES

Sidney D. Nelson and W. Perry Gordon

Department of Medicinal Chemistry
University of Washington
Seattle, WA   98195, U.S.A.

INTRODUCTION

Substituted hydrazines are receiving considerable attention
in chemical toxicology because of their widespread use as rocket
fuels, herbicides, intermediates in chemical synthesis and as
therapeutic agents for the treatment of tuberculosis, depression
and cancer. Hydrazines can produce many toxic responses including
CNS disturbances such as convulsions, hematological disturbances
such as methemoglobinemia and hemolysis, and tissue damage resulting from fat deposition and cellular necrosis as well as from
mutagenic and carcinogenic events (Juchau and Horita, 1972;
Druckrey, 1973). Although 20 million pounds of hydrazine itself
are produced annually in the USA, little is known of its metabolic
fate. Hydrazine is known to cause liver dysfunction and hematological and CNS disturbances in man and other animals exposed
acutely to high doses, and is lung tumorigenic in mice exposed
chronically to low doses (Toth, 1972). This report will summarize
our results to date on the metabolism of hydrazine both in vivo
in mice, and in vitro using liver microsomes and whole blood obtained from mice.

MATERIALS AND METHODS

Assay for Hydrazine

Hydrazine was assayed by gas chromatography (GC) as its
benzaldehyde azine adduct in a method similar to one already reported (Timbrell et al., 1977) using a nitrogen detector. Aliquots
of whole blood, urine, or homogenized tissue were acidified to
pH6 with phosphate buffer and extracted with methylene chloride.

To the remaining aqueous phase was added benzaldehyde in ethanol which was allowed to react for 30 min. at room temperature. The benzalazine formed was extracted with methylene chloride, and after drying (MgSO4), evaporation of solvent, and reconstitution in 50 μl of ethyl acetate, 1-2 μl was analyzed by GC using an HP 5840A e- equipped with NP detectors and a recording integrator; column, si- lanized glass 3.6 m x 2 mm i.d. packed with 3% OV-17 on 100/120 mesh Gas Chrom Q; oven, 200°C; injector, 250°C; detector, 280°C; He carrier flow, 35 ml/min; hydrogen, 3 ml/min; air, 50 ml/min. Sen- sitivity limits for the assay were approximately 0.5 μg/ml of hydrazine with a standard deviation of ± 0.2 μg/ml using peak area measurements and standard curves. Internal standardization did not significantly improve either the precision or accuracy of our assay.

Assay for Nitrogen

$^{15}N_2$ was analyzed as a metabolite of ^{15}N-hydrazine (Merck Isotopes) by gas chromatography-mass spectrometry and selected-ion-monitoring (GC-MS-SIM). Samples were analyzed on a HP 5985A GC-MS system using a 5 meter x 2 mm i.d. stainless steel column packed with molecular sieve (MS 5A) to separate other gases from nitrogen. $^{15}N_2$ was monitored at m/z 30 and the internal standard, isobutane, at m/z 57 (t-butyl cation). Complete details of the assay will be reported elsewhere. For studies in vivo, mice were placed in a Delmar-Roth all-glass metabolism cage modified to maintain an atmos- phere of oxygen and sulfur hexafluoride, a system suggested to us by Don Reed and Frank Dost at Oregon State University, Corvallis, Ore., USA. For in vitro studies, mouse liver microsomes or whole blood were incubated in septum-sealed vials in an atmosphere of helium and oxygen or modified as stated in the text.

RESULTS

Tissue Distribution of Hydrazine

Tissue distribution studies were carried out with single i.p. doses of 1 mmole/kg of hydrazine sulfate in male Swiss-Webster mice,

a strain of mouse that has been found to be susceptible to hydrazine-induced lung tumorigenesis (Toth, 1972). Hydrazine was distributed rapidly to most tissues with the highest levels appearing in
the kidney (Table 1). Loss of hydrazine from all tissues was extensive by 24 hours. Although the lungs are the primary target
organ for tumorigenicity, selective distribution or retention of
hydrazine by this tissue was not apparent from these studies.

Table 1.   Tissue Distribution of Hydrazine Sulfate
           in Swiss-Webster Mice[a]

Tissue	2 Hours	24 Hours
	$\mu g/g$[b]	
Liver	4.1 ± 0.7	0.6 ± 0.2
Spleen	7.8 ± 2.1	0.3 ± 0.1
Kidneys	22.8 ± 2.6	3.7 ± 0.5
Lungs	7.0 ± 1.0	0.8 ± 0.2
Brain	2.5 ± 0.4	0.3 ± 0.1
Heart	6.6 ± 1.1	0.5 ± 0.1
Stomach	3.7 ± 1.3	0.4 ± 0.2
Intestines	4.0 ± 1.5	0.4 ± 0.2

[a] Tissues removed from 5 male mice at the times
indicated after a dose of 1 mmole/kg of hydrazine
sulfate was administered i.p. in normal saline.
[b] Values represent means ± SE of hydrazine levels
in wet tissues as determined by GC.

After the acute i.p. dose, there was an initial rapid loss of hydrazine from the blood followed by a much slower decline over the next several hours (Figure 1). Approximately 15% of the dose was excreted in the urine over a period of 48 hours as either free hydrazine or labile conjugates, with an additional 25% appearing as hydrazine after heating with 1N HCl for 24 hours at 50°C (Figure 2).

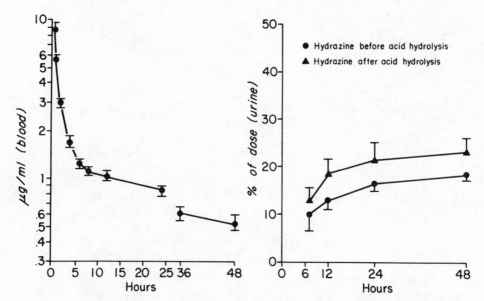

Fig. 1.   Time course of hydrazine loss from whole blood samples taken from Swiss-Webster mice after a dose of 1 mmole/kg i.p. of hydrazine sulfate. Each time point represents the mean ± SE of values obtained from 5 mice.

Fig. 2.   Time course of hydrazine excretion into mouse urine after a dose of 1 mmole/kg of hydrazine sulfate. Each point represents the mean ± SE of values obtained from 5 mice.

Nitrogen Formation from Hydrazine

    Within the first 1-2 hours after the administration of
^{15}N-hydrazine sulfate to male Swiss-Webster mice, approximately 20%
of the dose is expired as $^{15}N_2$ gas and by 48 hours another 10-15%
is eliminated as this metabolite (Figure 3). Much of the oxidation
appears to occur in the blood and requires the presence of oxyhemoglobin, methemoglobin, or carbon monoxyhemoglobin (Figure 4).

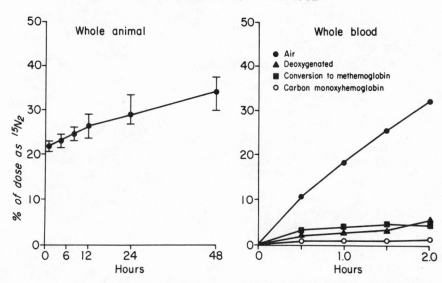

^{15}N-HYDRAZINE CONVERSION TO ^{15}N-NITROGEN

Fig. 3.   Time course for the ex-   Fig. 4.   Time course for the
          piration of $^{15}N_2$ by   formation of $^{15}N_2$ by
          Swiss-Webster mice after   1 ml of whole blood
          the administration of   obtained from Swiss-
          1 mmole/kg of ^{15}N-hydra-   Webster mice under the
          zine sulfate. Each time   conditions indicated.
          point represents the mean   Each time point repre-
          ± SE of values obtained   sents the average of
          from 5 mice.   duplicate determinations.

Mouse liver microsomes were also able to convert hydrazine to nitrogen in a reaction that required enzyme activity, NADPH, and oxygen (Table 2). However, this oxidation reaction was not inhibitable by a carbon monoxide/oxygen atmosphere. Evidence for diimide formation from hydrazine in both the oxyhemoglobin-mediated and microsomal oxygenase-mediated reactions was obtained by co-incubating styrene with hydrazine in the two systems. In both cases the reduced product, ethylbenzene was detected by GC analysis (10% SP-1000 on 80/100 Supelcoport column, 6 m x 2 mm i.d., with temp. programming 6 min at 100°C then 20°C/min to 140°C; retention time of ethylbenzene, 8.8 minutes).

Table 2. Microsomal Oxidation of Hydrazine to Nitrogen

Conditions	N_2 Formation	% of Control
	picomoles/mg protein/min[a]	
Control (He/O_2 8:2)	380 ± 29	100
Heat-denatured	78 ± 8	19
−NADPH	91 ± 8	24
−O_2	64 ± 10	17
CO/O_2 8:2	358 ± 36	94

[a]Values represent means ± SE of 4 incubations containing mouse liver microsomes (2 mg/ml) and NADPH.

Toxicity Studies

The single acute doses of hydrazine used in these studies, although producing substantial quantities of nitrogen presumably
through oxidation by oxyhemoglobin, had little effect on hematological
parameters (Table 3). No significant changes occurred to the red
cell count, total hemoglobin, methemoglobin or hematocrit 6 hours
after 1 mmole/kg doses of hydrazine sulfate to mice. Recently a
report has appeared which shows that high doses of hydrazine repeated over several days is required before significant hematological
disturbances are observed (Jain and Hochstein, 1979).

Table 3. Effects of Hydrazine on Hematological Parameters in Mice

Treatment	RBC Count	Total Hb	MetHb	Hematocrit
	$10^6/\mu l$	g%	% Hb	%
Controls[a]	4.7 ± 0.6	18 ± 2	0.5 ± 0.4	48 ± 1
Treated[b]	4.6 ± 0.5	16 ± 3	0.8 ± 0.3	47 ± 1

[a]Blood samples taken from 10 mice 6 hours after saline, i.p.
[b]Blood samples taken from 10 mice 6 hours after hydrazine sulfate,
1 mmole/kg i.p.

DISCUSSION

The metabolism of hydrazines as it relates to toxicity has
been the subject of several investigations. Studies on the metabolism of the antitubercular drug, isoniazid, and the antidepres-
sant drug, iproniazid, have indicated that the liver injury sus-
tained by some patients receiving these drugs results from their
hydrolysis to simple monosubstituted acyl- and alkyl-substituted
hydrazines which are further oxidized to reactive acylating and
alkylating agents (Mitchell et al., 1975; Nelson et al., 1976;
Nelson et al., 1978; Timbrell et al., 1980). These studies were
extensions of several investigations carried out on the metabolism
and toxicity of mono- and di-alkylhydrazines (Dost et al., 1966;
Prough et al., 1969; Druckrey, 1970).

However, few studies have been carried out on the fate of the
simplest and most prevalent hydrazine in our environment, hydrazine itself, despite several reports on toxicities associated with
exposure to hydrazine (Back and Thomas, 1970; Dambrauskas and
Cornish, 1964; Amenta and Johnston, 1962; Bodansky, 1924; Toth,
1972; Tosk et al., 1979; Wright and Tikkanen, 1980). The only
metabolites that have been detected are the mono- and di-acetylated

products of hydrazine (McKennis et al., 1955). One of the major
problems in determining the metabolic fate of hydrazine is the lack
of an appropriate radiolabeled analog for detection purposes, and a
lack of specific and sensitive assays for hydrazine itself. We,
therefore, employed gas chromatography with a nitrogen detector, and
a stable isotope analog of hydrazine, ^{15}N-hydrazine, coupled with
GC-MS-SIM for a more accurate assessment of the fate of hydrazine
in a mouse model for lung tumorigenesis caused by hydrazine.

The results of our studies to date show that hydrazine is
rapidly distributed to several tissues (Table 1) and is rapidly
metabolized in the Swiss-Webster mouse with 20-25% of the dose expired as nitrogen in the first 2 hours. Both oxyhemoglobin and
liver microsomal oxygenases are capable of oxidizing hydrazine to
nitrogen, and diimide, a powerful diazene reducing agent (Kondo et
al., 1977) has been indirectly detected as a probable intermediate
in the oxidation reaction (Figure 5). Previously, phenyldiazene,
has been shown to be formed in the oxyhemoglobin-mediated oxidation
of phenylhydrazine (Itano, 1975) and both radical and diazonium
products of this diazene have been implicated in hemolytic reactions
caused by phenylhydrazine (Goldberg and Stern, 1975; Misra and
Fridovich, 1976). However, hydrazine is a much less potent inducer

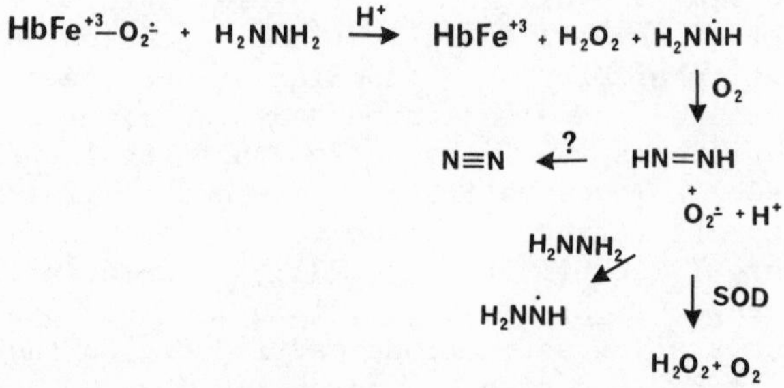

$$HbFe^{+3}{-}O_2^{\bar{\cdot}} \; + \; H_2NNH_2 \; \xrightarrow{\; H^+ \;} \; HbFe^{+3} + H_2O_2 + H_2N\dot{N}H$$

$$\downarrow O_2$$

$$N{\equiv}N \; \xleftarrow{\;?\;} \; HN{=}NH$$

$$\dot{O}_2^{\bar{\cdot}} + H^+$$

$$H_2NNH_2 \searrow \qquad \downarrow SOD$$

$$H_2N\dot{N}H \qquad \qquad$$

$$H_2O_2 + O_2$$

Fig. 5. Postulated scheme for the oxidation of hydrazine to
nitrogen by oxyhemoglobin.

of red blood cell hemolysis (Jain and Hochstein, 1979) in spite of
the fact that it is oxidized rapidly by oxyhemoglobin as shown in
this study. This would indicate that the alkyl substituent is the
moiety largely responsible for the hemolytic reaction.

After the initial rapid phase of nitrogen expiration from mice
dosed with hydrazine, a much slower rate of nitrogen evolution is
observed (Figure 3). This correlates with the disappearance of
hydrazine from the circulation (Figure 1). Administration of hydrazine by the i.v. route (Dr. Frank Dost, personal communication)
also leads to similar results indicating that the initial rapid
disappearance of hydrazine is probably caused by oxidation of hydrazine by oxyhemoglobin. This is then followed by a slower release
of hydrazine from tissues after the initial metabolism and distribution phase.

While nitrogen expiration accounts for approximately 35% of a
1 mmole/kg dose of hydrazine in the Swiss-Webster mouse, 10-15%
appears in the urine as hydrazine or conjugates hydrolyzable at
pH 6.0 (Figure 2). An additional 20-25% of the urinary metabolites
can be hydrolyzed to hydrazine at pH 1 with mild heating. We are
now in the process of determining the nature of these acid labile
conjugates and have detected the presence of both acetyl- and
diacetyl-hydrazine, the pyruvic and oxoglutaric acid hydrazones,
and a tetrahydro-oxo-pyridazine (1,4,5,6-tetrahydro-6-oxo-3-pyrida-
zine carboxylic acid) which could form by dehydrative cyclization
of the oxoglutaric acid hydrazone (Figure 6).

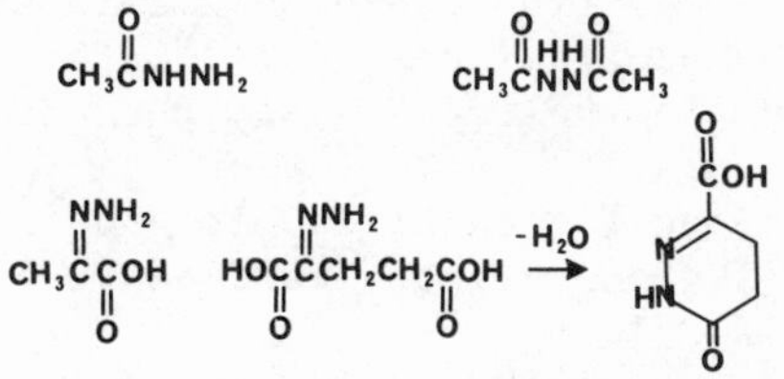

Fig. 6. Metabolites of hydrazine detected in mouse urine.

Thus, we have been able to account for approximately 70% of a dose of hydrazine in the mouse after an acute 1 mmole/kg i.p. dose as the sulfate. However, at the present time we do not know if the metabolism of hydrazine is related in any way to the toxicities caused by this widely used chemical compound. Reports have appeared suggesting that the nucleophilic nature of hydrazine itself may be responsible for some of the acute toxicities (McKennis et al., 1961; Springer et al., 1980). In order to better assess the relationship of hydrazine metabolism to its long-term toxic effects, such as carcinogenesis, we will have to more completely describe its metabolism and then carry out continuous dosing experiments coupled with distribution, metabolism and tumorigenesis studies.

ACKNOWLEDGEMENTS

This work was supported by NIEHS Young Environmental Scientist Health Research Grant No. ES01717 to S.D.N. and mass spectrometry support from grant No. NSF-CHE 7803068.

REFERENCES

Amenta, J. S. and Johnston, E. H., 1962, Hydrazine-induced alterations in rat liver, Lab. Invest., 11:956.
Back, K. C. and Thomas, A. A., 1970, Aerospace problems in pharmacology and toxicology, Ann. Rev. Pharmacol., 10:395.
Bodansky, M., 1924, The action of hydrazine and some of its derivatives in producing liver injury as measured by the effect on levulose tolerance, J. Biol. Chem., 58:799.
Dambrauskas, T. and Cornish, H. H., 1964, The distribution, metabolism, and excretion of hydrazine in rat and mouse, Tox. and Appl. Pharmacol., 6:653.
Dost, F. N., Reed, D. J., and Wang, C. H., 1966, The metabolic fate of monomethylhydrazine and unsymmetrical dimethylhydrazine, Biochem. Pharmacol., 15:1325.
Druckrey, H., 1970, Production of colonic carcinomas by 1,2-dialkyl-hydrazines and azoxyalkanes, in: "Carcinoma of the Colon and Antecedent Epithelium," W. J. Burdette, ed., Thomas, Springfield.
Druckrey, H., 1973, Specific carcinogenic and teratogenic effects of indirect alkylating methyl and ethyl compounds, Xenobiotica, 3:271.
Goldberg, B. and Stern, A., 1975, The generation of O_2^- by the interaction of the hemolytic agent, phenylhydrazine, with human hemoglobin, J. Biol. Chem., 250:2401.
Itano, H. A., Hirota, K., and Hosokawa, K., 1975, Mechanism of induction of haemolytic anaemia by phenylhydrazine, Nature, 256:665.
Jain, S. K. and Hochstein, P., Generation of superoxide radicals by hydrazine: its role in phenylhydrazine-induced hemolytic anemia, Biochim. Biophys. Acta, 586:128.

Juchau, M. and Horita, A., 1972, Metabolism of hydrazine derivatives of pharmacological interest, Drug Metab. Rev., 1:71.

Kondo, K., Murai, S., and Sonoda, N., 1977, Selenium catalyzed generation of diimide from hydrazine, Tet Lett., 42:3727.

McKennis, H. Jr., Weatherby, J. H., and Witkin, L. B., 1955, Studies on the excretion of hydrazine and metabolites, J. Pharmacol. Exp. Ther., 114:385.

McKennis, H. Jr., Yard, A. S., Adair, E. J., and Weatherby, J. H., 1961, L-γ-Glutamylhydrazine and the metabolism of hydrazine, J. Pharmacol. Exp. Ther., 131:152.

Misra, H. P. and Fridovich, I., 1976, The oxidation of phenylhydrazine: superoxide and mechanism, Biochemistry, 15:681.

Mitchell, J. R., Zimmerman, H. J., Ishak, K. G., Thorgeirsson, U. P., Timbrell, J. A., Snodgrass, W. R., and Nelson, S. D., 1975, Isoniazid-liver injury: clinical spectrum, pathology and probably pathogenesis, Ann. Intern. Med., 84:181.

Nelson, S. D., Mitchell, J. R., Timbrell, J. A., Snodgrass, W. R. and Corcoran, G. B., 1976, Isoniazid and iproniazid: activation of metabolites to toxic intermediates in man and rat, Science, 193:901.

Nelson, S. D., Mitchell, J. R., Snodgrass, W. R., and Timbrell, J. A., 1978, Hepatotoxicity and metabolism of iproniazid and isopropylhydrazine, J. Pharmacol. Exp. Ther., 206:574.

Prough, R. A., Wittkop, J. A., and Reed, D. J., 1969, Evidence for the hepatic metabolism of some monoalkylhydrazines, Arch. Biochem. Biophys., 131:369.

Springer, D. L., Broderick, D. J., and Dost, F. N., 1980, Effects of hydrazine and its derivatives on ornithine decarboxylase synthesis, activity, and inactivation, Tox. Appl. Pharmacol., 53:365.

Timbrell, J. A., Wright, J. M., and Smith, C. M., 1977, Determination of hydrazine metabolites of isoniazid in human urine by gas chromatography, J. Chromatogr., 138:165.

Timbrell, J. A., Mitchell, J. R., Snodgrass, W. R., and Nelson, S. D., 1980, Isoniazid hepatotoxicity: the relationship between covalent binding and metabolism in vivo, J. Pharmacol. Exp. Ther., 213:364.

Tosk, J., Schmeltz, I., and Hoffmann, D., 1979, Hydrazines as mutagens in a histidine-requiring auxotroph of Salmonella typhimurium, Mut. Res., 66:247.

Toth, B., 1972, Hydrazine, methylhydrazine, and methylhydrazine sulfate carcinogenesis in Swiss mice, Brit. J. Cancer, 9:109.

Von Wright, A. and Tikkanen, L., 1980, The comparative mutagenicities of hydrazine and its mono- and di-methyl derivatives in bacterial test systems, Mut. Res., 78:17.

METABOLISM OF PROCARBAZINE [N-ISOPROPYL-α-(2-METHYLHYDRAZINO)-p-TOLUAMIDE HCl]

R.A. Prough, M.W. Coomes, S.W. Cummings, and P. Wiebkin

Department of Biochemistry, The University of Texas

Health Science Center, Dallas, Texas 75235

INTRODUCTION

Procarbazine has been shown to be effective in the treatment of Hodgkin's disease and several other neoplastic disorders (Spivak, 1974). This hydrazine derivative has been shown to induce pulmonary tumors, leukemia, and mammary carcinoma in rodents (Kelly et al., 1964; Kelly et al., 1968) and myelogenous leukemia and malignant neoplasms in nonhuman primates (Sieber et al., 1978). The drug has been noted to cause nausea and central nervous system depression in humans (DeVita et al., 1965) and has been shown to suppress the immune system (Floersheim, 1963) to induce teratogenesis (Chaube and Murphy, 1969) and sterility (Fox and Fox, 1967).

During the last ten years, a number of research groups have endeavored to elucidate its metabolic disposition in rodents in an attempt to find plausible explanations for its biological activity (Baggiolini et al., 1969; Dewald et al., 1969; Prough et al., 1970; Weinkam and Shiba, 1978; Dunn et al., 1979). However, the relationship between procarbazine (PCZ) metabolism and its mode of action as an antitumor, carcinogenic, or toxic agent is not known. The present report will attempt to define the nature of two successive oxidative transformation steps in procarbazine metabolism and examine the possibility of other routes of metabolism. Of the reported metabolites of procarbazine (Reed and May, 1978), several products have been noted which represent compounds (azoxy derivatives) which might form reactive intermediates or compounds which may result from degradation of a free radical intermediate (methane; Dost and Reed, 1976). An attempt to relate these findings to the enzyme systems capable of handling hydrazine, azo, and azoxy compounds will be made in the ensuing pages.

MATERIALS AND METHODS

Hepatic microsomal protein from control or induced rats were prepared as described by Remmer et al. (1967). The formation of azoprocarbazine or azomethane was determined by measuring the aldazine of the methylhydrazine formed on addition of p-dimethylaminobenzaldehyde in 0.5 N H_2SO_4 to an aliquot of reaction mixture (Dunn et al., 1979). The methane formed from either methylhydrazine or the procarbazine derivatives was measured as reported by Prough et al. (1969). Covalent binding studies were performed as described by Wiebkin and Prough (1980). The high pressure liquid chromatography (h.p.l.c.) conditions developed to measure procarbazine metabolites consisted of an isocratic solvent system (acetonitrile:methylene chloride:hexane; 3:11:86) with a μBondapak CN column at 3.0 ml per min flow rate (Wiebkin and Prough, 1980). The metabolites could also be separated using reverse phase chromatography with a Waters' Assoc. Radial Compression Module fitted with a Radial PAK-A column. The separation utilized an isocratic solvent system consisting of 3% sodium acetate in 45% methanol at a flow rate of 2.0 ml per min. The separations were monitored by absorbance at 254 nm.

The azo, azoxy[1], and hydrazone derivatives of procarbazine were synthesized as reported by Dunn et al. (1979) and purified using the preceding h.p.l.c. systems. Radiolabeled procarbazine (^{14}C-methyl- and ^{14}C-ring-labeled) was obtained from The Drug Research and Development Program, Division of Cancer Treatment,

---

1. The nomenclature used for the azoxy isomers is based on IUPAC Tentative Rules which uses the infixes -NNO- or -ONN- to specify the position of the oxygen. For procarbazine, the two azoxy isomers would be N-isopropyl-α-(2-methyl-NNO-azoxy)-p-toluamide for the isomer with the oxygen closest to the benzyl group (I) and N-isopropyl-α-(2-methyl-ONN-azoxy)-p-toluamide for the isomer with the oxygen closest to the methyl group (II). For convenience, isomers I and II will be designated as the benzylazoxy isomer (I) and methylazoxy isomer (II), respectively. The nitrogen numbering system, N-1 or N-2, is based on the nomenclature of the parent compound, a 2-methyl-1-benzylhydrazine derivative. In the following structures, R equals $-C_6H_4-CO-NH-CH(CH_3)_2$.

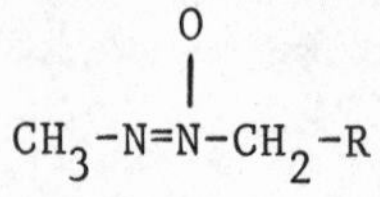

$$CH_3-N=N-CH_2-R \qquad\qquad CH_3-N=N-CH_2-R$$

(I) 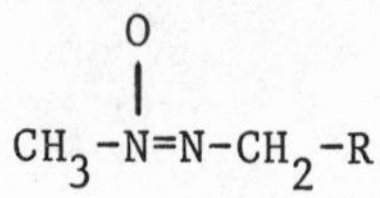   (II)

National Cancer Institute, Bethesda, MD. The ^{14}C-azo and azoxy
derivatives were prepared as described by Dunn et al. (1979) and
purified using the h.p.l.c. systems described above. Mass
spectra were obtained using a Finnegan 3100D mass spectrometer
coupled to a 6100 data system. UV and melting point measurements
were carried out as previously reported (Dunn et al., 1979).

RESULTS

Formation of Azoprocarbazine

        Recently, we reported that procarbazine is oxidized to its
azo derivative by liver microsomes in the presence of NADPH and
oxygen (Dunn et al., 1979). It was demonstrated that 90-95% of
the reaction must be due to the presence of an NADPH-dependent
reaction catalyzed by microsomal cytochrome P-450. The remaining
5-10% of the procarbazine oxidation activity was not dependent on
NADPH, but could be prevented by heat inactivating the microsomal
protein prior to the addition of procarbazine (Coomes and Prough,
unpublished results). This result suggests the presence of
another enzyme in liver microsomal protein, albeit one in low
activity.

        While looking at procarbazine oxidation in isolated
hepatocytes, Coomes et al. (1980) noted another enzyme activity
which is not dependent on reduced pyridine nucleotide and is
localized in the mitochondrial protein fraction. In Table 1, one
can see the effect of nicotinamide and N,N-dimethyl-
propargylamine (INACT) on the NADPH-dependent and independent
procarbazine oxidase activity of mitochondrial and microsomal
protein. The bulk of NADPH-independent activity is localized in
the mitochondrial fraction and is markedly inhibited by
nicotinamide and INACT. The presence of NADPH also decreased this
activity in mitochondrial fractions. These results suggest that
the mitochondrial activity is due to monoamine oxidase, a known
flavoprotein component of the outer mitochondrial membrane.

        Subsequent experiments involving enzyme purification,
inhibition of azo formation by benzylamine, and the exclusion of
the mitochondrial electron transport chain in this reaction
supported the preceding conclusion. As seen in Table 1, there was
little or no effect of nicotinamide or INACT on the
NADPH-dependent oxidation of procarbazine in microsomes, but a
measureable effect of these compounds on the minor
NADPH-independent procarbazine oxidation activity of microsomes.
This result indicated that the small amount of activity was due to
the contamination of microsomes with outer mitochondrial
membranes. A small amount of activity was also seen in the
cytosol and most likely is due to monoamine oxidase activity which
is normally released in the soluble fraction upon homogenization

Table 1. Relative Contribution of Microsomal and
Mitochondrial Procarbazine N-Oxidation

Source of enzyme[a]	PCZ Oxidation (nmoles/min/mg)	
	−NADPH	+NADPH
Homogenate	1.5	1.3
Mitochrondria	4.3	2.1
Mitochondria + 5 mM Nicotinamide	1.4	1.3
Mitochondria + 0.33 mM INACT[b]	0.3	0.2
Microsomes	1.3	9.9
Microsomes + 5 mM Nicotinamide	0.4	9.8
Microsomes + 0.33 mM INACT	0.1	9.6
Cytosol	0.7	0.8

a. The experiment was performed using liver mitochondria
and microsomes from phenobarbital-pretreated rats at pH
7.4 and 37°C in 0.1 mM potassium phosphate buffer and 1
mM EDTA.

b. INACT, N,N-dimethylpropargylamine.

of whole liver or mitochondria. The monoamine oxidase reaction
also functions when 1,2-dimethylhydrazine was used as a substrate;
the activity was 3.8 nmoles of azomethane formed per min per mg
mitochondrial protein (90% of that noted for PCZ).

These results collectively suggest that in hepatocytes, two
enzymes, monoamine oxidase and the cytochrome P-450-dependent
monooxygenase accounts for nearly all of the procarbazine
oxidation activity of the rat hepatocyte. The ratio of
NADPH-independent activity to the NADPH-dependent activity of rat
liver microsomal protein is approximately 0.4 and this ratio
decreased to 0.1 upon animal pretreatment with phenobarbital as
expected since monoamine oxidase activity is not increased by
animal pretreatment with barbituates.

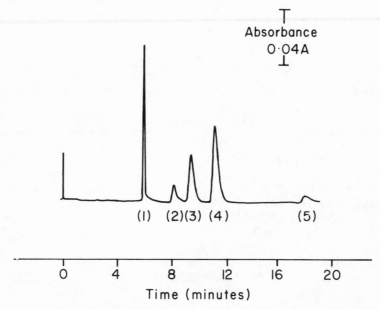

Figure 1. High Pressure Liquid Chromatographic Separation of Azoprocarbazine and Its Metabolites using a μBondapak CN column. 1. azoprocarbazine, 2. N-isopropyl-p-formylbenz-amide, 3. benzylazoxyprocarbazine, 4. methylazoxyprocarb-azine, and 5. N-isopropyl-p-formylbenzamide methylhydrazone.

Formation of Azoxyprocarbazine Derivatives.

One can separate a number of procarbazine metabolites with h.p.l.c. using cyano-bonded normal phase columns (Figure 1). When one subjects aliquots of the benzene extracts of reaction mixtures consisting of microsomal protein, NADPH, oxygen, and azoprocarbazine to h.p.l.c. analysis, the formation of at least three products can be noted: N-isopropyl-p-formylbenzamide and two additional metabolites which were subsequently shown to be azoxy derivatives. The two azoxy derivatives were chemically synthesized from azoprocarbazine and their physical properties compared with the microsomal metabolites (Table 2). The microsomal metabolites were identical in physical properties to

Table 2. Physical Characterization of the Chemically
          Synthesized Benzylazoxy- and Methylazoxy-
          procarbazine and the Microsomal Metabolites

Compound	Retention Time (min)[a]		Absorbance Maximum (nm)[b]	Molecular Weight (m/e)	Melting Point (^{o}C)
	A	B			
Benzylazoxy PCZ	10.4	6.8	226	235	124
Methylazoxy PCZ	12.1	7.4	227	235	138-9
Metabolite 3	10.3	6.8	226	235	123
Metabolite 4	12.1	7.3	227	235	138

a. These retention times are those obtained using a
μBondapak CN column with an isocratic solvent system of
hexane, methylene chloride, and acetonitrile (86:11:3,
v/v) at a flow rate of 3 ml per min (System A) and using
a Radial PAK-A column with an isocratic solvent system of
3% sodium acetate in 45% methanol at a flow rate of 2 ml
per min (System B). (See Figure 1 for metabolite
number).

b. Spectra were obtained with HPLC purified samples in
methanol.

the synthetic derivatives.   Based on our h.p.l.c.   and mass
spectral results and the proton magnetic resonance data of Weinkam
and Shiba (1978), we assume that the azoxy derivative denoted 3 in
the h.p.l.c.   trace (figure 1,2) is the benzylazoxy isomer
[N-isopropyl-α-(2-methyl-NNO-azoxy)-p-toluamide] and   that   the
derivative denoted 4 is the methylazoxy isomer [N-isopropyl-
α-(2-methyl- ONN-azoxy)-p-toluamide].

   The amount of methylazoxy derivative formed by   liver
microsomes was increased by animal pretreatment with either
5,6-benzoflavone or phenobarbital (Figure 2). However, the rate
of formation of the benzylazoxy derivative was increased only by
animal pretreatment with phenobarbital. These results suggested
that more than one form of the enzyme catalyzing the oxidation
reaction exists in rat liver.

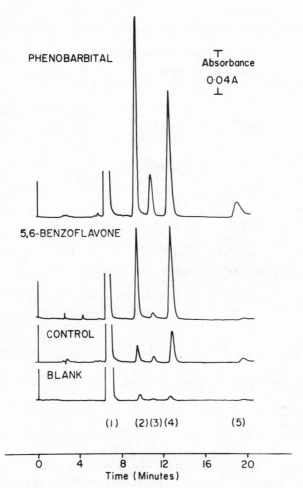

Figure 2.   Metabolism of Azoprocarbazine by Liver Microsomal
    Suspensions from Untreated, Phenobarbital-, and 5,6-benzo-
    flavone-Pretreated   Rats.   Microsomes   at   a   protein
    concentration   of   2.7   mg/ml   (untreated);   2.1   mg/ml
    (phenobarbital);   and   2.5   mg/ml   (5,6-benzoflavone)   were
    incubated with 0.25 mM azo-procarbazine for 10 mins at 37°C.
    (See Figure 1 for the metabolite numbering system.)

    Table 3 documents the effects of several inhibitors on
azoxyprocarbazine formation catalyzed by liver microsomes. It can
be seen that carbon monoxide, anti-NADPH-cytochrome c (P-450)

Table 3. The Effect of Some Inhibitors on the Rate of
          Formation of Benzylazoxy- and Methylazoxy-
          procarbazine

Addition	Benzylazoxy[a]	Methylazoxy	Total[b] Azoxy
None	0.25	1.08	1.33(100)
CO/O_2 (4:1, v/v)	0.11	0.31	0.42(32)
Anti-reductase globulin	0.02	0.11	0.13(10)
Non-immune globulin	0.27	1.13	1.40(105)
Octylamine (1.0 mM)	0.05	0.12	0.17(13)
SKF 525A (0.5 mM)	0.04	$-^c$	0.04(3)
Metyrapone (1 mM)	0.03	0.11	0.14(8)
Disulfiram (0.1 mM)	0.11	0.58	0.69(52)
Methimazole (1.0 mM)	0.06	0.10	0.16(12)

a. The rates are expressed as nmoles metabolite produced
per min per mg microsomal protein.

b. The numbers in parentheses represent the total
metabolite formation as a percentage of control.

c. The methylazoxyprocarbazine   metabolite   was   not
detected.

reductase globulin, octylamine, SKF 525A, and metyrapone are
extremely effective in inhibiting azoprocarbazine metabolism.
Methimazole and disulfiram also inhibited the reaction suggesting
that these compounds have a pronounced effect on some N-oxidation
reactions (Fiala et al., 1977), but have little affect on other
reactions of cytochrome P-450 in vitro (Prough and Ziegler, 1977).

    Azoprocarbazine has a far greater affinity for the
monooxygenase catalyzing its conversion to azoxy metabolites than
does the parent hydrazine, procarbazine. Procarbazine oxidation
exhibits a K_m value for procarbazine of 0.6 mM (Dunn et al.,
1979), while the K_m for azoprocarbazine ranged from 25-100 µM
(Wiebkin and Prough, 1980). It should be noted that preliminary
studies on the metabolism of the azoxy isomers indicate that the
compounds have much lower rates of metabolism at 100 µM azoxy
concentration by liver microsomes than azoprocarbazine (Wiebkin
and Prough, 1980). The low rates of metabolism of the azoxy

Table 4. Methane Production from Procarbazine   and   Its
          Metabolites   by   Liver   Microsomes   from
          Phenobarbital Pretreated Rats

Substrate	Methane Formed[a] (pmoles/min/mg)
Procarbazine (4 mM)	12 ± 5 (n=4)
Azoprocarbazine (0.25 mM)	39 ± 12 (n=11)
Benzylazoxy-PCZ (0.2 mM)	0 (n=4)
Methylazoxy-PCZ (0.2 mM)	0 (n=4)
Methylhydrazine (4 mM)	272 ± 63 (n=6)

a. The liver microsomes from phenobarbital pretreated
rats were prepared by the method of Remmer et al., 1967
and the values given are the rates obtained over a
fifteen min time course.

metabolites suggest that the kinetic studies reported by Wiebkin
and Prough (1980) measured the initial rates of azo metabolism,
not a steady-state rate of metabolism.

Formation of Methane from Procarbazine Metabolites.

     Since Reed and May (1978) have suggested that methane may
represent the product of a methyl radical obtained during PCZ
metabolism, we endeavored to reexamine the product-precursor
relation between the various metabolites of procarbazine and
methane formation. Table 4 shows the rates of methane formation
catalyzed by liver microsomes in the presence of oxygen and NADPH
by the postulated metabolites of procarbazine.   Azoprocarbazine
(0.25 mM) gave four times more methane than did procarbazine (4
mM). Other experiments demonstrate a lag in methane formation
(data not shown) suggesting that a product-precursor relationship
exists between the hydrazine and methane. These results are also
interpreted to mean that azoprocarbazine lies on the reaction
pathway between procarbazine and methane.   The   synthetic azoxy
derivatives (benzyl- and methylazoxy isomers) did not yield
methane upon incubation with NADPH, oxygen, and microsomal
protein.   At equal concentrations, the chemical decomposition
product of azoprocarbazine, methylhydrazine, gave methane as a
metabolite at a 6-fold higher rate than did the azo derivative.

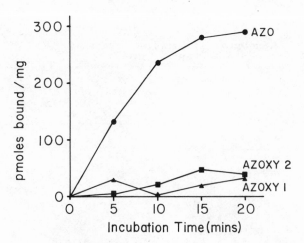

Figure 3. Time Course of Covalent Binding of ^{14}C-Methyl-labeled
    Azo- and the Azoxyprocarbazines to Liver Microsomal Protein
    from Phenobarbital-Pretreated Rats. Microsomes (approx.   2.0
    mg/ml)   were   incubated   at $37^{o}C$ with either 0.25 mM
    [14C]-methyl-labeled azo-, benzylazoxy-,   or   methylazoxy-
    procarbazine for the times indicated and the amount of
    covalently-bound material assessed radiometrically as
    described in Methods.   The values are means of at least 2
    separate experiments differing by less than 10%.
    Methyl-labeled   azoprocarbazine, ● ;   benzylazoxyprocarb-
    azine, ▲ ; Methylazoxy-procarbazine, ■ .

While the rates of methylhydrazine formed from azoprocarbazine due
to chemical decomposition are too small to account for the methane
formation (Dunn et al., 1979), we can not eliminate the
possibility   that   an   enzymatic   reaction   involving
tautomerization/hydrolysis exists to form methylhydrazine or that
the methylhydrazone derivative formed can also lead to methane
formation. Further experiments will have to be performed to
explore these possibilities.

Covalent Binding of Procarbazine Metabolites to Microsomal Protein

    Wiebkin   and   Prough   (1980)   have   demonstrated   that
azoprocarbazine can bind to microsomal protein when metabolically

activated by an NADPH- and oxygen-dependent reaction. It was also noted that the methyl portion of the molecule preferentially binds to protein under these conditions. This suggests that, prior to or during binding to microsomal protein, the reactive intermediate formed undergoes scission allowing the methyl end of the molecule to bind 3-4 times more readily than the benzyl end of the compound.

The ^{14}C-benzyl- and methyl-labeled azoxy derivatives were synthesized and the covalent binding of these derivatives compared to ^{14}C-methyl-labeled azoprocarbazine. Figure 3 shows the time course of covalent binding of ^{14}C-methyl-labeled azo, benzylazoxy, and methylazoxy derivatives of procarbazine. At 0.25 mM concentrations, the two azoxy derivatives did not bind covalently to microsomal protein to any appreciable extent. Compared to the methyl-labeled azoprocarbazine, the benzyl-labeled compounds did not lead to measureable covalent binding of the benzyl portion of the azoxy isomers (data not shown).

DISCUSSION

The results of the studies presented here suggest that procarbazine is metabolized to several distinct products which have been reported to form reactive intermediates (azoxy isomers; Weinkam and Shiba, 1978) or are the breakdown products of a reactive intermediate (methane). Our experiments demonstrate that methane results from metabolism of azoprocarbazine, but not from the azoxy isomers. In addition, the reactive species capable of covalently binding to microsomal protein also resulted from the metabolism of azoprocarbazine and not from metabolism of the azoxy derivatives as suggested for 1,2-dimethylhydrazine (Druckrey, 1970). These results are encapsulated in the scheme presented in Figure 4.

These studies have utilized liver microsomal protein as a metabolic activating system and as a trap for any reactive intermediate which might be formed. Additional metabolites might be formed by other enzyme systems capable of forming a chemically reactive molecule. It would be tempting to speculate that the reactive species which binds covalently to microsomal protein is a methyl radical which can also give rise to methane. We can not exclude entirely the role of azoxy derivatives in these reactions since these compounds can exist as either cis- or trans-isomers. The cis-isomers may be far more reactive and spontaneously degrade to give some reactive species. To date, we have no evidence for cis-isomers of the azo or azoxy isomers. Clearly, more information is required to unambiguously define the modes of procarbazine metabolism and potential reactive intermediates which may exist.

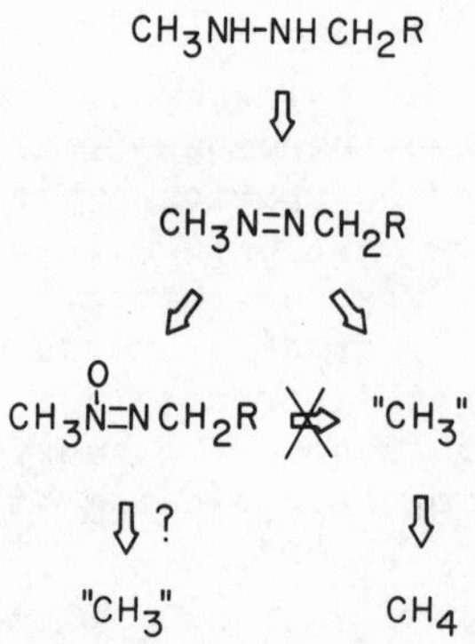

$$CH_3NH\text{-}NH\,CH_2R$$

$$\Downarrow$$

$$CH_3N\text{=}N\,CH_2R$$

$$\Swarrow \qquad \Searrow$$

$$\overset{O}{CH_3\overset{|}{N}\text{=}N\,CH_2R} \quad \cancel{\quad} \quad \text{"CH}_3\text{"}$$

$$\Downarrow ? \qquad\qquad \Downarrow$$

$$\text{"CH}_3\text{"} \qquad\qquad CH_4$$

Figure 4.   Scheme for Metabolism of Procarbazine
and Azoprocarbazine.

ACKNOWLEDGEMENTS

These studies were supported in   part   from   American   Cancer
Society   Grant   BC-336 and Robert A.   Welch Foundation Grant I-616.
RAP was a USPHS Research Career Development   Awardee   (HL   00255);
MWC and PW were Robert A.   Welch Pre- and Post-doctoral Fellows.

REFERENCES

Baggiolini, M., Dewald, B.,   and   Aebi,   H.,   1969,   Oxidation   of
    p-(N^1-methylhydrazino   methyl)-N-isopropyl benzamide (pro-
    carbazine)   to   the   methylazo   derivative   and   oxidative
    cleavage   of the N^2-C bond in the isolated perfused liver,
    Biochem.   Pharmacol., 18:2187.
Chaube, S., and Murphy, M.L., 1969, Fetal   malformations   produced
    in   rats   by N-isopropyl-α-(2-methylhydrazino)-p-toluamide
    hydrochloride (procarbazine), Teratology, 2:23.

Coomes, M.W., Wiebkin, P., and Prough, R.A., 1980, The metabolism of 1,2-disubstituted hydrazines in isolated hepatocytes, in: "Microsomes, Drug Oxidations, and Chemical Carcinogenesis", M.J. Coon, A.H. Conney, R.W. Estabrook, H.V. Gelboin, J.R. Gillette, and P.J. O'Brien, eds., Vol. 2, Academic Press, New York, p. 741.

DeVita, V.T., Hahn, M.A., and Oliverio, V.T., 1965, Monoamine oxidase inhibition by a new carcinostatic agent, N-isopropyl-α-(2-methylhydrazine)-p-toluamide (MIH), Proc. Soc. Exp. Biol. Med., 120:561.

Dewald, B., Baggiolini, M., and Aebi, H., 1969, N-Demethylation of p-(N¹-methylhydrazino methyl)-N-isopropyl benzamide (procarbazine), a cytostatically active methylhydrazine derivative, in the intact rat and in the isolated perfused rat liver, Biochem. Pharmacol., 18:2179.

Dost, F.N., and Reed, D.J., 1967, Methane formation in vivo from N-isopropyl-α-(2-methylhydrazino)-p-toluamide hydrochloride, a tumor inhibiting methylhydrazine derivative, Biochem. Pharmacol., 16:1741.

Druckrey, H., 1970, Production of colonic carcinomas by 1,2-dialkylhydrazines and azoxyalkanes, in: W.J. Burdette, ed., "Carcinoma of the Colon and Antedent Epithelium", Chapter 20, C.C. Thomas, Springfield, IL, p. 267.

Dunn, D.L., Lubet, R.A., and Prough, R.A., 1979, Oxidative metabolism of N-isopropyl-α-(2-methylhydrazino)-p-toluamide hydrochloride (procarbazine) by rat liver microsomes, Cancer Res., 39:4555.

Fiala, E.S., Bobotas, G., Kulakis, C., Wattenburg, L.W., and Weisburger, J.H., 1977, Effects of disulfiram and related compounds on the metabolism in vivo of the colon carcinogen, 1,2-dimethylhydrazine, Biochem. Pharmacol., 26:1763.

Floersheim, G.L., 1963, Prolonged survival time of skin homotransplants in mice with a methylhydrazine derivative, Experimentia (Basel), 19:546.

Fox, B.W., and Fox, M., 1967, Biochemical aspects of the actions of drugs on spermatogenesis, Pharmacol. Rev., 19:21.

Kelley, M.G., O'Gara, R.W., Gadekar, K., Yancey, S.T., and Oliverio, V.T., 1964, Carcinogenic activity of new antitumor agent, N-isopropyl-α-(2-methylhydrazino)-p-toluamide hydrochloride (NSC-77213), Cancer Chemother. Rep., 39:77.

Kelley, M.G, O'Gara, R.W., Yancey, S.T., and Botkin, C., 1968, Induction of tumors in rats with procarbazine hydrochloride, J. Natl. Cancer Inst., 40:1027.

Prough, R.A., Wittkop, J.A., and Reed, D.J., 1969, Evidence for the hepatic metabolism of some monoalkylhydrazines, Arch. Biochem. Biophys., 131:369.

Prough, R.A., Wittkop, J.A., and Reed, D.J., 1970, Further evidence on the nature of microsomal metabolism of procarbazine and related alkylhydrazines, Arch. Biochem. Biophys. 140:450.

Prough, R.A., and Ziegler, D.M., 1977, The relative participation of liver microsomal amine oxidase and cytochrome P-450 in N-demethylation reactions, Arch. Biochem. Biophys., 180:363.

Reed, D.J., and May, H.E., 1978, Cytochrome P-450 interactions with the 2-chloroethylnitrosoureas and procarbazine, Biochimie (Paris), 60:989.

Remmer, H., Greim, H., Schenkman, J.B., and Estabrook, R.W., 1967, Methods for the elevation of hepatic microsomal mixed function oxidase levels and cytochrome P-450, Methods in Enzymology, 10:703.

Sieber, S.M., Correa, P., Dalgard, D.W., and Adamson, R.H., 1978, Carcinogenesis and other adverse effects of procarbazine in nonhuman primates, Cancer Res., 38:2125.

Spivak, S.D., 1974, Procarbazine diagnosis and treatment-drugs five years later, Ann. Int. Med., 81:795.

Weinkam, R.J., and Shiba, D.A., 1978, Metabolic activation of procarbazine, Life Sciences, 22:937.

Wiebkin, P., and Prough, R.A., 1980, Oxidative metabolism of N-isopropyl-α-(2-methylhydrazino)-p-toluamide (azoprocarbazine) by rodent liver microsomes, Cancer Res., 40:3524.

METABOLISM OF MUTAGENIC AMINO-γ-CARBOLINES IN TRYPTOPHAN

PYROLYSATES

Ryuichi Kato[1], Yasushi Yamazoe[1], Kenji Ishii[1], Shiro Mita[1], Tetsuya Kamataki[1] and Takashi Sugimura[2]

1 Department of Pharmacology, School of Medicine, Keio University, Tokyo 160, Japan
2 National Cancer Center Research Institute Tokyo 104, Japan

Smoke condensates from cigarettes and charred parts of broiled fish and meat showed potent mutagenic activity to Salmonella typhimurium TA 98 and TA 100 in the presence of rat liver S-9 Mix (Nagao, et al., 1977). The mutagenic substances were formed also by pyrolysis of protein and amino acids (Nagao et al., 1977).

A number of new mutagenic compounds were isolated and identified from pyrolysates of amino acids, such as tryptophan, phenylalanine, glutamic acid and lysine (Sugimura et al., 1977; Kosuge et al., 1978; Sugimura and Nagao, 1979). Potent mutagens were isolated and also identified from pyrolysates of soybean globulin (Yoshida et al., 1978) and broiled sardine (Kasai et al., 1980). Among such compounds thus far isolated, two γ-carboline derivates, Trp-P-1 (3-amino-1,4-dimethyl-5H-pyrido [4,3-b] indole) and Trp-P-2 (3-amino-1-methyl-5H-pyrido [4,3-b] indole) are potently mutagenic. The mutagenicity of Trp-P-1 and Trp-P-2 as tested by a modified Ames method using the TA 98 strain of S.typhimurium, is

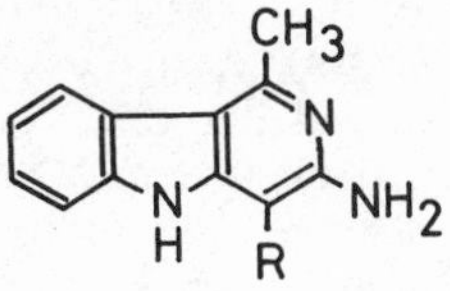

Fig. 1. Structures of Trp-P-1 and Trp-P-2

R = CH₃ Trp-P-1
R = H   Trp-P-2

about 120 and 320 times as high as that of benzo(a)pyrene, respectively (Sugimura and Nagao, 1979). The chemical structures of Trp-P-1 and Trp-P-2 are shown in Fig. 1.

Several lines of evidence obtained in vitro and in vivo experiments also indicate that Trp-P-1 and Trp-P-2 are carcinogenic in mammals (Takayama et al., 1979; Ishikawa et al., 1979). Recently Matsukura et al. (1980) found that Trp-P-1 and Trp-P-2 produced hepatocarcinomas in mice by ingesting diets containing these chemicals.

In this paper, we report the metabolic activation of Trp-P-1 and Trp-P-2 by hepatic microsomes and reconstituted cytochrome P-450 systems and the characterization of a mutagenic intermediate. Details of the procedures were as described in previous papers (Ishii et al., 1980a; Ishii et al.,1980b; Yamazoe et al., 1980).

METABOLISMS OF TRP-P-1 AND TRP-P-2 BY HEPATIC MICROSOMES

Since the activities converting Trp-P-1 and Trp-P-2 to mutagenic metabolites were found mainly in microsomal fractions but not in 105,000X g supernatant fractions of rat liver, we examined the metabolism by hepatic microsomes of Trp-P-1 and Trp-P-2. We mainly describe here the results obtained with Trp-P-2, but some similar results were also observed with Trp-P-1.

In accordance with the results of the mutagenesis seen with S. typhimurium TA 98, the activities to metabolize Trp-P-2 and the formation of the main active metabolite were found only in the microsomal fraction. The addition of 105,000X g supernatant fraction to microsomes did not increase the number of revertants and the formation of the main active metabolite.

To determine the nature of the mutagenic metabolite of Trp-P-2, we separated the metabolites using a high performance liquid chromatograph (HPLC) as described in a previous paper (Yamazoe et al., 1980). The major metabolite isolated, M-3, was mutagenic directly to S. typhimurium TA 98, without addition of S-9 Mix. The metabolism of Trp-P-2 in the incubation mixture containing microsomes from PCB-treated rats was fairly fast and the rates of disappearance of Trp-P-2 and formation of M-3 were 31 and 14 nmole/mg protein/10 min, respectively.

The chemical structure of M-3 was then analyzed by gas chromatography/mass spectrometry (GC/MS) using a JEOL D-300 mass spectrometer after the trimethylsilylation and dimethylisopropylsilyl-ation with N,O-bis-trimethylsilylacetamide and dimethylisopropylsilyl-(DMiPS)imidazole, respectively, and after the chlorination by heating with conc. hydrochloric acid (Razzouk et al., 1977). Table 1 shows GC/MS data for the dimethylisopropylsilyl derivatives of Trp-P-2 and M-3. The difference of 16 amu between the molecular ion

Table 1. GC/MS Data for the Dimethylisopropylsilyl (DMiPS)
         Derivatives of Trp-P-2 and the Active Metabolite (M-3)

Compound	Fragment ion				
Mono-DMiPS- Trp-P-2	297*(19.3)	282(3.1)	254(100)		
Mono-DMiPS- M-3	313*(82.2)	298(14.9)	297(17.8)	270(100)	254(37.8)

*
 Molecular ion
 Numbers in parentheses are relative intensities of the correspond-
 ing base peak.

peak of mono-DMiPS-M-3 and that of mono-DMiPS-Trp-P-2 indicates the
introduction of oxygen atom in Trp-P-2. The differences of 16 amu
between the ions formed by loss of isopropyl moiety (m/e 270 and m/e
254) and between those formed by loss of the methyl group (m/e 298
and m/e 282) also suggested the conversion of the amino-group of
Trp-P-2 to the hydroxyamino - group. A product showing a molecular
ion cluster at m/e 231 and m/e 233 with a relative ratio of 3:1,
indicating an introduction of chlorine atom into the molecule, was
found with M-3, but not with Trp-P-2. These results support the
presence of hydroxyamino-group in M-3. These plus further results
obtained with proton-NMR spectra at 200 MHz (Varian XL 200 spectrometer
operating in the Fourior-transform mode) indicate that the mutagenic
metabolite of Trp-P-2, M-3, is a 3-hydroxyamino derivative. Details
on the elucidation of structure of this mutagenic metabolite will be
published elsewhere.

METABOLIC ACTIVATION OF TRP-P-1 AND TRP-P-2 BY RAT HEPATIC MICROSOMES

        In a previous paper, we reported a sharp decrease in the number
of revertants induced by Trp-P-2 or Trp-P-1 when more than 10 µl per
plate of the liver 9,000 X g supernatant or microsomal fraction was
added (Ishii et al., 1980b). These results suggest the following
possibilities: decrease in the amount of Trp-P-2 or Trp-P-1, further
metabolism of mutagenic metabolite of Trp-P-2 or Trp-P-1 to nonmutagenic
metabolite(s) or killing effect due to the formation of a large
amount of toxic metabolite(s).

        We examined these possibilities by measuring the amounts of
Trp-P-2 and N-hydroxy-Trp-P-2 and the activity required to induce
the revertants, after the incubation. As shown in Fig. 2-A, the

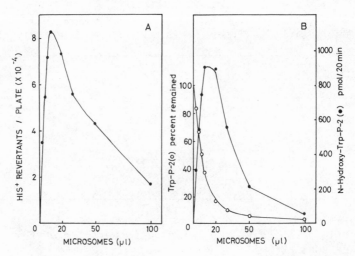

Fig. 2.   Effects of the amount of microsomes on the mutagenic
activation and N-hydroxylation of Trp-P-2. The mutation
assay was carried out as described by Ames et al., with
modifications. Liver microsomes from PCB-treated rats were
used. The concentration of microsomal protein was 5.6
mg/ml. The incubation mixture containing 1   g of Trp-P-2
(3.9 nmol) as a substrate and S. typhimurium TA 98 was
preincubated at 37°C for 20 min prior to pouring onto a
minimal agar plate. For HPLC analysis, the same volume of
acetonitrile was added to the reaction mixture and the
preparation centrifuged for 10 min at 2,500Xg. An
aliquot of the supernatant was injected onto the column.

number of revertants was maximum with 10 µl of liver microsomes from
PCB-treated rats and markedly decreased with 30, 50 and 100 µl.
Fig. 2-B shows the decrease in the amount of added Trp-P-2 during
the 20 min incubation. With increase in the amount of microsomes,
the remaining Trp-P-2 was markedly decreased and the amount of
N-hydroxy-Trp-P-2 was also decreased.   These results suggest the
occurrence of a rapid further metabolism of N-hydroxy-Trp-P-2 to
non-mutagenic metabolite(s) by microsomes.

The number of revertants of S. typhimurium TA 98 induced by
Trp-P-1 and Trp-P-2 in the presence of microsomes increased about
210 and 220 times or 60 and 90 times, respectively, when liver
microsomes from PCB- or 3-methylcholanthrene-treated rats were used.
On the other hand, liver microsomes from phenobarbital-treated rats
were only 3 to 4 times as active as those from control rats (Ishii
et al., 1980b).

In accordance with the results of the mutagenesis, the activity of N-hydroxylation of Trp-P-2 of microsomes was increased 200, 80 and 8 times by treatment of rats with PCB, 3-methylcholanthrene or phenobarbital, respectively (Yamazoe et al., 1980). These results suggest that Trp-P-1 and Trp-P-2 are activated by the P-448 form of the cytochrome and N-hydroxylation of Trp-P-2 is catalyzed by the P-448 form.

The induction of revertants by Trp-P-1 and Trp-P-2 and the formation of N-hydroxy-Trp-P-2 by rat hepatic microsomes required NADPH and oxygen and such were markedly inhibited by carbon monoxide and 7,8-benzoflavone but only slightly by metyrapone and SKF 525-A. These results indicate the involvement of the P-448 form of cytochrome P-450 for the metabolic activations of Trp-P-1 and Trp-P-2 and the formation of N-hydroxy-Trp-P-2.

In addition, about 97.7% of direct mutagenic activity was recovered in the M-3 fraction after application of the incubated mixture to the HPLC procedure. This means that the direct mutagen from the metabolism of Trp-P-2 is mainly located in M-3 fraction (N-hydroxy-Trp-P-2) (Yamazoe et al., 1980).

Table 2. Requirements for the Mutagenic Activation of Trp-P-1 and Trp-P-2 by a Purified and Reconstituted Cytochrome P-448 System

Incubation mixture[a]	His[+] revertants/plate	
	Trp-P-1[b]	Trp-P-2[b]
Complete	13,300(100%)	45,600(100%)
−Reductase	260(2.0%)	1,010(2.2%)
−P-448	22(0.2%)	0(0)
−NADPH	235(1.8%)	313(0.7%)
−DLPC[c]	7,700(57.9%)	30,400(69.9%)

[a] 0.1 nmol of cytochrome P-448 from PCB-treated rat liver microsomes (PCB P-448) was used.
[b] 3 μg of Trp-P-1 and 1 μg of Trp-P-2 were used.
[c] L-α-dilauroyl-3-phosphatidylcholine
From Ishii et al. (1980a), (permission granted).

Table 3. Mutagenic Activation and N-Hydroxylation of
         Trp-P-2 by Various Forms of Purified
         Cytochrome P-450

Preparation	Mutagenicity Rev./pmol P-450/10 min	N-Hydroxy-Trp-P-2 formation pmol/pmol P-450/10 min
PB P-450	10	< 0.18
MC P-448	1172	9.60
PCB P-450	60	0.57
PCB P-448	1029	7.94

PB P-450:   P-450 form purified from liver microsomes of
            phenobarbital-treated rats
MC P-448:   P-448 form purified from liver microsomes of
            3-methylcholanthrene treated rats
PCB P-450: P-450 form purified from liver microsomes of
            PCB-treated rats
PCB P-448: P-448 form purified from liver microsomes of
            PCB-treated rats

The incubation mixture consisted of 0.1 nmol of cytochrome
P-450(448), 0.5 unit of NADPH-cytochrome P-450 reductase,
30 µg of L-α-dilauroyl-3-phosphatidylcholine, 0.5 mM Trp-P-2
and cofactors in 0.5 ml of 50 mM phosphate (pH 7.4). After
10 min of incubation, a 200 µl portion of the mixture was
passed through a membrane filter (Yamazoe et al., 1980) and
then mixed with 20 µl of dimethylsulfoxide containing 0.3
nmol of menadione. An aliquot (20 µl) of the resultant
mixture was used for the mutation test. The formation of
N-hydroxy-Trp-P-2 was determined with HPLC using a supernatant
of acetonitrile-treated incubation mixture (Yamazoe et al.,
1980).

METABOLIC ACTIVATION OF TRP-P-1 AND TRP-P-2 BY PURIFIED CYTOCHROME
P-450 FROM RAT LIVER

    For further characterization of the biochemical mechanism, in
particular the involvement of P-448 form of the cytochrome in the
metabolic activation of Trp-P-2 and Trp-P-1, we carried out metabolic
studies using various purified forms of cytochrome P-450.

As shown in Table 2, the number of revertants in the complete reconstituted system was comparable with the number in the microsomal system, however, the omission of NADPH-cytochrome P-450 reductase, cytochrome P-448 or NADPH resulted in a complete decrease in the number of revertants (Ishii et al., 1980a). The formation of N-hydroxy-Trp-P-2 by these systems showed results identical to those observed with the number of revertants.

Concerning the capacities of various forms of cytochrome P-450 to produce the mutagenic metabolite, P-448 forms isolated from PCB- or MC-treated rats were very active, while P-450 forms isolated from PB- or PCB-treated rats were only about 0.4% or 4%, respectively, as high as those of the P-448 forms (Ishii et al., 1980a).

Table 3 shows the formation of N-hydroxy-Trp-P-2 and mutagenic activation by various forms of purified cytochrome P-450. The formation of N-hydroxy-Trp-P-2 was high with P-448 forms of the cytochromes isolated from PCB- or 3-methylcholanthrene-treated rats, whereas it was low in P-450 forms isolated from phenobarbital- or PCB-treated rats, in accordance with their capacities of mutagenic activation.

METABOLISM OF TRP-P-2 BY LIVER MICROSOMES FROM MICE GENETICALLY RESPONSIVE AND NON-RESPONSIVE TO 3-METHYLCHOLANTHRENE

Arylhydrocarbon hydroxylase (AHH: benzo(a)pyrene hydroxylase) is regulated by the Ah locus in mice and induced by treatment with 3-methylcholanthrene (Nebert et al., 1975). Since AHH activity is associated with some P-448 forms of cytochrome P-450, we used mice of C57BL/6N and DBA/2N strains to clarify the relationship between AHH and Trp-P-2 N-hydroxylase.

The formation of N-hydroxy-Trp-P-2 was markedly induced by 3-methylcholanthrene in C57BL/6N (B6) mice, but not in DBA/2N (D2) mice, in accordance with the induction of benzo(a)pyrene hydroxylase (Table 4 and Fig. 3). The number of revertants induced by Trp-P-2 was also increased by 3-methylcholanthrene treatment in B6 mice, but not in D2 mice, as reported previously (Nebert et al., 1979). The results suggest a close relationship between AHH and Trp-P-2 N-hydroxylase in mice.

SPECIES DIFFERENCES IN THE INDUCTION OF BENZO(A)PYRENE HYDROXYLASE AND TRP-P-2 N-HYDROXYLASE

3-Methylcholanthrene or PCB can induce benzo(a)pyrene hydroxylase in various species of animals except rabbits. Therefore, the induction of Trp-P-2 N-hydroxylase by PCB was studied in rabbits and data compared with findings in rats.

Table 4. Effects of Treatment of C57BL/6N (B6) and DBA/2N (D2)
         Mice with 3-Methylcholanthrene (3-MC) on Activities of
         Benzo(a)pyrene Hydroxylase, Trp-P-2 N-Hydroxylase
         and the Number of Revertants Induced by Trp-P-2

Strain	Treatment	Benzo(a)pyrene hydroxylase	Trp-P-2 N-Hydroxylase	Number of revertants[*]
		pmol/mg protein per min		
B6	corn oil	380 ± 31[**]	25 ± 2[**]	241 ± 33[**]
B6	3-MC	4857 ± 227	350 ± 61	20600 ± 3090
D2	corn oil	388 ± 13	27 ± 3	329 ± 75
D2	3-MC	391 ± 56	25 ± 3	494 ± 26

Each value represents the mean $\pm$ S.E.M. from 6-7 mice.
* The number of revertants is represented as revertants/20 μl
  of the filtrate.
** Significant difference from controls(B6) ($p < 0.01$).

Hepatic microsomes prepared from mice treated with 3-MC (80
mg/kg) or corn oil were used. Activity of benzo(a)pyrene hydroxylase was determined fluorometrically using 3-hydroxybenzo(a)pyrene
as the standard. Quantitation of Trp-P-2 N-hydroxylase activity
was conducted simultaneously with the mutation test.
From Yamazoe et el. (1980), with some modifications.

    As shown in Fig. 4, benzo(a)pyrene hydroxylase, Trp-P-2
N-hydroxylase and the number of revertants were increased by 6, 1150
and 810 times, respectively, in rats by treatment with PCB.   In
rabbits, the same treatment did not increase benzo(a)pyrene hydroxylase,
but did increase Trp-P-2 N-hydroxylase and the number of revertants
by 14 and 30 times, respectively. These results indicate that,
although benzo(a)pyrene hydroxylase and Trp-P-2 N-hydroxylase are
similar in many aspects, the both are clearly different in rabbits.

    Further studies on the similarity and difference between
benzo(a)pyrene hydroxylase and Trp-P-2 N-hydroxylase are underway.

MUTAGENIC ACTIVATION BY HUMAN LIVER MICROSOMES OF TRP-P-1 AND
TRP-P-2

    Trp-P-1 and Trp-P-2 would be mutagenic and carcinogenic to

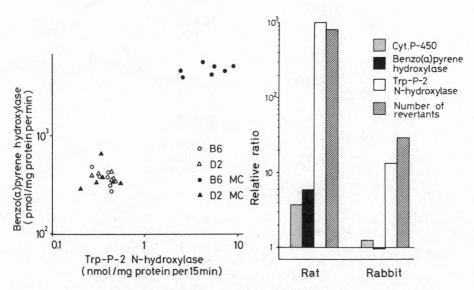

Fig. 3.   Relationship between Trp-P-2 N-hydroxylase and benzo(a)-
(left)   pyrene hydroxylase in mice. Both activities are represented
           on logarithmic scales and each point is the mean of dupli-
           cate determinations from an individual mouse. Experimental
           details are the same as described in Table 4.
           From Yamazoe et al. (1980), with a modification.

Fig. 4.   Effect of PCB-treatment on the content of cytochrome P-450,
(right)   activities of benzo(a)pyrene hydroxylase and Trp-P-2
           N-hydroxylase, and the number of revertants induced by
           Trp-P-2. Hepatic microsomes obtained from rats and rabbits
           treated with PCB (500 mg/kg.ip) or corn oil were used.
           Each column represents the ratio of values obtained with
           PCB-treated to corn oil-treated animals.

humans if the human liver microsomes are capable of activating these
chemicals to active mutagens. As can be seen in Fig. 5, microsomes
prepared from a healthy portion of a human liver, obtained at the
time of surgery for hepatic cancer, catalyzed the activation of
Trp-P-1 and Trp-P-2. The activities of human microsomes required to
produce mutagens, as determined by the number of revertants were
comparable to the activities of liver microsomes from the untreated
rats.

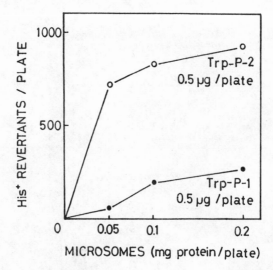

Fig. 5.   Mutagenic activation by human liver microsomes of Trp-P-1
          and Trp-P-2. To human microsomes and other necessary com-
          ponents was added 1 μg of Trp-P-1 or Trp-P-2, and incuba-
          tion carried out at 37°C for 20 min. The reaction mixture
          was passed through a Millipore filter (Yamazoe et al.,
          1980). A half of the original volume was mixed with S.
          typhimurium TA 98 and 2 ml of molten soft agar, and poured
          onto a minimal agar plate. From Kato et al. (1980).

N-HYDROXY-TRP-P-2 AS A POSSIBLE PROXIMATE OR ULTIMATE MUTAGEN AND
CARCINOGEN

     When Trp-P-2 and calf thymus DNA were incubated in the presence
of rat liver microsomes and NADPH-generating system, there was a
covalent binding of a metabolite of Trp-P-2 to DNA (Hashimoto et
al., 1978). Nemoto et al. (1979) reported that the covalent binding
of Trp-P-2 to calf thymus DNA was increased when ATP or GTP was
added to an S-9 mix. They also observed a low covalent binding in
the absence of 105,000 Xg supernatant fraction. These results
suggest that some further enzymatic activation of the Trp-P-2
metabolite(s) formed by microsomal cytochrome P-450 stimulates the
covalent binding of Trp-P-2 to DNA.

     Our recent studies also demonstrated the covalent binding of
1-C^{14}-Trp-P-2 to DNA when Trp-P-2 and calf thymus DNA were incubated
in the presence of various forms of purified cytochrome P-450
preparations, NADPH-cytochrome P-450 reductase, L-α-dilauroyl-3-phos-
phatidylcholine and NADPH (Mita et al., in preparation). The amount
of metabolite(s) of Trp-P-2 covalently bound to DNA varied with the
preparations but clearly correlated to the formation of N-hydroxy-

Trp-P-2.   We now have evidence for a direct covalent binding of N-hydroxy-Trp-P-2 from the M-3 fraction to calf thymus DNA (Mita et al., in preparation).

Hashimoto et al. (1980) demonstrated direct covalent binding of synthetic N-hydroxy-Trp-P-2 to calf thymus DNA, but the rate of binding was slow and was markedly accelerated after the acetylation of N-hydroxy-Trp-2.   In the case of N-acetyl-2-aminofluorene (AAF), N-hydroxy-AAF is considered to be a proximate carcinogen and N-O-acetyl-esters and N-O-sulfate of N-hydroxy-AAF are considered to be ultimate carcinogens.   On the other hand, N-hydroxy-2-aminofluorene, probably its free radical derivative, is considered to be an ultimate mutagen (Sakai et al., 1978).

Concerning Trp-P-2, all the available data indicate that N-hydroxy-Trp-P-2, probably the free redical derivative, is an ultimate mutagen, but further metabolic activations by enzymes in the supernatant fraction and in S. typhimurium TA 98 cannot be excluded.   In vitro and in vivo studies on the ultimate form of Trp-P-2 are underway to elucidate the mechanism of induction of hepatocarcinoma.

ACKNOWLEDGEMENTS

This work was supported by a Grant-in-Aid for Cancer Research from the Ministry of Education, Science and Culture, Japan.   We thank M. Ohara, Kyushu University for advice on the manuscript.

REFERENCES

Hashimoto, Y., Takeda, K., Shudo, K., Okamoto, T., Sugimura, T., and Kosuge, T., 1978, Rat liver microsome-mediated binding to DNA of 3-amino-1-methyl-5H-pyrido(4,3-b)indole, a potent mutagen isolated from tryptophan pyrolysate, Chem.-Biol. Interact., 23:137.
Hashimoto, Y., Shudo, K., and Okamoto, T., 1980, Metabolic activation of Trp-P-2 and Glu-P-1, Their reaction with DNA, Proc. Jpn. Cancer Assoc., 39th Annual Meeting, 33.
Ishii, K., Ando, M., Kamataki, T., Kato, R., and Nagao, M., 1980a, Metabolic activation of mutagenic tryptophan pyrolysis products (Trp-P-1 and Trp-P-2) by a purified cytochrome P-450-dependent monooxygenase system, Cancer Letters, 9:271.
Ishii, K., Yamazoe, Y., Kamataki, T., and Kato, R., 1980b, Metabolic activation of mutagenic tryptophan pyrolysis products by rat liver microsomes, Cancer Res., 40:2596.
Ishikawa, T., Takayama, S., Kitagawa, T., Kawachi, T., Kinebuchi, M., Matsukura, N., Uchida, E., and Sugimura, T., 1979, In vivo experiments on tryptophan pyrolysis products, in

"Naturally Occurring Carcinogens Mutagens and Modulators of Carcinogenesis" E. C. Miller, et al., eds., Japan Scientific Societies Press, Tokyo/ University Park Press, Baltimore.

Kasai, H., Yamaizumi, Z., Wakabayashi, K., Nagao, M., Sugimura, T., Yokoyama, S., Miyazawa, T., Spingarn, N. E., Weisburger, J. H., and Nishimura, S., 1980, Potent novel mutagens produced by broiling fish under normal conditions, Proc. Jap. Acad., 56(b):278

Kato, R., Yamazoe, Y., Kamataki, T., 1980, Metabolic activation of tryptophan pyrolysates, Trp-P-1 and Trp-P-2 by hepatic cytochrome P-450, in "Genetic and Environmental Factors in Experimental and Human Cancer" H. V. Gelboin et al., eds., Japan Scientific Societies Press, Tokyo

Kosuge, T., Tsuji, K., Wakabayashi, K., Okamoto, T., Shudo, K., Iitaka, Y., Itai, A., Sugimura, T., Kawachi, T., Nagao, M., Yahagi, T., and Seino, Y., 1978, Isolation and structures of mutagenic principles in amino acid pyrolysate, Chem. Pharm. Bull. (Tokyo), 26:611.

Matsukura, N., Kawachi, T., Uchida, E., Ogaki, H., Morino, K., Sugimura, T., and Takayama, S., 1980, Induction of liver tumors in CDF_1 mice by oral administration of Trp-P-1 and Trp-P-2 mutagenic principles in tryptophan pyrolysis products, Proc. Jpn. Cancer Asso., 39 th Annual Meeting, 32.

Nagao, M., Honda, M., Seino, Y., Yahagi, T., Kawachi, T., and Sugimura, T., 1977, Mutagenecities of protein pyrolysates, Cancer Lett., 2:335.

Nebert, D. W., Robinson, J. R., Niwa, A., Kumaki, K., and Poland, A. P., 1975, Genetic expression of arylhydrocarbon hydroxylase activity in the mouse, J. Cell Physiol., 85:393.

Nebert, D. W., Bigelow, S. W., Okey, A. B., Yahagi, T., Mori, Y., Nagao, M., and Sugimura, T., 1979, Pyrolysis products from amino acids and protein: Highest mutagenicity requires cytochrome P_1-450, Proc. Natl. Acad. Sci. USA, 76:5929.

Nemoto, N., Kusumi, S., Takayama, S., Nagao, M., and Sugimura, T., 1979, Metabolic activation of 3-amino-5H-pyrido(4,3-b)indole, a highly mutagenic principle in tryptophan pyrolysate, by rat liver enzymes, Chem.-Biol. Interact., 27:191.

Razzouk, C., Lhoest, G., Roberfroid, M., and Mercier, M., 1977, Subnanogram estimation of the proximate carcinogen N-hydroxy-2-fluorenylacetamide by gas-liquid chromatography, Anal. Biochem., 83:194.

Sakai, S., Reinhold, C. E., Wirth, P. J., and Thorgeirsson, S. S., 1978, Mechanism of in vitro mutagenic activation and covalent binding of N-hydroxy-2-acetylaminofluorene in isolated liver cell nuclei from rat and mouse, Cancer Res., 38:2058.

Sugimura, T., Kawachi, T., Nagao, M., Yahaigi, T., Seino, Y., Okamoto, T., Shudo, K., Kosuge, T., Tsuji, K., Wakabayashi, K., Iitaka, Y., and Itai, A., 1977, Mutagenic principles in tryptophan and phenylalanine pyrolysis products, Proc. Jpn. Acad., 53:58.

Sugimura, T., and Nagao, M., 1979, Mutagenic factors in cooked food, CRC Crit. Rev. Toxicol., 8:189.

Takayama, S., Hirakawa, T., Tanaka, M., Katoh, Y., and Sugimura, T., 1979, Transformation and neoplastic development of hamster embryo cells after exposure to tryptophan pyrolysis products in tissue culture, in "Naturally Occurring Carcinogens-Mutagens and Modulators of Carcinogenesis" E. C. Miller et al., eds., Japan Scientific Societies Press, Tokyo/ University Park Press, Baltimore.

Yamazoe, Y., Ishii, K., Kamataki, T., Kato, R., and Sugimura, T., 1980, Isolation and characterization of active metabolites of tryptophan-pyrolysate mutagen, Trp-P-2, formed by rat liver microsomes, Chem.-Biol. Interact., 30:125.

Yoshida, D., Matsumoto, T., Yoshimura, R., and Matsuzaki, T., 1978, Mutagenicity of amino-α-carbolines in pyrolysis products of soybean globulin, Biochem. Biophys. Res. Commun., 83:915.

METABOLIC ASPECTS OF THE COMUTAGENIC ACTION OF NORHARMAN

Takashi Sugimura, Minako Nagao and Keiji Wakabayashi

National Cancer Center Research Institute
Tsukiji, Chuo-ku, Tokyo, Japan

INTRODUCTION

Pyrolysis of amino acids, proteins and protein foods yields many mutagens (Sugimura et al., 1977b; Sugimura, 1979; Commoner et al., 1978; Yoshida et al., 1978; Pariza et al., 1979; Rappaport et al., 1979; Kasai et al., 1980a,b; Spingarn et al., 1980). We previously isolated very potent mutagens from a tryptophan pyrolysate, and these were identified as 3-amino-1,4-dimethyl-5H-pyrido-[4,3-b]indole (Trp-P-1) and 3-amino-1-methyl-5H-pyrido[4,3-b]indole (Trp-P-2) (Sugimura et al., 1977a). These two compounds are derivatives of γ-carboline. The nitrogen atom at the γ position was not derived from the α-amino nitrogen of the same molecule of tryptophan as the indole nucleus of these mutagens. During purification of Trp-P-1 and Trp-P-2, it was noticed that the recovery of mutagenic activity decreased abruptly at the step of removal of 9H-pyrido-[3,4-b]indole (norharman) and 1-methyl-9H-pyrido[3,4-b]indole (harman). Addition of the fraction containing norharman and harman to the fraction of Trp-P-1 and Trp-P-2 restored the mutagenic activity (Nagao et al., 1977c).

Norharman is β-carboline and harman is its methyl derivative. The nitrogen atom at the β position is possibly derived from the α-amino nitrogen of the same molecule of tryptophan as the indole nucleus of these substances. During pyrolysis, tryptamine was produced and one and two carbon fragments derived from other tryptophan molecules during pyrolysis may result in formation of the skeleton structures of norharman and harman, respectively, as shown schematically in Figure 1.

1011

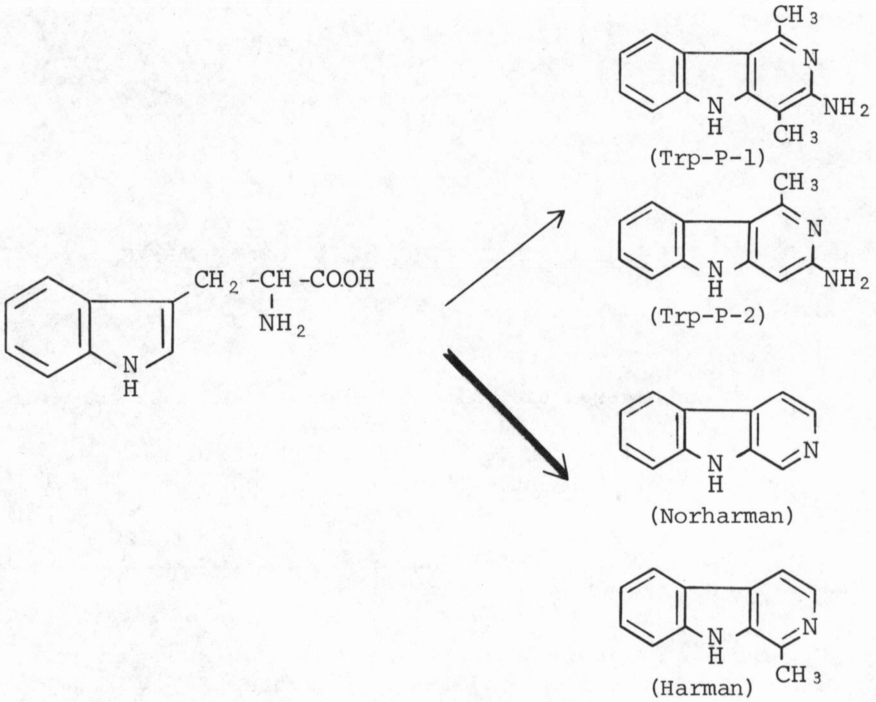

Fig. 1.   Formation of γ-carboline mutagens and β-carboline comutagens upon pyrolysis of tryptophan.

Trp-P-1 and Trp-P-2 are strong mutagens, as indicated in Table
1, and their specific mutagenic activities were higher than those
of most typical carcinogens, such as aflatoxin B_1, 4-nitroquinoline
1-oxide (4NQO), AF-2 (2-(2-furyl)-3-(5-nitro-2-furyl)acrylamide)
and benzo[a]pyrene (B[a]P) towards *Salmonella typhimurium* TA98.
With TA100 as a test strain, the specific mutagenic activities of
Trp-P-1 and Trp-P-2 were even higher than those of N-methyl-N'-
nitro-N-nitrosoguanidine (MNNG), B[a]P and N,N-dimethylnitrosamine
(DMN) (Sugimura, 1979).   Trp-P-1 and Trp-P-2 have potentials to
induce malignant transformation of Chinese hamster cells *in vitro*
(Takayama et al., 1977), and cells from the transformed colonies
could be transplanted into the cheek pouch of hamsters (Takayama et
al., 1978).   Local subcutaneous injection of Trp-P-1 yielded fibro-
sarcomas in rats (Ishikawa et al., 1979).   More recently it was
observed that oral administration of Trp-P-1 and Trp-P-2 to mice at
a concentration of 0.02% in their diet resulted in formation of
many malignant hepatomas, some of which metastasized to the lung.
Females were more susceptible than males (Matsukura et al., 1980).

Table 1. Comparison of the Sepcific Mutagenic Activities of
          the γ-Carboline Mutagens Trp-P-1 and Trp-P-2 from
          Tryptophan Pyrolysate with Those of Other Typical
          Mutagenic Carcinogens

TA98		TA100	
Trp-P-2	104,200	AF-2	41,900
Trp-P-1	39,000	Aflatoxin B_1	22,500
Aflatoxin B_1	6,000	4NQO	3,915
4NQO	965	Trp-P-2	1,750
AF-2	550	Trp-P-1	1,650
B[a]P	321	MNNG	870
DMN	0.00	B[a]P	644
MNNG	0.00	DMN	0.23
Norharman	0.00	Norharman	0.00
Harman	0.00	Harman	0.00

Unlike these γ-carboline derivatives, norharman and harman
were not mutagenic towards *Salmonella typhimurium* TA98 and TA100
even with metabolic activation, as also indicated in Table 1.
However, as mentioned before, norharman enhanced the mutagenicities
of Trp-P-1 and Trp-P-2. This enhancement of mutagenicity was
observed when excess S-9 was used, whereas suppression of the mutagenicity by norharman was observed with a relatively small amount
of S-9 (Nagao et al., 1978). Trp-P-2 can be converted to its
hydroxyamino derivative and norharman inhibits this *N*-hydroxylation
(Kato, personal communication). This inhibition of formation of
the *N*-hydroxyl compound could partly explain the suppression of the
mutagenicity of Trp-P-1 observed with a small amount of S-9. Norharman also enhanced the mutagenicity of *N,N*-dimethylaminoazobenzene
(DAB) with any amount of S-9 (Nagao et al., 1977a, 1978).

During studies on the effect of norharman on the mutagenicity
of DAB and related compounds, a most remarkable phenomenon was
observed. Aniline which itself is non-mutagenic towards *Salmonella*,
became mutagenic in medium with norharman (Nagao et al., 1977b).
This is shown in Table 2. Similarly, *o*-toluidine and yellow OB
were demonstrated to be mutagenic only in the presence of norharman
(Nagao et al., 1977b; Sugimura et al., 1977c). Aniline had been
suspected to be carcinogenic because of the high incidence of

Table 2. Mutagenicity of Aniline with Norharman

		Revertants/plate[a]
Aniline	100 µg	0
Norharman	200 µg	0
Aniline 100 µg and Norharman 200 µg		3,400

[a]*S. typhimurium* TA98 was used with S-9 mix.

tumors of the urinary bladder in workers in aniline industry (IARC Monographs 1974) and recently it was shown to be carcinogenic in a long-term animal test (U.S. DHEW Publ. 1978). *o*-Toluidine (IARC Monographs 1978) and yellow OB (IARC Monographs 1975) were previously reported to be carcinogenic.

The mutagenicities of several carcinogenic compounds could be demonstrated only in the presence of norharman. This report presents further recent progress on the action of norharman. Definitions of a comutagen and comutagenesis are proposed, and metabolic aspects of comutagenesis are also described.

METHODS AND MATERIALS

Microbial tests were carried out by the incubation method described by Yahagi et al. (1977) and by Sugimura and Nagao (1980). That is, the bacterial culture, S-9 mix and the test substance were *preincubated* for 20 minutes at 37°C, molten soft agar was added and the mixture was poured onto hard agar containing a limited amount of histidine and biotin. This incubation method is more efficient for detecting various mutagens, including pyrrolizidine alkaloids and DMN, than the original method of Ames, which has no preincubation step.

The bacterial strains *Salmonella typhimurium* TA98 and TA100, donated by Dr. Bruce N. Ames, were used throughout this work. TA98 is a frameshift mutant and TA100 is a base-pair change mutant. Both strains contain the pKM101 plasmid.

Norharman and harman were purchased from Aldrich Chemical Co., and their purity was checked by thin-layer chromatography using silica gel as a supporting layer and acetone-ethylacetate-1 N NH$_4$OH (20:20:3, v/v/v) as a solvent. Aniline and *o*-, *m*-, and *p*-toluidine were purified and provided by Dr. Y. Hashimoto, Tohoku University.

```
┌─────────────────────────────────────────────────────────────┐
│                                                               │
│   Non-mutagenic          +      Non-mutagenic                 │
│     compound A                    compound B,C,D or E          │
│                                                               │
│          +S-9 mix                                             │
│          ─────────────────────▶  Mutagenic                    │
│          Incubation                                           │
│                                                               │
│   A is defined as "COMUTAGEN"                                 │
│                                                               │
│   This phenomenon is defined as "COMUTAGENESIS"               │
│                                                               │
└─────────────────────────────────────────────────────────────┘
```

Fig. 2. Definition of a comutagen and comutagenesis

Yellow OB, DPhN and diphenylamine were from Tokyo Kasei Kogyo Co. Ltd., Tokyo. 3-Methylpyridine and its derivatives, N-nitrosonorharman and 8 carboline derivatives, were obtained from Nard Institute, Osaka.

S-9 mix, obtained from the liver of rats pretreated with poly-chlorinated biphenyls (KC-500) as described by Ames et al. (1975), was used throughout. Other conditions are described in the references cited.

COMUTAGENS AND COMUTAGENESIS

The concept of a comutagen is shown schematically in Figure 2. Various compounds, B, C, D, E, etc. none of which are mutagenic, but some of which are reported to be carcinogenic, were demonstrated to be mutagenic in the presence of a compound A. In this case, A is called a COMUTAGEN. The novel phenomenon that B, C, D, E, etc., are mutagenic in the presence of S-9 mix and compound A is defined as COMUTAGENESIS. Norharman corresponds to compound A. The carcinogenic process is classified into two steps, initiation and promotion. The step of initiation is caused by the initiator, and the step of promotion by the tumor promoter. Initiation is closely related to mutation which is caused by the initiator, namely the mutagen. Therefore, a *comutagen* and *comutagenesis* should correspond to a *coinitiator* and *coinitiation*, respectively, but further studies are required to prove this.

EFFECTS OF NORHARMAN ON ANILINE AND ITS METABOLITES

As shown in Figure 3, the number of revertants of TA98 increased linearly with up to 100 µg of aniline with a fixed amount

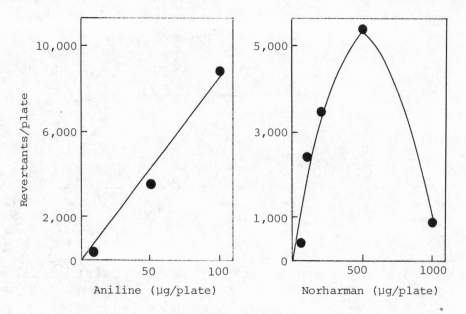

Fig. 3.   Comutagenic action of norharman with aniline. The amounts
           of norharman (a) and aniline (b) per plate were 200 µg
           and 50 µg, respectively. TA98 was used with S-9 mix.

of 200 µg of norharman. Figure 3 also shows that the number of
revertants of TA98 increased linearly with up to 500 µg of norharman with a fixed amount of 50 µg of aniline. Therefore, in most
experiments, 200 µg of norharman and 5Ò µg of aniline and other
test substances were used.

     Since the comutagenesis of aniline with norharman was observed
only in the presence of S-9 mix, the process probably involves
metabolism of aniline. To test this possibility, we examined the
effects of its metabolites, phenylhydroxylamine, nitrosobenzene, o-
and p-aminophenol and azoxybenzene. However, as shown in Table 3,
only aniline and the possible metabolites phenylhydroxylamine and
nitrosobenzene were mutagenic. Linear dose-response curves were
obtained for the mutagenicities of phenylhydroxylamine and nitrosobenzene with a fixed amount of norharman (200 µg).

COMUTAGENIC ACTIONS OF NORHARMAN AND ITS DERIVATIVES

     Norharman is β-carboline. It has been found in tobacco tar
(Poindexter et al., 1962), and also in toasted bread at 18 ng/g,
broiled beef at 39 ng/g, and broiled sardine at 158 ng/g (Yasuda et

Table 3. Comutagenic Actions of Norharman with Aniline and Its
         Metabolites

	Revertants/μg[a]		
	-S-9 mix	+S-9 mix	+S-9 mix + norharman[b]
Aniline	0	0	396
Phenylhydroxylamine	0	0	980
Nitrosobenzene	0	0	467
o-Aminophenol	0	0	0
p-Aminophenol	0	0	0
Azoxybenzene	0	1.3	1.6

[a] TA98 was used.
[b] Norharman was added at 200 μg per plate.

al., 1978). Some β-carboline derivatives are known to inhibit
amine oxidase and also to have hallucinogenic actions (Ho, 1972;
Lessin et al., 1967). Harman, 1-methyl-β-carboline, is found in
tobacco tar, toasted bread, broiled beef and broiled sardine
(Poindexter et al., 1962; Yasuda et al., 1978), and also in mushroom
and Japanese sake (Takase et al., 1966; Takeuchi et al., 1973).

Eleven compounds that are structurally related to norharman
were tested for comutagenicity. Of these, norharman was the strongest comutagen in the presence of S-9 mix and aniline. Harman had a
much weaker comutagenic action, and 9-methylnorharman and 3-methyl-
γ-carboline were only slightly comutagenic. Norharman was demonstrated to intercalate into double stranded DNA bases, but the more
efficient intercalator harman (Hayashi et al., 1977; Pezzuto et
al., 1980) had much less comutagenic action and the strongest
intercalator tested, ellipticine (Ashby et al., 1980), had no
comutagenic action. Thus intercalation may be partly responsible
for comutagenesis but it is not the only mechanism. α-Carboline,
γ-carboline, 1-aminonorharman, carbazole, fluorene and 4-azaflu-
orene had no comutagenic action.

Interaction of norharman with microsomal P-450 cytochrome(s)
was measured quantitatively by differential spectroscopy. Norharman interacts with the 6th ligand of ferric iron of cytochrome P-
450, but B[a]P binds to the active center of cytochrome P-450
(Fujino et al., 1980).

Table 4. Effect of Norharman on the Mutagenicities of Toluidine
          Isomers

| | µg | Revertants/plate[a] | |
		−Norharman	+Norharman[b]
o-Toluidine	40	26	1,250
	120	17	4,100
	200	21	5,460
m-Toluidine	40	22	47
	120	18	54
	200	22	45
p-Toluidine	40	34	35
	120	28	38
	200	29	35
Control		30	32

[a] TA98 was used with S-9 mix.
[b] 200 µg of norharman was added per plate.

COMUTAGENIC ACTIONS OF NORHARMAN ON O-TOLUIDINE AND ITS DERIVATIVES

     o-Toluidine was found to produce papillomatous change of the
urinary bladder of rats, but m- and p-toluidine were not carcinogenic (Yoshida et al., 1941). It is interesting that norharman was
comutagenic with only o-toluidine, not with m- or p-toluidines
(Table 4). The potential mutagenicities of toluidine isomers observed in the presence of norharman are apparently related to their
potential carcinogenic actions. Yellow OB, a carcinogenic azodye,
was mutagenic in the presence of norharman and S-9 mix. If yellow
OB is cleaved at its azobond by the action of azo-reductase in S-9
mix, o-toluidine should be released. But the azo compound Red
colour No. 2, which does not yield aniline or o-toluidine on treatment with azo-reductase, was not active even in the presence of
norharman and S-9 mix (Nagao, unpublished data).

COMUTAGENIC ACTIONS OF NORHARMAN WITH PYRIDINE DERIVATIVES

     As shown in Table 5, two of the aminopyridine derivatives
tested were demonstrated to be mutagenic in the presence of norharman and S-9 mix, but the others were not. Thus, as in the case of

Table 5.   Comutagenic Actions of Norharman with Aminopyridine
           Derivatives

	Revertants/100 µg[a]	
	-Norharman	+Norharman[b]
2-Aminopyridine	0	0
3-Aminopyridine	0	1,505
4-Aminopyridine	0	0
2,3-Diaminopyridine	0	0
2,6-Diaminopyridine	0	0
3,4-Diaminopyridine	0	0
2-Amino-3-methyl-pyridine	0	6,488
2-Amino-4-methyl-pyridine	0	0
2-Amino-5-methyl-pyridine	0	0
2-Amino-6-methyl-pyridine	0	0

[a] TA98 was used with S-9 mix.
[b] 200 µg of norharman was added per plate.

o-toluidine, there was a structural requirement for mutagenic
activity in the presence of norharman and S-9 mix.

COMUTAGENIC ACTION OF NORHARMAN ON N,N-DIPHENYLNITROSAMINE (DPhN)

     In short-term tests on carcinogens, conducted under the aus-
pices of the National Institute of Environmental Health Sciences of
the United States and Imperial Chemical Industry, England, DPhN was
first chosen as a negative control. We were asked to examine the
mutagenicity of DPhN in the *Salmonella* test. DPhN was not mutagenic towards TA98 or TA100 with and without S-9 mix. An additional
request was to test it with norharman. To our surprise, we found
that DPhN was mutagenic to *Salmonella typhimurium* TA98 in the
presence of norharman with S-9 mix (International Workshop, 1980;
Wakabayashi et al., 1980). Meanwhile, Cardy et al. (1979) reported
the carcinogenicity of DPhN in rats on oral administration. These
two findings were coincidental. Our finding indicates that the
inclusion of norharman in the preincubation medium broadens the
range of the microbial short-term test for detecting carcinogens as
positive mutagenic substances. In the absence of norharman, the

Table 6. Mutagenicities of Diphenylnitrosamine, Diphenylamine
          and N-Nitrosonorharman

	Revertants/μmol[a]			
	-Norharman[b]		+Norharman[b]	
	-S-9 mix	+S-9 mix	-S-9 mix	+S-9 mix
N,N-Diphenylnitrosamine	0	0	0	2,932
N,N-Diphenylamine	0	0	0	9,000
N-Nitrosonorharman	180	32		

[a] TA98 was used. [b] 200 μg of norharman was added per plate.

known carcinogens aniline, o-toluidine, yellow OB and DPhN were not
mutagenic.

     The possible mechanism of the comutagenic action of norharman
with DPhN was considered. An enzyme was found that catalyzed a
transnitrosation reaction in which DPhN acted as a donor of nitroso
groups (Stemmerman et al., 1980). Therefore, N-nitrosonorharman
was suspected to be formed as a mutagen from DPhN and norharman.
Studies showed, however, that although N-nitrosonorharman was a
direct acting mutagen, its action was very weak (Table 6). Furthermore diphenylamine was more mutagenic in the presence of norharman
and S-9 mix than DPhN under the same conditions, as also shown in
Table 6. Thus it seems likely that DPhN yields diphenylamine
metabolically, and that the latter exerts comutagenesis with norharman.

STRAIN DEPENDENCY OF COMUTAGENIC ACTION OF NORHARMAN

     The comutagenic actions of norharman with aniline and o-
toluidine were shown with TA98 (rfa, uvrB, pKM101) and also with
TA1538 (rfa, uvrB). However, TA1978 (rfa, uvr^+) was much less
susceptible to norharman and aniline in the presence of S-9 mix as
seen in Table 7. It is noteworthy that the comutagenic actions of
norharman with aniline and o-toluidine were not observed with
TA1535 (rfa, uvrB) or TA100 (rfa, uvrB, pKM101).

     Thus the biological intermediate produced by norharman and
aniline in the presence of S-9 mix seems to have a preferential
effect on the sequence in the frameshift type mutants TA1538 and
TA98. This strain dependency was observed for the comutagenic
actions of norharman with all other substances tested besides
aniline and o-toluidine.

Table 7. Strain Dependency of Comutagenic Action of Norharman

	Revertants/plate
Salmonella strain	Aniline 50 μg + Norharman 200 μg
TA98 (*rfa, uvr*B, pKM101)	2,180
TA1538 (*rfa, uvr*B)	2,460
TA1978 (*rfa, uvr*$^+$)	320
TA100 (*rfa, uvr*B, pKM101)	0
TA1535 (*rfa, vur*B)	0

With S-9 mix

REQUIREMENTS OF BOTH MICROSOMES AND SUPERNATANT FOR THE COMUTAGENIC ACTION OF NORHARMAN

S-9 mix, which is composed of microsomal and supernatant fractions, is required for the comutagenic action of norharman with aniline. Therefore, we incubated phenylhydroxylamine as a possible proximate metabolite of aniline with microsomes, microsomes plus NADPH, supernatant plus NADPH, and microsomes plus supernatant plus NADPH. The results are given in Table 8. It is evident that the microsomes alone had no ability to exert a comutagenic action by norharman in the presence of phenylhydroxylamine, but that further addition of NADPH resulted in formation of many revertants.

Table 8. Requirements for the Comutagenicity of Norharman
          with Phenylhydroxylamine

	Revertants/plate[a]
Microsomes	25
Microsomes + NADPH	2,060
Supernatant + NADPH	80
Microsomes + Supernatant + NADPH	12,060

200 μg norharman and 3 μg phenylhydroxylamine were used.
[a]TA98 was used.

The supernatant plus NADPH was ineffective for the comutagenicity of norharman with phenylhydroxylamine. A mixture of microsomes and supernatant plus NADPH was the most effective for the comutagenicity of norharman. Enzymes in the two fractions may thus act in collaboration to cause the comutagenicity of norharman.

In preliminary experiments seven metabolites were separated by thin layer chromatography after incubation of norharman with S-9 mix. However, none of them showed any comutagenic activity in the presence of phenylhydroxylamine with or without S-9 mix. Nevertheless, suggestive evidence is now available that a directly acting mutagen towards TA98 is produced during incubation of phenylhydroxylamine, norharman and S-9 mix. This compound appears to be labile, but more information on its nature should lead to clarification of the mechanism of the comutagenic action of norharman.

CONCLUDING REMARKS

Norharman, β-carboline, has a unique action, called comutagenicity. Some compounds, which are known or suspected to be carcinogenic, such as aniline, o-toluidine, yellow OB and DPhN, were found to be mutagenic towards TA98 in the presence of norharman and S-9 mix. Some other compounds, which have not yet been fully investigated for carcinogenicity, such as phenylhydroxylamine, nitrosobenzene, diphenylamine and 3-aminopyridine, were also demonstrated to be mutagenic in the presence of norharman and S-9 mix. The latter group of compounds should be subjected to further in $vivo$ carcinogenicity tests.

Since norharman is present in our environment, e.g., in tobacco tar, charred parts of foods and natural products, the effect of its comutagenic action should not be overlooked from the view point of environmental carcinogenesis.

At present, the mechanism of its comutagenic action is not fully understood, although the mechanism clearly involves metabolism. Very probably a reactive intermediate is formed during incubation of norharman with certain compounds plus S-9 mix.

Studies are required on whether this comutagenic action also occurs in mammalian cells and whether it is really related to coinitiation in the carcinogenic process.

REFERENCES

Ames, B. N., McCann, J., and Yamasaki, E., 1975, Methods for detecting carcinogens and mutagens with the $Salmonella$/mammalian-microsome mutagenicity test, Mutation Res., 31:347.

Ashby, J., Elliott, B. M., and Styles, J. A., 1980, Norharman and ellipticine: a comparison of their abilities to interact with DNA *in vitro*, Cancer Lett., 9:21.

Cardy, R. H., Lijinsky, W., and Hildebrandt, P. K., 1979, Neoplastic and nonneoplastic urinary bladder lesions induced in Fischer 344 rats and B6C3F₁ hybrid mice by N-nitrosodiphenylamine, Ecotoxicol. Environ. Safety, 3:29.

Commoner, B., Vithayathil, A. J., Dolara, P., Nair, S., Madyastha, P., and Cuca, G. C., 1978, Formation of mutagens in beef and beef extract during cooking, Science, 201:913.

Fujino, T., Matsuyama, A., Nagao, M., and Sugimura, T., 1980, Inhibition by norharman of metabolism of benzo[a]pyrene by the microsomal mixed function oxidase of rat liver, Chem.-Biol. Interactions, 32:1.

Hayashi, K., Nagao, M., and Sugimura, T., 1977, Interactions of norharman and harman with DNA, Nucl. Acids Res., 4:3679.

Ho, B. T., 1972, Monoamine oxidase inhibitors, J. Pharm. Sci., 61:821.

Int. Agency Res. Cancer, 1974, "IARC Monographs on the Evaluation of Carcinogenic Risk of Chemicals to Man", vol. 4, Aniline, Int. Agency Res. Cancer, Lyon, pp. 27.

Int. Agency Res. Cancer, 1975, " IARC Monographs on the Evaluation of Carcinogenic Risk of Chemicals to Man", vol. 8, Yellow OB, Int. Agency Res. Cancer, Lyon, pp. 287.

Int. Agency Res. Cancer, 1978, "IARC Monographs on the Evaluation of Carcinogenic Risk of Chemicals to Man", vol. 16, *ortho*-Toluidine, Int. Agency Res. Cancer, Lyon, pp. 349.

International Program for the Evaluation of Short-Term Tests for Carcinogenicity, 1980, Elsevier, Amsterdam, in press.

Ishikawa, T., Takayama, S., Kitagawa, T., Kawachi, T., Kinebuchi, M., Matsukura, N., Uchida, E., and Sugimura, T., 1979, *In vivo* experiments on tryptophan pyrolysis products, in: "Naturally Occurring Carcinogens-Mutagens and Modulators of Carcinogenesis", E. C. Miller, J. A. Miller, I. Hirono, T. Sugimura and S. Takayama, ed., Japan Sci. Soc. Press, Tokyo/Univ. Park Press, Baltimore, pp. 159.

Kasai, H., Yamaizumi, Z., Wakabayashi, K., Nagao, M., Sugimura, T., Yokoyama, S., Miyazawa, T., and Nishimura, S., 1980a, Structure and chemical synthesis of Me-IQ, a potent mutagen isolated from broiled fish, Chem. Letter, in press.

Kasai, H., Yamaizumi, Z., Wakabayashi, K., Nagao, M., Sugimura, T., Yokoyama, S., Miyazawa, T., Spingarn, N. E., Weisburger, J. H., and Nishimura, S., 1980b, Potent novel mutagens produced by broiling fish under normal conditions, Proc. Jpn Acad., 56B:278.

Lessin, A. W., Long, R. F., and Parkes, M. W., 1967, The central stimulant properties of some substituted indolylalkylamines and β-carbolines and their activities as inhibitors of monoamine oxidase and the uptake of 5-hydroxytryptamine, Br. J. Pharmac. Chemother., 29:70.

Matsukura, N., Kawachi, T., Uchida, E., Morino, K., Ohgaki, H.,
     Sugimura, T., and Takayama, S., 1980, Induction of liver
     tumors in CDF$_1$ mice by oral administration of Trp-P-1 and
     Trp-P-2 mutagenic principles in tryptophan pyrolysis products,
     Proc. Jpn Cancer Assoc., in press.
Nagao, M., Yahagi, T., Honda, M., Seino, Y., Kawachi, T., and
     Sugimura, T., 1977a, Comutagenic actions of norharman deriv-
     atives with 4-dimethylaminoazobenzene and related compounds,
     Cancer Lett., 3:339.
Nagao, M., Yahagi, T., Honda, M., Seino, Y., Matsushima, T., and
     Sugimura, T., 1977b, Demonstration of mutagenicity of aniline
     and o-toluidine by norharman, Proc. Jpn Acad., 53B:34.
Nagao, M., Yahagi, T., Kawachi, T., Sugimura, T., Kosuge, T., Tsuji,
     K., Wakabayashi, K., Mizusaki, S., and Matsumoto, T., 1977c,
     Comutagenic action of norharman and harman, Proc. Jpn Acad.,
     53:95.
Nagao, M., Yahagi, T., and Sugimura, T., 1978, Differences in effects
     of norharman with various classes of chemical mutagens and
     amounts of S-9, Biochem. Biophys. Res. Commun. 83:373.
Pariza, M. W., Ashoor, S. H., Chu, F. S., and Lund, D. B., 1979,
     Effects of temperature and time on mutagen formation in
     pan-fried hamburger., Cancer Lett., 7:63.
Pezzuto, J. M., Lau, P. P., Luh, Y., Moore, P. D., Wogan, G. N.,
     and Hecht, S. M., 1980, There is a correlation between the
     DNA affinity and mutagenicity of several 3-amino-1-methyl-
     5H-pyrido[4,3-b]indoles, Proc. Natl. Acad. Sci. USA, 77:1427.
Poindexter, E. H., Carpenter, Jr., and Carpenter, R. D., 1962, The
     isolation of harmane and norharmane from tobacco and cigarette
     smoke, Phytochemistry, 1:215.
Rappaport, S. M., McCartney, M. C., and Wei, E. T., 1979, Volatil-
     ization of mutagens from beef during cooking, Cancer Lett.,
     8:139.
Spingarn, N. E., Kasai, H., Vuolo, L. L., Nishimura, S., Yamaizumi,
     Z., Sugimura, T., Matsushima, T., and Weisburger, J. H., 1980,
     Formation of mutagens in cooked foods. III. Isolation of a
     potent mutagen from beef, Cancer Lett., 9:177.
Stemmermann, G. N., Mower, H., Ichinotsubo, D., Tomiyasu, L., Mandel,
     M., and Nomura, A., 1980, Mutagens in extracts of human
     gastric mucosa, J. Natl. Cancer Inst., 65:321.
Sugimura, T., 1979, Naturally occurring genotoxic carcinogens, in:
     "Naturally Occurring Carcinogens-Mutagens and Modulators of
     Carcinogenicity", E. C. Miller, J. A. Miller, I. Hirono, T.
     Sugimura, and S. Takayama, ed., Japan Sci. Soc. Press, Tokyo/
     Univ. Park Press, Baltimore, pp. 241.
Sugimura, T., Kawachi, T., Nagao, M., Yahagi, T., Seino, Y., Okamoto,
     T., Shudo, K., Kosuge, T., Tsuji, K., Wakabayashi, K.,
     Iitaka, Y., and Itai, A., 1977a, Mutagenic principle(s) in
     tryptophan and phenylalanine pyrolysis products, Proc. Jpn
     Acad., 53:58.

Sugimura, T., and Nagao, M., 1980, Modification of mutagenic activity, in: "Chemical Mutagens", F. J. de Serres, and A. Hollaender, ed., vol. 6, Plenum Press, New York and London, pp. 41.

Sugimura, T., Nagao, M., Kawachi, T., Honda, M., Yahagi, T., Seino, Y., Sato, S., Matsukura, N., Matsushima, T., Shirai, A., Sawamura, M., and Matsumoto, H., 1977b, Mutagen-carcinogens in food with special reference to highly mutagenic pyrolytic products in broiled foods, in: "Origin of Human Cancer", H. H. Hiatt, J. D. Watson, and J. A. Winsten, ed., Cold Spring Harbor Lab., Cold Spring Harbor, NY., pp. 1561.

Sugimura, T., Nagao, M., Matsushima, T., Yahagi, T., and Hayashi, K., 1977c, Recent findings on the relation between mutagenicity and carcinogenicity, Nucl. Acds Res., Spec. Publ., No.3, 41.

Takase, S., and Murakami, H., 1966, Studies on the fluorescence of Sake. Part 1. Fluorescence spectrum of Sake and identification of harman, Agr. Biol. Chem., 30:869.

Takayama, S., Hirakawa, T., and Sugimura, T., 1978, Malignant transformation in vitro by tryptophan pyrolysis products, Proc. Jpn Acad., 53B:418.

Takayama, S., Katoh, Y., Tanaka, M., Nagao, M., Wakabayashi, K., and Sugimura, T., 1977, In vitro transformation of hamster embryo cells with tryptophan pyrolysis products, Proc. Jpn Acad., 53B:126.

Takeuchi, T., Ogawa, K., Iinuma, H., Suda, H., Ukita, K., Nagatsu, T., Kato, M., and Umezawa, H., 1973, Monoamine oxidase inhibitors isolated from fermented broths, J. Antibiotics, 26: 162.

U. S. DHEW Publ. No. (NIH) 78-1385, 1978, Bioassay of aniline hydrochloride for possible carcinogenicity, Public Health Serv., Natl. Inst. Health, Bethesda.

Wakabayashi, K., Nagao, M., Kawachi, T., and Sugimura, T., 1980, Comutagenic effect of norharman with N-nitrosamine derivatives, Mutation Res., in press.

Yahagi, T., Nagao, M., Seino, Y., Matsushima, T., Sugimura, T., and Okada, M., 1977, Mutagenicities of N-nitrosamines on Salmonella, Mutation Res., 48:121.

Yasuda, T., Yamaizumi, Z., Nishimura, S., Nagao, M., Takahashi, Y., Fujiki, H., and Sugimura, T., 1978, Detection of comutagenic compounds, harman and norharman in pyrolysis products of proteins and food by GCMS, Med. GCMS Soc., vol. 3, pp. 97.

Yoshida, D., Matsumoto, T., Yoshimura, R., and Matsuzaki, T., 1978, Mutagenicity of amino-α-carbolines in pyrolysis products of soybean globulin, Biochem. Biophys. Res. Commun., 83:915.

Yoshida, T., Shimauchi, T., and Kin, C., 1941, Experimentalle Studien über die Entwicklung des Harnblasentumors, I. Mitt. Gann, 35:272.

REACTIVE INTERMEDIATES FROM NITROSAMINES

Michael C. Archer

Department of Medical Biophysics
University of Toronto
Toronto, Canada, M4X 1K9

Carcinogenic nitrosamines are believed to undergo enzymatic
α-hydroxylation (Figure 1) as a first step in a reaction sequence
leading to the production of the ultimate alkylating agent (Magee
and Barnes, 1967; Druckrey et al., 1967). Spontaneous cleavage
of the carbon-nitrogen bond in the α-hydroxynitrosamine (II) leads
to production of an aldehyde fragment and the alkyldiazohydroxide
(III). The latter may then produce either the diazoalkane (IV), or
cationic products (V and VI) which may then either be trapped by
water as alcohols, or react at a nucleophilic site on a biomolecule
such as DNA.

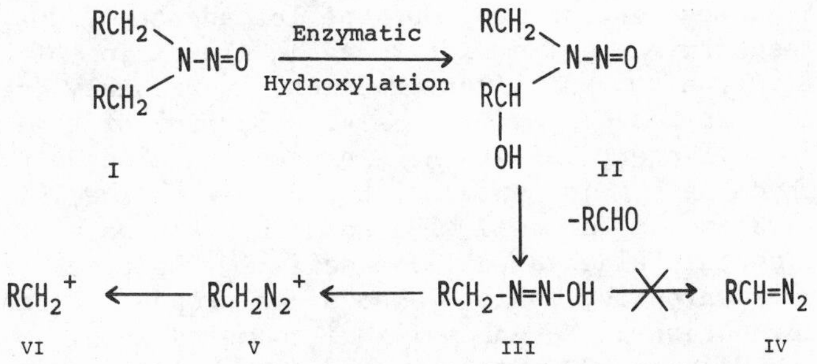

Fig. 1.   Reaction scheme for production of alkylating agent fol-
           lowing microsomal metabolism of N-nitrosodialkylamine.

Lijinsky and his co-workers have performed experiments with
nitrosamines labeled with deuterium atoms which suggest that alkylation of nucleic acids does not proceed via the diazoalkane
(Lijinsky et al., 1968; Ross et al., 1971). Evidence for uni-

molecular reactivity or participation of cationic species during
alkylation reactions by nitrosamines, however, is only indirect
(Lawley, 1976). In an effort to provide explicit evidence for the
nature of the ultimate alkylating agent for nitrosamines, we have
relied on the well known property of primary alkyl carbonium ions
to rearrange via a hydride or an alkyl shift to form a more stable
secondary or tertiary ion (Collins, 1971; March, 1977):

$$RCH_2\overset{+}{C}H_2 \longrightarrow R\overset{+}{C}HCH_3$$

The n-propyl cation constitutes the simplest system in which
a non-degenerate rearrangement can occur. Thus, metabolism of
N-nitrosodi-n-propylamine by isolated rat liver preparations *in
vitro* with water as the nucleophile, should lead to production of
isopropanol if the n-propyl carbonium ion is produced. Hepatic
nucleic acids containing bases modified by an isopropyl group
should be observed following administration of N-nitrosodi-n-
propylamine to a rat *in vivo* if the n-propyl cation is the ultimate alkylating agent. We have tested both of these possibilities.

Table 1 shows the analysis of reaction mixtures for n-propanol and isopropanol following a 90 minute incubation of N-
nitrosodi-n-propylamine with various rat liver fractions in the
presence of an NADPH-generating system (Park et al., 1977; Park
and Archer, 1978a). Both n-propanol and isopropanol were formed
in reaction mixtures containing the microsomal fraction, but there
was no evidence for propanol formation in the soluble enzyme
fraction or when SKF-525A, a potent cytochrome P450 inhibitor, was
added to the microsomal fraction. The combined yield of pro-
panols from 280 µmoles of N-nitrosodi-n-propylamine was 6.1 µmoles
and 28.5 µmoles for the 105,000 g pellet and the 9000 g supernatant respectively. The difference in the ratios of n- to isopropanol in the two rat liver fractions is probably due to
differences in the composition (e.g. polarity) of the reaction
mixtures (Zollinger, 1961). As expected, we also detected propionaldehyde as a major product, but only with the 105,000 g
pellet (8.3 µmoles formed). The soluble fraction of rat liver is
known to contain aldehyde oxidase activity (Deitrich, 1966). For
comparative purposes, Table 1 also shows propanol formation following incubation of N-n-propyl-N-nitrosourea in 0.1 M Tris
buffer, pH 7.4, at 37°. As expected (Moss, 1974; Kirmse, 1976)
both propanol isomers were formed, the total yield in this case
being almost quantitative.

These results indicate that microsomal metabolism of N-
nitrosodi-n-propylamine *in vitro* leads to production of propanol
through the intermediary of a carbocation as illustrated in
Figure 1.

Table 1

Analysis of n-propanol and isopropanol obtained by
incubation of N-nitrosodi-n-propylamine (NDPA) with
rat liver fractions and incubation of N-n-propyl-N-
nitrosourea (PNU) in 0.1 M tris buffer, pH 7.4 at 37°C.

| System[a] | Proportion of n- and iso- forms in propanol mixture (%) | |
	n-propanol	isopropanol
NDPA + 105,000 g pellet	61	39
NDPA + 9,000 g supernatant	82	18
NDPA + 105,000 g supernatant	ND	ND
NDPA + 105,000 g pellet + SKF 525A	ND	ND
PNU + buffer	61	39

ND = none detected

[a] Reaction mixtures contained the following components in a final
volume of 20 ml: 15 ml of rat liver fraction prepared from a 25%
(w/v) liver homogenate;   0.1 M Tris buffer, pH 7.4;   61.5 mM $MgCl_2$;
2.1 mM NADP;   21.8 mM glucose-6-phosphate; and 50 units of
glucose-6-phosphate dehydrogenase.   0.28 mmole of N-nitrosodi-n-
propylamine were added to start the reaction.   After 90 min., the
reaction was terminated by adding 20 ml of 20% $ZnSO_4$ followed by
20 ml of sat. $Ba(OH)_2$.   Following removal of the precipitate by
centrifugation, the supernatant was distilled at atmospheric pressure through a short path micro-distillation head.   The distillate
which was collected on ice was analyzed for propionaldehyde, isopropanol and n-propanol by gas chromatography with a 2m Porapak-N
column at 135°.   Product identities were confirmed by gas chromatography-mass spectrometry.

We next investigated whether N-nitrosodi-n-propylamine, when
administered to animals, would lead to formation of isopropylated,
as well as n-propylated bases in DNA (Park et al., 1980).   In
1971, Krüger demonstrated formation of 7-propylguanine in rat
liver following administration of N-nitrosodi-n-propylamine, but
no attempt was made in that study to distinguish between 7-n-
propyl and 7-isopropyl guanines.

N-[2,3-^{3}H]nitrosodi-n-propylamine was injected intraperitoneally into each of four male Sprague-Dawley rats.   After 12
hours, the animals were sacrificed, the livers removed, and DNA

isolated by the method of Swann and Magee (1968). Unlabeled
authentic 7-n-propylguanine and 7-isopropylguanine were added to
the DNA before hydrolysis with perchloric acid at 100°. The
hydrolysate was examined for the presence of the labeled propylguanine isomers by reverse phase liquid chromatography using a μ-
Bondapak C18 column. Figure 2 shows that at a flow rate of 1 ml/
min. 0.05 M ammonium formate pH 3.5 containing 3% methanol gave
separation of the propylguanine isomers. The isomers were also
well separated from the major bases and 7-methylguanine, which
eluted close to the solvent front. When the eluant was 0.01 M
potassium phosphate, pH 6.8, containing 5% methanol, the two
propylated isomers eluted as a single peak with a retention time of
10.5 min, which was again well separated from the major bases and
7-methylguanine.

Figure 2 shows that the DNA hydrolysate contained only 7-n-
propylguanine. In a similar experiment, we showed that rat liver

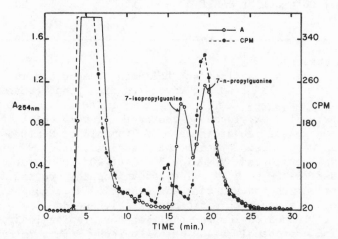

Fig. 2.   Chromatographic profile of DNA hydrolysate from rat
           liver 12 h after application of 133 mg/18.5 mCi/kg N-
           [2,3-^{3}H]-nitrosodi-n-propylamine. Conditions included:
           column, 30-cm μ-Bondapak-C18; eluant, 3% methanol/0.05
           M ammonium formate (pH 3.5); flow rate, 1 ml/min; and
           sample, 10 mg hydrolyzed DNA containing unlabeled 7-n-
           propylguanine (41 μg/mg DNA) and 7-isopropylguanine (34
           μg/mg DNA), which was injected five times in five equal
           portions. Reprinted with permission from Park et al.,
           1980.

RNA did contain a small amount of the rearranged adduct, 7-iso-
propylguanine, but the amount represented less than 5% of the 7-n-

propylguanine formed (Park et al., 1980). The results of our *in vitro* experiments supported a reaction sequence, initiated by microsomal oxidation of the nitrosamine, leading to formation of carbonium ions. The results of the *in vivo* experiment therefore suggest that in the intact cell, this reaction sequence can be intercepted by nucleophilic sites in DNA, exemplified here by the N7 position of guanine, before a carbocation is formed. The ultimate alkylating agent in the cell is probably one of the earlier electrophilic intermediates such as the α-hydroxynitrosamine alkyldiazohydroxide, or alkyldiazonium ion (II, III or IV, Figure 1) that reacts with a nucleophilic site on the DNA molecule in a concerted biomolecular reaction.

The results of the *in vivo* experiment could also be explained by formation of 7-isopropylguanine in DNA followed by loss of this adduct at a much faster rate than 7-n-propylguanine. Although this possibility cannot be ruled out, especially if enzymatic repair is involved, the combined steric and electronic effects of the n-propyl group would not be expected to differ significantly from those of the isopropyl group in either a unimolecular or nucleophile-assisted excision reaction (Lawley and Brookes, 1963; Wiberg, 1964).

An additional and somewhat surprising feature of the reactivity of N-nitrosodi-n-propylamine *in vivo* is that, in addition to direct transfer of a propyl group to DNA or RNA, the nitrosamine also acts as a methylating agent (Krüger, 1971). 7-methylguanine is in fact the major alkylation product in hepatic nucleic acids when N-nitrosodi-n-propylamine or its β-hydroxy or β-oxo derivatives are administered to rats (Krüger, 1973; Krüger and Bertram, 1973). Krüger suggested that N-nitrosodi-n-propylamine is metabolically degraded via two β-oxidation reactions followed by cleavage of the acyl fragment to yield N-nitrosomethylpropylamine which then acts as the methylating agent.

During our investigation of the *in vitro* metabolism of N-nitrosodi-n-propylamine, we were able to show that, in addition to oxidation at the carbon atom α to the N-nitroso group, β-oxidation can also take place (Park and Archer, 1978a). Thus, N-nitroso-2-hydroxypropylpropylamine (NHPPA) was isolated and characterized following incubation of N-nitrosodi-n-propylamine with the microsomal fraction of rat liver. We next showed that NHPPA could be oxidized further to N-nitroso-2-oxopropylpropylamine (NOPPA) by the microsomal preparation (Park and Archer, 1978a). Finally, with NOPPA as substrate, we showed that either a reduction reaction may take place with the microsomal fraction or the soluble fraction from rat liver to yield NHPPA, or NOPPA may undergo microsomal α-oxidation (Park and Archer, 1978b).

The results of these experiments therefore provide direct evidence for the hypothesis formulated by Krüger (1971) that N-nitrosodi-n-propylamine may be metabolized by two consecutive β-oxidation reactions.

In further experiments designed to confirm and extend Krüger's observations on the methylating properties of N-nitrosodi-n-propylamine and its β-oxidized derivatives, we have shown that administration of highly purified NOPPA to rats indeed leads to formation of 7-methylguanine in hepatic DNA (Leung et al., 1980). As shown in Figure 3, O^6-methylguanine was additionally formed.

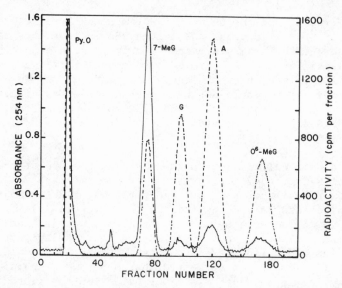

Fig. 3.   Sephadex G-10 chromatography of a hydrolysate of 7 mg DNA from rat liver 12 hr following intraperitoneal injection of 310 mg/23 mCi/kg N-nitroso-2-oxo[1,3-^{3}H] propylpropylamine;   ● ——— ●, Radioactivity; ● - - - - ●, absorbance (254 nm);   Py.O, pyrimidine oligonucleotides; 7-MeG, 7-methylguanine;   A, adenine: G, guanine; O^6-MeG, O^6-methylguanine;   eluant, 0.05 M ammonium formate, pH 6.8, containing 0.02% sodium azide;   flow rate 33.6 ml/hr;   fraction size, 2.8 ml.   Unlabeled 7-methyl-guanine and O^6-methylguanine were added to the isolated DNA prior to hydrolysis.   Reprinted with permission from Leung et al., 1980.

The ratio of O^6-methylguanine : 7-methylguanine was 0.07, which is very similar to the ratio of these two methylated guanines found following N-nitrosodimethylamine administration at compar-

able dose levels and times (0.05 - 0.12, Nicoll et al., 1975).
Since Lawley (1976) has shown that the ratio of O^6-methylguanine:
7-methylguanine in DNA is indicative of the reactivity of the
methylating agent, our result suggested that the methylating
agent formed from NOPPA is similar in reactivity to that formed
from N-nitrosodimethylamine, and supported the hypothesis of
Krüger that N-nitrosomethylpropylamine may be the methylating
agent formed from NOPPA.

In subsequent experiments, however, we have been unable to
provide any evidence for the formation of N-nitrosomethylpropylamine from NOPPA. Although NOPPA is converted into N-nitrosomethylpropylamine in a base-catalyzed, non-enzymatic reaction,
this takes place only at high pH; we observed no detectable
reaction at physiological pH at 37° even after 16 hours (Leung
et al., 1980). In an extensive search, we have yet found no rat
liver fraction capable of catalyzing the conversion of NOPPA to
N-nitrosomethylpropylamine (Leung et al., 1980; Leung and Archer,
unpublished observations), although we have possibly not investigated appropriate conditions to allow detection of such an
enzyme. Thus the mechanism by which NOPPA acts as a methylating
agent for DNA in the rat remains to be determined.

SUMMARY

Metabolism of N-nitrosodi-n-propylamine by isolated rat liver
fractions yielded both n-propanol and isopropanol, providing evidence for a reaction sequence, initiated by formation of the α-
hydroxynitrosamine, that leads to formation of carbonium ions.
Administration of N-nitrosodi-n-propylamine to rats led to formation of 7-n-propylguanine but not 7-isopropylguanine in hepatic
DNA, showing that in the intact cell, the reaction sequence is
intercepted by nucleophilic sites in DNA before a carbocation is
formed. Alkylation of DNA therefore appears to occur primarily by
a bimolecular reaction not involving free alkyl cations.

N-nitrosodi-n-propylamine is also oxidized by rat liver
microsomes by two consecutive β-oxidation reactions to yield N-
nitroso-2-oxopropylpropylamine. The latter agent leads to production of 7-methylguanine and O^6-methylguanine in liver DNA
following its administration to rats. The chemical nature of the
ultimate methylating agent is unclear.

ACKNOWLEDGEMENTS

These investigatio..s were supported by Grants CA 26651 and CA
21951 awarded by the National Cancer Institute, DHEW, Research Career Development Award ES 00033 awarded by the National Institute of
Environmental Health Sciences, DHEW, and the Ontario Cancer

Research and Treatment Foundation. The collaboration of Kwan-Hang
Leung, Kwanghee K. Park and John S. Wishnok is gratefully ack-
nowledged.

REFERENCES

Collins, C. J., 1971, Reactions of primary aliphatic amines with
        nitrous acid, Accts. Chem. Res., 4:315.
Deitrich, R. A., 1966, Tissue and subcellular distribution of
        mammalian aldehyde-oxidizing capacity, Biochem. Pharmacol.,
        15:1911.
Druckrey, H., Preussmann, R., Ivankovic, S., and Schmäl, D., 1967,
        Organotrope carcinogene wirkungen bei 65 verschiedenen N-
        nitroso-Verbindungen an BD-Ratten, Z. Krebsforsch., 69:103.
Kirmse, W., 1976, Nitrogen as leaving group: aliphatic diazonium
        ions, Angew. Chem. Int. Ed. Engl., 15:251.
Krüger, F. W., 1971, Metabolismus von Nitrosaminen in vivo, I.
        Über die β-Oxidation Aliphatischer Di-n-alkylnitrosamine:
        Die Bildung von 7-Methylguanin neben 7-Propyl-bzw. 7-
        Butylguanin nach Applikation von Di-n-propyl- oder Di-n-
        butylnitrosamin. Z. Krebsforsch., 76:145.
Krüger, F. W., 1973, Metabolism of nitrosamines in vivo, II. On
        the methylation of nucleic acids by aliphatic di-n-alkyl-
        nitrosamines in vivo, caused by β-oxidation: The in-
        creased formation of 7-methylguanine after application of
        β-hydroxypropyl-propyl-nitrosamine compared to that after
        application of di-n-propyl-nitrosamine. Z. Krebsforsch.,
        79:90.
Kruger, F. W., and Bertram, B., 1973, Metabolism of nitrosamines
        in vivo, III. On the methylation of nucleic acids by ali-
        phatic di-n-alkyl-nitrosamines in vivo resulting from β-
        oxidation: The formation of 7-methylguanine after ap-
        plication of 2-oxo-propyl-propyl-nitrosamine and methyl-
        propyl-nitrosamine. Z. Krebsforsch., 80:189.
Lawley, P. D., and Brookes, P., 1963, Further studies on the
        alkylation of nucleic acids and their constituent nucleo-
        tides, Biochem. J., 89:127.
Lawley, P. D., 1976, Carcinogenesis by alkylating agents, in:
        "Chemical Carcinogens", C. E. Searle, ed., American
        Chemical Society Monograph 173, Washington D.C.
Leung, K. H., Park, K. K., and Archer, M. C., 1980, Methylation of
        DNA by N-nitroso-2-oxopropylpropylamine: Formation of O^6
        and 7-methylguanine and studies on the methylation mechan-
        ism. Tox. Appl. Pharmacol., 53:29.
Lijinsky, W., Loo, J., and Ross, A. E., 1968, Mechanism of alkyl-
        ation of nucleic acids by nitrosodimethylamine, Nature,
        218:1174.
Magee, P. N., and Barnes, J. M., 1967, Carcinogenic nitroso com-
        pounds, Adv. Cancer Res., 10:164.

March, J., 1977, "Advanced organic chemistry: reaction mechanisms
    and structure", McGraw-Hill, New York.

Moss, R. A., 1974, Some chemistry of alkanediazotates, Acc. Chem.
    Res., 7:421.

Nicoll, J. W., Swann, P. F., and Pegg, A. E., 1975, Effect of di-
    methylnitrosamine on persistence of methylated guanines in
    rat liver and kidney DNA, Nature, 254:201.

Park, K. K., Wishnok, J. S., and Archer, M. C., 1977, Mechanism of
    alkylation by N-nitroso compounds: Detection of re-
    arranged alcohol in the microsomal metabolism of N-
    nitrosodi-n-propylamine and base-catalyzed decomposition
    of N-n-propyl-n-nitrosourea, Chem. -Biol. Interact., 18:
    349.

Park, K. K., and Archer, M. C., 1978a, Microsomal metabolism of
    N-nitrosodi-n-propylamine: Formation of products result-
    ing from α- and β-oxidation, Chem. -Biol. Interact., 22:
    83.

Park, K. K., and Archer, M. C., 1978b, Metabolism of N-nitroso-2-
    oxopropylpropylamine by rat liver: Formation of products
    resulting from both oxidation and reduction, Cancer
    Biochem. Biophys. 3:37.

Park, K. K., Archer, M. C., and Wishnok, J. S., 1980, Alkylation
    of nucleic acids by N-nitrosodi-n-propylamine: Evidence
    that carbonium ions are not significantly involved, Chem.
    -Biol. Interact., 29:139.

Ross, A. E., Keefer, L., and Lijinsky, W., 1971, Alkylation of
    nucleic acids of rat liver and lung by deuterated N-
    nitrosodiethylamine in vivo, J. Natl. Cancer Inst., 47:
    789.

Swann, P. F., and Magee, P. N., 1968, Nitrosamine-induced carcino-
    genesis. The alkylation of nucleic acids of the rat by N-
    methyl-n-nitrosourea, dimethylnitrosamine, dimethylsulfate
    and methylmethane sulfonate, Biochem. J., 110:39.

Wiberg, K. B., 1964, "Physical Organic Chemistry", John Wiley &
    Sons, New York.

Zollinger, H., 1961, "Diazo and Azo Chemistry: Aliphatic and
    Aromatic Compounds", Interscience, New York.

FORMATION AND FATE OF REACTIVE INTERMEDIATES OF PARATHION

James Halpert and Robert A. Neal

Center in Environmental Toxicology, Department of
Biochemistry, Vanderbilt University School of Medicine
Nashville, Tennessee 37232

INTRODUCTION

A number of compounds containing the thiono-sulfur group (P=S
or C=S) exhibit toxic properties when administered to laboratory
animals or man. These effects include inhibition of thyroid hormone synthesis, induction of neoplasia, blood dyscrasias, liver
damage, lung damage, and inhibition of liver, lung, and brain monooxygenases. The major effort in this laboratory has been directed
towards elucidating the mechanisms by which thiono-sulfur compounds
cause inactivation of the cytochrome P-450 dependent monooxygenase
system in vitro and liver and lung damage in vivo. There is substantial evidence based on in vivo and in vitro studies utilizing
inhibitors and inducers of metabolism that the inactivation of
cytochrome P-450 and organ damage are caused by metabolites formed
during the oxidative desulfuration of thiono-sulfur compounds.
However, the oxygen analogs, which represent the stable oxidative
metabolites, cause neither inhibition of cytochrome P-450 nor
organ damage. This finding strongly implicates reactive intermediates formed during the desulfuration of thiono-sulfur compounds in mediating their adverse effects. Consistent with this
interpretation is the observation of covalent binding to target
organ macromolecules when thiono-sulfur compounds are administered
in vivo and of covalent binding to microsomes when the compounds
are incubated in vitro in the presence of NADPH. For a review of
the toxicology and metabolism of thiono-sulfur compounds see Neal
(1980).

INITIAL IN VITRO STUDIES WITH PARATHION

The insecticide parathion (diethyl p-nitrophenyl phosphoro-

thionate) has served as a model compound for investigating the
mechanism by which thiono-sulfur compounds cause inactivation of
cytochrome P-450 in vitro. Parathion is converted by the cytochrome P-450 dependent monooxygenase system in insects and mammals
to paraoxon, which is a potent inhibitor of the enzyme acetylcho-
linesterase. The high acute toxicity of parathion precludes the
administration to experimental animals of doses sufficient to
inhibit cytochrome P-450 in vivo. However, the parathion analog
diethyl phenyl phosphorothionate, which is much less acutely toxic,
has been found to decrease cytochrome P-450 levels both in vivo
and in vitro, and to cause centrilobular hydropic degeneration of
the liver in phenobarbital-pretreated rats in a manner analogous
to carbon disulfide (Seawright et al., 1976). Thus it was considered that elucidation of the mechanism by which parathion causes
inhibition of cytochrome P-450 in vitro, in addition to shedding
light on cytochrome P-450 structure and function, should have
relevance for an understanding of how thiono-sulfur compounds
exert adverse effects on the liver when administered in vivo.

    Based on studies with a chemical model system (Ptashne and
Neal, 1972; Herriott, 1971) and on oxygen-18 studies using liver
microsomes (Ptashne et al., 1971) a chemical mechanism for the
formation of paraoxon from parathion was proposed. According to
this mechanism (Figure 1) the initial reaction in the metabolism
of parathion by the cytochrome P-450 containing monooxygenase
system is the transfer of a singlet oxygen atom from cytochrome
P-450 to one of the unshared electron pairs on the thiono-sulfur
group of parathion to form an intermediate S-oxide. The S-oxide
of parathion cannot be isolated and is shown here in brackets.
It was proposed that the S-oxide then cyclizes into a tricyclic
structure, referred to as a phosphooxythirane, which on electron
rearrangement loses atomic sulfur and forms the product paraoxon.

    If the sulfur atom released from parathion as proposed in
Figure 1 is in its singlet state, it would be a highly reactive
electrophile which would bind readily to nucleophiles near the site
of its release. The thiono-sulfur group of parathion has been
found to bind covalently to tissue macromolecules following
administration of $[^{35}S]$ parathion in vivo (Poore and Neal, 1972),
and on incubation with hepatic microsomes in vitro (Norman et al.,
1974). When double-labeled $[^{32}P, \ ^{35}S]$ parathion was incubated in
the absence of NADPH with liver microsomes from phenobarbital-
treated rats, only a trace of radioactivity could be found bound
to the microsomes. However, in the presence of NADPH, a substantial amount of sulfur and a small but significant amount of the
phosphorous-containing portion of the parathion molecule became
covalently bound. The ^{32}P-binding could not be accounted for by
the binding of the paraoxon formed during the incubations, suggesting binding of one or more of the intermediate S-oxides shown in
Figure 1. The results also demonstrated that the majority of the

sulfur bound to the microsomes was free of the phosphorous-containing part of the molecule and thus must be atomic sulfur released
in the metabolism of parathion to paraoxon. This was further
substantiated by the finding that the amount of sulfur bound under
these conditions is equivalent to the amount of paraoxon formed
(Norman et al., 1974).

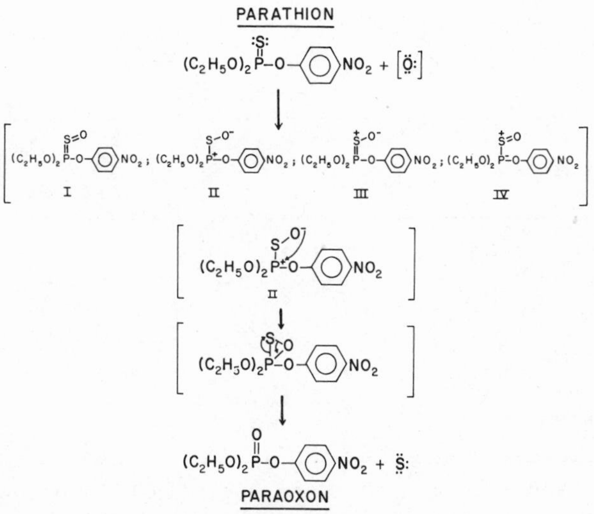

Fig. 1:   Chemical mechanism for the metabolism of parathion to
          paraoxon by the mammalian hepatic cytochrome P-450 depen-
          dent monooxygenase system.

Metabolism of parathion by liver microsomes was found to cause
a decrease in the level of cytochrome P-450 detectable as its
carbon monoxide complex and a decrease in the rate of metabolism
of benzphetamine. However these effects could not be attributed
to paraoxon or any of the other stable metabolites formed during
the incubations. These findings strongly implicated binding of
atomic sulfur and/or the S-oxide of parathion as being responsible
for the inhibition of the cytochrome P-450. These results set the
stage for an investigation of the mechanism of the inactivation of
cytochrome P-450 during parathion metabolism using a reconstituted
monooxygenase system containing the major form of cytochrome P-450
from liver microsomes of phenobarbital-treated rats.

INACTIVATION OF PURIFIED RAT LIVER CYTOCHROME P-450 BY PARATHION

Covalent Binding of ^{35}S to Cytochrome P-450

    With the use of a partially purified antibody to the cyto-
chrome P-450, covalent binding of a ^{35}S-containing metabolite of
[^{35}S] parathion to the cytochrome P-450 of a reconstituted system
could be demonstrated (Kamataki and Neal, 1976). The results of
such an experiment are shown in Table 1.

Table 1. Binding of the Sulfur Atom of Parathion
to Rat Liver Cytochrome P-450

Conditions[a]	^{35}S or ^{14}C Binding (nmol/nmol P-450/5 min)
[^{35}S] Parathion	
Complete System	1.02 ± 0.02
-NADPH	0.04 ± 0.01
-Reductase	0.04 ± 0.00
[^{14}C] Parathion	
Complete System	0.02 ± 0.00

The data are taken from Kamataki and Neal (1976) and represent the
mean $\pm$ standard deviation of duplicate determinations.

[a]The complete incubation mixture contained 0.5 nmole of cytochrome
P-450, 0.25 unit of NADPH-cytochrome c reductase, 100 μg of
dilauroyl-L-3-phosphatidylcholine, 0.05 M HEPES buffer (pH 7.8),
0.015 M $MgCl_2$, 100 μg of sodium deoxycholate, 0.1 mM EDTA, 0.1 mM
[35S]- or [ethyl-^{14}C] parathion, and).1mM NADPH in a total volume of 1.0 ml. The reaction mixtures were incubated at 37° for
5 min. The reactions were stopped by placing the incubation in
an ice-water bath. After approximately 5 min. 3 mg of partially
purified antibody to cytochrome P-450 were added, and the mixtures were allowed to stand at 0-4° for 16 hr. All of the cytochrome P-450 was precipitated under these conditions. The precipitate representing cytochrome P-450 combined with its antibody
was separated by centrifugation and washed extensively prior to
scintillation counting.

    In the absence of NADPH or NADPH-cytochrome P-450 reductase,
less than 4% as much binding to the cytochrome P-450 was observed
as in the presence of a complete system. This low amount of ^{35}S
may represent non-covalently bound material. Furthermore, when
the incubation was carried out in the presence of NADPH but using
[ethyl-^{14}C] rather than [^{35}S] parathion, less than 2% as much

radioactivity was bound to the cytochrome P-450. Again, this low
amount of binding could represent non-covalently bound material.
These findings showed that metabolism of parathion is required for
covalent binding of sulfur to cytochrome P-450 and that essentially
all of the sulfur bound to the P-450 is free of the remainder of
the parathion molecule.

Three different experimental approaches were employed to
determine whether cytochrome P-450 was the sole site of attack of
atomic sulfur from parathion. Using the partially purified antibody to the P-450, approximately 90% of the ^{35}S covalently bound
to the proteins of the reconstituted system could be precipitated
along with all of the cytochrome P-450 and approximately 30% of
the reductase activity (Kamataki and Neal, 1976). These results
demonstrated that the majority of the bound sulfur was associated
with the P-450. It remained unclear, however, whether some of the
radioactivity in the precipitate represented ^{35}S bound to reductase
which had co-precipitated with the P-450 and whether the low amount
of radioactivity in the supernatant represented ^{35}S bound to reductase or ^{35}S which had been released from the P-450 during the 16
hours allowed for precipitation of the P-450-antibody complex.

Therefore the ^{35}S-labeled proteins of the reconstituted system
were submitted to SDS gel electrophoresis (Kamataki and Neal, 1976).
The majority of the radioactivity was associated with high molecular weight protein which remained at the origin. However, of
the protein which migrated into the gel, only the cytochrome P-450
contained any substantial amount of radioactivity. No accumulation
of protein at the origin was observed when the reconstituted system
was incubated with parathion in the absence of NADPH. Furthermore,
when the ^{35}S labeled proteins were treated with CN$^-$ prior to SDS
gel electrophoresis, the accumulation of both protein and radioactivity at the origin could be prevented. The CN$^-$ treatment released approximately 50% of the ^{35}S as ^{35}SCN$^-$, in good agreement
with results obtained with C^{35}S$_2$ using whole microsomes (Catignani
and Neal, 1975). This suggested the presence of hydrodisulfide
linkages (R-S-^{35}SH), presumably formed by attack of atomic sulfur
on cysteine residues in the cytochrome P-450. The efficacy of CN$^-$
and other nucleophiles such as β-mercaptoethanol and dithiothreitol
in dissociating the high molecular weight aggregates formed during
the metabolism of parathion by the reconstituted system or intact
microsomes, suggested that the aggregates were stabilized by inter-
molecular polysulfide bonds between hydrodisulfide linkages on one
protein molecule and sulfhydryl groups on other molecules. These
results provided important information about the physico-chemical
state of the cytochrome P-450 after metabolism of parathion and
confirmed the conclusion based on the antibody experiments that
P-450 is the major site of attack of ^{35}S from parathion. It remained unclear, however, whether any of the labeled protein at the
origin represented aggregates containing cytochrome P-450 and the

reductase. Cross-linking of P-450 with the reductase might offer
one explanation for the co-precipitation of P-450 and reductase
activity upon treatment of the ^{35}S labeled reconstituted system
protein with the antibody to the P-450.

Final confirmation that cytochrome P-450 is the predominant if
not sole target of ^{35}S released from [^{35}S] parathion by the reconstituted system was provided by studies using a modified system
where the NADPH-cytochrome P-450 reductase was replaced by cumene
hydroperoxide (Yoshihara and Neal, 1977). With both the standard
and modified reconstituted systems, 65% of the sulfur released
during the metabolism of parathion to paraoxon became bound to
protein. In subsequent experiments using the standard reconstituted system total protein-bound radioactivity was equated with
radioactivity bound to cytochrome P-450 (Halpert and Neal, 1980).
These experiments were carried out at a low reductase:P-450 ratio
(1:10 unit/nmol), so that the reductase constituted less than 10%
of the total protein by weight.

Nature of the Non-dissociable ^{35}S Bound to Cytochrome P-450

The results of the experiments with CN$^-$ did not reveal whether
that portion of the bound ^{35}S not released as ^{35}SCN$^-$ represented
hydrodisulfide linkages not accessible to the reagent or sulfur
bound in linkages chemically stable to CN$^-$ treatment (Kamataki and
Neal, 1976). If the sulfur atom released from parathion were in
the singlet state, it could be expected to participate in carbon-
hydrogen insertion reactions in a manner similar to carbenes and
nitrenes. A carbon-hydrogen insertion reaction would give rise to
a mercaptan, which would be resistant to treatment with CN$^-$.
Considering the reactivity of singlet atomic sulfur, it might be
expected to bind at or near the site of its release. Thus an
investigation was undertaken of the nature of the non-dissociable
sulfur bound to the cytochrome P-450 in order to gain information
about both the state of the sulfur released from parathion and
about the cytochrome P-450 itself (Halpert and Neal, 1980).

The presence of two chemically different classes of sulfur
covalently bound to cytochrome P-450 was demonstrated by several
different approaches. Treatment with 5 mM dithiothreitol for 1 hr
at room temperature readily removed 75% of the bound ^{35}S. However,
no further radioactivity could be released even if the dithiothreitol treatment was performed in the presence of 1% SDS or sodium
cholate at 37° or 1% SDS at 100°. Similar results were obtained
when the ^{35}S-labeled protein was treated with performic acid.
Such treatment would be expected to convert hydrodisulfide link-
ages to non-labeled cysteic acid plus ^{35}SO$_4^{-2}$ (R-S-^{35}SH--->R-SO$_3^-$
+ ^{35}SO$_4^{-2}$) and to convert any amino acids derivatized by carbon-
hydrogen insertion to ^{35}S containing sulfonic acid derivatives
R-^{35}SH----->R-^{35}SO$_3^-$). The oxidized protein, which was insoluble

in all media tested, was treated for 24 hr with trypsin at an enzyme: substrate ratio of 1:30 (w/w), and the digest was chromatographed on a calibrated column of Sephadex G-10. Approximately 25% of the radioactivity eluted in the void volume, apparently representing ^{35}S bound to peptide material, whereas the remainder eluted in the position of standard [35]SO$_4^{-2}$. The identity of the second peak was confirmed by chromatography in two other systems. The peptide-bound ^{35}S could be resolved into two components by column electrophoresis at pH 1.9. When the performic acid oxidized protein was subjected to acid hydrolysis and then column electrophoresis, three ^{35}S-containing components could be resolved (Figure 2).

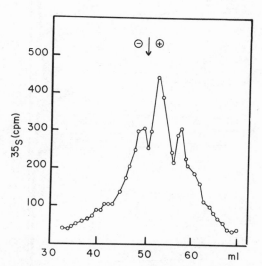

Fig. 2: Column electrophoresis at pH 1.9 of an acid hydrolysate of performic acid oxidized ^{35}S-labeled protein of a reconstituted system after incubation with [^{35}S] parathion. Electrophoresis was performed for 7.5 hr at 1000 V and 13.8 mA. The [35]SO$_4^{-2}$ was not recovered. The incubation with parathion was performed as described by Halpert and Neal (1980).

The heterogeneity of the bound ^{35}S was confirmed by acid hydrolysis of the ^{35}S-labeled protein without previous performic acid oxidation, which yielded four labeled peaks upon chromatography on a Beckman 121 amino acid analyzer (Fig. 3). The first peak appeared at the break-through and constituted approximately

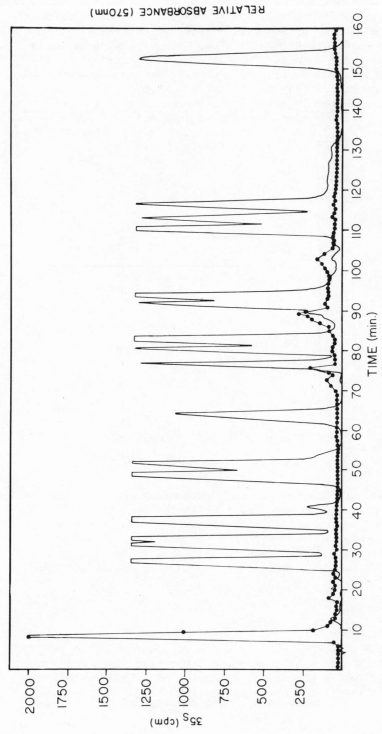

Fig. 3: Ion-exchange chromatography on a Beckman 121 amino acid analyzer of an acid hydrolysate of a cytochrome P-450 containing reconstituted system incubated with [^{35}S] parathion. One aliquot of the sample corresponding to 10 nmoles protein was monitored by ninhydrin (smooth curve). An identical aliquot was then applied to the column, and fractions of 1-min were collected and monitored for radioactivity (●——●).

70% of the radioactivity recovered, apparently representing material derived from the sulfur bound to cysteine residues. The remaining ^{35}S appeared as three ninhydrin-positive peaks. From their elution positions on the amino acid analyzer and on Sephadex G-10, two would appear to be hydrophobic and one aromatic. Experiments are now in progress to positively identify the structure of these derivatized amino acids.

Loss of Heme, Cytochrome P-450, and Enzymatic Activity During Parathion Metabolism

Metabolism of parathion by a reconstituted system was found to be accompanied by a considerable loss of heme and cytochrome P-450 (Table 2). In a series of six experiments the heme loss was calculated to be $81 \pm 10\%$ of the cytochrome P-450 loss. Catalase offered little or no protection against the loss of heme, suggesting that it was not due to hydrogen peroxide formation during the metabolism of parathion. A loss of heme was also observed in intact microsomes incubated with parathion in the presence of NADPH. The only change in the oxidized spectrum of the reconstituted system was a decrease in the absorbance at 417 nm; no new spectral bands indicative of modified heme appeared.

Table 2. Parathion-mediated loss of cytochrome P-450
and heme in a reconstituted system

System[a]	P-450 (nmol/ml)	heme (nmol/ml)[b]
-NADPH	0.96 ± 0.05	1.22 ± 0.13
Complete	0.22 ± 0.01	0.49 ± 0.03
+Catalase[c]	0.24 ± 0.00	0.59 ± 0.06

All results represent the mean $\pm$ standard deviation of duplicate or triplicate determinations.

[a]The complete system contained 1.6 nmoles cytochrome P-450, 0.16 units NADPH-cytochrome P-450 reductase, 0.1 mM parathion, 50 µg dilauryl L-3 phosphatidylcholine, 50 µg sodium deoxycholate, 15 mM $MgCl_2$, 0.1 mM EDTA, 0.2 mM NADPH, and 0.05 M HEPES buffer, pH 7.5 in a volume of 0.5 ml. After incubation for 30 min at 37°C, 0.75 ml 0.17 M Tris-HCl, pH 7.4, containing 33% glycerol was added, and the samples were placed on ice until assay for heme and cytochrome P-450.
[b]Determined as pyridine hemochromagen. Not corrected for any contribution due to catalase.
[c]500 units.

Table 3. Parathion-mediated loss of cytochrome P-450 and enzymatic
activity in a reconstituted system

System[a]	P-450 (nmol/ml)	Benzphetamine[b] (nmol/min/ml)	(turnover #)	Ethoxycoumarin[b] (nmol/min/ml)	(turnover #)
-NADPH	3.41	730	214	41.5	12.2
Complete	1.59	240	150 (70%)	10.2	6.4 (53%)

[a] The complete system contained 4 nmoles cytochrome P-450, 0.4 units reductase, 100 μg dilauryl L-3 phosphatidylcholine, 100 μg sodium deoxycholate, 0.05 M HEPES buffer (pH 7.5), 15 mM $MgCl_2$, 0.1 mM EDTA, 0.05 mM parathion, and 0.2 mM NADPH in a final volume of 1.0 ml. Incubation was for 30' at 37°C. Prior to assay the samples were dialyzed for 48 hr at 4°C against 4 x 1 liter portions of 0.1 M Tris-HCl, pH 7.4, containing 20% glycerol.
[b] Aliquots corresponding to 0.1 nmole P-450 were taken and assayed for benzphetamine or ethoxycoumarin metabolism in the presence of a complete system containing a saturating amount of the reductase.

The loss of enzymatic activity of the cytochrome P-450 after
incubation with parathion exceeded the loss of cytochrome P-450
detectable as its carbon monoxide complex, as evidenced by a
decrease in the turnover number of the modified   protein relative
to controls (Table 3). The relative turnover numbers of the modified enzyme towards benzphetamine and ethoxycoumarin, respectively,
were 70% and 53%. The relationship between the loss of cytochrome
P-450 and of benzphetamine demethylase activity remained constant
in a series of samples ranging in P-450 loss from 10-50%.

Correlation Between ^{35}S-binding and P-450 Loss

In a series of samples ranging from 0.5 to 2 nmoles ^{35}S bound/
nmole P-450, 75% of the covalently bound radioactivity could be
dissociated with 5 mM dithiothreitol without any restoration of
cytochrome P-450 detectable as its carbon monoxide complex or of
monooxygenase activity. This finding suggested that the activity
losses might be due to that component of the bound sulfur remaining after dithiothreitol treatment or to irreversible structural
changes in the protein, whether related or unrelated to covalent
binding of sulfur to cysteine residues. In a series of 10 samples,
a 1:1 correlation was found between the loss of cytochrome P-450
and the degree of non-dithiothreitol dissociable sulfur binding.
However by including 1 mM dithiothreitol in the incubation mixture,
up to 60% of the dithiothreitol-stable sulfur binding could be prevented without any protection against the loss of the P-450 or of
the enzymatic activity. Thus, the apparent correlation between
DTT-stable sulfur binding and P-450 loss observed when DTT was
omitted from the incubation may be only a fortuitous one, both
parameters reflecting the degree of parathion metabolism, and
hence of inactivation.

Model Studies with Sulfhydryl Compounds

These were carried out to investigate whether some of the
inhibitory effects of covalent binding could be mimicked by modification of cysteine residues. Purified rat liver cytochrome P-450
from phenobarbital-treated rats was found to contain 3 free sulfhydryl groups readily titratable with 4,4'dipyridinedisulfide (PDS).
In the presence of 0.5% SDS all seven sulfhydryl groups indicated by
amino acid analysis could be titrated. Modification with PDS or
p-bromophenacyl bromide of the 3 sulfhydryl groups accessible in
the native protein (absence of SDS) had no effect on the benzphetamine demethylase activity. The reconstituted system containing
p-bromophenacylated cytochrome P-450 (3 nmoles/nmole P-450) also
metabolized parathion and became labeled with ^{35}S in a manner
similar to the native, i.e., 75% of the ^{35}S was dissociable with
dithiothreitol. In the presence of 0.5% SDS all seven sulfhydryl
groups could be alkylated with p-bromophenacyl bromide. Although
all enzyme activity was lost, no loss of heme was observed nor was

the cytochrome P-420 level, detectable as its carbon monoxide complex, in the alkylated protein lower than that of controls treated
with only SDS. It thus does not appear that derivatization of the
cysteine groups of cytochrome P-450 per se causes a loss of heme.
These results also suggest that those cysteine residues modified
by binding of atomic sulfur released during parathion metabolism
reside in the interior of the cytochrome P-450 molecule.

SUMMARY

     Approximately 50% of the loss of monooxygenase activity during
parathion metabolism by a reconstituted system can be attributed to
a loss of heme from the cytochrome P-450. The mechanism of the
loss of heme is not yet known. However, the only change in the
oxidized spectrum accompanying it is a decrease in the absorbance
at 417 nm. The additional 50% loss of activity does not appear to
involve mere derivatization of essential amino acids in the P-450
by the covalent binding of the atomic sulfur released during the
metabolism of parathion to paraoxon. On the one hand, the 75% of
the sulfur bound to the P-450 which appears to be in the form of a
hydrodisulfide can be readily removed with no restoration of cytochrome P-450 or monooxygenase activity. On the other hand, the
amount of irreversibly bound sulfur can be reduced by inclusion of
dithiothreitol during the incubation with parathion without any
protection of the enzyme. Furthermore, this irreversibly bound
sulfur appears to be heterogeneously distributed (Figures 2-3).
Thus the most likely explanation for the loss of monooxygenase
activity not accounted for by the heme loss is that the structural
changes accompanying the binding of atomic sulfur to cysteine
residues cause irreversible inhibition. This interpretation is
based on two findings. First, even in the presence of SDS, not all
the modified cytochrome P-450 behaves electrophoretically as a
monomer even after treatment with dithiothreitol (Halpert and Neal,
1980). Second, the cysteine residues attacked by atomic sulfur
appear to be those inaccessible to sulfhydryl or alkylating agents
in the absence of SDS. Thus even after the disruption of the polysulfide covalent linkages by dithiothreitol treatment of the modified cytochrome P-450, a portion of the enzyme molecules may remain
in an inactive aggregational state or conformation under the conditions of the enzyme assay (absence of denaturing agent).

     It appears doubtful that the atomic sulfur which causes inac-
tivation of P-450 in vitro is responsible for the liver damage
observed upon in vivo administration of the parathion analog
diethylphenyl phosphorothionate or other thiono-sulfur compounds
to rats. In vitro experiments suggest that cytochrome P-450 is
the predominant if not sole site of attack of atomic sulfur in
intact microsomes (Kamataki and Neal, 1976) as well as in the
reconstituted system. If a similar situation exists in vivo, the
only likely damage to the endoplasmatic reticulum caused by atomic

sulfur is the inactivation of the cytochrome P-450 itself. Although
inactivation of cytochrome P-450 may have important consequences
for the metabolism of other compounds, it is unlikely to cause liver
damage per se. Furthermore, since 75-90% of the atomic sulfur
released during the in vitro metabolism of parathion to paraoxon
becomes bound to the microsomal protein (P-450), the amount of
atomic sulfur which could conceivably react with cytosolic proteins
is extremely limited. It appears more likely that binding of intermediate S-oxides similar to those shown in Figure 1 is responsible
for liver damage in vivo caused by diethylphenyl phosphorothionate
and related compounds.

REFERENCES

Catignani, C. L. and Neal, R. A., 1975, Evidence for the formation
        of a protein bound hydrodisulfide resulting from microsomal
        mixed-function oxidase catalyzed desulfuration of carbon
        disulfide, Biochem. Biophys. Res. Commun., 65:629.
Halpert, J. and Neal, R. A., 1980, Inactivation of purified rat
        liver cytochrome P-450 during the metabolism of parathion
        (diethyl p-nitrophenyl phosphorothionate), J. Biol. Chem.,
        255:1080.
Herriott, A. W., 1971, Peroxy acid oxidation of phosphinothioates,
        a reversal of stereochemistry, J. Amer. Chem. Soc., 93:3304.
Kamataki, T. and Neal, R. A., 1976, Metabolism of diethyl p-nitro-
        phenyl phosphorothionate (parathion) by a reconstituted
        mixed-function oxidase enzyme system: studies of the covalent
        binding of the sulfur atom, Mol. Pharmacol., 12:933.
Neal, R. A., 1980, Microsomal metabolism of thiono-sulfur compounds:
        mechanism and toxicological significance, in: "Reviews in
        Biochemical Toxicology 2," E. Hodgson, J. R. Bend, and R.
        Philpot, eds., Elsevier North Holland, New York.
Norman, B. J., Poore, R. E., and Neal, R. A., 1974, Studies of the
        binding of sulfur released in the mixed-function oxidase-
        catalyzed metabolism of diethyl p-nitrophenyl phosphoro-
        thionate (parathion) to diethyl p-nitrophenyl phosphate
        (paraoxon), Biochem. Pharmacol., 23:1733.
Poore, R. E. and Neal, R. A., 1972, Evidence for extrahepatic meta-
        bolism of parathion, Toxicol. Appl. Pharmacol., 23:759.
Ptashne, K. A. and Neal, R. A., 1972, Reaction of parathion and
        malathion with peroxytrifluoroacetic acid, a model system
        for the mixed function oxidases, Biochemistry, 11:3224.
Ptashne, K. A., Wolcott, R. M., and Neal, R. A., 1972, Oxygen-18
        studies on the chemical mechanisms of the mixed-function
        oxidase catalyzed desulfuration and dearylation reactions
        of parathion, J. Pharmacol. Exp. Ther., 179:380.
Seawright, A. A., Hrdlicka, J., and De Matteis, F., 1976, The
        hepatotoxicity of O,O-diethyl, O-phenyl phosphorothionate
        (SV_1) for the rat, Br. J. Exp. Path., 57:16.

Yoshihara, S. and Neal, R. A., 1977, Comparison of the metabolism of
    parathion by a rat liver reconstituted mixed-function oxidase
    enzyme system and by a system containing cumene hydroperoxide
    and purified rat liver cytochrome P-450, Drug Metab. Dispos.,
    5:191.

DISCUSSION

Following Nelson's talk, Prough asked what concentrations of
hydrazine were added to the microsomal reaction mixtures. Nelson
answered that both 0.1 mM and 1.0 mM concentrations had been used
but that the data presented were only those with 1.0 mM hydrazine.
Prough commented that hydrazine is metabolized by the FAD-containing
monooxygenase (Ziegler's enzyme) but that it is a very poor substrate. Cottrell questioned the use of sulfur hexafluoride (SF_6)
as a carrier gas rather than helium or argon in the *in vitro* experiments. Nelson referred the question to Reed who had suggested
the SF_6; Reed stated that it has a great advantage for low temperature use since it is a condensible gas. Nelson added that it is an
excellent carrier for their gas chromatographic system since it is
very volatile, comes off as rapidly as helium and is very unreactive.
Gorrod asked whether they had used either octylamine or n-methyl-
imipramine (NMI) to differentiate between Ziegler's enzyme and the
P-450 enzyme system. Nelson replied that they had not used either
one; he added that NMI did not differentiate well between the two
enzyme systems.

Gorrod commented after Prough's presentation that Prough's
previous work with the simple hydrazines had indicated that they
were oxidized by Ziegler's enzyme, while his current report with
these substituted hydrazines indicated that P-450 was the oxidizing
enzyme. Prough replied that there are still unexplained results
with the alkyl hydrazines; former work from Reed's laboratory indicated that the methyl hydrazines were not oxidized by P-450, but
earlier work of both Mitchell and Nelson suggested that P-450 was
involved in such reactions; he emphasized that there is no clear
understanding of these oxidations at present. He added that the
chemistry of the hydrazines makes it easy to visualize either a
heme protein or a flavin-dependent monooxygenase carrying out a
very easy oxidation of the acyl group. Juchau asked what percentage
of the covalent binding of hydrazines would be inhibited by blocking
the P-450-mediated reaction, and Prough said that he could not
answer that question as yet but such work was in progress in his
laboratory.

Gorrod challenged Kato after his talk to explain why the point of metabolic attack in his pyrolyzed tryptophan compounds was on the pyridine nitrogen rather than on the amino nitrogen, as seen with most other aminopyridine compounds described in the literature. The possible tautomeric forms of the ring nitrogen, the amine or imine tautomers, could presumably influence the point of metabolic attack. Kato, however, had no explanation for the discrepancy. He also replied that he had not determined in which tautomeric form the pyrolysates were found. Parke asked whether any of the hydroxy-amino metabolites form complexes with cytochrome P-450; Kato replied that the pyrolysates do form such complexes but he had no information about the metabolites. He responded negatively to Parke's question as to whether the P-450 appeared to change to P-448 on addition of the pyrolysates.

Following Sugimura's talk, Wareing asked why it was that 3'-aminopyridine was a co-mutagen whereas the 2'- and 4'-compounds were not. Sugimura did not know the reason. J. Miller then asked whether they had performed "co-carcinogenesis" experiments with those pairs of compounds which were individually non-carcinogenic. Sugimura said that they had done co-carcinogenic studies with the mutant harman compound and the results suggest that it is a co-initiator, however the complete data are still not all available. In response to a question by Boyland as to the mechanism of action of his "co-initiator" compound, Sugimura responded that it does promote intercalation. He also said that it is possible that nor-harman is an inhibitor of many enzymes. Green asked whether nor-harman is extensively metabolized and the reply was that it is converted to at least three or four metabolites, none of which is co-mutagenic. Remmer then suggested that perhaps this is why the Japanese prefer raw fish! Hildebrandt asked if the addition of catalase and methanol decreased the number of revertant colonies in the Ames test. The response was that they had never done this experiment.

After Archer's talk, Parkinson asked if the observed formation of isopropanol in Archer's experiments might not be due to a reduction of the acetone that commonly contaminates the pyridine nucleotide cofactors, since the oxidation of a 2-hydroxypropyl compound to the corresponding 2-oxypropyl compound is reversible. Archer replied that the controls run with the pyridine nucleotides seemed to eliminate such a possibility. Furthermore, n-propanol did not produce isopropanol nor did isopropanol isomerize to n-propanol, so that n-propanol could not be considered a source of the isopropanol produced in their experiments. Magee then asked Archer how his data supporting the much more selective alkylation of DNA on the oxygens by the nitrosomethylating agents than by dimethylsulfate or methyl methanesulfonate could be acceptable if Lawley's proposed mechanism is incorrect. Archer responded that

since nitrosocompounds are much stronger alkylating agents than
dimethylsulfate or methyl methanesulfonate, the latter compounds
are less discriminating in the nucleophilic centers which they
attack. His only difference with Lawley is that the ultimate electrophile is not the ionic species that was first suggested. Also
Lawley proposed that the two classes of alkylating agents are S_N2
and S_N1 reagents, and it now appears that the term S_N1 is perhaps
a misnomer. Regardless of the ionic species, it seems that these
compounds are still very reactive alkylating agents. Boyland asked
what other members of this class of relatively selective alkylating
agents there are, i.e. which react like nitrosamines as distinct
from reactions like dimethylsulfate (other than diazomethane).
Archer replied that he was not postulating another reaction scheme
but only suggesting that the ultimate electrophile was further
back in the scheme, and may be the alpha hydroxynitrosamine itself,
the alpha diazotate or the carbonium ion. Boyland then asked
whether diazomethane doesn't alkylate uridine differently than
dimethylsulfate. Archer replied that he did not know.

    After the paper of Halpert and Neal (presented by Guengerich),
Netter asked whether the activity of the P-450 which remains after
partial destruction of this enzyme by parathion has undergone
qualitative changes, possibly allosteric in nature, which could be
evidenced as increase in specific activity towards selected substrates. He cited some previous work done in his laboratory (Legrum
et al., Toxicol. Appl. Pharmacol., 48:195, 1979) in which they
showed that CoCl$_2$ pretreatment of mice resulted in a 4-fold increase
in 7-ethoxycoumarin metabolism. Guengerich responded that there
are selective changes in the specific activity of the remaining
P-450; one of his figures indicated an overall decrease in P-450
with a drop of benzphetamine demethylase activity to only 1/3 of
the original activity and a drop in the 7-ethoxycoumarin demethylase activity to about 1/4 the original activity. However, this
system is a purified cytochrome P-450, and the enzyme may have undergone some changes during the purification process. Franklin asked
Guengerich about the stoichiometry of the S atoms and the P-450
molecules to which he replied that 3 of the 4 sulfur atoms become
bound in hydrosulfide linkages. Franklin then asked if it is
correct to say that only one of the 4 sulfurs is attacking the
P-450; Guengerich replied that that is conjecture as the mechanism
of P-450 destruction by the sulfur is still unclear. Gilette asked
whether the remaining activities of the P-450 had been separated
and the systems purified so as to determine any effect of sulfur
on these activities; Guengerich said that these experiments had
not been done.

FACTORS AFFECTING ACETYLHYDRAZINE HEPATOTOXICITY

A K Bahri, C S Chiang, J A Timbrell

Clinical Toxicology Unit
Department of Clinical Pharmacology
Royal Postgraduate Medical School, London, UK

INTRODUCTION

Acetylhydrazine is the metabolite of isoniazid thought to be responsible for the hepatotoxicity of the drug (Mitchell et al, 1976). Acetylhydrazine, identified as a urinary metabolite of isoniazid in man (Timbrell et al, 1977) causes extensive hepatic necrosis in phenobarbital pretreated rats (Mitchell et al, 1976). Various studies have indicated that metabolic activation by the microsomal enzymes is responsible for the production of a reactive acylating species which covalently binds to liver protein (Nelson et al, 1976; Timbrell et al, 1980) (Fig. 1).

However, there are other pathways of metabolism for acetyl-hydrazine such as acetylation, hydrazone formation and possibly hydrolysis to hydrazine and acetate (Timbrell et al, 1980; Wright & Timbrell, 1978). The relative importance of all these pathways will be important in the development of hepatotoxicity. Isoniazid, the parent drug from which acetylhydrazine is formed in vivo is an inhibitor of acetylation in vivo (Wright & Timbrell, 1979) and the microsomal activation in vitro (Timbrell & Wright, 1979) of acetyl-hydrazine. These effects may be important in determining the toxicity of isoniazid.

Rifampicin, a drug often given with isoniazid and thought to potentiate its hepatotoxicity, is known to be an acute inhibitor but chronic inducer of the microsomal enzymes (Pessayre & Mazel, 1976). In vivo studies in human subjects however failed to show any effects on the metabolism of isoniazid consistent with enzyme induction (Timbrell et al, unpublished results). Therefore the possibility that rifampicin may influence the hepatotoxicity of

1055

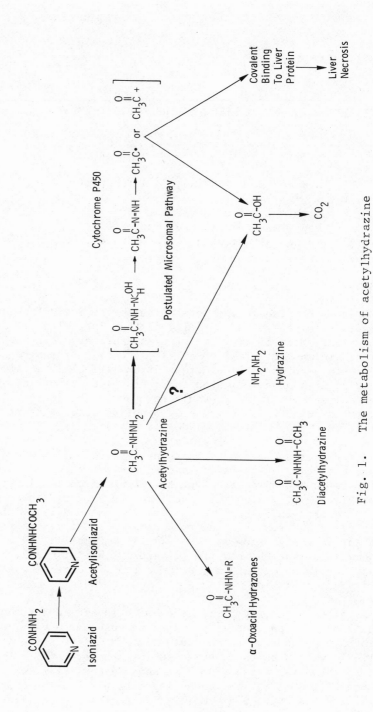

Fig. 1.   The metabolism of acetylhydrazine

acetylhydrazine and hence isoniazid in some other way was investigated in rats, a species not susceptible to microsomal enzyme
induction by rifampicin. Consequently the hepatotoxicity of
acetylhydrazine has been studied in normal and phenobarbital
pretreated rats and the effect of isoniazid and rifampicin on the
hepatotoxicity and disposition of acetylhydrazine has also been
studied.

MATERIALS AND METHODS

^{14}C-Acetylhydrazine hydrochloride was synthesised and purified
as previously described (Wright & Timbrell, 1978) and acetylhydrazine
hydrochloride was prepared from acetylhydrazine base by lyophilisa-
tion in 2M HCl followed by recrystallisation from methanol/ether.

All other compounds were obtained commercially.

Methods

Animals

Male, Sprague-Dawley rats (180-200 g) were used. All compounds
were given by i.p. injection in water. Rifampicin was dissolved
in water by adjusting the pH with dilute HCl to pH 3. Controls
received vehicle only.

Phenobarbitone was given daily (75 mg/kg) for 4 days prior to
acetylhydrazine; isoniazid was given simultaneously with acetylhydrazine and rifampicin was given either 30 minutes before or
daily for 6 days prior to treatment with acetylhydrazine.

Controls received treatment with isoniazid, phenobarbitone or
rifampicin only.

Histology

Animals were killed by cervical dislocation 24 hours after
treatment. Livers were removed and samples of tissue taken for
histology. Details of quantitation of liver necrosis will be
described elsewhere.

Determination of Hepatic Cytochrome P-450

Hepatic cytochrome P-450 was determined by the method of
Omura and Sato (1964) in the microsomal fraction from liver homogenates, with a final protein concentration of 2-3 mg/ml. Protein
concentration was determined by the method of Lowry et al (1951).

Metabolic Studies

     Animals given ^{14}C-acetylhydrazine hydrochloride were placed
in metabolism cages which allowed the separate collection of
expired CO_2, urine and faeces. Radioactivity in expired CO2 and
urine was determined as previously described by scintillation
counting (Wright & Timbrell, 1978). Urinary metabolites of ^{14}C-
acetylhydrazine were determined by thin layer radiochromatography
as previously described (Wright & Timbrell, 1978).

RESULTS

Histology

     The livers of normal animals 24 hours after the administration
of acetylhydrazine in doses of 0.09 to 0.9 mmol/kg showed no
evidence of necrosis. Portal tracts showed some infiltration by
inflammatory cells and mitotic figures were present particularly
after higher doses of acetylhydrazine.

     However, in phenobarbital pretreated animals marked centri-
lobular necrosis was evident after doses of 0.09 mmol/kg and the
extent of this damage increased in a dose dependent manner (Fig.
2).

     After doses of rifampicin alone mild portal tract inflammation
was apparent and after the highest doses (50 and 100 mg/kg/day for
6 days) there were also some sites of focal necrosis. Animals
pretreated with rifampicin followed by acetylhydrazine showed no
hepatic necrosis however and no potentiation of the lesion compared
with animals treated with rifampicin alone.

     Isoniazid in normal and phenobarbital pretreated animals did
not cause any hepatic damage (Table 1). However phenobarbital
pretreated animals given isoniazid and acetylhydrazine together
showed no hepatic necrosis (Table 1).

Biochemical Studies

     Measurement of hepatic cytochrome P-450 levels in treated
animals 24 hours after dosing revealed that in animals given
acetylhydrazine after phenobarbital pretreatment the level was
significantly reduced in a dose dependent manner (Fig. 2) (Table
1), but not by acetylhydrazine in normal, non-induced animals.
Rats treated with rifampicin for 6 days also showed a dose dependent
decline in cytochrome P-450 (Fig. 3).

Metabolic Studies

     Treatment of rats with rifampicin either in single doses or

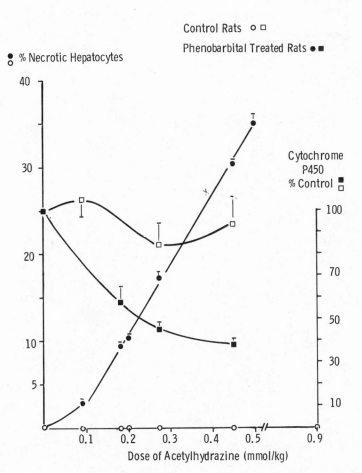

Fig. 2.   Relationship between dose of acetylhydrazine,
          hepatic necrosis (○,●) and cytochrome P-450
          (□,■) in normal and phenobarbital pretreated
          rats.

TABLE 1: The effect of pretreatments on the hepatotoxicity of acetylhydrazine

Dose of Acetylhydrazine (mmol/kg)	% Necrosis				Cytochrome P-450 (nmol/mg protein)	
	Pretreatment				Pretreatment	
	None	Phenobarbital (Pb)	Isoniazid (INH) (0.9 mmol/kg)	Pb + INH (0.4 mmol/kg)	None	Phenobarbital
0	0	0	0	0	0.33 ± 0.44	0.92 ± 0.02 *
0.09	0	3.1 ± 0.5 *			0.35 ± 0.03	
0.18	0	9.1 ± 0.5				0.53 ± 0.04
0.2	0	10.3 ± 0.6		0		
0.27	0	17.3 ± 0.4			0.28 ± 0.03	0.42 ± 0.02
0.45	0	30.9 ± 0.9			0.31 ± 0.04	0.35 ± 0.02
0.9	0		0			
		$n \geqslant 5$	$n \geqslant 3$	$n \geqslant 3$	$n = 3$	$n \geqslant 4$

* $p < 0.001$

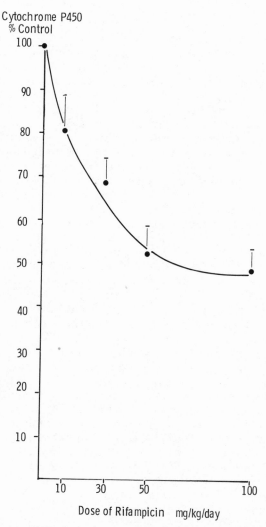

Fig. 3.   Effect of daily dosing with rifampicin
           for 6 days on hepatic cytochrome P-450
           in rats

chronically for 6 days resulted in a significant reduction in the acetylation of acetylhydrazine as indicated by the ratio diacetyl-hydrazine:acetylhydrazine. However no effect on excretion of $^{14}CO_2$ was discernable (Table 2).

DISCUSSION

The results presented, clearly indicate that acetylhydrazine is hepatotoxic in phenobarbital pretreated animals but not in control animals even over a 10 fold dose range. In the pheno-barbital pretreated animals where the level of cytochrome P-450 is increased 2-3 times there is a dose dependent, linear increase in necrosis (Fig. 2). The large increase in toxicity caused by phenobarbital pretreatment suggests that this pretreatment is not simply increasing the amount of existing microsomal enzyme responsible for the toxic pathway but may be increasing a normally minor form of cytochrome P-450 and thereby diverting a major proportion of the metabolism via the toxic pathway.

If the toxic pathway was a major route of metabolism in non-induced animals then hepatic necrosis should be observed at a dose level 2-3 times higher than that required in induced animals, reflecting the difference in the amount of cytochrome P-450. The dose response curve for the non-induced animals should therefore simply be offset up the dose scale, with the dose threshold 2-3 times higher than that in induced animals, as is the case with paracetamol induced hepatic necrosis (Mitchell et al, 1973). This is clearly not the case with acetylhydrazine and supports the conclusion that in normal animals the toxic pathway is a relatively minor one being increased to a large extent by phenobarbital pretreatment.

The dose dependent decrease in hepatic cytochrome P-450 seen in those animals suffering hepatic necrosis (Fig. 2) may simply be the result of cellular destruction and loss of enzyme and membrane integrity rather than a specific attack on the enzyme. The decline in cytochrome P-450 in rifampicin treated animals however is not associated with any significant hepatocellular damage. The mechanism of this loss is therefore unexplained at present.

It is clear that isoniazid reduces the hepatotoxicity of its metabolite acetylhydrazine when administered simultaneously to phenobarbital pretreated animals. Previous studies showed that isoniazid inhibited both the acetylation of acetylhydrazine in vivo (Wright & Timbrell, 1979) and the microsomal enzyme mediated covalent binding of acetylhydrazine to protein in vitro (Timbrell & Wright, 1979). Although these effects would be expected to work in opposition the decrease in hepatotoxicity caused by isoniazid reported here is presumably the result of inhibition of the micro-somal enzyme mediated activation. This finding may explain the

TABLE 2: Effect of Rifampicin on the Metabolism of ^{14}C-Acetylhydrazine in Rats

Pretreatment	Urine (0-6 hr)	$^{14}CO_2$	% Dose (0-6 hr urine)		R
			Acetylhydrazine	Diacetylhydrazine	
None	31.9 ± 1.3	27.2 ± 1.1	19.1 ± 1.4	11.3 ± 1.2	0.6 ± 0.08
Rifampicin 1 x 50 mg/kg	29.5 ± 1.3	33.1 ± 1.3	17.7 ± 0.8	7.2 ± 0.5	0.41 ± 0.03
Rifampicin 6 x 50 mg/kg	23.8 ± 5.3	27.8 ± 0.6	15.1 ± 2.6	4.3 ± 1.2	0.28 ± 0.06*

R = Diacetylhydrazine:Acetylhydrazine

Acetylhydrazine (0.4 mmol/kg) and Rifampicin were given by i.p. injection

Results are means of at least 3 animals ± S.E.

*$p < 0.05$

A. K. BAHRI ET AL.

fact that single large doses of isoniazid do not cause hepatic
damage in phenobarbital induced rats but several smaller doses do
(Mitchell et al, 1976).

It is clear that rifampicin pretreatment does not potentiate
the hepatotoxicity of acetylhydrazine nor is there an additive
effect. Even though rifampicin inhibited the acetylation which
could divert more acetylhydrazine through the microsomal pathway
(Fig. 1) no increase in $^{14}CO_2$ was observed (Table 2). This may be
due to the decrease in cytochrome P-450 also caused by rifampicin
pretreatment. The mechanism underlying this reduction of cytochrome
P-450, not observed by Pessayre and Mazel (1976) and the inhibition
of acetylation is as yet unexplained but could be related to the
cholestatic effects of rifampicin.

In conclusion acetylhydrazine hepatotoxicity, described as a
centrilobular necrosis, depends on metabolism through a microsomal
enzyme mediated pathway which is inducible by phenobarbital.
Isoniazid blocks the hepatic necrosis caused by acetylhydrazine,
possibly by inhibiting this pathway. Rifampicin does not influence
acetylhydrazine hepatotoxicity although it does inhibit acetylation
to diacetylhydrazine and causes a reduction in cytochrome P-450.

REFERENCES

Lowry, O.H.. Rosebrough, N.J., Farr, A.L. & Randall, R.J. (1951)
    Protein measurement with the Folin phenol reagent. J. Biol. Chem.,
    193: 265
Mitchell, J.R., Jollow, D.J., Potter, W.Z., Davis, D.C.,
    Gillette, J.R., & Brodie, B.B. (1973) Acetaminophen induced
hepatic necrosis. 1. Role of drug metabolism. J. Pharmacol. exp.
    Ther., 187: 185
Mitchell, J.R., Zimmerman, H.J., Ishak, K.G., Thorgeirsson, S.S.,
    Timbrell, J.A., Snodgrass, W.R. and Nelson, S.D. (1976) Isoniazid
    liver injury : Clinical spectrum, pathology and probable patho-
    genesis. Ann. Intern. Med., 84: 181
Nelson, S.D., Mitchell, J.R., Timbrell, J.A., Snodgrass, W.R. &
    Corcoran, G.B. (1976) Isoniazid and iproniazid : Activation of
    metabolites to toxic intermediates in man and rat. Science,
    193: 901
Omura, T. & Sato, R. (1964) The carbon monoxide binding pigment of
    liver microsomes. II. Solubilisation, purification and properties
    J. Biol. Chem., 239: 2370
Pessayre, D. & Mazel, P. (1976) Induction and inhibition of hepatic
    drug metabolising enzymes by rifampicin. Biochem. Pharmac., 25: 94
Timbrell, J.A., Wright, J.M. & Baillie, T.A. (1977) Monoacetyl-
    hydrazine as a metabolite of isoniazid in man. Clin. Pharmacol.
    Ther., 22: 602
Timbrell, J.A., Mitchell, J.R., Snodgrass, W.R., Nelson, S.D. (1980)
    Isoniazid hepatotoxicity : The relationship between covalent

binding and metabolism in vivo. J. Pharmacol. exp. Ther., 213: 364

Wright, J.M. & Timbrell, J.A. (1978) Factors affecting the metabolism of ^{14}C-acetylhydrazine in rats. Drug Metab. Disp., 6: 561

INHIBITION OF METABOLISM--MEDIATED CYTOTOXICITY BY 1,1-DISUBSTITUTED HYDRAZINES IN MOUSE MASTOCYOTOMA CELLS (LINE P815)

P. Wiebkin, R. N. Hines, R. E. Nelson, and R. A. Prough

Department of Biochemistry, University of Texas Health

Science Center at Dallas, Dallas, TX 75235

INTRODUCTION

A number of cell culture systems have been used to assess the cytotoxicity of chemicals (Rofe, 1971, Dawson, 1972, Worden, 1974), employing a wide range of criteria for the expression of cytotoxicity within these cells. It is now recognized that the majority of chemicals require metabolic activation for their toxic potential to be realized. Since this specialized cellular function is not usually retained in the normal state in culture, most cell culture cytotoxicity tests have been restricted to those compounds that are directly cytotoxic and do not require prior metabolic activation. We have made use of a modification of the mixed cell culture system first developed by Fry and Bridges (1977b) and then extended by Wiebkin et al. (1978), employing freshly isolated hepatocytes, possessing the full complement of drug metabolizing enzymes as the "metabolizing" component and cultured mouse mastocytoma cells (Line P815) as the "response" component.

Although hydrazine derivatives have been shown to be toxic and/or carcinogenic in vivo (Toth, 1975, Sieber et al., 1978), the demonstration of similar biological activity with these compounds using in vitro model systems has not been very successful. Indeed, our results (data not shown) indicate that of the hydrazines tested, none has the ability to inhibit the growth of the mouse mastocytoma cells, either in the presence or absence of a metabolic activating system. However, it has been reported (Hines and Prough, 1980) that the addition of 1,1-disubstituted hydrazines to microsomal suspensions in the presence of NADPH and oxygen results in the formation of a metabolite complex with cyto-

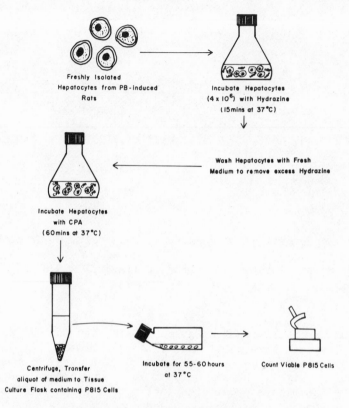

Figure 1. Protocol of the Hepatocyte/Mouse Mastocytoma (Line P815) Cell Toxicity System.

chrome P-450 that significantly inhibits the mixed-function oxidase activity of this enzyme system. In this study we have focused our attention, using this hepatocyte/mastocytoma system, on the ability of N-aminopiperidine (NAP) to inhibit the cytotoxicity of the antitumor agent, cyclophosphamide (CPA), a compound known to require metabolic activation before its full toxic potential is expressed.

MATERIALS AND METHODS

Male Sprague-Dawley CD rats (80-100g body weight) were used throughout this study. They were all allowed free access to food and water. Rats were given a daily ip. dose of either sodium phenobarbital in 0.9% saline (80 mg/kg body weight for 4 days) or 5,6-benzoflavone in corn oil (50 mg/kg body weight for 4 days). Following the last injection, the animals were starved 18 hours prior to sacrifice. Tissue culture medium, serum and reagents were obtained from Gibco Biocult N.Y. CPA was obtained from Mead Johnson and Co., Evansville, IN and NAP was purchased from Aldrich Chemical Company, MA. "Falcon" Tissue culture plasticware was purchased from American Scientific Products.

Hepatocytes were isolated from rat liver by collagenase/hyaluronidase digestion as described by Fry et al. (1976) using sterile techniques throughout, and after assessment of viability (routinely 85-95%) by the dye exclusion test (Cummings, 1970), they were diluted to 2×10^6 viable hepatocytes/ml of culture medium comprised of 5% (v/v) fetal calf serum and 10% (v/v) tryptose phosphate broth in Leibovitz L-15 medium. Two ml samples of this cell suspension were pipetted into 10 ml conical flasks and incubated with either CPA or NAP as the protocol of the experiment dictated at 37°C in a shaking water bath (approx 100 oscillations/minute). The CPA and NAP were dissolved in phosphate buffered saline 'A' (PBS'A') at 10 times the desired final concentration and sterilized by membrane filtration prior to use. The protocol used for the studies reported here is shown in Figure 1. Determination of the number of viable mouse mastocytoma P815 cells was carried out by their ability to exclude trypan blue using an improved Neubaur hemocytometer.

RESULTS

When CPA was incubated with freshly isolated hepatocytes from phenobarbital-treated rats, there was a marked decrease in the number of viable P815 cells after 55 hours in culture. The effect of varying the CPA concentration on the inhibition of P815 cell growth in the presence and absence of hepatocytes was then studied, and the results are shown in Figure 2. In the presence of hepatocytes there was a sharp decrease in the number of viable P815 cells counted after 55 hours in culture as the concentration of CPA increased such that at a concentration of 0.27 mg/ml the viable cell number was only 6% of that present in the control cultures. When hepatocytes were omitted from the incubation mixture, no toxicity to the P815 cells could be detected even at the highest concentration of CPA used. The pre-incubation of hepatocytes with NAP (4.2 mM), followed by incubation with differ-

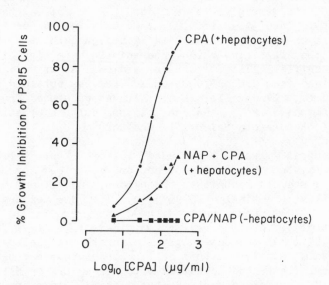

Figure 2. The Effect of N-Aminopiperidine (NAP) on the Dose
Response of the Metabolism-Mediated Cytotoxicity of Cyclo-
phosphamide (CPA) in Mouse Mastocytoma Cells (Line P815).
The materials and methods are as described in Figure 1.

ent concentrations of cyclophosphamide resulted in a significant
change in the dose-response curve for the toxicity of activated
CPA to P815 cells (Figure 2). At all concentrations of CPA used
4.2 mM NAP reduced the inhibition of P815 cell growth by
approximately 60%.

The effect of varying the NAP concentrations on the reduction
of activated CPA toxicity to P815 cells was also studied, and the
results shown in Figure 3. From this dose-response we have
calculated a ID_{50} value (defined as that dose of NAP which reduces
the cytotoxicity of CPA by 50%) of 1.0 mM.

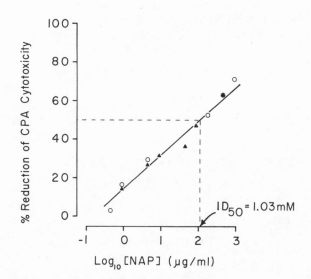

Figure 3.   Dose Response of the Ability of N-Aminopiperidine (NAP) to Inhibit the Metabolism-Mediated Cytotoxicity of Cyclophosphamide (CPA) in Mouse Mastocytoma Cells (Line P815). The materials and methods utilized in this experiment are as described in Figure 1.

The results shown in Table 1 indicate that in the case of 1,1-disubstituted hydrazines there is a correlation between the extent of complex formation with cytochrome P-450 observed in microsomal suspension and the degree of inhibition of the metabolite-mediated cytotoxicity of CPA. 1,2-Disubstituted hydrazines were unable to form the inhibitory complex in microsomal suspensions and also failed to inhibit significantly the CPA metabolite toxicity in the hepatocyte/P815 cell-culture system.

The ability of NAP to inhibit the CPA metabolite toxicity of the P815 cells was significantly reduced when hepatocytes from untreated or 5,6-benzoflavone rats were used as the activating component in this system (Table 1). A similar reduction in the ability of NAP to form the inhibitory metabolite complex with cytochrome P-450 was also noted.

TABLE 1

Comparison of the Ability of Certain Hydrazine Derivatives to Reduce the Metabolism-mediated Cytotoxicity of Cyclophosphamide in Mouse Mastocytoma Cells and Their Ability to Form the Inhibitory Metabolite-Complex with Cytochrome P-450 in Microsomal Suspensions.

Hydrazine[a] Derivative	Ability to Inhibit Toxicity[b]	Ability to Form Complex[b]
PHENOBARBITAL PRETREATED ANIMALS[c]		
1,1-Disubstituted		
N-Aminopiperidine	100	100
1,1-Dimethylhydrazine	69	23
1,2-Disubstituted		
1,2-Dimethylhydrazine	18	0
Procarbazine	55	0
5,6-BENZOFLAVONE PRETREATED ANIMALS[c]		
N-Aminopiperidine	44	11
CONTROL ANIMALS[c]		
N-Aminopiperidine	53	18

[a] Hydrazine concentrations used were 1 mM and the cyclophosphamide concentration used was 0.27 mg/ml.

[b] The ability of the various hydrazines to inhibit the metabolism-mediated toxicity of cyclophosphamide or form the metabolite-complex is expressed relative to the data obtained with N-aminopiperidine.

[c] Hepatocytes used for the metabolic activation system were isolated from rats pretreated as indicated.

DISCUSSION

A number of in vitro mixed cell culture systems have been developed in recent years, for the detection of cytotoxicity, mutagenesis and carcinogenicity of xenobiotics, particularly those requiring metabolic activation before their biological effect is expressed. Isolated hepatocytes, retaining the capacity for metabolizing a wide range of xenobiotics similar to that found in the in vivo situation (Fry and Bridges, 1977a, 1978), have been used as a metabolizing component in those systems as freshly isolated cell suspensions (Fry and Bridges, 1977b, Wiebkin et al., 1978, Green et al., 1977) and in short term primary culture (San and Williams, 1977, Michalopoulos et al., 1978, Langenbach et al., 1978, Jones and Huberman, 1980).

Using the mixed hepatocyte/mouse mastocytoma cell culture system described here, we have confirmed the studies of Fry and Bridges (1977b), who reported that CPA toxicity to the response component is only fully expressed in the presence of isolated rat hepatocytes. Several hydrazine derivatives were therefore tested in this system, and found to be almost non-toxic to the mouse mastocytoma cells, as reflected by their inability to inhibit cell growth. However, employing a slight modification of the original Fry and Bridges method, it was noted that pre-incubation of freshly isolated rat hepatocyte suspension with N-aminopiperidine (NAP), a 1,1-disubstituted hydrazine, inhibited the metabolism-mediated cytotoxicity of CPA to mouse mastocytoma cells in a dose dependent manner.

Since the effect of animal pretreatment, the concentration dependence, and the substrate specificity on the formation of the inhibitory complex in vitro and inhibition of CPA toxicity in culture are similar, one might conclude that a cytochrome P-450 metabolite complex is formed in isolated hepatocytes upon incubation with 1,1-disubstituted hydrazines and this complex prevents the metabolism of cyclophosphamide to form toxic products. It is well established that the activation of CPA occurs primarily in the liver and is mediated by the cytochrome P-450 linked mixed-function oxygenase system (Sladek, 1971, Hales and Jain, 1980). This study highlights the versatility such in vitro mixed cell culture systems possess. They can not only be utilized for the evaluation of the biological effects of xenobiotics, particularly those requiring activation by microsomal monooxygenases such as cytochrome P-450, but also drug-drug interactions that could occur at the site of modification within the cell.

ACKNOWLEDGEMENTS

Supported in part by grants from the American Cancer Society (BC-336) and the Robert A. Welch Foundation (I-616). RNH was a Trainee on USPHS Training Grant T 32/GM 07062 and RAP is the recipient of a USPHS Research Career Development Awardee, HL 00255.

REFERENCES

Cummings, H., 1970, in: "Virology-Tissue Culture", First Edition, Butterworths, London, p. 26.
Dawson, M., 1972, in: "Cellular Pharmacology", Charles C. Thomas, Springfield, IL.
Fry, J.R., Jones, C.A., Wiebkin, P., Bellemann, P., and Bridges, J.W., 1976, The enzymic isolation of adult rat hepatocytes in a functional and viable state, Anal. Biochem., 71:, 341.
Fry, J.R., and Bridges, J.W., 1977a, The metabolism of xenobiotics in cell suspensions and cell cultures, in: "Progress in Drug Metabolism", J.W. Bridges and L.F. Chasseaud, eds., Vol. 2, Wiley and Sons, Chichester, p. 71.
Fry, J.R., and Bridges, J.W., 1977b, A novel mixed hepatocyte-fibroblast culture system and its use as a test for metabolism-mediated cytotoxicity, Biochem. Pharmacol., 26:969.
Fry, J.R., and Bridges, J.W., 1978, Use of primary hepatocyte cultures in biochemical toxicology, in: "Reviews in Biochemical Toxicology", Vol 1, Elsevier/North Holland, New York, p. 201.
Green, M.H.L., Bridges, B.A., Rogers, A.M., Horspool, G., Muriel, W.J., Bridges, J.W, and Fry, J.R., 1977, Mutagen screening by a simplified bacterial fluctuation test: Use of microsomal preparations and whole liver cells for metabolic activation, Mutation Research, 48:287.
Hales, B.F., and Jain, R., 1980, Characteristics of the activation of cyclophosphamide to a mutagen by rat liver, Biochem. Pharmacol., 29:256.
Hines, R.N., and Prough, R.A., 1980, The characterization of an inhibitory complex formed with cytochrome P-450 and a metabolite of 1,1-disubstituted hydrazines, J. Pharmac. Exp. Therap., 214:80.
Jones, C.A., and Huberman, E., 1980, A sensitive hepatocyte-mediated assay for the metabolism of nitrosamines to mutagens for mammalian cells, Cancer Res., 40:406.
Langenbach, R., Freed M.J., and Huberman, E., 1978, Liver cell-mediated mutagenesis of mammalian cells by liver carcinogens, Proc. Nat. Acad. Sci. U.S.A., 75:2864.

Michalopoulos, G., Sattler, G.L., O'Connor, L., and Pitot, H.C., 1978, Unscheduled DNA synthesis induced by procarcinogens in suspension and primary cultures of hepatocytes on collagen membranes, Cancer Res., 38:1866.

Rofe, P.C., 1971, Tissue culture and toxicology, Fed. Cosmet. Toxicol., 9:683.

San, R.H.C., and Williams, G.M., 1977, Rat hepatocyte primary cell culture-mediated mutagenesis of adult rat liver epithelial cells by procarcinogens, Proc. Soc. Exp. Biol. Med., 156:534.

Sieber, S.M., Correa, P., Dalgard, D.W., and Adamson, R.M., 1978, Carcinogenic and other adverse effect of procarbazine in non-human primates, Cancer Res., 38:2125.

Sladek, N.E., 1971, Metabolism of cyclophosphamide by rat hepatic microsomes, Cancer Res., 31:901.

Toth, B., 1975, Synthetic and naturally occurring hydrazines as possible cancer causative agents, Cancer Res., 35:3693.

Wiebkin, P., Fry, J.R., and Bridges, J.W., 1978, Metabolism-mediated cytotoxicity of chemical carcinogens and non-carcinogens, Biochem. Pharmacol., 27:1849.

Worden A.N., 1974, The use of cell culture in toxicology, in: "Modern Trends in Toxicology", E. Boyland and R. Goulding, eds., Vol. 2,, Butterworths, London, p. 216.

EPOXIDES FROM PARACETAMOL - A POSSIBLE EXPLANATION FOR PARACETAMOL TOXICITY

Lidia J. Notarianni, Harriet G. Oldham, P.N. Bennett,
C.C.B. Southgate and R.T. Parfitt

School of Pharmacy and Pharmacology
University of Bath, Claverton Down, Bath BA2 7AY
England

Paracetamol (acetaminophen) is a commonly used mild analgesic which is considered safe in normal therapeutic doses. Large doses however are known to cause liver necrosis, which can be fatal, in man and experimental animals. This toxicity is associated with a highly reactive, unstable metabolite of paracetamol and although the identity of the toxic metabolite(s) is uncertain, several lines of investigation have led to three entities: $\underline{N}$-hydroxypara-cetamol[1], N-acetylbenzoquinonimine[1] and paracetamol-2,3-dihydro 2,3-oxide[2], being proposed. Most of the evidence amassed to date has been based on *in vitro* studies utilising radiolabelling techniques, the results being assumed to mimic the biochemical transformations occurring in the intact liver. Our approach has been to study the products of *in vivo* metabolism by examining in more detail the urinary metabolites of paracetamol in man with the aim of obtaining useful information on the nature of the toxic intermediate metabolite.

The metabolism of paracetamol was examined in (a) adult volunteers given a therapeutic dose (1-2g) of paracetamol and (b) patients self-poisoned with paracetamol. Urinary metabolites were assayed by high performance liquid chromatography (HPLC) using a modification of the method of Adriaenssens[3]. A typical chromatogram is shown in Figure 1. All the well known metabolites of paracetamol were identified with the use of standard compounds: paracetamol-4-glucuronide, paracetamol-4-sulphate, 3-$\underline{S}$-cysteinyl-paracetamol, and 3-$\underline{S}$-($\underline{N}$-acetylcysteinyl)-paracetamol. In addition to these metabolites, several other drug-related chromatographic peaks were observed which, with the use of specific enzyme

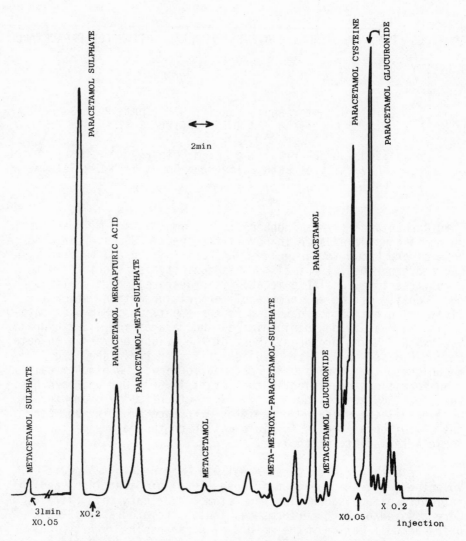

Fig. 1.   HPLC CHROMATOGRAM OF 0-8 hour urine after paracetamol (2g)

hydrolases (sulphatase, β-glucuronidase) and reference samples,
were identified as metacetamol (N-acetyl-m-aminophenol), paracet-
amol-3-sulphate and m-methoxyparacetamol. There is evidence to
suggest that two further compounds were metacetamol-3-glucuronide
and metacetamol-3-sulphate. The paracetamol used in the
volunteer studies was free from any contamination by metacetamol.
Since no acetanilide was found in the urines it is unlikely that
the observed metacetamol arose as a result of paracetamol reduction to acetanilide and subsequent oxidation via the 3,4-dihydro-
3,4-oxide intermediate. Figure 2 depicts the above and other
less well-known metabolites of paracetamol isolated by other
workers[4, 5, 6].

     The finding of metacetamol and its conjugates, and those
metabolites of paracetamol bearing an oxygen substituent in the
2- or 3- positions on the ring poses two questions:-
          1. Which pathways are involved in their formation?
          2. Do these pathways involve a reactive intermediate
               likely to cause hepatotoxicity?
Figure 3 shows possible routes obligatorily involving an epoxide
intermediate. Figure 4 shows a scheme in which the central
intermediate is the N-acetylbenzoquinonimine. Our assertion is
that paracetamol degradation proceeds via at least one of the
epoxide intermediates shown in Figure 3. Our evidence for this
consists of:-
          (i) The discovery of conjugates of 2-hydroxyparacetamol
               reported by Mrochek[4] and confirmed by results from
               our own laboratory.
          (ii) The various metabolites detected with oxygen in the
               3- position.
Although as can be seen from Figure 4 the benzoquinonimine could
possibly give rise to the latter metabolites, it is difficult to
conceive how the 2-hydroxy metabolites can be obtained by this
route without the intervention of an epoxide.

     From inspection of the range of metabolites in Figure 2 it
seems highly probable that 4-hydroxy-2,3-dihydroacetanilide-2,3-
oxide is an intermediate in paracetamol metabolism. Epoxides
are well-known in biological systems (some such compounds are
known carcinogens[7]) and are highly reactive, unstable entities
capable of reacting with macromolecules or with liver glutathione.
This compound is therefore a likely candidate for the hepatotoxic
metabolite of paracetamol, but as shown in Figures 3 and 4 other
epoxides as well as the benzoquinonimine may also be involved.

     In summary, we have shown that the urine of humans treated
with paracetamol contains a range of 2- and 3- substituted
metabolites. Our proposal is that they arise via one or more
epoxide intermediates which are responsible for the hepatotoxicity

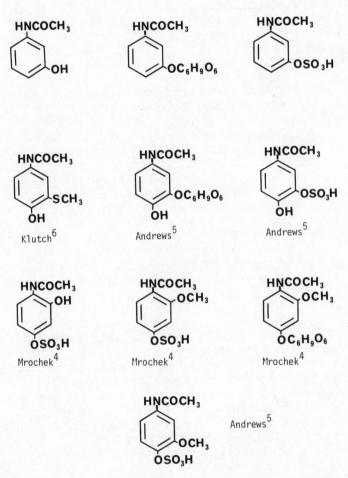

FIGURE 2.   Minor metabolites of paracetamol found in human urine

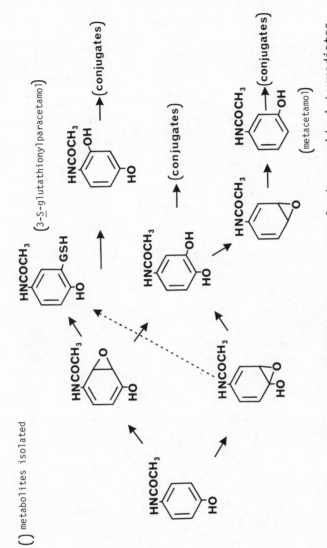

() metabolites isolated

Figure 3. Possible metabolism of paracetamol via epoxide intermediates

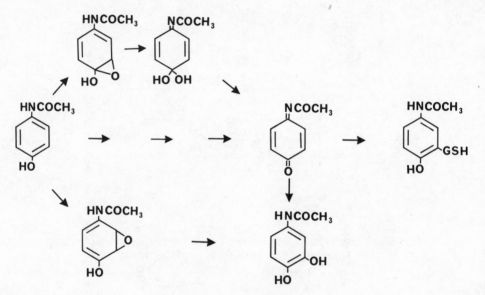

Figure 4.   Possible metabolism of paracetamol
via N-acetylbenzoquinonimine

of paracetamol. Further experiments are in progress to examine
this in more detail.

We are indebted to Dr. R.S. Andrews of Sterling Winthrop for his
generous gifts of paracetamol metabolites.

HGO is in receipt of a studentship from the Arthritis and
Rheumatism Council.

References
1. W.Z. Potter, S.S. Thorgeirsson, D.J. Jollow and J.R.
    Mitchell, Acetaminophen-induced hepatic necrosis
    V Correlation of hepatic necrosis, covalent
    binding and glutathione depletion in hamsters.
    Pharmacol. 12: 129 (1974).
2. J.A. Hinson, L.R. Pohl and J.R. Gillette, N-hydroxy-
    acetaminophen: a microsomal metabolite of N-hydro-
    xyphenacetin but apparently not of acetominophen.
    Life Sci. 24: 2133 (1979).
3. P.I. Adriaenssens and L.F. Prescott, High performance
    liquid chromatographic estimation of paracetamol
    metabolites in plasma. Br.J.Clin.Pharmacol. 6(1):
    87 (1978).
4. J.E. Mrochek, S. Katz, W.H. Christie and S.R. Dinsmore,
    Acetominophen metabolism in man, as determined by
    high resolution liquid chromatography. Clin.Chem.
    20: 1086 (1974).
5. R.S. Andrews, C.C. Bond, J. Burnett, A. Saunders and
    K. Watson, Isolation and identification of para-
    cetamol metabolites. J.Int.Med.Res. 4(4): 34
    (1976).
6. A. Klutch, W. Levin, R.L. Chang, F. Vane and A.H.
    Conney, Formation of a thiomethyl metabolite of
    phenacetin and acetominophen in dogs and man.
    Clin.Pharmacol.Ther. 24(3): 287 (1978).
7. I.A. Smith and P.G. Seybold, Methyl(a)anthracenes:
    correlations between theoretical reactivity
    indices and carcinogenicity. Int.J.Quantum
    Chem: quantum biology symposium 5, 311 (1978).

EVIDENCE FOR REDOX CYCLING OF ACETAMINOPHEN AND ITS REACTIVE

METABOLITE BY ENDOGENOUS MICROSOMAL SYSTEMS

George B. Corcoran[1] and Jerry R. Mitchell[2]

Department of Medicine and Institute for Lipid Research
Baylor College of Medicine, Houston,TX 77030 U.S.A. and
Department of Pharmacology, The George Washington Uni-
versity, Washington, D.C. 20037 U.S.A.

INTRODUCTION

The metabolic activation of acetaminophen to an arylating
metabolite was originally postulated to occur via N-oxidation to
the hydroxamic acid (Mitchell et al., 1972, 1973). Evidence supporting this hypothesis was provided by subsequent studies of
species differences and of the effects of mixed-function oxygenase
inducers and inhibitors on the N-oxidation of 2-acetylaminofluorene
(Thorgeirsson et al., 1972, 1973), 4-chloroacetanilide (Thorgeirsson
et al., 1972; Hinson et al., 1975) and phenacetin (Hinson and
Mitchell, 1976) to hydroxamic acids. Because of the formation of
meta-substituted sulfhydryl adducts of acetaminophen, we proposed
that N-hydroxy-acetaminophen rapidly dehydrated to the reactive
N-acetyl-p-benzoquinoneimine and this was the ultimate toxic species
form in vivo (Mitchell et al., 1974; Jollow et al., 1974).

To investigate more directly the possible role of N-acetyl-p-
benzoquinoneimine in acetaminophen-induced hepatic and renal necrosis, we prepared N-hydroxy-acetaminophen by removing the benzyl
protecting group from 4-benzoxy-N-hydroxy-acetanilide by hydrogen-
olysis, as originally suggested by R. S. Andrews (Sterling Winthrop
Laboratories, United Kingdom). It was hoped that N-hydroxyacetaminophen would serve as a convenient chemical source of the toxic
acetaminophen metabolite and would facilitate characterization of
this arylating species. We then examined: 1) the rate of decomposi-

---

1. From a Dissertation presented to the Department of Pharmacology,
   The Graduate School of Arts and Sciences, The George Washington
   University, in partial fulfillment of requirements for the Ph.D.
2. Burroughs Wellcome Scholar in Clinical Pharmacology.

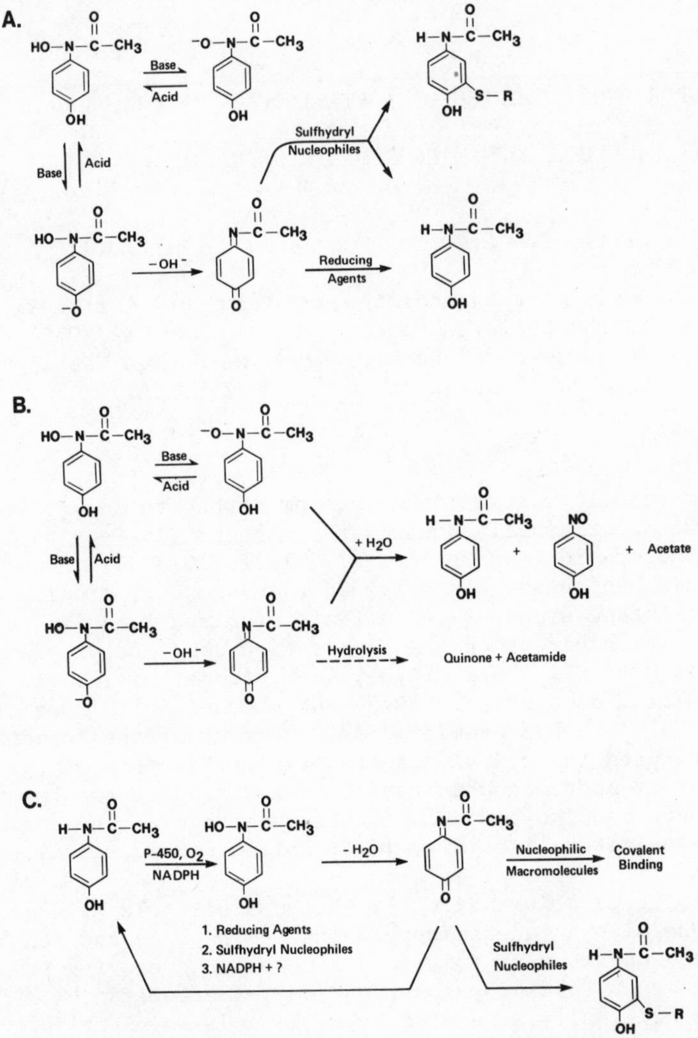

Figure 1.   Postulated Reactions of Sulfhydryl Compounds and Reduc-
            ing Agents With the Reactive Intermediates Formed From
            N-Hydroxy-Acetaminophen and Acetaminophen.

tion of synthetic N-hydroxy-acetaminophen, which was markedly decreased by acid conditions and the presence of sulfhydryl agents;
2) the reaction of the compound with glutathione, N-acetylcysteine
and cysteine, which yielded adducts that were identical to those
formed following metabolic activation of acetaminophen itself; and
3) the toxicity of the compound, which produced centrilobular hepatic necrosis in mice at doses below the hepatotoxic dose of
acetaminophen (Corcoran et al., 1978a, 1978b, 1979). N-Hydroxy-
acetaminophen and its reaction products with sulfhydryl compounds
were assayed by gas chromatography, high pressure liquid chromatography and gas chromatography-mass spectrometry. In the presence
of cysteine, N-hydroxy-acetaminophen decomposed to yield acetaminophen and the acetaminophen-cysteine adduct, accounting totally for
the disappearance of N-hydroxy-acetaminophen. Conversely, in the
absence of a sulfhydryl compound, the rate of decomposition of N-
hydroxy-acetaminophen was much faster and only 50% of its disappearance could be accounted for as acetaminophen (Corcoran et al.,
1978 a,b).

Two other groups concomitantly reported the successful synthesis of N-hydroxy-acetaminophen by other approaches (Calder et al.,
1978; Healey et al; 1978; Gemborys et al., 1978). They too found
that the decomposition of N-hydroxy-acetaminophen was pH and temperature dependent but they did not examine the effect of sulfhydryl compounds on rate of decomposition or on reaction products.
Calder and coworkers (1978), however, did observe marked depletion
of hepatic and renal glutathione and extensive hepatic and renal
necrosis after N-hydroxy-acetaminophen was administered to mice
and rats.

Results from preceding studies of the reaction of N-hydroxy-
acetaminophen with sulfhydryl compounds (Corcoran et al., 1978a,
1978b, 1979) have been consistent with the hypothesis that the
hydroxamic acid undergoes a base-catalyzed dehydration to N-acetyl-
p-benzoquinoneimine, which is the arylating species (Fig. 1A).
The investigations of Calder et al. (1978), Healey et al. (1978)
and Gemborys et al. (1978) demonstrating a pH optimum of 9-10 for
decomposition of N-hydroxy-acetaminophen are also consistent with
this hypothesis. We now report studies testing whether products
formed from the decomposition of N-hydroxy-acetaminophen in the
absence of sulfhydryl compounds also could arise from the N-
acetyl-p-benzoquinoneimine intermediate (Fig. 1B). The investigations are extended to an analysis of the reactive intermediate
formed from acetaminophen by microsomal mixed-function oxygenases
to test whether the intermediate has reaction properties similar
to N-acetyl-p-benzoquinoneimine (Fig. 1C). Finally, based upon
these properties, the potential for redox cycling of acetaminophen
and its reactive metabolite by endogenous microsomal systems is
examined.

Table 1

Identification and Quantitation of N-Hydroxy-acetaminophen and its
Decomposition Products by Mass Spectrometry and Gas Chromatography.
Acidified aliquots (0.2 ml) of incubation solutions were extracted
with ethyl acetate. After extracts were evaporated, residues were
treated with N,O-bis-(trimethylsilyl)trifluoroacetamide in CH_3CN,
and heated at 120°C for 10 minutes. Two-microliter samples were injected on a 6 ft., 1% OV-17 gas chromatographic column and the temperature was programmed from 100° to 200°C at 9°/minute (flame
ionization detection; carrier gas 70 cc N_2/min). Peak area ratios
of p-nitrosophenol, p-nitrophenol, acetaminophen and N-hydroxy-
acetaminophen to p-chloroacetanilide (retention time 5.4 min) were
proportional to their concentrations (r = 0.96, 0.99, 0.98 and 0.98
respectively) over the range of concentrations encountered experimentally (0.01 to 20 mM).

SILYLATED DER-IVATIVES OF	RETENTION TIME (MIN)	MAJOR MASS SPECTRAL IONS (m/z)
QUINONE[1, 2]	0.8	$108(M^+)$, 82, 80, 54, 53, 52
p-NITROSOPHENOL[2]	4.0	$195(M^+)$, 180, 166, 150, 91, 73
p-NITROPHENOL[2]	6.6	$211(M^+)$, 196, 182, 150, 91, 73,
ACETAMINOPHEN[2]	7.7	$295(M^+)$, 280, 254, 237, 206, 181, 180, 165, 116, 114, 73
N-HYDROXY-ACETAMINOPHEN[2]	10.7	$311(M^+)$, 296, 269, 268, 223, 206, 192, 179, 147, 73, 45, 43
N-HYDROXY-ACETAMINOPHEN[3]	–	352, 340, $312(MH^+)$, 296, 222
ACETAMINOPHEN-CYSTEINE AS PENTAFLUOROPROPIONYL METHYL ESTER DERIVATIVE[4]	–	$576(M^+)$, 413, 398, 354, 342, 329, 312, 300, 287, 247, 234, 215, 183, 100, 45, 43

[1]Quinone does not form a silylated derivative. [2]GC-electron impact
mass spectrum obtained with an LKB type 9000 Gas Chromatograph-Mass
Spectrometer (conditions: ionizing voltage 70 eV, ionizing current
60 μA, accelerating voltage -3.5 kV, source temperature 270°C,
separator temperature 270°C). [3]GC-chemical ionization mass
spectrum obtained with a Varian Model 1700 Gas Chromatograph interfaced with a Finnigan Model 1015 Mass Spectrometer, on-line with a
Systems Industries Model 150 Data System (conditions: ionizing
voltage 140 eV, source temperature ambient, methane served as the
carrier and reagent gas, source pressure 1 Torr). [4]Electron
impact mass spectrum obtained by direct probe insertion of the
pentafluoropropionyl methyl ester, rather than the silyl derivative,
into a Varian MAT CH7 Mass Spectrometer (conditions: ionizing
voltage 20 eV, ionizing current 100 μA, accelerating voltage -3kV,
source temperature 220°C, probe temperature 100°C).

MATERIALS AND METHODS

All reagents and materials were the highest grade commercially available. ^{14}C-(Carbonyl)- and ^{3}H-(general)-4-hydroxy-acetanilide were purchased from New England Nuclear Corp., Boston, MA and purified as previously described (Mitchell et al., 1974).

N-Hydroxy-acetaminophen (4.5 mM) was incubated in 0.1 M phosphate buffer at varying pH and at 37° with shaking. Incubations (2 ml) were started by adding N-hydroxy-acetaminophen at zero time as a distilled water solution, prepared immediately before use and maintained at 0°. All incubations contained 0.1 mg/ml p-chloroacetanilide and 0.4 mg/ml p-fluorophenol as internal standards. For a sample analysis, see Table 1.

The acetaminophen-cysteine conjugate was measured as follows. Aliquots (0.1 ml) were withdrawn from incubations, added to 0.1 ml ice cold methanol and analyzed on the same day. Determinations were performed with a Waters Associates (Milford, MA) high pressure liquid chromatographic (HPLC) system: Model U6K Injector, Model 6000A Solvent Delivery System, Model 440 Absorbance Detector (254 nm filters), 0.4 by 30 cm   Bondapak C_{18} column, and a water: methanol:acetic acid (86.5:12.5:1.0) mobile phase at a flow rate of 2.0 ml/min. The following retention times were observed: acetaminophen, 4.9 min; acetaminophen-cysteine conjugate, 4.1 min; p-fluorophenol, 10.8 min. Acetaminophen-cysteine conjugate concentrations were routinely determined by HPLC separation and by detection of absorbance at 254 nm. The relationship between concentration and the peak area ratio of the conjugate to internal standard (p-fluorophenol) was established by incubating ^{14}C-(carbonyl)-acetaminophen (sp. act. 1720 DPM/nmole) with hepatic microsomes (3.22 mg protein/ml), NADPH and 1 mM cysteine. Incubations (37° for 30 min) were terminated by addition of equal volumes of ice cold methanol. After samples were centrifuged supernatants were filtered (0.5 micron filter) and injected directly onto the HPLC column. Column effluent was collected at 10 second intervals and counted in a Packard Model 3003 liquid scintillation spectrometer. After corrections were made for quench (channels ratio) and counting efficiency, DPM were converted to nanomoles of acetaminophen-cysteine conjugate. Conjugate absorptivity was 83.1 $\pm$ 3.6% (n=6) that of acetaminophen at 254 nm.

RESULTS

N-Hydroxy-acetaminophen and its decomposition products were identified by analysis of full scan GC-electron impact mass spectra and by comparison with synthetic reference standards (Table 1). The identity of N-hydroxy-acetaminophen was further verified by gas chromatography-chemical ionization mass spectrometry of the bis-trimethylsilyl ether (Table 1). Thereafter, the rate of decomposi-

Table 2

Effect of Cysteine on Decomposition of N-Hydroxy-acetaminophen and Appearance of Products. N-Hydroxy-acetaminophen was incubated for 15, 60 or 120 minutes at 37° in 0.1 M phosphate buffers of differ-pH, with equimolar cysteine (+), 10 mM cysteine (++) or no cysteine at all (0). Analyses are described in Table 1 and Methods. ABBREVIATIONS: N-Hydroxy-acetaminophen (NOH-PHAA); Acetaminophen (PHAA); Acetaminophen-Cysteine (PHAA-CYS); p-Nitrosophenol (p-NOP); p-Nitrophenol (p-NO$_2$P); Cysteine (CYS).

	CONDITIONS			NOH-PHAA		PHAA[a]	PHAA-CYS	p-NOP	p-NO$_2$P	TOTAL
	CYS	pH	MIN	mM	(%)	(%)	(%)	(%)	(%)	(%)
A.	+	5.4	0	11.53	99.5	0.0	0.4			100.0
			15	11.27	97.3	0.0	0.5	N.D.[b]	N.D.	97.8
			60	10.20	88.1	0.0	0.6			88.7
	+	7.4	0	11.87	95.6	0.0	4.4			100.0
			15	10.67	86.0	3.5	6.5	N.D.	N.D.	96.0
			60	5.34	43.0	23.3	12.2			78.5
	++	7.7	0	7.18	99.3	0.0	0.7	0.0	0.0	100.0
			60	4.92	68.0	9.0	18.9	0.0	0.0	95.9
	+	8.2	0	10.19	97.4	0.0	2.6			100.0
			15	8.63	82.5	8.3	13.1	N.D.	N.D.	103.9
			60	4.08	39.0	26.0	43.8			108.8
	++	9.5	0	8.15	99.1	0.0	0.9	0.0	0.0	100.0
			60	0.24	2.9	27.7	67.0	0.0	0.0	97.7
	+	10.4	0	8.10	98.9	0.0	1.1			100.0
			15	4.70	53.4	5.0	30.2	N.D.	N.D.	92.6
			60	2.34	28.6	13.8	58.1			100.5
	++	11.4	0	5.86	99.5	0.0	0.5	0.0	0.0	100.0
			60	5.25	89.1	2.5	9.7	0.0	0.0	101.3
	+	13.0	0	8.14	98.1	0.0	1.9			100.0
			15	7.83	94.3	0.0	1.1	N.D.	N.D.	95.4
			120	7.87	94.8	0.0	4.3			99.1
B.	0	7.7	0	4.51	100.0	0.0	0.0	0.0	0.0	100.0
			60	0.69	15.3	43.7	0.0	40.6	3.7	103.3
	0	9.0	0	4.85	100.0	0.0	0.0	0.0	0.0	100.0
			60	0.01	0.2	46.0	0.0	51.1	4.9	102.2
	0	11.7	0	5.35	100.0	0.0	0.0	0.0	0.0	100.0
			60	4.14	77.4	12.6	0.0	10.5	1.5	102.0

[a]Values are corrected to account for a small amount of acetaminophen (5-10%) present as an impurity in N-hydroxy-acetaminophen.
[b]N.D. = Not Determined.

tion and the decomposition products of N-hydroxy-acetaminophen were determined by gas chromatography (Table 1, 2).

Without the addition of other agents, N-hydroxy-acetaminophen

decomposed to give approximately equal amounts of acetaminophen and
nitrosophenol and small amounts of benzoquinone and nitrophenol
(Table 2B). The small concentration of benzoquinone (less than
1%) was not routinely quantitated, but the disappearance of
N-hydroxy-acetaminophen always could be accounted for by the formation of acetaminophen, nitrosophenol and nitrophenol. Loss of
N-hydroxy-acetaminophen was maximal at pH 9-10 and minimal at pH
5 and 13 (Table 2), as previously reported (Corcoran et al., 1978a,
b; Healey et al., 1978; Gemborys et al., 1978).

The addition of sulfhydryl compounds such as cysteine markedly
slowed the decomposition of N-hydroxy-acetaminophen (Corcoran et al.,
1978a, 1978b), and altered product formation (Table 2). Again the
rate of decomposition was fastest near pH 9.5. The distribution of
products formed in the presence of cysteine was also influenced by
pH. The amount of acetaminophen exceeded the amount of acetaminophen-cysteine adduct at pH 7.4 after 60 minutes of incubation
whereas the reverse relationship occurred at higher pH values.

Inclusion of other reducing agents such as ascorbic acid markedly the slowed decomposition of N-hydroxy-acetaminophen and altered
product formation as well (Table 3). As seen with cysteine, 10 mM
ascorbic acid inhibited completely the formation of nitrosophenol
and nitrophenol. However, ascorbate did not add to the reactive
intermediate of N-hydroxy-acetaminophen to form a conjugate that
could be detected by HPLC analysis. The disappearance of N-hydroxy-
acetaminophen was accounted for stoichiometrically by the formation
of acetaminophen (Table 3F).

We compared the enzymatic formation of the reactive metabolite
of acetaminophen by microsomal monooxygenases with the base-catalyzed (pH 7.4) formation of the reactive metabolite of N-hydroxy-
acetaminophen by investigating the effects of cysteine and ascorbate
on these processes. Addition of cysteine alone trapped the reactive
intermediate derived from the NADPH-dependent metabolism of acetaminophen (Table 4A) and from the chemical decomposition of N-
hydroxy-acetaminophen (Table 4B) as acetaminophen-cysteine adducts.
Cysteine simultaneously inhibited the arylation of microsomal proteins by acetaminophen (Table 4C) and diminished the acetaminophen-
stimulated oxidation of NADPH during microsomal metabolism (Table
4D).

When increasing amounts of ascorbic acid were introduced into
microsomal incubations containing acetaminophen and cysteine or
into aqueous solutions containing N-hydroxy-acetaminophen and
cysteine, the amount of acetaminophen-cysteine formed from acetaminophen or from N-hydroxy-acetaminophen decreased in a concentration-dependent manner (Table 4A, B). Addition of ascorbic acid
had a similar concentration-dependent effect on the arylation of
microsomal proteins by the reactive acetaminophen metabolite (Table

Table 3

Effects of Ascorbic Acid on Decomposition of N-Hydroxy-acetaminophen
and Formation of Products.   Ascorbic acid and equimolar amounts of
cysteine (+) were added to incubations of N-hydroxy-acetaminophen in
0.1 M phosphate buffer.   Analyses are described in Table 1 and in
Methods.   Abbreviations are given in Table 2 except for ascorbic aci
(ASC).

	CONDITIONS			NOH-PHAA		PHAA[a]	PHAA-CYS	p-NOP	p-NO$_2$P	TOT
	CYS	ASC	pH	MIN	mM	(%)	(%)	(%)	(%)	(%)	(%
A.	0	0	7.4	0	5.58	100.0	0.0	0.0			100
				10	2.73	48.9	15.6	0.0	N.D.[b]	N.D.	64
				30	1.48	26.5	22.6	0.0			49
B.	0	1	7.4	0	5.58	100.0	0.0	0.0			100
				10	5.24	93.9	1.4	0.0	N.D.	N.D.	95
				30	4.38	78.5	10.8	0.0			89
C.	+	0	7.4	0	5.71	99.7	0.0	0.3			100
				10	5.66	98.8	1.9	3.3	N.D.	N.D.	104
				30	5.10	89.0	5.4	7.3			101
D.	+	1	7.4	0	6.20	99.7	0.0	0.3			100.
				10	6.12	98.4	0.5	2.2	N.D.	N.D.	101.
				30	5.30	85.2	4.2	6.1			95.
E.	0	0	9.4	0	4.24	100.0	0.0	0.0	0.0	0.0	100.
				10	0.05	1.2	49.3	0.0	37.7	5.2	93.
				30	0.00	0.0	50.0	0.0	44.3	3.1	97.
F.	0	10	9.4	0	4.43	100.0	0.0	0.0	0.0	0.0	100.
				10	1.36	30.7	65.5	0.0	0.0	0.0	96.
				30	0.41	9.2	97.3	0.0	0.0	0.0	106.

a,b see Table 2.

4C) but did not decrease acetaminophen-stimulation of NADPH oxidation more than that caused by addition of cysteine alone (Table
4D).

Although NADPH oxidation was stimulated 16.4 nmols/mg/5 min by
the metabolism of acetaminophen, only a small fraction of oxidized
acetaminophen (0.223 nmols/mg/5 min) became covalently bound to
microsomal proteins (Table 4C).   Addition of cysteine to incubations resulted in much more of the oxidized, arylating metabolite
of acetaminophen becoming trapped as the cysteine conjugate (2.39
nmols/ mg/5 min), in spite of the apparently contradictory action
of cysteine to decrease acetaminophen-stimulated NADPH oxidation.

Finally, ascorbic acid abolished acetaminophen-cysteine conjugate formation without influencing the rate of acetaminophen-stimulated NADPH oxidation in the presence of cysteine, the basal rate of NADPH oxidation in the absence of drug, or the rate of ethylmorphine N-demethylation (43.7 $\pm$ 1.5 nmols HCHO formed /mg/5min).

DISCUSSION

In aqueous solutions, N-hydroxy-acetaminophen decomposes to acetaminophen and nitrosophenol and small amounts of nitrophenol and benzoquinone. Presumably, the nitrophenol arises from further oxidation of nitrosophenol, and benzoquinone arises from hydrolysis of N-acetyl-p-benzoquinoneimine (Fig. 1B). Decomposition of N-hydroxy-acetaminophen in buffered solutions is maximal around pH 9.5, which is consistent with N-acetyl-p-benzoquinoneimine formation being the rate-determining step (Fig. 1B). Nitrosophenol and acetaminophen are formed from N-hydroxyacetaminophen in an approximately 1:1 ratio. The appearance of equal amounts of both an oxidized and a reduced product suggests a 2-electron redox couple in which the reactive N-acetyl-p-benzoquinoneimine formed by dehydration of N-hydroxy-acetaminophen is reduced to acetaminophen by a second N-hydroxy-acetaminophen molecule which becomes oxidized to nitrosophenol (Fig. 1B).

In reactions containing cysteine, the loss of N-hydroxy-acetaminophen is markedly diminished and the formation of nitrosophenol but not of acetaminophen is abolished. Under these conditions reactive N-acetyl-p-benzoquinoneimine is either reduced or conjugated before it can react with N-hydroxy-acetaminophen (Fig. 1A). This prevents paired nitrosophenol and acetaminophen formation by the bimolecular reaction pathway shown in Figure 1B. In this system the formation of acetaminophen versus the acetaminophen-cysteine conjugate is pH dependent. Under acid conditions the reduction of N-acetyl-p-benzoquinoneimine to acetaminophen is predominant whereas under basic conditions adduct formation is greater (Fig. 1A). Presumably, this occurs because cysteine is most nucleophilic as an anion and anion formation is suppressed by acidic conditions.

As with cysteine, addition of ascorbic acid decreases the rate of decomposition of N-hydroxy-acetaminophen and inhibits formation of nitrosophenol. Ascorbic acid, which lacks the nucleophilic but not the reducing action of cysteine, stoichiometrically converts N-hydroxy-acetaminophen to acetaminophen by reducing the N-acetyl-p-benzoquinoneimine intermediate (Fig. 1A). Apparently, neither cysteine nor ascorbic acid reduce N-hydroxy-acetaminophen directly to acetaminophen because both agents decrease the overall rate of decomposition of N-hydroxy-acetaminophen. Moreover, they do not promote the formation of acetaminophen at acid and basic pH's which are not favorable for dehydration to the N-acetyl-p-benzo-quinoneimine intermediate.

Table 4

Effect of Ascorbic Acid on Acetaminophen-Cysteine Conjugate Formation
Covalent Binding and NADPH Oxidation from Microsomal Metabolism of
Acetaminophen. Metabolic studies (Parts A, C and D) were conducted i
0.1 M phosphate at pH 7.4 and 37°C for 5 minutes with hepatic microsomes (1 mg protein/ml) from male Swiss mice, 1 mM acetaminophen, 0.
mM NADPH and 1 mM cysteine. Values reported are means and standard
deviations of three determinations. In Part B, N-hydroxy-acetaminophen (4 mM) was reacted with cysteine (4 mM) at pH 7.4 and 37°C for
120 min. Products are expressed as % of initial N-hydroxy-acetaminophen concentration. Covalent binding (^{3}H-acetaminophen, sp.act. 1060
DPM/nmol) and NADPH oxidation were determined as described previously
(Potter et al., 1973; Sasame et al., 1973). Other analyses and abbreviations are described in Tables 1 and 2 and Methods.

CONDITIONS		PART A FORMATION OF ACETAMINO-PHEN-CYSTEINE FROM MICROSOMAL METABOLISM OF ^{3}H-ACETAMINOPHEN.			PART B FORMATION OF ACETAMINOPHEN-CYSTEINE (%) AND ACETAMINO-PHEN (%) FROM N-HYDROXY-ACETAMINOPHEN (% REMAINING).		
CYS	ASC (mM)	nmol/mg/5min		(%)	NOH-PHAA	PHAA	PHAA-CYS
+	0	2.39 + .28		(100)	4.2	65.0	33.0
+	1	1.45 + .21		(60)	7.1	71.4	28.0
+	5	0.84 + .06		(35)	8.5	82.7	18.9
+	10	0.16 + .03		(7)	9.6	89.0	5.8
+	30	0.02 + .01		(1)	9.2	96.0	2.2

		PART C COVALENT BINDING FROM MICROSOMAL METABOLISM OF ^{3}H-ACETAMINOPHEN.			PART D NADPH OXIDATION FROM MICROSOMAL METABOLISM OF ^{3}H-ACETAMINOPHEN.		
		nmol/mg/5 min		(%)	nmol/mg/5 min		(%)
0	0	.223 + .025		(100)	16.4 + .66		(100)
+	0	.009 + .002		(4)	8.4 + .90		(51)
0	1	.048 + .006		(21)	11.4 + .46		(70)
0	5	.014 + .002		(6)	9.5 + 1.50		(58)
0	10	.019 + .002		(9)	9.0 + 1.50		(55)
0	30	.003 + .001		(1)	7.1 + 1.20		(43)

Both cysteine and ascorbic acid inhibit the arylation of
microsomal proteins by the toxic metabolite of acetaminophen formed
oxidatively by cytochrome P-450 monooxygenases (Table 4C). The
scheme in Figure 1C predicts this behavior when the toxic metabolite
of acetaminophen is a reactive quinoneimine species, as postulated.
An important observation is the ability of increasing concentrations
of ascorbic acid to decrease formation of the acetaminophen-cysteine
adduct derived from the microsomal oxidation of acetaminophen and

from the base-catalyzed decomposition of N-hydroxy-acetaminophen
(Table 4A,B). This effect of ascorbic acid on the adduct derived
from oxidation of acetaminophen is explained by the reduction of
the N-acetyl-p-benzoquinoneimine intermediate to acetaminophen
before conjugation with cysteine can occur (Fig. 1C). It apparently
does not result from inhibition of microsomal oxidation of acetaminophen because acetaminophen-stimulated NADPH oxidation in the
presence of cysteine is not decreased further by addition of ascorbic acid (Table 4D). Furthermore, ascorbic acid does not inhibit
the oxidation of other cytochrome P-450 substrates such as ethyl
morphine.

These results also explain why the amount of reactive metabolite formed from acetaminophen always appears to be greater when
the metabolite is quantitated by measuring the amount of sulfhydryl
adduct formed rather than by measuring the amount of covalent
binding as an index of metabolite formation (Table 4A,C). In the
absence of a high concentration of nucleophile, the strong reducing
environment of the microsomal system probably favors the reduction
of most of the N-acetyl-p-benzoquinoneimine back to acetaminophen
before it has time to react with the nucleophilic sites present in
low concentration in protein macromolecules. Conversely, when
present in high concentrations in microsomal incubations, sulfhydryl
nucleophiles add efficiently to N-acetyl-p-benzoquinoneimine before
much reduction can occur, resulting in greater estimates of the rate
which the arylating intermediate is formed. Both measurements,
however, likely underestimate the amount of acetaminophen that is
oxidized to N-acetyl-p-benzoquinoneimine.

These conclusions are further strengthened by examining the
stoichiometry of NADPH utilization during the microsomal oxidation
of acetaminophen. Far more NADPH is oxidized than can be accounted
for by the oxidation of acetaminophen to its arylating metabolite,
when the amount of toxic species is estimated by covalent binding
or by sulfhydryl trapping measurements. Furthermore, acetaminophen-stimulated NADPH oxidation is decreased by reducing agents
such as cysteine and ascorbate. These findings suggest that in
the absence of these reducing agents, some NADPH is utilized to
reduce N-acetyl-p-benzoquinoneimine back to acetaminophen. For
each molecule of acetaminophen that undergoes cyclic metabolism
(i.e. the oxidation of acetaminopohen to N-acetyl-p-benzoquinoneimine followed by reduction of the quinoneimine back to acetaminophen), 4 molecules of NADPH would be consumed. The best evidence
for the cyclic metabolism of acetaminophen can be seen after
ascorbic acid is added to incubations containing acetaminophen.
As the concentration of ascorbic acid is increased, the covalent
binding of acetaminophen is abolished, indicating that all N-
acetyl-p-benzoquinoneimine being produced is reduced back to acetaminophen by ascorbate. Concomitantly, concentrations of ascorbic
acid that decrease acetaminophen-stimulated NADPH consumption by

about half do not influence the metabolism of other substrates
such as ethylmorphine. It is likely that reduction of the quinone-
eimine by NADPH is competitively abolished by ascorbic acid but
this reducing agent has no effect on the NADPH-dependent oxidation
of acetaminophen.

New studies applying electrochemical techniques have also provided evidence that N-acetyl-p-benzoquinoneimine is the arylating
metabolite formed from acetaminophen. Kissinger recently reported
(1978) that N-acetyl-p-benzoquinoneimine can be generated electro-
chemically from acetaminophen and that the intermediate reacts
with nucleophiles to produce sulfhydryl adducts that are indistin-
guishable from those formed during the microsomal metabolism of
acetaminophen.

Thus, the collective data would appear to establish that the
reactive intermediate resulting from P-450-mediated oxidation of
acetaminophen is N-acetyl-p-benzoquinoneimine (Mitchell et al.,
1978; Jollow et al., 1974). Furthermore, variations in the hepatotoxicity of acetaminophen with animal species and strains (Davies
et al., 1974), enzyme induction and inhibition (Mitchell et al.,
1972, 1973; Potter et al., 1973) and genetic responsiveness (Thorgeirsson et al., 1978) correlate directly with differences in the
ability to N-hydroxylate other drugs (Thorgeirsson et al., 1972,
1973; Hinson et al., 1975; Hinson and Mitchell, 1976).

However, one may speculate about alternative pathways that
could be involved in acetaminophen-induced liver injury. Isolation
of 3-methoxy-acetaminophen from urine (Knox and Jurrand, 1977) raise
the possibility that the toxicity of acetaminophen was due to a
3,4-epoxide metabolite or a 3,4-catechol derived from the epoxide.
However, this pathway is not likely to be involved in the toxicity
of acetaminophen. When the toxic arylating metabolite of acetaminophen is generated enzymatically under an $^{18}O_2$ atmosphere and
trapped with glutathione, no ^{18}O is found in the adduct, indicating
that it did not arise from 3,4-epoxidation (Hinson et al., 1977).
Moreover, the 3,4-catechol metabolite of acetaminophen contains
^{18}O and therefore arises from a pathway other than the toxifying
one (Hinson et al., 1979). In addition, the trace amount of 3,4-
catechol formed in microsomal incubations is not decreased by
addition of ascorbic acid or sulfhydryl agents (Hinson et al.,
1979), yet ascorbic acid and sulfhydryls abolish the covalent
binding of acetaminophen (Table 4C).

Acetaminophen might also be oxidized to N-acetyl-p-benzoquinoneimine directly, or through two one-electron steps. However,
there is no precedent for reactions of this kind mediated in vivo
by the cytochrome P-450 monooxygenases at the present time.

REFERENCES

1.   Calder, I. C., Healey, K., Yong A. C., Crowe, A. C., Ham, K.
         N., and Tange, J. D. 1978, N-Hydroxyparacetamol - its
         role in the metabolism of paracetamol. In: "Biological
         Oxidation of Nitrogen," (J. W. Gorrod, ed.), pp. 309-
         318, Elsevier/ North Holland Biomedical Press, Oxford.
2.   Corcoran, G. B., Vaishnav, Y. N., and Mitchell, J. R. 1978
         Reactivity of the N-hydroxy metabolite of acetaminophen
         including sulfhydryl adduct formation. Proc. Seventh
         Intl. Congr. Pharmacol., p. 651, (abstract).
3.   Corcoran, G. B., Mitchell, J. R., Vaishnav, Y. N., Horning,
         E. C., and Nelson, S. D. 1978, Formation of chemically
         reactive metabolites from drugs. In: Advances in Pharma-
         cology and Therapeutics, Vol. 9 (Y. Cohen, ed.), pp.
         103-111, Pergamon Press, Oxford.
4.   Corcoran, G. B., Mitchell, J. R., Vaishnav, Y. N., and Horning,
         E. C. 1979, Sulfhydryl adduct and acetaminophen forma-
         tion from N-hydroxy-acetaminophen: Mechanistic implica-
         tions. Pharmacologist 21:220, (abstract).
5.   Davis, D. C., Potter, W. Z., Jollow, D. J., and Mitchell, J. R.
         1974, Species differences in hepatic glutathione depletion,
         covalent binding and hepatic necrosis after acetaminophen.
         Life Sci. 14:2099-2109.
6.   Gemborys, M. W., Gribble, G. W., and Mudge, G. H. 1978, Synthe-
         sis of N-hydroxyacetaminophen, a postulated toxic metabo-
         lite of acetaminophen, and its phenolic sulfate conjugate.
         J. Med. Chem. 21:649-652.
7.   Healey K., Calder, I. C., Yong, A. C., Crowe, C. A., Funder,
         C. C., Ham, K. N., and Tange, J. D. 1978, Liver and kidney
         damage induced by N-hydroxyparacetamol. Xenobiotica
         8:403-411.
8.   Hinson, J. A., Mitchell, J. R., and Jollow, D. J. 1975, Micro-
         somal N-hydroxylation of p-chloroacetanilide. Molec.
         Pharmacol. 11:462-469.
9.   Hinson, J. A., and Mitchell, J. R. 1976, N-Hydroxylation of
         phenacetin by hamster liver microsomes. Drug Metab.
         Dispos. 4:430-435.
10. Hinson, J. A., Nelson, S. D., and Mitchell, J. R. 1977, Studies
         on the microsomal formation of arylating metabolites of
         acetaminophen and phenacetin. Molec. Pharmacol. 13:
         625-633.
11. Hinson, J. A., Pohl, L. R., Monks, T. J., and Gillette, J. R.
         1979, 3-Hydroxyacetaminophen: A microsomal metabolite
         of acetaminophen. Pharmacologist 21:219 (abstract).
12. Jollow, D. J., Thorgeisson, S. S., Potter, W. Z., Hashimoto,
         M., and Mitchell, J. R. 1974, Acetaminophen-induced hepatic
         necrosis. VI. Metabolic disposition of toxic and nontoxic
         doses of acetaminophen. Pharmacology 12:251-271.

13. Kissinger, P. T. 1978, Application of electrometric techniques
      to pharmacological problems. Pharmacologist 20:139.
14. Knox, J. H., and Jurand, J. 1978, Determination of paracetamol
      and its metabolites in urine by high-performance liquid
      chromatography using reversed-phase bonded supports.
      J. Chromatogr. 142:651-670.
15. Mitchell, J. R., Potter, W. Z., Jollow, D., Davis, D., Gillette
      J. R. 1974, Acetaminophen-induced hepatic necrosis. I.
      Potentiation by inducers and protection by inhibitors of
      drug-metabolizing enzymes. Fed. Proc. 31:539 (abstract).
16. Mitchell, J. R., Jollow, D., Gillette, J. R. and Brodie, B. B.
      1973, Drug metabolism as a cause of drug toxicity. Drug
      Metab. Dispos. 1:418-423.
17. Mitchell, J. R., Thorgeirsson, S. S., Potter, W. Z., Jollow,
      D. J., Keiser, H. 1974, Acetaminophen-induced hepatic in-
      jury. Protective role of glutathione in man and ration-
      ale for possible therapy. Clin. Pharmacol. Therap. 16:
      676-684.
18. Potter, W. Z., Davis, D. C., Mitchell, J. R., Jollow, D. J.,
      Gillette, J. R. and Brodie, B. B. 1973, Acetaminophen-
      induced hepatic necrosis. III. Cytochrome P-450 mediated
      covalent binding in vitro. J. Pharmacol. Exp. Therap. 187
      203-210.
19. Sasame, H. A., Mitchell, J. R., Thorgeirsson, S. S., and
      Gillette, J. R. 1973, Relationships between NADH and NADPH
      oxidation during drug metabolism. Drug Metab. Dispos.
      1:150-155.
20. Thorgeirsson, S. S., Jollow, D., Potter, W., Sasame, H.,
      Mitchell, J. R., Gillette, J. R. and Brodie, B. B. 1972,
      Acetaminophen-induced hepatic necrosis. III. Role of
      cytochrome P-450 in the toxicity of N-acetylarylamines.
      Fifth Intl. Congr. Pharmacol., p. 223 (abstract).
21. Thorgeirsson, S. S., Jollow, D. J., Sasame, H. A., Green, I.,
      and Mitchell, J. R. 1973, The role of cytochrome P-450 in
      N-hydroxylation of 2-acetylaminofluorene. Molec. Pharm-
      acol. 9:398-404.
23. Thorgeirsson, S. S., Felton, J. S., and Nebert, D. W. 1975,
      Genetic differences in the aromatic hydrocarbon-inducible
      N-hydroxylation of 2-acetylaminofluorene- and acetamino-
      phen-produced hepatotoxicity in mice. Molec. Pharmacol.
      11:159-165.

PROSTAGLANDIN SYNTHETASE CATALYZED ACTIVATION OF PARACETAMOL

Peter Moldéus, Anver Rahimtula*,
Bo Andersson and Margareta Berggren
Department of Forensic Medicine
Karolinska Institutet
104 01 Stockholm 60

INTRODUCTION

Paracetamol (acetaminophen) is a widely used analgesic and antipyretic drug which is considered to be nontoxic at therapeutic concentration, but produces liver and renal damage when taken in overdoses (Proudfoot and Wright, 1970; Mitchell et al. 1973). It is currently believed that the microsomal cytochrome P-450-linked monooxygenase system is responsible for activating paracetamol to an electrophilic intermediate that can bind covalently to cellular macromolecules to produce cell damage (Jollow et al., 1973). In the presence of reduced glutathione (GSH), the reactive species is trapped as the corresponding glutathione conjugate (Mitchell et al., 1973).

Marnett and coworkers (1975,1977) have shown that during prostaglandin biosynthesis in sheep seminal vesicle microsomes, several xenobiotics, including benzo(a)pyrene, undergo substantial cooxygenation.

In this manuscript we report that sheep seminal vesicle and also rat kidney microsomes can catalyze arachidonic acid-dependent metabolism of paracetamol to a reactive metabolite which may conjugate with glutathione or bind covalently to protein. Preliminary evidence indicates that this metabolite is similar if not identical to that produced during NADPH dependent microsomal metabolism of paracetamol.

---

* Present address: Department of Biochemistry, Memorial University of New Foundland, St. Johns, New Foundland, Canada A1B 3X9.

METHODS AND MATERIAL

Microsomes were prepared as previously described (Wlodawer and Samuelsson, 1973) from sheep seminal vesicles that had been stored frozen at -80° and from rat kidneys as described by Ernster et al. (1962).

Incubations were performed at 25° in 0.1 M phosphate buffer, pH 8.0, at a protein concentration of 1 mg per ml of incubate. The final concentration of arachidonic acid was 0.3 mM and that of GSH 2.5 mM. Reactions were started with microsomes and terminated by the addition of 0.2 ml 3 N perchloric acid per ml of incubation mixture. Incubations with kidney microsomes were performed at 37°.

Protein concentration was measured according to the method of Lowry et al.

Paracetamol glutathione conjugate formation was determined by high-performance liquid chromatography as described earlier (Moldéus, 1978).

Covalent binding to protein was determined by the method of Jollow et al. (1973) using ^{3}H-paracetamol as substrate.

Arachidonic acid and GSH were purchased from Sigma Chemical Co., St. Louis, Mo and ^{3}H-paracetamol from NEN, Boston, Mass. Linolenic acid hydroperoxide was produced from linolenic acid in the presence of soybean lipoxygenase as described by Funk et al. (1976). N-OH paracetamol was a gift from Dr R. Andrews Sterling Winthrop. Other chemicals were obtained from local commercial suppliers.

RESULTS

In the presence of arachidonic acid, paracetamol and GSH, microsomes isolated from sheep seminal vesicles (SSV) catalyzed the formation of a paracetamol glutathione conjugate (Table I). The glutathione conjugate was identified by high-performance liquid chromatography and found to cochromatograph with the paracetamol glutathione conjugate formed during aerobic incubation of liver microsomes with paracetamol and GSH in the presence of a NADPH-regenerating system (Moldéus, 1978). The reaction catalyzed by SSV microsomes was enzymatic in nature, since no activity was detected in the absence of microsomes, or with boiled microsomes. Neither was activity detected if either of GSH, paracetamol or arachidonic acid was absent from the incubation. Indomethacin, a potent inhibitor of prostaglandin synthetase, inhibited the formation of the GSH conjugate of paracetamol by 70% at a concentration of 100 μM (Table I).

Table 1. Formation of Paracetamol Glutathione Conjugate
Catalyzed by SSV Microsomes

Incubation conditions	Paracetamol Glutathione conjugate nmol/mg protein per min
Complete	22.0
- microsomes	1.3
- arachidonic acid	0.9
+ indomethacin, 100 µM	6.5
Complete (boiled microsomes)	0

Paracetamol concentration was 1 mM.

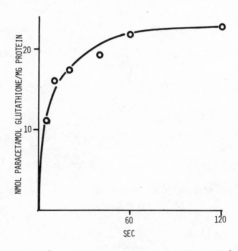

Fig. 1. Rate of formation of paracetamol glutathione conjugate
in SSV microsomes. Paracetamol concentration was 0.2 mM.

As demonstrated in Fig. 1, the formation of paracetamol glutathione conjugate by SSV microsomes was linear with time for
only a few seconds and levelled off after only ten seconds of incubation. After two minutes, the reaction had almost terminated.
It is of interest to note that the initial reaction velocity was
much higher than that usually observed with mouse liver microsomes in presence of NADPH. The affinity for paracetamol was also
quite high in the reaction catalyzed by SSV microsomes, since the
reaction was saturated at about 200 µM paracetamol (Fig. 2).

If GSH was omitted from the reaction mixture covalent bin-

ding of paracetamol to protein was observed (Fig. 3). Similar to
the formation of the glutathione conjugate (Fig. 1), the rate of
covalent binding was initially rapid but levelled off after about
one minute of incubation. The amount of covalently bound paracetamol was however much less than the amount of glutathione conjugate
formed.

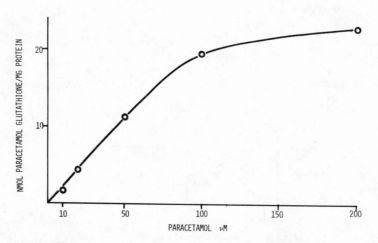

Fig. 2. Glutathione conjugate formation in SSV microsomes at
        different paracetamol concentrations. Time of incuba-
        tion was 1 min.

Rat kidney microsomes were also able to catalyze the formation of a paracetamol glutathione conjugate in the presence of
arachidonic acid and GSH (Table 3). The activity, which was considerably lower than with SSV microsomes, was inhibited by indomethacin, indicating the involvement of prostaglandin synthetase
in the reaction. The activity in the presence of arachidonic acid
was considerably higher than what was found with NADPH (Table 3).

Formation of N-OH paracetamol has been suggested to be the
first step in the formation of the reactive metabolite of paracetamol in liver (Mitchell et al., 1973). N-OH paracetamol then spontaneously forms N-acetylimidoquinone which is believed to be the
reactive species of paracetamol. When N-OH paracetamol was incubated together with SSV microsomes, arachidonic acid and GSH, a glutathione conjugate was rapidly formed; a conjugate with different
retention time on high performance liquid chromatography compared
to that formed from paracetamol. The rate of formation of this
conjugate corresponded well to that with paracetamol and the activity levelled off after a few seconds (Fig. 4).

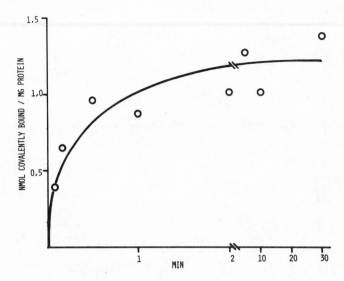

Fig. 3. Covalent binding to protein of paracetamol in SSV microsomes. Incubations were performed as described in Methods
but in the absence of GSH and at a paracetamol concentration of 0.2 mM.

Table 3. Formation of Paracetamol Glutathione Conjugate
Catalyzed by Rat Kidney Microsomes

Incubation conditions	nmol/mg per 30 min
Complete	7.1
+ indomethacin, 200 μM	4.1
+ NADPH generating system, no arachidonic acid	1.3

In order to prevent breakdown of paracetamol glutathione
incubations were performed in the presence of serine bora-
te (20 mM). Protein concentration was 2 mg/ml of incubate
and the concentration of paracetamol 1 mM.

In the incubation medium N-OH paracetamol rapidly undergoes
degradation to the N-acetylimidoquinone which may spontaneously
conjugate with GSH. Formation of this conjugate was also observed
during incubation with N-OH paracetamol (Fig. 4). This glutathione conjugate had similar retentiontime on high performance
liquid chromatography to that formed from paracetamol incubated
with SSV or kidney microsomes.

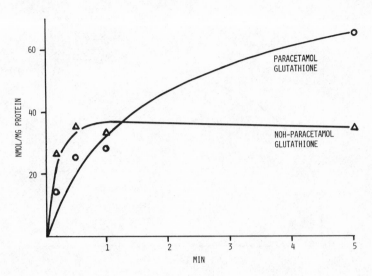

Fig. 4. Formation of glutathione conjugates of N-OH paracetamol in
        SSV microsomes. Incubations were performed as described in
        Methods at an N-OH paracetamol concentration of 0.2 mM.

DISCUSSION

    The results of the present study clearly demonstrate the abi-
lity of prostaglandin synthetase to catalyze the metabolic activation of paracetamol. This activity is observed in microsomes from
both sheep seminal vesicles and rat kidney and is demonstrated
both as covalent binding to protein and formation of a paracetamol
glutathione conjugate. The reactive metabolite formed is apparently similar if not identical to that formed by liver microsomes in
the presence of NADPH. This is supported by similar retention
times of the glutathione conjugates on high-performance liquid
chromatography. It is also of interest to note that both the initial reaction velocity and the affinity for paracetamol are much
higher in the SSV microsomal system as compared to the NADPH-dependent monooxygenase reaction in liver microsomes. The apparent
K_m-value for paracetamol is only about 60 µM (calculated from Fig.
2) in the SSV microsomal system, whereas it is known to be in the
millimolar range in liver microsomes.

    Prostaglandin synthetase, which is present in high concentra-
tions in the seminal vesicles and kidneys exhibits both cyclooxy-
genase activity, catalyzing the oxygenation of arachidonic acid to
the hydroperoxyendoperoxide (PGG_2) (Hamberg et al., 1974; Nugteren
and Hazelhof, 1973), and peroxidase activity, catalyzing the reduction of PGG_2 to the hydroxyendoperoxide (PGH_2) (Miyamoto et al.,

1976; O'Brien and Rahimtula, 1976). It is the latter activity that appears to be responsible for the metabolic activation of paraceta-mol. This conclusion is supported by the observation that linole-nic acid hydroperoxide could support formation of the paracetamol glutathione conjugate with SSV microsomes (Moldéus and Rahimtula, 1980) and is also in accordance with previous results by Marnett and Reed (1979) demonstrating cooxidation of benzo(a)pyrene to quinones during metabolism of PGG_2 or 15-hydroperoxy-5,8,11,13-eicosatetraenoic acid by SSV microsomes.

The identity of the electrophilic metabolite of paracetamol has not yet been established. However, since it can be formed as a result of prostaglandin synthetase peroxidase function, its for-mation appears to involve a one-electron oxidation reaction which could lead to hydrogen abstraction to yield the phenoxy radical of paracetamol. This radical may in turn be further oxidized to the N-acetylimidoquinone prior to reacting with GSH.

Evidently N-hydroxylation is not involved in the activation since N-OH paracetamol could be activated to a reactive metabolite different from that formed from paracetamol. The activation of N-OH paracetamol may thus also involve an hydrogen abstraction.

The toxicological implications of the metabolic activation of paracetamol as a result of prostaglandin synthetase function are presently unclear. Obviously, this mechanism is of particular interest for the interpretation of the renal toxicity of the drug and may well be of importance for the production of reactive para-cetamol metabolites. This is supported by the observation that kidney microsomes are able to catalyze arachidonic acid dependent activation of paracetamol (cf Table 3).

SUMMARY

Microsomes isolated from sheep seminal vesicles (SSV) were found to catalyze the metabolic activation of paracetamol as evi-denced by formation of paracetamol glutathione conjugate when SSV microsomes were incubated with paracetamol in the presence of arachidonic acid and GSH. In the absence of GSH covalent binding of paracetamol to protein was observed. The activity was inhibited by indomethacin indicating the involvement of prostaglandin syn-thetase in the reaction. The initial activity was very rapid, and the affinity for paracetamol in the reaction was high, in fact, much higher than with microsomes from mouse liver using an NADPH generating system.

N-OH paracetamol was also activated by SSV microsomes in the presence of arachidonic acid to a metabolite apparently different from that formed from paracetamol since the retention times of the respective glutathione conjugates differed significantly.

Finally it was shown that rat kidney microsomes were also able to catalyze the formation of a paracetamol glutathione conjugate in the presence of arachidonic acid and GSH. The activity was however considerably less than with SSV microsomes.

ACKNOWLEDGEMENTS

This study was supported by grants from the Swedish Council for planning and coordination of research and Swedish Medical Research Council (No. B 81-03P-5636-02 5 021 05636-G918).

REFERENCES

Ernster, L., Siekevitz, P., and Palade, G.E., 1962, Enzyme-structure relationship in the endoplasmic reticulum of rat liver, J. Cell. Biol., 15:541.

Funk, M.O., Isaac, R., and Porter, N.A., 1976, Preparation and purification of lipid hydroperoxides from arachidonic and γ-linolenic acids, Lipids, 11:113.

Hamberg, M., Svensson, J., Wakabayashi, T., and Samuelsson, B., 1974, Isolation and structure of two prostaglandin endoperoxides that cause platelet aggregation, Proc. Nat. Acad. Sci. USA, 71:345.

Jollow, D.J., Mitchell, J.R., Potter, W.Z., Davis, D.C., Gillette, J.R., and Brodie, B.B., 1973, Acetaminophen induced hepatic necrosis.II. Role of covalent binding in vivo, J. Pharmacol. Exp. Ther., 187:195.

Marnett, L.J., and Reed, G.A., 1979, Peroxidatic oxidation of benzo(a)pyrene and prostaglandin biosynthesis, Biochemistry, 18:2923.

Marnett, L.J., Reed, G.A., and Johnson, J.T., 1977, Prostaglandin synthetase dependent benzo(a)pyrene oxidation: Products of the oxidation and inhibition of their formation by antioxidants, Biochem. Biophys. Res. Commun., 79:569.

Marnett, L.J., Wlodawer, P., and Samuelsson, B., 1975, Co-oxygenation of organic substrates by the prostaglandin synthetase of sheep vesicular gland, J. Biol. Chem., 250:8510.

Mitchell, J.R., Jollow, D.J., Potter, W.Z., Davis, D.C., Gillette, J.R., and Brodie, B.B., 1973, Acetaminophen-induced hepatic necrosis. I. Role of drug metabolism, J. Pharmacol. Exp. Ther., 187:185.

Miyamoto, T., Ogino, N., Yamamoto, S., and Hayaishi, O., 1976, Purification of prostaglandin endoperoxide synthetase from bovine vesicular gland microsomes, J. Biol. Chem., 251:2629.

Moldéus, P., 1978, Paracetamol metabolism and toxicity in isolated hepatocytes from rat and mouse, Biochem. Pharmacol., 27:2859.

Moldéus, P., and Rahimtula, A., 1980, Metabolism of paracetamol to a glutathione conjugate catalyzed by prostaglandin synthetase. Biochem. Biophys. Res. Commun., In press.

Nugteren, D.H., and Hazelhof, E., 1973, Isolation and properties of intermediates in prostaglandin biosynthesis, Biochim. Biophys. Acta, 326:448.

O'Brien, P.J., and Rahimtula, A.D., 1976, The possible involvement of a peroxidase in prostaglandin biosynthesis. Biochem. Biophys. Res. Commun., 70:832.

Proudfoot, A.T., and Wright, N., 1974, Acute paracetamol poisoning, Brit. Med. J., 3:557.

Wlodawer, P., and Samuelsson, B., 1973, On the organization and mechanism of prostaglandin synthetase, J. Biol. Chem. 248:5673.

ROUTES TO THE FORMATION OF N-METHYL-4-AMINOPHENOL, A METABOLITE

OF N,N-DIMETHYLANILINE

N.J. Gooderham and J.W. Gorrod

Department of Pharmacy
Chelsea College (University of London)
Manresa Road, London, SW3 6LX, U.K.

INTRODUCTION.

We have recently shown the in vitro metabolism of N,N-dimethyl-
aniline to include N-oxidation, N-demethylation and ring
hydroxylation (Gorrod and Gooderham 1980). The major metabolites
(fig1) were N-methylaniline, N,N-dimethylaniline-N-oxide and
N,N-dimethyl-4-aminophenol; aniline, N-methyl-4-aminophenol and
4-aminophenol were produced to a lesser extent.

The formation of N-methyl-4-aminophenol could have arisen via
N-demethylation of N,N-dimethyl-4-aminophenol and/or by 4-hydroxy-
lation to N-methylaniline, (fig.1). It was the purpose of this study
to examine this proposal and establish the route or routes of
formation of this compound during the in vitro metabolism of
N,N-dimethylaniline.

We have also examined the effect of a number of enzyme
inhibitors on these reactions in an attempt to elucidate the
enzymology involved.

MATERIALS AND METHODS

SKF 525A (2-diethylaminoethyl-2,2-diphenylvalerate)was a gift from
Smith, Kline and French Ltd (Welwyn Garden City, Herts, U.K.), DPEA
(2,6-dichloro-6-phenylphenoxyethylamine) from Eli Lilly and Co Ltd
(Erlwood, Surrey, U.K.) and MMI (Methimazole) from Professor
D.M. Ziegler (University of Texas, Austin, Texas, U.S.A.).
Catalase and n-Octylamine were purchased from BDH Ltd (Poole, Dorset,
U.K.) and superoxide dismutase from Sigma Laboratories (Kingston-
upon-Thames, Surrey, U.K.). All other materials were obtained and
used as previously described, (Gorrod and Gooderham 1980).

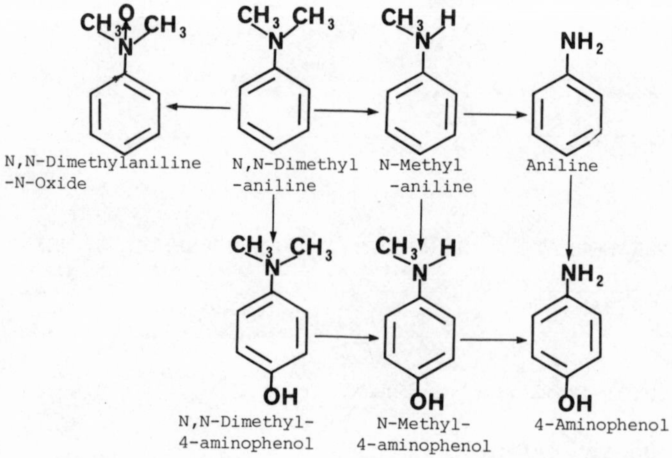

Fig.1   The Metabolism of N,N-Dimethylaniline

Tissue Preparations

Adult male Dunkin Hartley Albino guinea pigs (600g) were used as sources of tissue. Washed microsomes were prepared using the methods of Gorrod et al.,(1975).

Incubation Procedures

Incubations were carried out aerobically in 25ml Erlenmeyer flasks fitted with ground glass stoppers, at 37°C using a shaking water bath. A typical incubate contained substrate 5 μmoles, tissue 1ml ($\equiv$ 0.5g original tissue), NADP 2 μmoles, glucose-6-phosphate 10μmoles, glucose-6-phosphate dehydrogenase 3 units, magnesium chloride 20μmoles and phosphate buffer (Sorensens 0.2m pH 7.4). Inhibitors, when employed, were made up in phosphate buffer and added to the incubate. In all experiments, incubates had a final volumes of 3.5ml. Incubations were terminated after 15 mins unless otherwise indicated.

Analysis of Anilines, N,N-Dimethylaniline-N-oxide and Aminophenols.

The G L C methods of Gorrod et al.,(1975) were used to determine anilines and N,N-dimethylaniline-N-oxide. Aminophenols were determined using the method of Gorrod and Patterson (1980).

Table 1.   Incubation of the Major Metabolites of N,N-Dimethyl-
           aniline with Fortified Guinea Pig Hepatic Microsomal
           Preparations.

Substrate / Metabolite	N,N-Dimethyl aniline	N,N-Dimethyl aniline-N-oxide	N,Methyl aniline	N,N-Dimethyl-4-aminophenol
NN-Dimethyl aniline-N-oxide	65.4	-	-	0
NN-Dimethyl aniline	-	82.6	-	0
N-Methyl aniline	73.0	25.9	-	0
Aniline	0	0	ND	0
NN-Dimethyl-4-aminophenol	9.8	1.7	-	-
N-Methyl-4-aminophenol	4.6	2.4	21.7	24.3
4-Aminophenol	1.7	0	0	0

ND - Not determined

All values are nmoles of product formed per mg microsomal protein
per 15 min incubation.

RESULTS

     Incubation of N,N-dimethylaniline with fortified guinea pig
hepatic microsomal preparations produced N,N-dimethylaniline-
N-Oxide, N-methylaniline and N,N-dimethyl-4-aminophenol as primary
metabolites. N-Methyl-4-aminophenol, 4-aminophenol and aniline are
secondary products formed after an incubation lag time (Gorrod and
Gooderham 1980). To determine from which reactions N-methyl-4-
aminophenol can arise, incubation of the primary metabolites of
N,N-dimethylaniline with fortified hepatic microsomes were performed.
The results are summarized in table 1.

The incubation of both N-methylaniline and N,N-dimethyl-4-aminophenol apparently leads to the formation of appreciable amounts of N-methyl-4-aminophenol.

Effect of Incubation Time and Substrate Concentration on the in vitro Metabolism of N,N-Dimethylaniline, N-Methylaniline and N,N-Dimethyl-4-aminophenol

The effect of incubation time and substrate concentration on the in vitro metabolism of N,N-dimethylaniline, N-methylaniline and N,N-dimethyl-4-aminophenol is shown in figs 2 and 3. It is obvious from fig 2 that N-methylaniline is a better substrate for 4-hydroxylation than is N,N-dimethylaniline.

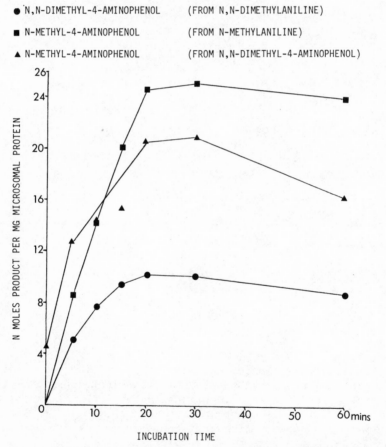

● N,N-DIMETHYL-4-AMINOPHENOL   (FROM N,N-DIMETHYLANILINE)

■ N-METHYL-4-AMINOPHENOL   (FROM N-METHYLANILINE)

▲ N-METHYL-4-AMINOPHENOL   (FROM N,N-DIMETHYL-4-AMINOPHENOL)

Fig 2   The Effect of Incubation Time on the 4-Hydroxylation of N,N-Dimethylaniline and N-Methylaniline and the N-Demethylation of N,N-Dimethyl-4-aminophenol.

● N,N-DIMETHYL-4-AMINOPHENOL   (FROM N,N-DIMETHYLANILINE)

■ N-METHYL-4-AMINOPHENOL   (FROM N-METHYLANILINE)

▲ N-METHYL-4-AMINOPHENOL   (FROM N,N-DIMETHYL-4-AMINOPHENOL)

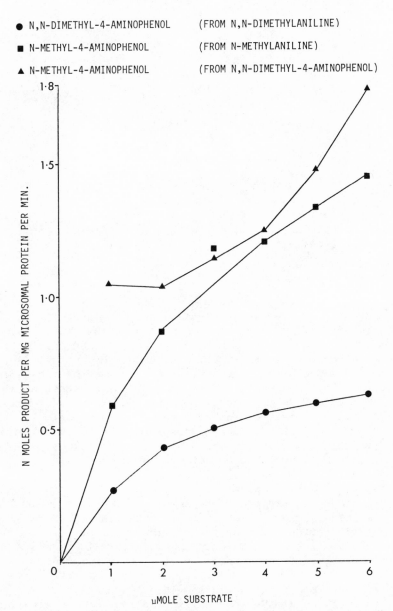

Fig 3   The Effect of Substrate Concentration on the 4-Hydroxylation
        of N,N-Dimethylaniline and N-Methylaniline and the
        N-Demethylation of N,N-Dimethyl-4-aminophenol.

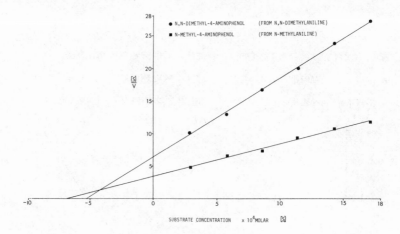

Fig 4   Hanes Plots for the 4-Hydroxylation of N,N-Dimethylaniline
        and N.Methylaniline by Guinea Pig Hepatic Microsomal
        Preparations.

It would seem that N,N-dimethyl-4-aminophenol is readily
N-demethylated to form N-methyl-4-aminophenol. However, this
reaction appears to be predominantly non-enzymic, an observation
confirmed by additional experiments in which N,N-dimethyl-4-
aminophenol was incubated in buffer in the absence of tissue and/
or cofactors. Incubation in buffer alone produced substantial
amounts of N-methyl-4-aminophenol, whilst the presence of tissue
and particularly cofactors inhibited this N-demethylation.
Incubation of N,N-dimethyl-4-aminophenol for longer than 15 mins
resulted in large amounts of 4-aminophenol being formed especially
in the absence of tissue and cofactors.

     Hanes plots were used in the kinetic analyses of the enzymic
4-hydroxylations and are presented in fig.4. Apparent Km and V. max
values for the 4-hydroxylation of N,N-dimethylaniline were 5.3×10^{-4}M
and 0.83 nmoles product formed per mg microsomal protein per min,
and for 4-hydroxylation of N-methylaniline 6.9×10^{-4}M and 2.01
nmoles product formed per mg microsomal protein per min.

Effect of some Inhibitors on the in vitro 4-Hydroxylation of
N,N-Dimethylaniline and N-Methylaniline and on the N-Demethylation
of N,N-Dimethyl-4-aminophenol

     Table 2 illustrates the effect of a number of inhibitors on the
4-hydroxylation of N,N-dimethylaniline and N-methylaniline and the
N-demethylation of N,N-dimethyl-4-aminophenol.

Table 2. The Effect of Various Inhibitors on the Formation of
          N,N-Dimethyl-4-aminophenol from N,N-Dimethylaniline and
          N-Methyl-4-aminophenol from N-Methylaniline and
          N,N-Dimethyl-4-aminophenol using Guinea Pig Hepatic
          Microsomes.

Substrate ⟍ Inhibitor	Inhibitor Concentration	N,N-Dimethyl aniline	N-Methyl aniline	N,N-Dimethyl -4-aminophenol
Control	NIL	100	100	100
SKF 525A	10^{-5}M	92	100	117
	10^{-4}M	87	84	93
	10^{-3}M	69	56	102
DPEA	10^{-5}M	49	83	82
	10^{-4}M	31	69	71
	10^{-3}M	6	31	43
n-Octyla mine	10^{-5}M	89	88	118
	10^{-4}M	61	58	81
	10^{-3}M	17	24	43
MMI	10^{-5}M	92	71	123
	10^{-4}M	87	51	104
	10^{-3}M	93	36	100
Carbon monoxide, Oxygen, Nitrogen Ratio	$O_2:N_2$ (1:9)	100	100	100
	$CO:O_2:N_2$(5:1:4)	65	47	16

Values are percentages of controls

Effect of SKF 525A, DPEA and n-Octylamine

     SKF 525A DPEA and n-octylamine are known inhibitors of many
cytochrome P-450 mediated reactions (Anders and Mannering 1966,
Kiese 1966 and Testa and Jenner 1976, respectively). However, both
DPEA and n-octylamine have also been shown to enhance N-oxidation
by the microsomal flavoprotein amine oxidase, (Ziegler, Poulsen and
McKee 1971 and Ziegler, McKee and Poulsen 1973). All three compounds
were observed to reduce the 4-hydroxylation of both N,N-dimethyl-
aniline and N-methylaniline, the inhibition in all cases being
concentration dependant. DPEA and n-Octylamine, but not SKF 525A,
were observed to inhibit the N-demethylation of N,N-dimethyl-4-
aminophenol (see Table 2).

Effect of Carbon Monoxide

Carbon monoxide inhibited the 4-hydroxylation of both
N,N-dimethylaniline and N-methylaniline (see table 2), indicative
of Cytochrome P-450 involvement (Estabrook et al 1970). Carbon
monoxide also inhibited the N-demethylation of N,N-dimethyl-4-
aminophenol suggesting that this reaction is partially catalysed
by cytochrome P-450 as well as the non-enzymic demethylation
described above.

Effect of M.M.I.

Methimazole (MMI) has been reported as a specific inhibitor
of the microsomal flavoprotein amine oxidase enzyme, having no
effect on cytochrome P-450 (Ziegler and Poulsen 1979). MMI had no
effect on the 4-hydroxylation of N,N-dimethylaniline but did inhibit
the 4-hydroxylation of N-methylaniline. Furthermore this effect

Table 3   The Effect of Sodium Chloride, Superoxide Dismutase and
Catalase on the Formation of N,N-Dimethyl-4-aminophenol
from N,N-Dimethylaniline and N-Methyl-4-aminophenol from
N-Methylaniline and N,N-Dimethyl-4-aminophenol using
Guinea Pig Hepatic Microsomes.

Substrate / Incubate	N,N-Dimethyl aniline	N-Methylaniline	N,N-Dimethyl-4-aminophenol
A	100	100	100
B	71	55	106
C	76	47	153
D	73	59	167
E	68	54	107

Values are percentages of controls

A -   Substrate (Control)
B -   Substrate + Sodium chloride (0.2 M)
C -   Substrate + Sodium chloride (0.2 M) + Superoxide dismutase
      (0.06mg/ml)
D -   Substrate + Sodium chloride (0.2 M) + Superoxide dismutase
      (0.06mg/ml) +Catalase (2.3μl/ml).
E   Substrate + Sodium chloride (0.2 M) + Catalase (2μl/ml).

was accentuated with increasing inhibitor concentration. The
N-demethylation of N,N-dimethyl-4-aminophenol was unaffected by
M.M.I. (see Table 2)

Effect of Superoxide Dismutase

It has been proposed that superoxide anion is the active
oxygenating species in cytochrome P-450 oxidations (Strobel and
Coon 1971). However, superoxide dismutase, with and without
catalase, produced no inhibition, other than that due to the presence
of sodium chloride, on the 4-hydroxylation of N,N-dimethylaniline
and N-methylaniline. (see Table 3). Therefore we conclude that the
superoxide anion does not play a role in these reactions.

Interestingly the effect of superoxide dismutase, particularly
in the presence of catalase, produced a pronounced enhancement of
the N-demethylation of N,N-dimethyl-4-aminophenol.

DISCUSSION

Our earlier studies on the in vitro metabolism of N,N-dimethyl-
aniline established that ring hydroxylation of the substrate occurred;
the enzyme system responsible was shown to be microsomal and located
predominantly in the liver (Gorrod and Gooderham 1980). We
proposed that the finding of N-methyl-4-aminophenol following in
vitro N,N-dimethylaniline metabolism was due to (a) 4-hydroxylation
of N-methylaniline, (b) N-demethylation of N,N-dimethyl-4-aminophenol
both of which are metabolically formed from the parent tertiary
amine.

Incubation of the primary metabolites (N-methylaniline, N,N-
dimethylaniline-N-oxide and N,N-dimethyl-4-aminophenol) with
fortified guinea pig hepatic microsomes showed this hypothesis to
be essentially correct. Small amounts of N-methyl-4-aminophenol
and N,N-dimethyl-4-aminophenol were found after incubating
N,N-dimethylaniline N oxide with microsomes, however, we believe
these arise from reduction of the substrate to the amine followed
by 4-hydroxylation. In any case the formation of N-methyl-4-amino-
phenol from N,N-dimethylaniline-N-oxide is not a major in vitro
route to the secondary aminophenol.

The conversion of N,N-dimethyl-4-aminophenol to N-methyl-4-
aminophenol is largely non-enzymic, however the presence of
cytochromes may partially catalyse the reaction, indicated by
carbon monoxide inhibition. These results are consistent with the
findings of Eyer et al., (1974) who has described the autocatalytic
oxidation of N,N-dimethyl-4-aminophenol under aerobic conditions
at pH 7.4. One of the products of this reaction was shown to be
N-methyl-4-aminophenol. Eyer et al proposed that this compound
was a breakdown product of an unstable quinoneimine.

The formation of the quinoneimine could be reversed by the presence of reducing equivalents. This would explain the inhibitory effect that cofactors have on the formation of M-methyl-4-amino-phenol from N,N-dimethyl-4-aminophenol. It has been suggested (Eyer 1980) that the equilibrium for the formation of the quinoneimine lies towards N,N-dimethyl-4-aminophenol to favour the back reaction. Therefore removing O_2^-, for example with superoxide dismutase, would stimulate formation of the quinoneimine which would lead in turn to enhanced N-methyl-4-aminophenol formation. The inhibition of N-methyl-4-aminophenol formation from N,N-dimethyl-4-aminophenol in the presence of DPEA and n-octylamine may be an artifact, the lower amounts of secondary aminophenol observed being due to the reactivity of the quinoneimine with the primary aliphatic amine inhibitors (Nagasawa and Gutmann 1959).

The evidence presented strongly suggests that the in vitro 4-hydroxylation of N,N-dimethylaniline by fortified guinea pig hepatic microsomal preparations is cytochrome P-450 mediated. This is also true for the 4-hydroxylation of N-methylaniline although the inhibition by MMI could indicate the involvement of the microsomal flavoprotein amine oxidase probably via an N-oxidised intermediate (Beckett and Belanger 1976). However, if the contribution of this enzyme had been significant the reaction should have been enhanced by n-octylamine and DPEA in line with the observations of Ziegler et al., (1971) and (1973). Since this is clearly not the case, we must conclude that the 4-hydroxylation of N-methylaniline is predominantly cytochrome P-450 mediated.

ACKNOWLEDGEMENTS

N.J.G. thanks the Science Research Council for the award of a studentship. We are grateful to Mr. L. Disley for his help with preparation of tissue.

REFERENCES

Anders M.W. and Mannering G.J., 1966,   Inhibition of drug
metabolism. 1. Kinetics of N-demethylation of ethylmorphine by
2-diethylaminoethyl 2,2-diphenylvalerate HCl (SKF525A) and related
compounds, Mol. Pharmacol, 2:319

Beckett A.H. and Bélanger P.M., 1976, Metabolic N-oxidation of
secondary and primary aromatic amines as a route to ring
hydroxylation to various N-oxygenated products and to dealkylation
of secondary amines, Biochem. Pharmacol., 25:211

Estabrook R.W., Franklin M.R. and Hildebrandt A.G., 1970. Factors
influencing the inhibitory effect of carbon monoxide on cytochrome
P-450 catalysed mixed function oxidation reactions, Annal. New
York Acad. Sci., 174:218

Eyer P., Kiese M., Lipowsky G. and Weger N., 1974,   Reactions of
4-dimethylaminophenol with haemoglobin and autooxidation of
4-dimethylaminophenol, Chem. Biol. Interact., 8:41

Eyer P. 1980,   Private Communication

Gorrod J.W. and Gooderham N.J., 1980, The in vitro metabolism
of N,N-dimethylaniline by guinea pig and rabbit tissue preparations
Eur. J. Drug Metab. In Press

Gorrod J.W. and Patterson L.H., 1980,   Evidence for p-hydroxylation
of N-ethyl-N-methylaniline, Xenobiotica, 10:603

Gorrod J.W., Temple D.J. and Beckett A.H., 1975, The metabolism
of n-ethyl,N-methylaniline by rabbit liver microsomes. The
measurement of metabolites by gas-liquid chromatography,
Xenobiotica, 5:453

Kiese M., 1966, The biochemical production of ferrihaemoglobin
forming derivatives from aromatic amines and mechanisms of
ferrihaemoglobin formation, Pharmacol. Revs., 18:1091.

Nagasawa H.T. and Gutmann H.R., 1959, The oxidation of
o-aminophenols by cytochrome C and cytochrome oxidase, J.Biol.
Chem, 234:1593

Strobel H.W. and Coon M.J., 1971, Effect of superoxide generation
and dismutation on hydroxylation reactions catalysed by liver
microsomal cytochrome P-450, J.Biol.Chem., 246:7826

Testa B. and Jenner P., 1976, Chapter 2.1., in: Drug Metabolism, Chemical and Biochemical aspects, B. Testa and P. Jenner eds., Pub. Marcel Dekker inc., N.Y.

Ziegler D.M., McKee E.M. and Poulsen L.L., 1973, Microsomal flavoprotein catalysed N-oxidation of arylamines, Drug Metab & Disposit., 1:314

Ziegler D.M. and Poulsen L.L., 1979, The liver microsomal FAD-containing monooxygenase, J.Biol.Chem 254:6449

Ziegler D.M., Poulsen L.L. and McKee E.M.,1971, Interaction of primary amines with a mixed function amine oxidase isolated from pig liver microsomes, Xenobiotica, 1:523

NICOTINE Δ1'(5)' IMINIUM ION : A REACTIVE INTERMEDIATE

IN NICOTINE METABOLISM**

A.R. Hibberd* and J.W. Gorrod

Department of Pharmacy, Chelsea College
University of London, Manresa Road
London SW3 6lx U.K.

INTRODUCTION

Nicotine Δ1'(5)' iminium ion, postulated as an intermediate in the metabolism of nicotine to cotinine (Murphy, 1973) was subsequently shown to be a discrete entity in this pathway, in vivo and in vitro (Hibberd and Gorrod, 1978; Brandange and Lindblom, 1979). The enzyme responsible for the conversion of this iminium ion to cotinine, in vitro, being "soluble" aldehyde oxidase (Brandange and Lindblom, 1979; Hibberd, 1979).

The expected reactivity of nicotine Δ1'(5)' iminium ion may allow it to react with tissue nucleophiles which could have severe consequences. This reactivity could account for the interaction of nicotine with receptor sites by either ionic or covalent binding, or it could produce further metabolic products, by reaction with compounds such as glutathione (GSH), followed by enzymic degradation to a mercapturic acid (fig 1). Thus nucleophiles may be expected to inhibit cotinine production from nicotine or the iminium ion when they are incorporated into in vitro system, and such inhibition has been observed with CN, (Murphy,1973 ; Hucker et al., 1960 ; Booth & Boyland, 1971; Gorrod et al., 1971). GSH and cysteine (Hibberd, 1979; Gorrod et al., 1971).

The aim of this study was to examine the possibility that nicotine Δ1'(5)' iminium ion could react with GSH as proposed by Gorrod and Jenner,(1975) and with cyanide as proposed by Murphy(1973).

---

* Present address Northwick Park Hospital, Harrow and The School of Pharmacy, (University of London), Brunswick Square, London, WC1 U.K.

** Supported by grant No.1194 from the The Tobacco Research Council U.S.A., inc.

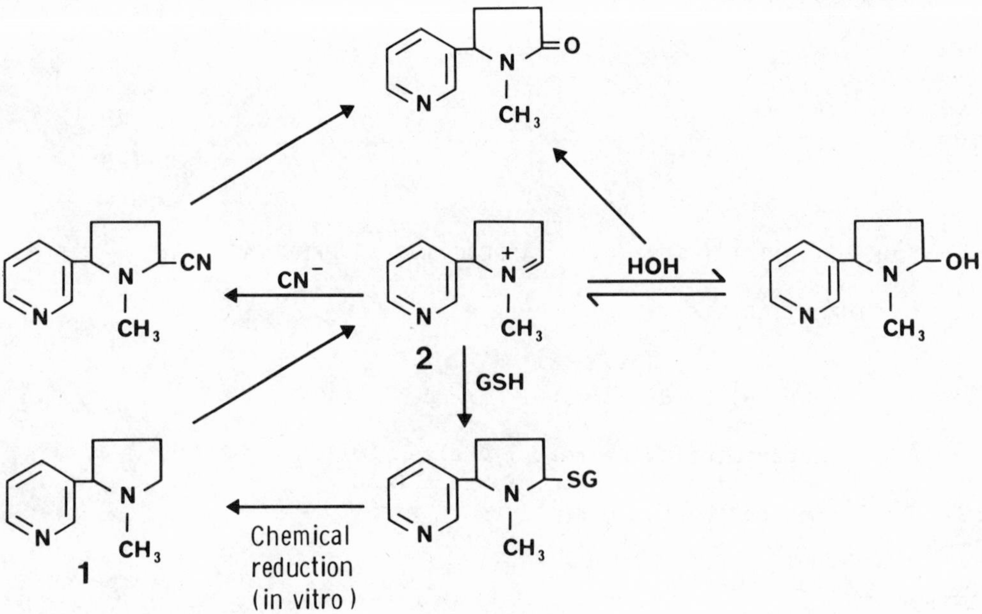

Fig.1   Some proposed reaction involved in the metabolism of
        nicotine (1) and nicotine Δ1'(5)' iminium ion (2).

METHODS

All materials were obtained as previously described   (Hibberd, (1979).

Preparation of the 5'cyano adduct

To an aqueous solution (pH<3)of crystalline nicotine Δ1'(5)' iminium
ion diperchlorate in an ice bath was added KOH (5N) until the pH
was 7.0.   An amount of KCN equimolar to the iminium ion was added
and the solution stirred for one hour, during which it was allowed
to come to room temperature.   The solution was then extracted over
NaCl (1g/5ml of solution) six times with an equal volume of
dichloromethane.   The pooled extracts were dried over anhydrous
sodium sulphate and concentrated under reduced pressure to leave
an orange brown oil.   The 5'-cyanonicotine distilled at 135°(1-2mm).
The compound was examined by mass spectrometry (MS) (Murphy, 1973 ;
Nguyen et al., 1979) and by infra-red spectroscopy (i.r.) as a film
on potassium bromide discs and by gas chromatography   (GLC).

     Incubations were prepared employing 5'cyanonicotine (1.25μmoles)
in water (0.5ml): guinea pig or rabbit hepatic 140,000 x g
supernatant (4-9mg protein/ml) (1ml); phosphate buffer pH 7.4 (2ml).
Incubations were carried out at 37° in shaking water bath for up to
60 minutes.   Control incubates without tissue were prepared to test
for any chemical formation of cotinine at each of the incubation
times.

Extraction and assay of cotinine by GLC followed previously
described methods (Beckett and Triggs, 1966) except that
extraction was carried out after the addition of 1g NaCl. Some
concentrated extracts were examined by thin layer chromatography
and areas corresponding to authentic cotinine characterised by GLC
and MS (Hibberd, 1979).

Preparation of the GSH Iminium Ion Adduct

Aqueous solutions, 2-3ml of the iminium ion (2mmoles) and 20ml
GSH (8mmoles), adjusted to pH 6.8 with NaOH (5N) whilst in an ice
bath, were mixed, the pH adjusted to 7.5-7.8 with NaOH(5N), and
constantly stirred for an hour during which time the reaction
mixture was allowed to come to room temperature. This was then
reduced to approximately 1ml under reduced pressure at 30°- 35° to
yield a concentrate which applied to two sheets of Whatman 3M paper
and the chromatograms developed (descending) in an argon filled
tank with butanol: propranol: ammonia solution 2N (2:1:1) as
solvent. The air dried chromatogram was viewed under U.V. light
(366 nm), fluorescent bands were marked and lengthwise strips were
sprayed with detecting reagents as follows; (i) Ninhydrin (0.2%w/v)
in acetone, (ii) $K_2Cr_2O_7$ 0.1M plus CH_3COOH 0.1M (equal volumes)
followed by $AgNO_3$ 0.1M; all aqueous solutions (Knight, 1958; Booth
et al., 1960; Boyland & Sims, 1962), (iii) Koenig reaction
(modified) (Papadopoulous and Kintzios, 1963). The band R_F 0.16
to 0.20 which fluoresced (blue) strongly under U.V. light also
reacted positively to all three detecting reagents as follows;
(i) mauve purple, (ii) yellow orange on red brown background,
(iii) orange. The band was removed, shredded and eluted with
4x20 ml of methanol containing ammonia solution (S.G. 0.88) (5%v/v).
Pooled eluates were filtered, concentrated at 20° and again subjected
to the same chromatographic and detection techniques as before.
The eluted compound from band R_FO. 16-0.20 was concentrated to near
dryness under reduced pressure. After addition of anhydrous
methanol (0.5 ml) the precipitated creamy brown solid was collected
and stored stored in a vacuum desiccator. Portions of the solid
were examined by (a) Paper Chromatography - following HBr (S.G.1.49)
hydrolysis under reflux at 120° for 10 hours. The hydrolysate was
evaporated under reduced pressure at 70° and the brown solid
remaining was dissolved in methanol containing 5% ammonia (S.G.0.88)
(0.25ml) and chromatographed on Whatman 3M paper using butanol-1-ol
acetic acid (glacial): water (2:1:1).   The following compounds
were applied to the same chromatogram; nicotine, cotinine, cystine,
glycine, glutamic acid, glutathione oxidised (GSSG). The detecting
reagents described above were again employed; (b) Mass spectrometry
- electron impact (EI-MS), gas chromatography EI mass spectrometry
(GC-MS) and chemical ionisation mass spectrometry (CI-MS). Mass
spectra were also obtained of nicotine, GSH, and nicotine $\Delta1'(5')$
iminium ion diperchlorate: (c) GLC - Analysis of the adduct in
aqueous methanol and following reduction (30minutes) with

acidified sodium borohydride. The ether extracts of the basified solution of the latter were concentrated to 20-30μl at 40°, both FID and N detection was used (GLC conditions were as previously described (Hibberd, 1979); (d) Melting Point - Determination (e) NMR Spectroscopy - using D_2O as solvent. Spectra were also obtained of GSH and nicotine Δ1'(5') iminium ion $diClO_4$1n D_2O using a Perkin Elmer R32 NMR Spectrometer before and after mixing. (f) U.V. Spectroscopy - Using a methanolic solution. Spectra were also obtained of nicotine Δ1'(5') iminium ion $diClO_4$ nicotine and GSH in methanolic solution using a Pye-Unicam SP800 Spectrophotometer and Pye-Unicam SP21 recorder.
Additional experiments were preformed using tandem silica cells. GSH and nicotine Δ1'(5') iminium ion dichlorate (0.3mg/ml) were separately dissolved in phosphate buffer (pH 7.0). A base line was established, the contents of the tandem cells mixed and a spectrum recorded using a 10 x scale expansion.

RESULTS AND DISCUSSION

Evidence for the formation of 5'cyanonicotine

The GLC conditions and retention times for nicotine, cotinine and the cyano compound were seen to be quite different (table 1) and clearly showed that a new compound was formed.

The GC-MS and CI-MS data for the cyano compounds plus literature values for GC-MS are recorded in table 2.

The GC-MS characteristics of the synthesised cyanonicotine was in close agreement with literature values for 5'cyanonicotine (Nguyen et al., 1979). The i.r. spectrum of the cyano compound displayed an absorption at 2400 cm^{-1}, consistent with the nitrile moiety (Murphy,1973; Willard, 1974) indicating the formation of the 5'cyano-adduct from the iminium ion. This further supports Murphy's implication of this iminium ion as an intermediate in nicotine metabolism.

Metabolic Experiments

During a study of CN^- inhibition of cotinine formation from nicotine Δ1'(5') iminium ion and hepatic "soluble" preparations decreasing inhibition with increasing incubation time was observed. This suggested that 5'cyanonicotine may have been a substrate for the soluble enzyme producing cotinine. Results from the present work showed that guinea pig, rabbit hepatic "soluble" are able to produce cotinine from 5'cyanonicotine. (fig 2)

Table 1 - GLC conditions and retention times for nicotine,
         cotinine and 5'cyanonicotine.

COMPOUND	TEMPERATURE (°C) OVEN	INJ.PORT	RETENTION TIME (MINUTES)
NICOTINE	125	250	2.5
5' CYANO-NICOTINE	125	250	3.3
COTININE	190	250	5.3

Column packed with carbowax 20M 2%, KOH 5% on chromosorb W
(AW-DMDCS; 80-100 MESH); carrier gas - nitrogen (10 psi).

Table 2   Major fragments in the mass spectra
          of 5 cyanonicotine

	eV m/e (% relative abundance)
GC-MS (i)	70 187(M+,18), 160(65), 159(37), 109(100) 108 (62), 82 (86), 80 (45)
(ii)	70 187 (M+, 23,7), 160 (65). 109 (100), 82(82.5)
CI-MS	188 (M+1$^+$, 35), 161 (100)

GC-MS   (i) Sample   Prepared from iminium ion.
MS Instrument and conditions: VG 12F mass spectrometer with VG
   Digispec 8 data system. Source temperature 140°C. Linked to
   a Pye 104 GLC, with 1 metre column as in table 1.

Helium at a flow rate of 20 ml /min. was the carrier gas; oven
temperature was 135°C and the injection block at 250°C.

(ii) Literature values (Nguyen et al, 1979) using an AEI MS-12
interfaced to a PDP8/1 computer using DS-30 software.

CI-MS = Direct inlet using isobutane reagent gas
        Instrument, VG MM 12F Emission current - 500 microamp
        Source temperature - 140°C All m/e Values are +1

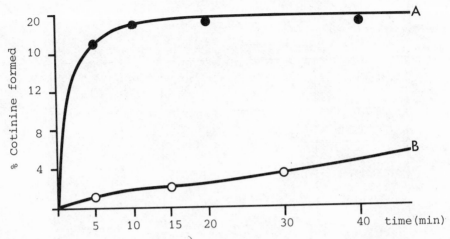

Fig.2 In vitro conversion of 5'cyanonicotine to cotinine by
(A) guinea pig and (B) rabbit 140,000xg hepatic supernatant

Guinea pig tissue produced cotinine maximally after 10 minutes
incubation, approximately 26 nmoles cotinine/mg protein (19.5% of
substrate) being formed. With rabbit tissue cotinine production
rose over 60 minutes reaching approximately 20 nmoles cotinine/mg
protein (7.4% of substrate). It can be concluded that a metabolic
pathway to cotinine, mediated by an hepatic soluble enzyme exists
via 5'cyanonicotine.

Evidence for Nicotinyl-GSH Adduct

Detection techniques provided the following evidence (i) -
amino acid (purple-mauve with ninhydrin (Smith, 1960), (ii) S-
containing (Knight, 1958; Booth et al.,1960; Boyland and Sims, 1962)
or an easily oxidised compound, (iii) pyridine containing compound
(orange with the Koenig reaction (Papadopoulous and Kintzios, 1963)
The chromatographic tank was filled with argon to minimise the
oxidation of the adduct. Chromatographic separation of the products
of HBr hydrolysis yielded the following, on the basis of their R_F
values and response to detecting reagents; cystine (due to cysteine
oxidation); GSH: GSSH: glycine and glutamic acid. Koenig positive
material was found from R_F0.48 almost to the solvent front, thus
demonstrating several pyridine compounds were present. The
tripeptide (GSH) adduct would have been expected to yield the amino
acids found after hydrolysis as described. The nicotinyl moiety
covalently bound would have been expected to release pyridinyl
compounds under the above conditions.

The MS evidence for the adduct rests on the presence of both GSH
and nicotinyl fragments in the spectra, viz: El-MS, nicotinyl
fragments, m/e 163,161,82, GSH fragments, m/e 126,83,56. Fragments
common to both entities were also seen viz m/e 85,84,57. Cl-MS,
nicotinyl fragments, m/e 177,163,121, GSH fragments, m/e 130
(Hibberd, 1979). Results from GC-MS demonstrated nicotine fragments
viz,, m/e 162 133,119,84 (Pillotti et al.,1976) at the appropriate
retention time. The mass spectrum of this peak provided good
evidence for the presence of nicotine fragments from the adduct.

    Direct injection of the compound onto GLC produced small
peaks consistent with the R_T of authentic nicotine using FID and
N detector. However, following borohydride reduction the peak
(FID and N detector) corresponding to nicotine was much larger.
This added strongly to the idea of the adduct being a nicotinyl
compound.

    The NMR spectra of GSH and the iminium ion are shown in fig.3a
and b. After mixing the solutions the H-5 (Brandange and Lindblom
1979) signal (δ9.5) of the iminium ion was lost (fig 3c) thus
suggesting that reaction at this position had occurred. Solutions
of the adduct in D_2O displayed a large singlet consistent with the
N-CH_3 group (δ1.95 - fig 3d) but no signals attributable to
pyridine protons were detected above the background. This is
consistent with a similar observation reported regarding the NMR
spectra of a GSH conjugate of Cambendazole (Wolf et al., 1980) and
may be attributable to trace amounts of paramagnetic metal ions
causing selective line-broadening (Ihnab and Bersohn 1970). Signals
attributable to GSH were, however, present (fig 3a and d)
(Lindblom, 1978). Perhaps the pyridine protons of the adduct were
in such close proximity to those of GSH (which may have wrapped
around the nicotinyl moiety) that great reduction in relaxation
times occurred so preventing the signals being seen (Shoolery,
1972). The N-CH_3 signal, however, shows that the compound
included all, or part, of the iminium ion molecule. Following
mixing of the GSH and iminium in tandem cells a change in the
UV spectra was observed between 265 and 300nm; the greatest
increase in absorbance occurred at 270nm. This indicates that
reaction between the two molecules had occurred. UV spectra
recorded for the adduct, iminium ion, nicotine and GSH in methanol
showed maximal absorbances of 225 and 255 nm for the adduct, 217
and 259 for nicotine $\Delta1'(5')$ iminium ion $diClO_4$, 215 and 260 nm
for nicotine and 210nm for GSH.

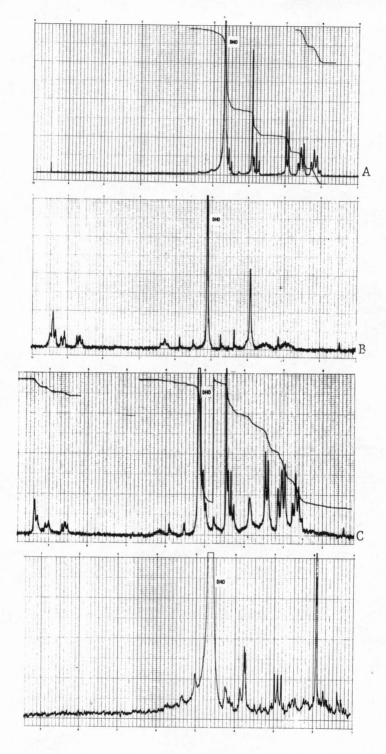

Figure 3   NMR spectra of (a) GSH, lock signal TSP. (b) Nicotine $\Delta1'(5')$ iminium ion dichlorate, lock signal external TMS. (c) Nicotine $\Delta1'(5')$ iminium ion plus freshly added GSH, lock signal external TMS, (d) isolated GSH adduct, lock signal DSS. All spectra were measured in D_2O.

◄──

   The melting point determination on the adduct showed charring from 188-264 and fusing of the tarry masses from 298-330. The melting points of the iminium ion $diClO_4$ and GSH were found to be sharp and are 230-231 and 194 respectively.

   The production of the cyano and GSH iminium ion adducts by chemical reaction in vitro show that, covalent bonding with tissue nucleophiles might occur during nicotine metabolism. The enzymic as well as chemical formation of GSH conjugates with reactive species is known to occur (Daly et al., 1972) and this may also be the case with nicotine iminium species and nucleophiles. It is thought that many reactive intermediates play important roles in the causation of toxic reactions and carcinogenesis due to their ability to react with nucleophiles e.g. in tissue proteins, DNA or RNA (Gorrod, 1979, Sims & Grover, 1974., Miller, 1970). Thus, by analogy, the reactivity of the nicotine $\Delta1'(5')$ iminium ion may imply a toxicological threat in people exposed to nicotine.

References

Beckett A.H. and Triggs E.J., 1966, Determination of nicotine and its metabolite, cotinine, in urine by gas chromatography, Nature 211:1415-1417

Booth J. and Boyland E., 1971, Enzymic oxidation of (-)-nicotine by guinea pig tissues in vitro. Biochem.Pharmacol 20:407-415

Booth J., Boyland E. and Sims P., 1960, Metabolism of polycyclic compounds: 15. The conversion of naphthalene into a derivative of glutathione by rat-liver slices. Biochem.J. 74: 117-122

Boyland E. and Sims P., 1962, Metabolism of polycyclic compounds: 20. The metabolism of phenanthrene in rabbits and rats: mercapturic acids and related compounds. Biochem.J. 84:564-570.

Brandange S. and Lindblom L., 1979a, The enzyme 'aldehyde oxidase' is an iminium oxidase. Reaction with nicotine $\Delta1'(5')$ iminium ion. Biochem. Biophys. Res. Com. 91:991-996.

Brandange S. and Lindblom L., 1979b, Synthesis, structure and stability of nicotine $\Delta1'(5')$ iminium ion, - an intermediary metabolite of nicotine. Acta. Chem. Scand. B33:187-191

Daly J.W., Jerina D.M. and Witkop B., 1972, Arene oxides and the
NIH shift: The metabolism, toxicity and carcinogenicity of aromatic
compounds. Experientia 28:1129-1264

Gorrod J.W., 1979, Toxic products produced during metabolism of
drugs and foreign compounds in Drug Toxicity. J.W. Gorrod ed., Pub
Taylor and Francis, London.

Gorrod J.W. and Jenner P., 1975, The metabolism of tobacco
alkaloids in Essays in Toxicology vol 6 W.J. Hayes ed. Pub Academic
Press.

Gorrod J.W.,Jenner P.J., Keysell G. and Beckett A.H., 1971,
Selective inhibition of alternative oxidative pathways of nicotine
metabolism in vitro. Chem. Biol. Interact. 3:269-270

Hibberd A.R.,1979. Studies on the metabolism and excretion of
nicotine and related compounds. Ph.D. thesis, University of London.

Hibberd A.R. and Gorrod J.W. 1978. The conversion of nicotine
Δ1'(5') iminium ion to cotinine in vivo and in vitro. Proc. 7th
Intern. Pharmac. Conf. Paris p.651 (Abstract No.2082)

Hucker H.B., Gillette J.R. and Brodie B.B., 1960. Enzymic pathway
for the formation of cotinine, a major metabolite of nicotine in
rabbit liver. J.Pharmacol. Exp. Therap. 129:94-100

Ihnab M. and Bersohn R., 1970, 'H Nuclear magnetic resonance study
of the copper (II) - carnosine complex in aqueous solution.
Biochemistry 9:4555-4566

Knight R.H. and Young L., 1958, Biochemical studies of toxic agents
11. The occurence of premercapturic acids. Biochem.J. 70:111-119

Lindblom L., 1978, Personal communication.

Miller J.A., 1970, Carcinogenesis by chemicals : An overview
Cancer Res. 30:559-576

Murphy P.J., 1973, Enzymatic oxidation of nicotine to nicotine
Δ1'(5') iminium ion. J.Biol. Chem. 248:2796-2800.

Nguyen T.L., Gruenke L.D. and Castagnoli N., 1979, Metabolic
oxidation of nicotine to chemically reactive intermediates.
J.Med.Chem. 22:259-263.

Papadopoulous N.M. and Kintzios J.A., 1963, Formation of metabolite
from nicotine by a rabbit liver preparation. J. Pharmacol. Exp.Ther
140:269-277

Pilotti A., Enzell C.R., McKennis H., Bowman E.R., Dufra E. and
Holmstedt B., 1976.
Studies on the identification of tobacco alkaloids, their mammalian
metabolites and related compounds by gas chromatography - mass
spectrometry. Beitr. Tabakforsch 8:339-349

Shoolery J., 1972. A basic guide to NMR pub. Varian Associates,
California U.S.A.

Sims P. and Grover P.L., 1974, Epoxides in polycyclic aromatic
metabolism and carcinogenesis. Advn. Cancer Res. 20:165-274

Smith I.,1960, Chromatographic and Electrophoretic Techniques,
vol.1 Pub. Heinman, London.

Willard H.H., Merritt L.L. and Dean J.A., 1974. Instrumental
methods of analysis. Pub. Van Nostrand Co. London U.K.

Wolf D.E., Vandenheuvel W.J.A., Tyler T.R., Walker R.W.,
Koniuszy F.R., Gruber V., Arison B.H., Rosegay A., Jacob T.A. and
Wolf F.J., 1980, Identification of a glutathione conjugate of
cambendazole formed in the presence of liver microsomes. Drug
Metab. and Disp. 8:131-138

ENZYME CHARACTERISTICS OF IMINE N-HYDROXYLATION*

J.W. Gorrod and M. Christou

Department of Pharmacy, Chelsea College University of
London,
Manresa Road,   London, SW3 6LX U.K.

INTRODUCTION

Imines have been implicated as intermediates in the metabolic
deamination of aliphatic amines such as d-amphetamine, (Hucker et
al., 1971, Hucker 1973), being subsequently oxygenated to oximes.
Oxygen-18 studies reported by Parli et al., (1971a) and (1973)
further supported the possible role of imines as intermediates in
oxime and ketone formation during the deamination of a number of
primary amines. Hucker et al., (1978), proposed that metabolic
N-oxidation of an imine may lead to the formation of a hydrocarbon.
Previously, Parli et al., (1971b) showed that 2,4,6, trimethylacetophenoneimine was converted metabolically to the corresponding
oxime.

The purpose of the present study was to extend these observations
by studying the metabolic N-hydroxylation of a series of stable
acetophenoneimines to the corresponding oximes using hepatic microsomal
fractions of various mammalian species. Furthermore, it was desirable
to elucidate the enzymology of this type of N-hydroxylation by
consideration of species differences and influence of inducing agents,
inhibitors and activators and thus confirm the findings of Parli et
al., (1971b) who were the first to report that the N-hydroxylation
of 2,4,6 trimethylacetophenoneimine to an oxime was mediated via
cytochrome P-450.

---

*   This work was supported by a postgraduate research studentship
    to M. Christou from Chelsea College, (University of London).

EXPERIMENTAL

Materials

NADP, G6P and G6PD were purchased from Boeringer Mannheim.

Chemical Syntheses

The various imines and oximes used in this study, namely: 2,6 dimethyl-; 2,4,6 trimethyl-; 2,3,5,6 tetramethyl-; 2,3,4,5,6 pentamethyl-and 2,6 dichloro-acetophenoneimines and oximes were synthesised in our laboratory using experimental procedures based on methods described by Pearson and Keaton (1963) and Hauser and Hoffenberg (1955) and will be described in detail elsewhere.

Preparation of Enzymes and Pretreatment of Animals

Livers were obtained from New Zealand White rabbits, Wistar strain rats, Laboratory animal centre Albino mice and male Dunkin-Hartley Albino guinea pigs. Hepatic microsomes and other cell fractions were prepared at 0°C by standard sedimentation procedures (Gorrod et al., 1975). The washed microsomal fractions were resuspended in isotonic tris KCl buffer pH 7.4 equivalent to 1g of original liver per 2ml of suspension. For studies on enzyme induction microsomes were prepared on day 5 from the livers of female rats receiving (80mg/kg ip) Phenobarbitone for 3 days or a single dose of Arochlor 1254 (100mg/kg ip) on day 1. Similarly microsomes were prepared on day 2 after a single dose of 3-methylcholanthrene (40mg/kg ip). The protein and cytochrome P-450 content of microsomal tissues were determined by standard methods.

Tissue Incubations

The 3.5 ml incubation mixture contained 5 μmol.of substrate in 0.5ml distilled water, 2μmol. of NADP, 10μmol. of G-6-P, 1 unit of G-6-P-D, 20μmol. $MgCl_2$ in 2.0ml of phosphate buffer pH 7.4 and 1 ml of tris KCl buffer containing microsomes equivalent to 0.5g of liver or other tissue. The reaction mixture was then incubated with shaking in air at 37°C for 30 minutes. For qualitative experiments, the flasks were incubated at 37°C for 1 hour. In all experiments, suitable controls were carried out.

Analytical Methods

All reactions were terminated by the addition of 1.0ml of 1N HCl. The reaction mixtures were extracted with 3 x 4 ml twice distilled diethyl ether and the combined extracts were evaporated to 10-20 μl. Under these conditions any unreacted imine was not extracted into the organic phase.

Thin layer chromatography (T.L.C.) was used for the separation and identification of the oximes formed metabolically; by directly comparing the Rf values of the metabolites with those of authentic oximes. The oximes were localised by UV 254 nm light and with suitable detection reagents.

For quantitative determinations direct specific G.L.C. methods were used. At the end of the incubation period, and after the enzymic activity was terminated, 5µl of a methanolic solution of internal marker, (equivalent to 250nmoles of another oxime with similar extraction properties), were introduced into the reaction mixture and the extractions were carried out as described earlier. The concentrated extract (1µl) was injected into a Perkin Elmer F33 gas Chromatograph containing a 2m coiled glass column packed with either (a) 3% OV 17 on chromosorb G or (b) 3% UC W 98 on Diataport S and the peak height ratio of the test compound to internal marker obtained. The amount of oxime formed metabolically was estimated from calibration curves, previously obtained using incubates spiked with known amounts of authentic oximes. All determinations were of total oxime formation (syn + anti).

RESULTS

General   Preliminary qualitative experiments showed that liver microsomes from various mammalian species, including the guinea pig, readily catalysed the N-hydroxylation of all imines to their corresponding oximes.

Table 1. N-hydroxylation of 2,6 Dimethylacetophenoneimine by
          Various Whole Organ Homogenates from Rabbit

Whole organ homogenate	nmoles/ /incubate	% oxime formation	% of liver activity
Liver	553	11.10	100
Lung	240	4.96	44.8
Kidney	23	0.46	4.2
Intestine	22	0.43	4.0
Spleen	20	0.40	3.6
Brain	12	0.24	2.2
Heart	6	0.12	1.1
Bladder	5	0.11	1.0

Table 2. N-hydroxylation of 2,6 Dimethylacetophenoneimine by
various Cell Fractions from Rabbit Liver Homogenate

Liver cell fraction	nmoles/ incubate	% oxime formation	% liver activity
Whole liver homogenate	553	11.1	100
Nuclei	300	6.0	54.2
Mitochondria	48	0.9	8.7
9,000 xg supernatant	578	11.6	104
140,000xg soluble fraction	17	0.3	3.1
microsomes	460	9.2	83.2
washed microsomes	485	9.7	87.7

Organ and Cellular Distribution of Imine N-hydroxylase.

     This was investigated using tissue preparations of homogenates
of various organs from the rabbit. The results outlined in tables
1 and 2 establish that imine N-hydroxylation occurs mainly in the
liver, although considerable activity has been found in the lung and
that imine N-hydroxylase resides in the microsomal fraction.

The Effect of Various Potential Inhibitors, Activators and Inducing
Agents.

     The effect of incorporation of potential metabolic inhibitors
and activators in the incubation medium is summarised in table 3.
The effect of incubation in an atmosphere containing Carbon monoxide
is illustrated in fig. 1.

     The effect of pretreatment of animals with inducing agents is
represented diagrammatically in fig. 2, showing the effect of the
inducers on the in vitro N-hydroxylation of imines in female rats
and also their effect on protein and cytochrome P-450 content of the
microsomal preparations. Phenobarbitone was found to induce
N-hydroxylation and there was a parallel rise in protein concentration
and cytochrome P-450 levels. 3-Methylcholanthrene had no appreciable
effect on N-hydroxylation and minimal effect on protein and
cytochrome P-450. Arochlor 1254 was the strongest inducer of oxime
formation and had the greatest effect on protein and cytochrome
P-450 levels.

Table 3. Effect of Incorporating Potential Inhibitors or
Activators in the Incubation Mixture.

N.D. (Not Determined)

Compound added to incubation mixture	Conc. in moles	% N-hydroxylation (Control = 100%)	
		2,6 dimethyl-acetophenoneimine	2,6 dichloro-acetophenoneimine
Control		100	100
DPEA	10^{-3}	26	46
SKF 525A	10^{-3}	53	76
N-Octylamine	10^{-3}	44	75
1-Naphthylthiourea	10^{-3}	104	115
Methimazole	10^{-3}	60	70
Quinuclidine	10^{-3}	78	88
Bromazepam	10^{-3}	94	80
Dabco	10^{-3}	84	92
Iproniazid	10^{-3}	84	76
α-Naphthoflavone	10^{-3}	84	87
β-Naphthoflavone	10^{-3}	N.D.	405
Aniline	10^{-3}	88	80
Acetone	0.5	96	90

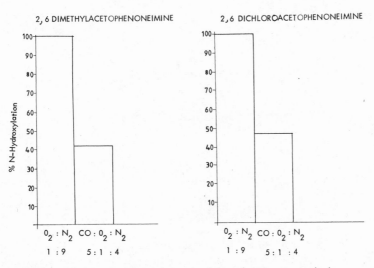

Fig.1 Effect of incubation in an atmosphere containing carbon
monoxide.

DISCUSSION

This paper presents evidence that hepatic microsomes of different species catalyse the conversion of a series of acetophenoneimines to the corresponding oximes. These findings confirm and extend the observations of Parli et al., (1971b) who were the first to report that an imine can be N-hydroxylated to an oxime, by rat and rabbit liver microsomes.

Although our present results do not provide any further direct evidence regarding the role of imines as intermediates of oxime formation from aliphatic amines, the fact that several stable imines have been established as oxime precursors (enzymatically) does support the proposals of earlier workers (Hucker et al., 1971; Hucker 1973, Parli et al., 1971a,1971b, 1973) that imines are possible intermediates in the deamination of primary aliphatic amines.

The importance of our present findings lies in the elucidation of the enzymology of imine N-hydroxylation. The fact that the guinea pig readily catalysed the N-hydroxylation of the imines suggests that cytochrome P-450 and not P-448 is involved in this conversion.

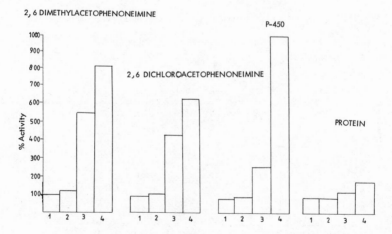

Fig. 2   The effect of 3-methylcholanthrene; phenobarbitone and
         Arochlor 1254 on the in vitro N-hydroxylation of imines in
         female rat.
         (1= Control 2=3-MC pretreatment 3=phenobarbitone pretreatment
         (4= Arochlor pretreatment)

It was confirmed that the reaction was a typical mixed function mono-oxygenase requiring oxygen and an NADPH regenerating system, in agreement with Parli et al., (1971b).

Studies directed towards the characterisation of the imine N-hydroxylase activity of microsomes using different inhibitors and activators produced some interesting results.

DPEA was found to be a potent inhibitor of this reaction in agreement with the findings of Parli et al., (1971b). Ziegler et al., (1971) reported that DPEA has an activating effect on the N-oxidation of tertiary N,N dialkylanilines; thus imine N-hydroxylation appears to be distinct from the N-oxidation of dialkylanilines which has been shown by Ziegler et al., (1969) & Patterson and Gorrod (1977) to be catalysed by a microsomal flavoprotein.

SKF 525A was also found to inhibit this N-hydroxylation reaction. Again, this compound has been shown to inhibit cytochrome P-450 dependent metabolic reactions, (Anders and Mannering 1966), but to have no effect on metabolic N-oxidation reactions involving the flavoprotein containing amine oxidase (Ziegler et al., 1966). This compound has also been reported to inhibit the N-oxidation of pyridines (Gorrod and Damani 1979), a reaction which has now been established as being cytochrome P-450 mediated.

N-octylamine, which was shown to enhance N-oxidation of tertiary dialkylanilines (Ziegler et al., 1969), was found to be a potent inhibitor of imine N-hydroxylation, in agreement with the findings of Smith and Gorrod (1978) on the N-hydroxylation of anilines, and Gorrod and Damani (1979) on the N-oxidation of pyridines, both reactions being mediated via cytochrome P-450.

Quinuclidine had only a small inhibitory effect. This compound according to Gorrod's classification (Gorrod 1973), belongs to group 1C and its N-oxidation is thought to be predominantly mediated via the flavoprotein containing amine oxidase. Since the evidence accumulated so far suggests that N-hydroxylation is mediated via cytochrome P-450, the relatively small effect is to be expected.

1-Naphthylthiourea has been reported to be a specific inhibitor of the flavoprotein containing amine oxidase (Ziegler and Mitchell (1972). This compound failed to inhibit imine N-hydroxylation. However, MMI another inhibitor of the flavoprotein amine oxidase, caused some inhibition of imine N-hydroxylation.

Aniline was used as a potential inhibitor since it is thought that aniline N-hydroxylation is mediated via cytochrome P-450. At a concentration of 10^{-3} it did cause a small inhibition.

α-Naphthoflavone caused some inhibition, but β-Naphthoflavone greatly activated the N-hydroxylation of 2,6 dichloroacetophenone-imine. β-Naphthoflavone has previously been reported to be a cytochrome P-450 inducer.

Bromazepam and acetone failed to significantly alter the N-hydroxylation of the imines studied.

It was also shown that the reaction rate of imine N-hydroxylation is decreased to about 40% when incubations were carried out in an atmosphere containing Nitrogen: oxygen: carbon monoxide 4:1:5 by volume. Again, N-oxidation of N,N diethylaniline is not affected by carbon monoxide, whereas the N-hydroxylation of aniline has been reported to be inhibited by about 40% in the presence of carbon monoxide (Hlavica et al., 1969). This observation further substantiates the suggestion that imine N-hydroxylation is mediated via cytochrome P-450.

In addition, enzyme activity was found to be substantially increased by previous treatment of animals with phenobarbitone, whereas treatment with 3-methylcholanthrene had no effect.

All these observations are consistent with the conclusion that the conversion of imine to oxime is catalysed by a typical cytochrome P-450 dependent hepatic microsomal N-oxidase and not the 3-methylcholanthrene inducible P-448 or the flavoprotein containing amine oxidases, and emphasise the multiplicity of microsomal N-oxidases (Gorrod 1978).

REFERENCES

Anders M.W. and Mannering G.J., 1966, Inhibition of drug metabolism 1. Kinetics of N-demethylation of ethylmorphine by 2-diethylamino-ethyl 2,2-diphenylvalerate HCl (SKF 525A) and related compounds, Molec.Pharmacol., 2:319

Gorrod J.W., 1973, Differentiation of various types of biological oxidation of nitrogen in organic molecules, Chem.Biol.Interact., 7:289

Gorrod J.W., 1978, On the Multiplicity of microsomal N-oxidase systems, in "Mechanisms of oxidizing enzymes" T.P. Singer and R.N. Ondavza, ed., Elsevier North-Holland Inc. N.Y.

Gorrod J.W., Temple D.J. and Beckett A.H., 1975, The metabolism of N-ethyl- N-methylaniline by rabbit liver microsomes: The measurement of metabolites by gas-liquid chromatography, Xenobiotica, 5:453

Gorrod J.W. and Damani L.A., 1979, The effect of various potential inhibitors, activators and inducers on the N-oxidation of 3-substituted pyridines in vitro, Xenobiotica, 4:219

Hauser C.R., and Hoffenberg D.S., 1955, Formation of certain mesityl ketoximes from ketimines. Beckmann rearrangements, J.Am.Chem.Soc., 77:4885

Hlavica P., Kiese M., Lange G. and Mor G., 1969, The effect of carbon monoxide on the N-hydroxylation of aniline by microsomes of the rabbit liver, Arch. Expt. Pathol. and Pharmacol. 263

Hucker H.B., 1973, Phenylacetoneoxime - An intermediate in amphetamine deamination, Drug Metabolism and Disp., 1:332

Hucker H.B., Michniewicz B.N.I. and Rhodes R.E., 1971, Phenyl-2-propanoneoxime, an intermediate in the oxidative deamination of d-(+)-amphetamine, Biochem.Pharmacol., 20:2123

Hucker H.B., Balletto A.J., Christy M.E. and Anderson P.S.,1978, A novel deamination reaction: Metabolism of 9,10-dihydro-9,10,11, trimethylanthracen-9,10-imine to 9,10-dimethylanthracene, in "Biological Oxidation of Nitrogen." J.W. Gorrod ed, Elsevier North-Holland Inc.NY.

Parli C.J., Wang N. and Mahon R.E., 1971a, The mechanism of the oxidation of d-amphetamine by rabbit liver oxygenase. Oxygen-18 studies, Biochem.Biophys.Res.Commun. 43:1204

Parli C.J., Wang N., and McMahon R.E., 1971b, The enzymatic N-hydroxylation of an imine. A new cytochrome P-450-dependent reaction catalysed by hepatic microsomal monooxygenases, J.Biol. Chem. 246:6953

Parli C.J., and McMahon R.E., 1973, The mechanism of microsomal deamination: Heavy isotope studies, Drug Metabolism and Disp. 1:337

Patterson L.H., and Gorrod J.W., 1977, In vitro C- and N-oxidation of 4-substituted unsymmetrical aromatic amines, Naunyn-Schmiedeberg's Arch.Pharmac., 297 (Suppl.II) R5.

Pearson D.E., and Keaton O.D., 1963, Lethargic reactions I. The preparation of hindered oximes, J.Org.Chem. 28:1557

Smith M.R. and Gorrod J.W., 1978, The microsomal N-oxidation of some primary aromatic amines in "Biological Oxidation of Nitrogen", J.W. Gorrod, ed. Elsevier North-Holland. Inc.NY.

Ziegler D.M., Mitchell C.H., and Jollow D., 1969, The properties
of a purified hepatic microsomal mixed function amine oxidase, in
"Microsomes and Drug Oxidations", J.R. Gillette, A.H. Conney,
G.J. Cosmides, R.W. Estabrook, J.R. Fouts and G.J. Mannering ed.
Acad.Press. NY.

Ziegler D.M., Poulsen L.L. and McKee E.M., 1971, Interaction of
primary amines with a mixed function amine oxidase isolated from
pig liver microsomes, Xenobiotica 1:523

Ziegler D.M. and Mitchell C.H., 1972, Microsomal oxidase IV.
Properties of a mixed function amine oxidase isolated from pig
liver microsomes, Arch.Biochem.Biophys. 150:116

THE EFFECT OF SUBSTRATES AND INHIBITORS OF MONOAMINE OXIDASE ON HEPATIC DIMETHYLNITROSAMINE METABOLISM AND MUTAGENICITY

Brian G. Lake, Ian R. Rowland, Rosalyn A. Harris, Michael A. Collins, John C. Phillips, Richard C. Cottrell and Sharat D. Gangolli

The British Industrial Biological Research Association Woodmansterne Road, Carshalton, Surrey SM5 4DS, U.K.

Dimethylnitrosamine (DMN) is a potent hepatotoxin and hepato-carcinogen in the rat (Magee and Barnes, 1967) and has also been shown to be a bacterial mutagen (Malling and Frantz, 1974). It is generally recognised that DMN requires metabolic activation to exert its biological effects (Heath, 1962; Magee and Barnes, 1967). Although it has been suggested that the hepatic metabolism of DMN is catalysed by enzymes of the cytochrome P-450 dependent mixed function oxidase (m.f.o.) complex (Argus et al., 1976; Czygan et al., 1973), recent studies have indicated that several pathways of DMN metabolism, including some independent of cytochrome P-450, exist in rat liver (Godoy et al., 1978; Kroeger-Koepke and Michejda, 1979; Lake et al., 1976). In this communication we demonstrate that substrates and inhibitors of monoamine oxidase (MAO, EC 1.4.3.4) are potent inhibitors of both DMN metabolism and DMN-induced mutagenesis.

Hepatic postmitochondrial supernatant fractions from male Sprague-Dawley rats (80-150g) were prepared in 0.154M KCl containing 50mM Tris-HCl buffer pH 7.4 as described previously (Lake et al., 1976). Assays of DMN demethylase, ethylmorphine N-demethylase and aniline 4-hydroxylase were determined at a substrate concentration of 5mM in postmitochondrial supernatant fractions as described else-where (Lake et al., 1976, 1979). The assay of DMN -, aflatoxin B_1 - and benzo(a)pyrene-induced mutagenicity was a modification of the mutagenesis test in liquid culture devised by Ames et al., (1975) as described elsewhere (Rowland et al., 1980).

The addition of aminoacetonitrile and isoxazole, which are known inhibitors of MAO (Riceberg et al., 1975; Bolt et al., 1974),

markedly inhibited the metabolism of DMN to formaldehyde by rat
hepatic postmitochondrial supernatant fractions (Table 1). However,
both compounds had little effect on the activities of the two m.f.o.
enzymes measured, namely ethylmorphine N-demethylase and aniline
4-hydroxylase. Similarly, benzimidazole and indazole, which are
inhibitors of rat hepatic microsomal MAO (B.G. Lake, unpublished
observations), inhibited the metabolism of DMN but not that of the
two m.f.o. substrates. DMN metabolism was also inhibited by the
addition of the MAO substrates 2-phenylethylamine, tryptamine and
tyramine (Table 1) but not by cadaverine which is a substrate of
diamine oxidase (Hill and Bardsley, 1975).

Table 1. Effect of Some Inhibitors on DMN Demethylase, Ethylmorphine
          N-Demethylase and Aniline 4-Hydroxylase Activities in
          vitro.

Compound	Concn. (mM)	% Control Activity[a]		
		DMN demethylase	Ethylmorphine N-demethylase	Aniline 4-hydroxylase
Aminoacetonitrile	10	30[b]	100	90
Isoxazole	10	55[b]	95	100
Benzimidazole	0.5	70[b]	95	100
Indazole	0.0025	40[b]	100	90
2-Phenylethylamine	0.05	50[b]	100	85
Tryptamine	0.05	50[b]	105	85
Tyramine	0.1	60[b]	105	105
Cadaverine	1	105	120	110

[a]Results are presented as the mean of 4 determinations in the
presence of inhibitor expressed as a percentage of the mean in the
absence of inhibitor.
[b]$P < 0.01$ (Student's t test).

     Both MAO substrates and inhibitors were potent inhibitors of
DMN-induced mutagenesis in the Ames test, whereas the diamine
oxidase substrate cadaverine had no effect (Table 2). In contrast
to the inhibition of DMN-induced mutagenesis neither MAO substrates
nor MAO inhibitors had any marked effect on the mutagenesis of
either aflatoxin B_1 or benzo(a)pyrene. Both these carcinogens are
considered to be metabolically activated by enzymes of the m.f.o.
complex (Krieger et al., 1975; Wood et al., 1976).

Table 2. Effect of Some MAO Substrates and Inhibitors on the
          Mutagenicity of DMN, Aflatoxin B_1 and Benzo(a)pyrene in
          the Ames Test.

Compound	Concn. (mM)	Mutagenicity (% of Control)[a]		
		DMN	Aflatoxin B_1	Benzo(a)pyrene
2-Phenylethylamine	0.1	10	95	85
Tryptamine	0.1	0	105	95
Tyramine	0.1	15	80	95
Benzimidazole	0.1	20	90	95
Indazole	0.02	15	85	110
Cadaverine	1	95	n.d.[b]	n.d.

[a]Results are presented as the number of mutants induced by either
DMN (50mM), aflatoxin B_1 (0.0013 mM) or benzo(a)pyrene (0.02 mM)
in the presence of inhibitor expressed as a percentage of the
bacterial mutants in the absence of inhibitor. None of the
inhibitors affected the survival of the Salmonella typhimurium
TA 100 tester strain.
[b]n.d., not determined.

     The results of these studies demonstrate that a number of
compounds of diverse chemical structure inhibit the metabolism and
mutagenesis of DMN at concentrations which had little effect on
several cytochrome P-450 dependent processes. As these compounds
are either known substrates or inhibitors of MAO the results may
suggest the participation of an enzyme(s) with MAO-like properties
in the hepatic metabolism/bioactivation of this nitrosamine.

ACKNOWLEDGEMENT

     This work forms part of a research project sponsored by the
U.K. Ministry of Agriculture, Fisheries and Food to whom our thanks
are due. The results of the research are the property of the
Ministry of Agriculture, Fisheries and Food and are Crown Copyright.

REFERENCES

Ames, B.N., McCann, J., and Yamasaki, E., 1975, Methods for
          detecting carcinogens and mutagens with the Salmonella/
          mammalian-microsome mutagenicity test, Mutation Res., 31:347.

Argus, M.F., Arcos, J.C., Pastor, K.M., Wu, B.C., and Venkatesan, N., 1976, Dimethylnitrosamine - demethylase : absence of increased enzyme catabolism and multiplicity of effector sites in repression. Hemoprotein involvement, Chem.-Biol. Interactions, 13:127.

Bolt, A.G., Ghosh, P.B., and Sleigh, M.J., 1974, Benzo-2,1,5-oxadiazoles - a novel class of heterocyclic monoamine oxidase inhibitors, Biochem. Pharmacol., 23:1963.

Czygan, P., Greim, H., Garro, A.J., Hutterer, F., Schaffner, F., Popper, H., Rosenthal, O., and Cooper, D.Y., 1973, Microsomal metabolism of dimethylnitrosamine and the cytochrome P-450 dependency of its activation to a mutagen, Cancer Res., 33:2983.

Godoy, H.M., Diaz Gomez, M.I., and Castro, J.A., 1978, Mechanism of dimethylnitrosamine metabolism and activation in rats, J. Natl. Cancer Inst., 61:1285.

Heath, D.F., 1962, The decomposition and toxicity of dialkylnitrosamines in rats, Biochem. J., 85:72.

Hill, C.M., and Bardsley, W.G., 1975, Histamine and related compounds as substrates of diamine oxidase (histaminase), Biochem. Pharmacol., 24:253.

Krieger, R.I., Salhab, A.S., Dalezios, J.I., and Hsieh, D.P.H., 1975, Aflatoxin B_1 hydroxylation by hepatic microsomal preparations from the rhesus monkey, Food Cosmet. Toxicol., 13:211.

Kroeger-Koepke, M.B., and Michejda, C.J., 1979, Evidence for several demethylase enzymes in the oxidation of dimethylnitrosamine and phenylmethylnitrosamine by rat liver fractions, Cancer Res., 39:1587.

Lake, B.G., Phillips, J.C., Harris, R.A., and Gangolli, S.D. 1979, The effect of ammonium sulfate on the metabolism of dimethylnitrosamine and other xenobiotics by rat hepatic microsomes, Drug Metab. Dispos., 7:181.

Lake, B.G. Phillips, J.C., Heading, C.E., and Gangolli, S.D., 1976, Studies on the in vitro metabolism of dimethylnitrosamine by rat liver, Toxicology, 5:297.

Magee, P.N., and Barnes, J.M., 1967, Carcinogenic nitroso compounds, Adv. Cancer Res., 10:163.

Malling, H.V., and Frantz, C.N., 1974, Metabolic activation of dimethylnitrosamine and diethylnitrosamine to mutagens, Mutation Res., 25:179.

Riceberg, L.J., Simon, M., Van Vunakis, H., and Abeles, R.H., 1975, Effects of aminoacetonitrile, an amine oxidase inhibitor, on mescaline metabolism in the rabbit, Biochem. Pharmacol., 24:119.

Rowland, I.R., Lake, B.G., Phillips, J.C., and Gangolli, S.D., 1980, Substrates and inhibitors of hepatic amine oxidase inhibit dimethylnitrosamine-induced mutagenesis in Salmonella typhimurium, Mutation Res., 72:63.

Wood, A.W., Levin, W., LU, A.Y.H., Yagi, H., Hernandez, O., Jerina,
        D.M., and Conney, A.H., 1976, Metabolism of benzo(a)pyrene
        and benzo(a)pyrene derivatives to mutagenic products by
        highly purified hepatic microsomal enzymes, J. Biol. Chem.,
        251:4882.

DIMETHYLNITROSAMINE ADDUCTS EXCRETED IN RAT URINE

Kari Hemminki and Harri Vainio

Department of Industrial Hygiene
and Toxicology
Institute of Occupational Health
SF-00290 Helsinki 29, Finland

INTRODUCTION

Biological exposure tests are available for a few carcinogens, but the low exposures generally encountered require high sensitivity, which may limit the applicability of the tests. Urinary excretion products of some carcinogens have been described (Jones, 1973; Chasseud, 1976; Fishbein, 1979). It is however not clear what the biological relevance of the excreted metabolites is to the genotoxic action of chemicals. We have therefore been interested in developing an assay to detect the excreted covalent reaction products of carcinogens with nucleic acids and protein. Although the amounts of such adducts may be small, the availability of urine should help to make the approach feasible. In the development of the techniques we describe here some urinary excretion products found after the administration of dimethylnitrosamine, a carcinogen requiring metabolic activation for its effects (Magee et al., 1976).

RESULTS

Wistar stain rats were injected intraperitoneally with 25 µCi of di-(^{14}C) methylnitrosamine (specific radioactivity 12 mCi/mmol, Amersham). The rats were kept in urine collection cages and urine was collected daily. Protein and particulate material were removed by ethanol precipitation (50 % ethanol, 0° for 10 min,

followed by centrifugation). The ethanol supernatants
were concentrated to about 1/5 of the volume and used
for Sephadex G 10 chromatography on a 0.6 x 90 cm
column. The column was eluted with 0.1 M ammonium formate buffer, pH 6.5, and 30 drop (about 2 ml) fractions
were collected. An aliquot of each fraction was used
for the determination of radioactivity. Each sample
of urine analysed contained, three main peaks of radioactivity (I to III, in the order of elution, Fig. 1).
The peaks of radioactivity were pooled and their identification was attempted.

Each of the three peaks of radioactivity were
analysed in thinlayer and high pressure liquid chromatography (HPLC). Thin-layer chromatography was
performed on cellulose plates with two different solvent
systems (Fig. 2). A number of methylated nucleic
acid bases, nucleosides and amino acid derivatives were
used as standards. The radioactivity of fraction I,
isolated by Sephadex G 10 chromatography, had an identical mobility to S-methylcysteine. Fraction II could
not be identified with the help of the standards
available. Fraction III resembled 7-methylguanine by
its mobility.

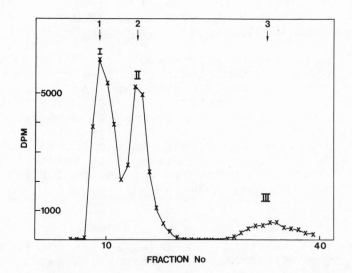

Fig. 1.   Sephadex G 10 chromatography of concentrated
          rat urine collected 16 h after the admin-
          istration of (^{14}C)dimethylnitrosamine.

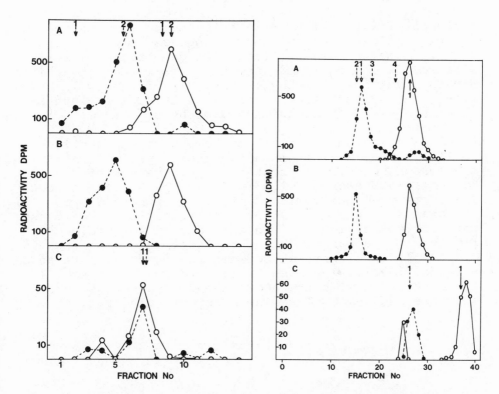

Fig. 2 and Fig. 3.   Cellulose TLC (Fig. 2) and HPLC
        (Fig. 3) separation of fraction I (A),
        fraction II (B) and fraction III (C) isolated
        as described in Fig. 1.   TLC was developed in
        butanol:ethanol:H$_2$O (85:10:25, ---●---) or in
        methanol:HCl:H$_2$O (70:20:20, ——o——).   The
        HPLC eluent was H$_2$O (-●-) or 20 % methanol
        (--o-) in a µBondaPak C$_{18}$ column.   The figures
        on top of each panel refer to the elution of
        the referent compound either in H$_2$O (---) or
        in 20 % methanol (——).   Fig. 2A: 1 = Є-methyl-
        lysine, 2 = S-methylcysteine; Fig. 2C: 1 =
        7-methylguanine.   Fig. 3A: 1 = S-methyl-
        cysteine, 2 = Є-methyllysine, 3 = 1-methyl-
        histidine, 4 = 3-methylhistidine; Fig. 3C:
        1 = 7-methylguanine.

Three solvent systems were used in HPLC in order
to confirm the identity of the urine excretion fractions.
The separations with two solvent systems using a
μBondaPak C_{18} column are shown in Fig. 3. The designation of fraction I as S-methylcysteine and that of
fraction III as 7-methylguanine appears to hold. Fraction II remains unidentified as it does not comigrate
with any of the following standards: S-methylglutathione,
S-methylcysteine, ϵ-methyllysine, 1- and 3-methylhisti-
dine, 7-methylguanosine, 7-methylxanthosine, 7-methyl-
guanine, 7-methylxanthine and 7-methyluric acid.

Dissociation constants of compounds in fractions I
to III were determined in a two-phase system. The
aqueous phase (0.5 ml) contained phosphate buffers
(50 mM), NaOH and HCl with indicated pH's. The aqueous
phase was vigorously mixed with 1 ml of butanol, and
the phases were separated by centrifugation. The phases
were collected, neutralized and counted for radioactivity. The apparent pK's were determined from the
mid-points of transition between the phases (Moore and
Koreeda, 1976). The results with fractions I and III
are consistent with the above identification as S-methylcysteine and 7-methylguanine, respectively (Fig. 4).

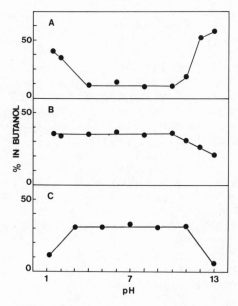

Fig. 4.   Estimation of pK values for fractions I (A),
          II (B) and III (C). The partition was between
          butanol (1 ml) and an aqueous phase (0.5 ml).

Fraction II shows hardly any transitions or possibly a weak one in the alkaline region (Fig. 4B).

The relative turnover of each fraction was estimated and compared to our previous results with the disappearance of (^{14}C)-methyl groups from rat liver nucleic acids and protein (Hemminki and Savolainen, 1979) and from rat serum proteins (Hemminki and Savolainen, 1980). Fig. 5 shows semilogarithmic plots of the disappearance of radioactivity from liver RNA (1, $T_{1/2}$ = 2.3 days), DNA 2, $T_{1/2}$ = 4.2 days) and protein (3, $T_{1/2}$ = 2.8 days), and serum protein (4, $T_{1/2}$ = 2.3 days).

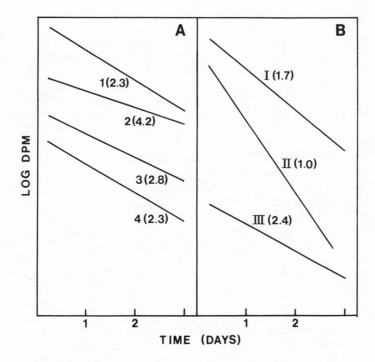

Fig. 5. Turn-over rates of (^{14}C)-methylated rat liver RNA (1), DNA (2), protein (3), of serum protein (4), and of fractions I to III. The figures in the parentheses are approximate half-lives in days.

The apparent turnover rates of urine fraction I, II and III are 1.7, 1.0 and 2.4 days, respectively.

DISCUSSION

The determination of carcinogen-adducts in urine is a potential method for the monitoring of a known chemical exposure. Moreover the urinary excretion products may be valuable in providing information about an unknown exposure in as far as the structures of the modified nucleic acid bases and of amino acids can be determined. Some urinary excretion products of carcinogens have been previously described. Literature is available on the appearance in urine of S-alkylated cysteine or its derivatives after an exposure to ethyl methanesulfonate (Roberts and Warwick, 1958), other alkyl methanesulfonates (Pillinger et al., 1964; Jones and Edwards, 1973; Jones, 1973; Chasseud, 1976), and dimethylnitrosamine (Craddock, 1965) in experimental animals, and to methyl chloride in man (van Doorn et al., 1980). Furthermore excessive excretion of 7-methylguanine has been detected in rat urine after the administration of dimethylnitrosamine (Craddock and Magee, 1967).

In the present study three main fractions of radioactivity were detected in rat urine after the administration of dimethylnitrosamine. Fraction I was thought to consist of S-methylcysteine. It had an apparent half-life slightly shorter than that of methylated proteins in vivo. Urinary S-methylcysteine may be derived from the degradation products of methylated proteins or glutathione, or both. Fraction II remained unidentified and its properties suggested that it was probably not a methylated degradation product of nucleic acids nor of protein. Fraction III was identified as 7-methylguanine, which may be derived from methylated RNA and DNA.

The present results with a model compound provided evidence that degradation products of alkylated proteins and nucleic acids may be detected in urine. Further studies on the urinary adducts of carcinogens are carried out with more complex alkylating agents.

REFERENCES

Chasseud, L. F., 1976, Conjugation with glutathione and
    mercapturic acid excretion, in: Glutathione:
    metabolism and functions, M. Arias and W. B.
    Jakoby, eds., Raven Press, New York, pp. 77-114.
Craddock, V. M., 1965, Reaction of the carcinogen
    dimethylnitrosamine with proteins and with thiol
    compounds in the intact animal, Biochem. J.,
    94:323.
Craddock, V. M., and Magee, P. N., 1967, Effect of admin-
    istration of the carcinogen dimethylnitrosamine
    on urinary 7-methylguanine, Biochem. J.,   104:
    435.
Van Doorn, R., Borm, P. J. A., Leijdekkers, Ch.-M.,
    Henderson, P. Th., Reuvers, J., and van Bergen,
    T. J., 1980, Detection and identification of
    S-methylcysteine in urine of workers exposed to
    methyl chloride, Int. Arch. Occup. Environ.
    Health, 46:99.
Fishbein, L., 1979, Potential industrial carcinogens
    and mutagens, Studies in Environmental Science,
    No. 4, Elsevier, Amsterdam, p. 534.
Hemminki, K., and Savolainen, H., 1980, Alkylation of
    rat serum proteins by dimethylnitrosamine and
    acetyl aminofluorene, Toxicol. Lett., in press.
Hemminki, K., and Savolainen, H., 1979, Methylation of
    neuronal and glial macromolecules by methyl-
    nitrosourea and dimethylnitrosamine in vivo,
    Toxicol. Lett. 4:287.
Jones, A. R., 1973, The metabolism of biological alky-
    lating agents, Drug Metab. Rev., 2:71.
Jones, A. R., and Edwards, K., 1973, Alkylating esters,
    VII. The metabolism of isopropyl methanesul-
    phonate and iso-propyl iodide in the rat,
    Experientia, 29:538.
Magee, P. N., Montesano, R., and Preussmann, 1976,
    N-Nitroso compounds and related carcinogens,
    in: Chemical Carcinogens (ACS Monographs 173),
    C. E. Searle, ed., American Chemical Society,
    Washington D.C., pp.491-625.
Moore, P. D., and Koreeda, M., Application of the change
    in partition coefficient with pH to the struc-
    ture determination of alkyl substituted
    guanosines, Biochem. Biophys. Res. Comm., vol. 73,
    2:459.

Pillinger, D. J., Fox, B. W., and Craig, A. W., 1964,
        Metabolic studies in rodents with C^{14}-labelled
        methyl methanesulfonate, in: Isotopes in Exper-
        imental Pharmacology, L. J. Roth, ed., University
        of Chicago Press, Chicago, pp. 415-432.
Roberts, J. J., and Warwick, G. P., 1958, Studies on the
        mode of action of tumour-growth-inhibiting alkyl-
        ating agents, I, The fate of ethyl methanesul-
        phonate ("Half Myleran") in the rat, Biochem.
        Pharmacol., 1:60.

A NOVEL PATHWAY OF NITROSAMINE METABOLISM IN LIVER MICROSOMES:

DENITROSATION OF NITROSAMINES BY CYTOCHROME P-450

D. Schrenk, M. Schwarz, H.A. Tennekes and W. Kunz

The German Cancer Research Centre, Institute of
Biochemistry
Im Neuenheimer Feld 280, D-6900 Heidelberg, Germany

INTRODUCTION

It is generally accepted that activation of carcinogenic nitrosamines results from oxidative dealkylation of the parent compound (figure 1). This initial step in nitrosamine metabolism –
catalysed by microsomal mono-oxygenase(s) – leads to the generation of highly reactive alkylating intermediates which bind covalently to nucleophilic centres in cellular macromolecules (Magee,
1962).

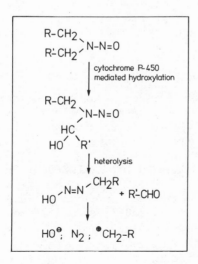

Figure 1 : Activation of N-nitrosamines

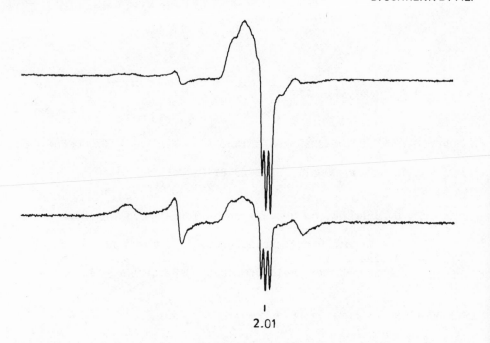

2.01

Figure 2 : EPR spectra of samples containing reduced liver microsomes
and di-n-propylnitrosamine or diphenylnitrosamine.
Upper spectrum was taken of reduced liver microsomes and
diphenylnitrosamine(1 mg/0.3 ml microsomal suspension).
Lower spectrum was taken of reduced liver microsomes and
di-n-propylnitrosamine(3 μl/0.3 ml microsomal suspension).
EPR spectra were taken at 100°K,microwave power 30 mW,
modulation amplitude 10 G,Gain 1000.
Liver microsomal suspension was prepared from phenobarbital-pretreated male NMRI mice and contained 86 mg protein
per ml and 1.8 nmol cytochrome P-450 per mg protein.

However, results of recent studies, described in this paper,
indicate that the mono-oxygenase system may also mediate an alternative reductive pathway of nitrosamine metabolism which leads to
denitrosation of the parent compound.

DENITROSATION OF NITROSAMINES IS CATALYZED BY CYTOCHROME P-450

EPR spectra with samples containing di-n-propylnitrosamine or
diphenylnitrosamine and reduced liver microsomes have been found to
show the three-line absorption characteristic of the ferrous NO-
hemoprotein complex (figure 2). This phenomenon is accompanied by
the formation of nitrite in the incubation medium, as indicated by
the appearance of an absorption peak at approximately 443 nm (Ebel

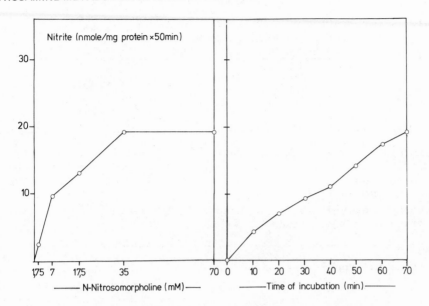

Figure 3 : Concentration- and time-dependent formation of nitrite
from N-nitrosomorpholine in liver microsomes from
untreated male NMRI mice.
Effects of NNM concentration were analysed following
a 50-minute incubation period at 37°C.The impact of
incubation time on nitrite formation was studied using
a 35 mM NNM concentration.One ml incubation mixture(total
volume 5 ml) contained approximately 2mg microsomal protein
and 1.6 nmol cytochrome P-450.Incubations were performed
in the dark under aerobic conditions in the presence of
an NADPH-regenerating system.

et al., 1975).

Nitrite concentration in the medium can be assayed using
Griess' reagent, as previously reported (Appel et al., 1980). The
generation of nitrite in liver microsomes in vitro was shown to depend on NNM concentration and incubation time (figure 3), and the
reaction was markedly accelerated in liver microsomes from phenobarbital-pretreated animals (figure 4). Nitrite formation was strongly inhibited by the addition of mono-oxygenase inhibitors, or by omission
of NADPH (table 1). It would thus appear virtually certain that the
denitrosation of nitrosamines is catalysed by the microsomal monooxygenase system.

However, pre-treatment of the animals with 3-methylcholanthrene
(3-MC) did not enhance denitrosation of nitrosamines in vitro. This
result suggests that 3-MC inducible mono-oxygenase species do not
catalyze this reaction.

Table 1. Generation of nitrite from dimethylnitrosamine is catalysed
          by microsomal mono-oxygenase(s)

	Denitrosation (nmole nitrite/ mg protein x 50 min)	% inhibition
Basic system with 35 mM substrate	8.3 + 1.3	–
Boiled microsomes	0.02+ 0.01	99
NADPH omitted	0.05+ 0.03	99
Under CO (30 sec)	5.4 + 0.5	34
Piperonyl butoxide (0.2 mM) added	3.8 + 0.3	54
Metyrapone (0.1 mM) added	4.7 + 0.5	43

Experimental conditions as in figure 3

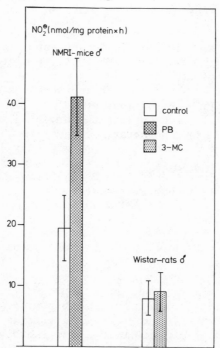

Figure 4 : Nitrite formation from NNM(35 mM)by rat and mouse liver
           microsomes following pre-treatment of the animals:pheno-
           barbital(0.1 % in the drinking water for 5 days)or 3-MC
           (40 mg/kg i.p. for 2 days). Animals were killed 24 hrs
           after last treatment. Assay procedures as in figure 3.
           PB microsomes contained 1.9 nmol P-450/mg protein.
           3-MC microsomes contained 1.4 nmol P-448/mg protein.

MECHANISTIC ASPECTS OF NITROSAMINE DENITROSATION

A strong correlation could be demonstrated between the rate of denitrosation of six different nitrosamines and their partition coefficients in CH_2Cl_2/H_2O (figure 5). Physico-chemical properties of the nitrosamine molecule thus would appear to influence formation of the enzyme-substrate-complex or cleavage of the N-N bond.

The generation of nitrite from DMN was strongly inhibited by the addition of superoxide scavengers, such as superoxide dismutase and epinephrine (table 2). Consequently, it would appear that superoxide radicals play a critical role in the generation of nitrite from nitrosamines. Superoxide radicals are probably released from the oxy-cytochrome P-450-substrate-complex (Estabrook et al., 1979) and it can be speculated that these radicals might cleave the N-N bond of free or enzyme-associated nitrosamines.

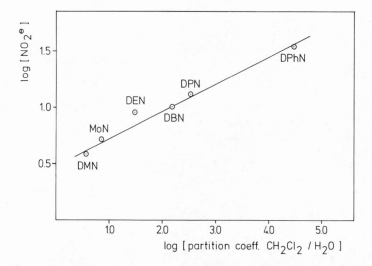

Figure 5 : Relationship between the rate of nitrite formation from various nitrosamines and their partition coefficients in CH_2Cl_2/H_2O.Partition coefficients as reported by Singer et al.(1977),except for DPhN which was determined in this lab. Assay procedures as in figure 3.Linear regression curve after logarithmic transformation: $\log NO_2^- = 0.23(\pm0.05)x$ $\log$ P.C.$(CH_2Cl_2/H_2O) + 0.54(\pm0.13)$;n = 6, r = 0.99, s = 0.06, F = 138.
Abbreviations: P.C. = partition coefficient ;DMN = dimethyl nitrosamine;MoN = N-nitrosomorpholine;DEN = diethylnitros-amine;DBN = di-n-butylnitrosamine;DPN = di-n-propylnitros-amine;DPhN = diphenylnitrosamine.Substrate concentration was 13.5 mM.

Table 2.   Effects of superoxide- and OH-radical scavengers and of
           hydrogen peroxide on nitrite formation from dimethylnitro-
           samine in mouse liver microsomes.

Incubation system	Denitrosation (nmole nitrite/ mg protein x 50 min)	% inhibition
Basic system	3.19 + 0.28	–
+ superoxide dismutase (10μg/ml)	1.70 + 0.15	47
+ H$_2$O$_2$ (2.5 mM)	3.06 + 0.23	4
+ epinephrine (0.2 mM)	1.09 + 0.17	66
+ benzoate (0.8 mM)	3.10 + 0.27	3
+ benzoate (80 mM)	1.61 + 0.20	50
+ mannitol (0.8 mM)	2.98 + 0.21	7
+ mannitol (80 mM)	1.79 + 0.16	44

Substrate concentration: 1.35 mM.
Experimental conditions as in figure 3.

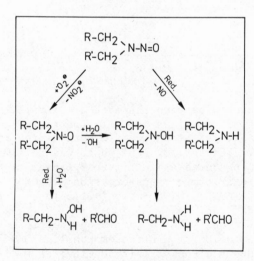

Figure 6 : Proposed pathways of nitrosamine denitrosation.

Nitrite formation was affected only by the addition of a very high dose (80 mM) of OH scavengers, such as mannitol or benzoate. Similarly, we could not detect significant changes in nitrite production in hydrogenperoxide-supplemented liver microsomes.

As reported earlier from this laboratory (Appel et al., 1979), NO-cytochrome P-450 complex formation has also been observed in samples of nitrosamines and dithionite-supplemented liver microsomes. Thus under anaerobic conditions, reductive cleavage of the N-N bond might also occur.

Proposed pathways of nitrosamine denitrosation are shown in figure 6. These pathways ultimately result in the generation of primary amines and primary and secondary hydroxylamines. The observation that methylamine and N-methylhydroxylamine are products of dimethylnitrosamine metabolism in rat liver microsomes (Grilli et al., 1975), is, therefore,in agreement with our postulations. Methylamine has also been detected in liver and urine of mice exposed to dimethylnitrosamine (Heath et al., 1958). In addition, it has also been reported that the postulated free OH-radicals are, in fact, generated in samples of liver microsomes and nitrosamines (Floyd et a., 1975).

SUMMARY AND CONCLUSIONS

The results of this study indicate that nitrosamines are denitrosated in liver microsomes in vitro. This reaction is catalysed by phenobarbital-inducible cytochrome P-450 species. Superoxide radicals play a critical role in nitrosamine denitrosation. Proposed pathways are shown in figure 6.

REFERENCES

Appel, K.E., Ruf, H.H., Mahr, B., Schwarz, M., Rickart, R., and
        Kunz, W., 1979, Binding of nitrosamines to cytochrome P-450
        of liver microsomes, Chem.-Biol. Interact., 28:17.
Appel, K.E., Schrenk, D., Schwarz, M., Mahr, B., and Kunz, W., 1980,
        Denitrosation of nitrosamines by liver microsomes; possible
        role of cytochrome P-450, Cancer Lettres, 9:13.
Ebel, R.E., O'Keefe, D.H., and Peterson, J.A., 1975, Nitric oxide
        complexes of cytochrome P-450, FEBS Lettres, 55:198.
Estabrook, R.W., Kawano, S., Werringloer, J., Kutnan, H., Tsuji, H.,
        Graf, H., and Ullrich, V., 1979, Oxycytochrome P-450: its
        breakdown to superoxide for the formation of hydrogen per-
        oxide, Acta biol. med. germ., 38:423.
Floyd, R.A., Lailing, M.S., Stuart, M.A., and Reigh, D.L., 1978,
        Spin trapping of free radicals produced from nitrosamine
        carcinogens, Photochem. and Photobiol., 28:857.

Grilli, S., and Prodi, G., 1975, Identification of dimethylnitrosa-
    mine metabolites in vitro, Gann, 66:473.
Heath, D.F., and Dutton, A., 1958, The detection of metabolic pro-
    ducts from dimethylnitrosamine in rats and mice, Biochem. J.,
    70:619.
Magee, P.N., and Farber, E., 1962, Toxic liver injury and carcino-
    genesis methylation of rat liver nucleic acids by dimethyl-
    nitrosamine in vivo, Biochem. J., 83:114.
Singer, G.M., Taylor, H.W., and Lijinsky, W., 1977, Liposolubility
    as an aspect of nitrosamine carcinogenicity: quantitative
    correlations and qualitative observations, Chem.-Biol.
    Interact., 19:133.

INVESTIGATIONS OF THE BIOACTIVATION OF N-NITROSOPYRROLIDINE IN THE
RAT

Richard C. Cottrell, Brian G. Lake, John C. Phillips,
Philip J. Young, David G. Walters, Andrew J. Allars and
Sharat D. Gangolli
The British Industrial Biological Research Association
Carshalton, Surrey SM5 4DS, U.K.

The alicyclic nitrosamine N-nitrospyrrolidine , (NPYR), like
its simpler dialkyl analogue N-nitrosodimethylamine (NDMA), is a
potent hepatocarcinogen on chronic administration to the rat
(Druckrey et al., 1967). There are, however, some significant
differences in the biological effects of these two compounds. Not
only is NPYR appreciably less potent as a carcinogen than NDMA
(Preussmann et al., 1977; Keefer et al., 1973; Terracini et al.,
1967; Hoch-ligeti et al., 1968) it is also markedly less hepatotoxic
(Hendy and Grasso, 1977).

We have been interested for some time in rationalizing these
differences in behaviour in terms of the metabolic bioactivation and
degradation processes which the nitrosamines undergo. Unfortunately,
the information available to us at the beginning of our studies was
not sufficient to allow such a rationalisation. We consequently set
out to study the metabolism of NPYR in the expectation that the
knowledge gained would lead to an explanation for its biological
effects.

Our initial results, however, only seemed to emphasize the
similarities between the two. NPYR appears to competitively inhibit
the conversion of NDMA to formaldehyde by rat liver 10,000g supernatant preparations (Cottrell et al., 1977). The metabolism of
[2,5 - 14C] NPYR to $^{14}CO_2$ is inhibited by a variety of compounds
whose most conspicuous common feature seemed to be that they also
inhibited the degradation of [14C] NDMA to $^{14}CO_2$ in vivo (Cottrell
et al., 1979). The rate of $^{14}CO_2$ exhalation from either [14C]
labelled nitrosamine was not effected by inhibitors of the hepatic
cytochrome P 450 dependent mixed function oxidase system (Cottrell
et al., 1979). Moreover the rate of metabolism of the two nitro-

samines in vivo was similar at similar molar dose levels (Cottrell et al., 1977; Phillips et al., 1975).

The first major difference that we observed was in their metabolism to $^{14}CO_2$ by the isolated perfused liver (Cottrell et al., 1979). While the rate of conversion of NDMA was sufficiently rapid to account for the rate observed in the intact animal the rate of degradation of NPYR was consistently much slower in the isolated liver. This work provided a clear hint of the importance, at least pharmacokinetically, of extra-hepatic metabolism in the biodegradation of NPYR. This hint was subsequently confirmed by whole-body auto-radiographic studies (Creasy, D.M., Young, P.J. and Cottrell, R.C., unpublished results) of the time dependance of the distribution of NPYR metabolites and by investigations involving in situ intestinal loops (Cottrell et al., 1979).

The reason for the specificity of NPYR as a carcinogen to the rat liver is, therefore, likely to be more subtle than merely that the liver is the major site of metabolism. Either a particular bioactivation process occurs in the liver which does not take place to the same extent in other organs, or the liver is more susceptible to a particular injury than other tissues which may be exposed to a similar insult.

About this time evidence appeared in the literature to suggest that the metabolism of NPYR was essentially analagous to NDMA (Hecht et al., 1978). 4-Hydroxybutanol was observed (by trapping as its 2,4 dinitrophenylhydrazone) both in the urine of rats treated with NPYR and in liver microsomal incubations. The metabolic consequences expected from such a degradative process would be the production of 4-hydroxybutyric acid by further oxidation and of 4-butyrolactone either by lactonisation of the acid or as a result of direct oxidation of the cyclic hemiacetal form in which the aldehyde exists in water.

These expectations were subsequently confirmed in a study of the metabolism of NPYR by rat liver fractions (Hecker et al., 1979) in which 1,4-butanediol (a reduction product) was also observed. Our concurrent investigations of the urinary metabolites of NPYR (Cottrell et al., 1980) have also provided evidence for 4-hydroxy-butyric acid and its lactone. We were unable to detect 1,4-butanediol in these urine samples by G.L.C. (detection limit less than 1 p.p.m.).

When we examined the radiolabelled metabolites of $[2,5 -^{14}C]$ NPYR in the urine, following a range of dose levels, we were a little surprised to find an obviously complex mixture. The use of $[2,2,5,5 -^3H ; 2,5 -^{14}C]$ NPYR allowed us to concentrate our attention on the isolation and identification of those metabolites which appeared to be nearest in structure to the parent nitrosamine. We

were able to identify a number of metabolites and rapidly realised
that the metabolic degradation of NPYR must be more complex than
previously supposed.

An examination of the effects on the H.P.L.C. urinary metabolite
profiles of some inhibitors of NPYR metabolism to CO_2 (Cottrell et
al., 1980) enabled us to group the urinary metabolites and suggested
possible pathways that could lead to the observed end-point products.

As a result of acquiring these forms of evidence we came to the
conclusion that at least three routes of biodegradation of NPYR
occur in the rat. At this stage, we cannot say which organs are
involved in which routes.

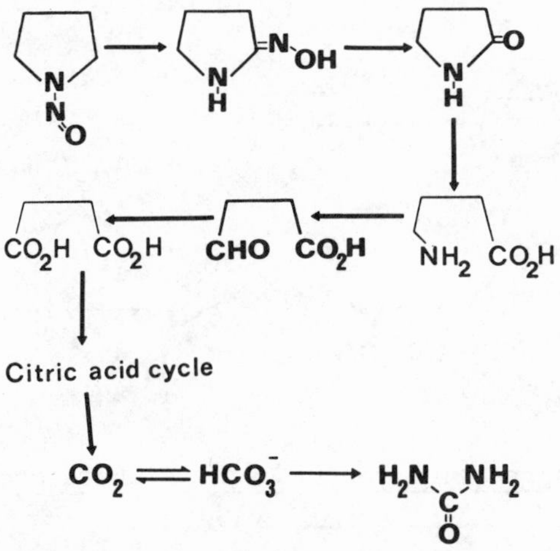

Fig. 1.   One possible biodegradation route of NPYR in the rat.
          Compounds shown in bold type have been identified radio-
          labelled following administration of L2,5 -^{14}C] NPYR

One of the routes postulated involves an initial rearrangement reaction (or apparent rearrangement) to yield pyrrolidinone-2-oxime which was detected radiolabelled in the urine of rats given a high dose of [2,5 -^{14}C] NPYR by a two-dimensional T.L.C. method. Subsequent (enyzmatic?) hydrolysis of this amide oxime could yield pyrrolidin-2-one. Ring opening to 4-aminobutyric acid followed by transamination to succinic semialdehyde and oxidation to succinic acid could lead into the citric acid cycle and thus to CO_2. Products of bicarbonate anabolism, such as urea, would also be expected to acquire radiolabel (Fig. 1).

We established that pyrrolidinone-2-oxime, pyrrolidin-2-one, succinic semialdehyde, CO_2 and urea were radiolabelled following the administration of [2,5 -^{14}C] NPYR, strongly suggesting that such a route does operate in the rat.

A number of inhibitors of NPYR degradation to CO_2 also gave rise to a dramatic increase in the urinary excretion of pyrrolidin-2-one; consistent with the hypothesis that the inhibition of CO_2 production is at least partly due to the blocking of this pathway at a point after the generation of the lactam. We were unable to find an efficient inhibitor of CO_2 production which did not cause this increase in pyrrolidin-2-one excretion. Thus our current evidence leads us to think that this route is pharmacokinetically dominant in the rat.

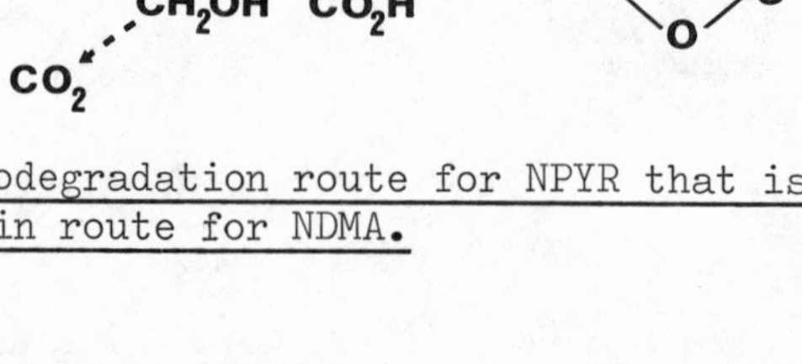

Fig. 2.   The biodegradation route for NPYR that is analagous to the main route for NDMA.

A second route involves the generation of 4-hydroxybutanol which may be oxidized to 4-hydroxybutyric acid or its lactone (Figure 2). Both of these oxidation products were identified, as products of NPYR metabolism, in rat urine. We were surprised to find that pretreatment with SKF 525A, which does not effect the rate of CO_2 generation from NPYR and does not change the urinary excretion of pyrrolidin-2-one, leads to a marked increase in urinary levels of 4-butyrolactone. This effect could be due to the blocking of this pathway only if this route does not contribute significantly to the overall in vivo rate of CO_2 production. A less attractive hypothesis, however, is that the drug merely alters the rate of formation or hydrolysis of the lactone.

A third route yields 3-hydroxy-N-nitrosopyrrolidine and, we suggest, via further metabolism, to C-1 pool precursors and CO_2 (Figure 3). The identification of radiolabel in urinary dimethylamine but not in methylamine, together with evidence from $3H/^{14}C$ labelling experiments, indicates the generation of some C-1 pool precursor. A proportion of the CO_2 exhaled from NPYR appears to arise from this route since none of the inhibitors we have studied so far completely eliminate CO_2 production and none effect the urinary excretion of dimethylamine or 3-hydroxy-N-nitrosopyrrolidine.

Clearly the questions of interest to this symposium are "which pathways are bioactivating and which are the biologically reactive intermediates?".

The only tools readily available to us to answer these questions were to chemically synthesise putative reactive intermediates and determine their mutagenicity in the "Ames" bacterial test system.

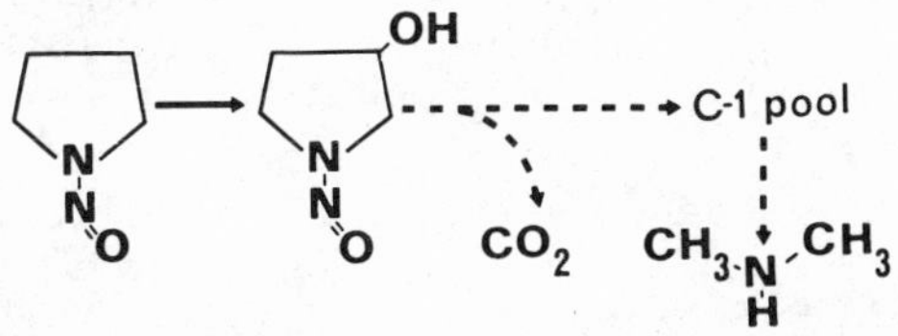

Fig. 3   Suggested outcome of the 3-hydroxylation pathway of NPYR degradation

Pyrrolidinone-2-oxime, the earliest identified intermediate of one route, was synthesised and found to be relatively non-toxic to rats (and Salmonella) and inactive in the Ames system. The route of biodegradation of NPYR which involves this amide oxime must there-fore, on current evidence, be considered a detoxifying route. It is of interest that such a role is assigned to a major biodegradation route since the difference in potency between NDMA and NPYR may thus be explained. The other products of this route are naturally occurring in the rat and therefore, presumably, unlikely to be toxic or carcinogenic.

The route via 4-hydroxybutanal clearly has the potential to be an activating route (Hecht et al., 1978). As yet the mechanism of generation of the hydroxyaldehyde is unresolved (Cottrell et al., 1980) and so the significance of this pathway to the carcinogenic process is somewhat speculative. Again, subsequent products of this route are naturally occurring intermediary metabolites in the rat.

3-Hydroxy-N-nitrosopyrrolidine is of interest since it is a mutagen in the "Ames" system without metabolic activation by rat liver "S-9" (Stoltz and Sen, 1977). Indeed, in our hands it seems to be more active alone than NPYR with "S-9" activation. It is thus difficult to exclude the possibility that the generation of this intermediate is a bioactivation process. We also observed that the inclusion of "S-9" in the Ames test considerably reduced the mutagenicity of 3-hydroxy-N-nitrosopyrrolidine which is consistent with our suggestion that it can be further degraded.

Thus the questions must remain open. We do not yet know with any degree of rational certainty what are the bioactivation routes of NPYR or the identity of the reactive intermediates formed. Further work is required if speculation is to be replaced by information.

By way of conclusion may I add a note of caution. It is perhaps worth mentioning in a symposium of this kind, that biological reactivity in a compound does not necessarily imply any extraordinary chemical reactivity. We need only to look at the efficiency and specificity of active-site-directed irreversible enzyme inhibitors to realise how a compound may have profound biological effects without displaying any unusual general chemical reactivity. It is the interaction of the "intermediate" with a particular intra-cellular target which will determine its biological reactivity. It is also possible, at least in principle, for profound biological consequences to follow an interaction between a chemical and a cellular component in which there are no changes in covalency at all. Clearly processes of this type will be particularly difficult to detect and study.

(This work forms part of a research project sponsored by the U.K. Ministry of Agriculture, Fisheries and Food to whom our thanks are due. The results of the research are the property of the Ministry of Agriculture, Fisheries and Food and are Crown Copyright).

References

Cottrell, R.C., Walters, D.G., Young, P.J., Phillips, J.C., Lake, B.G., and Gangolli, S.D., 1980, Studies of the urinary metabolites of N-nitrosopyrrolidine in the rat, Toxicol. Appl. Pharmacol., 54:368.

Cottrell, R.C., Young, P.J., Herod, I.A., Lake, B.G., Phillips, J.C., and Gangolli, S.D., 1977, Preliminary studies of the metabolism of N-nitroso [2,5 -^{14}C] pyrrolidine in the rat, Biochem. Soc. Trans., 5:1011.

Cottrell, R.C., Young, P.J., Walters, D.G., Phillips, J.C., Lake, B.G., and Gangolli, S.D., 1979, Studies of the metabolism of N-nitrosopyrrolidine in the rat, Toxicol. Appl. Pharmacol., 51:101.

Druckrey, H., Preussmann, R., Ivankovic, S., and Sohmahl, D., 1967, Organotrope Carcinogene Wirkungen bei 65 Verscheidenen N-nitroso-Verbindungen an B-D Ratten, Z. Krebsforsch., 69:103.

Hecht, S.S., Chen, C.-H.B., and Hoffmann, D., 1978, Evidence for metabolic α-hydroxylation of N-nitrosopyrrolidine, Cancer Res., 38:215.

Hecker, L.I., Farrelly, J.G., Smith, J.H., Saavedra, J.E., and Lyon, P.A., 1979, Metabolism of the liver carcinogen N-nitroso-pyrrolidine by rat liver microsomes, Cancer Res., 39:2679.

Hendy, R., and Grasso, P., 1977, Hepatotoxic response to single or repeated injections of N-nitrosopyrrolidine in the rat, Chem.-Biol. Interactions, 18:309.

Hoch-ligeti, C., Argus, M.F., and Arcos, J.C., 1968, Combined carcinogenic effects of dimethylnitrosamine and 3-methyl-cholanthrene in the rat., J. Natl. Cancer Inst., 40:535.

Keefer, L.K., Lijinsky, W., and Garcia, H., 1973, Deuterium isotope effect on the carcinogenicity of dimethylnitrosamine in rat liver, J. Natl. Cancer Inst., 51:299.

Phillips, J.C., Lake, B.G., Heading, C.E., Gangolli, S.D. and Lloyd, A.G., 1975, Studies on the metabolism of dimethylnitrosamine in the rat.1. Effect on dose, route of administration and sex, Food Cosmet. Toxicol., 13:203.

Preussmann, R., Schmahl, D., and Eisenbrand, G., 1977, Carcino-genicity of N-nitrosopyrrolidine:dose response study in rats, Z. Krebsforsch., 90:161.

Stoltz, D.R., and Sen, N.P., 1977, Mutagenicity of five cyclic N-nitrosamines : assay with Salmonella typhimurium, J. Natl. Cancer Inst., 54:393.

Terracini, B., Magee, P.N., and Barnes, J.M., 1967, Hepatic pathology in rats on low dietary levels of dimethylnitrosamine, Brit. J. Cancer., 21:559.

REACTIONS OF AROMATIC NITROSO COMPOUNDS WITH THIOLS

Chr. Diepold, P. Eyer, H. Kampffmeyer,
and K. Reinhardt

Institut für Pharmakologie und Toxikologie der
Medizinischen Fakultät der Universität München
Nussbaumstr. 26, 8000 München 2, FRG

INTRODUCTION

Since the discovery of N-oxigenation in vivo by KIESE[1] and the MILLER's[2] this reaction is generally accepted to be the first activation step in toxication of aromatic amines leading to methaemoglobinaemia and carcinogenesis. Despite the wealth of literature concerning further activation steps and detoxication reactions as well, detailed reports on the influence of thiols on those reactions are rare[3-6]. Recently we presented a scheme of the reactions of nitrosobenzene with reduced glutathione[7] and NEUMANN and his group[8,9] extended this reaction scheme for 4-nitrosotoluene in the reaction with glutathione and other thiols. Thereby it became obvious that different substituents of the nitrosoarenes markedly influence the reaction pathway. Thus we decided to compare the reactions of different p-substituted nitrosoarenes (ArNO) with glutathione (GSH) and other thiols since these novel type reactions are of biological significance and occur in the liver and red cells with significant alterations of the glutathione status.

MATERIALS and METHODS

Nitrosobenzene, nitrobenzene, aniline, 4-nitroacetophenone, 4-aminoacetophenone, 4-chloraniline, 3,4-dichloroaniline, 4-phenetidine, 4-N,N-dimethylphenylenediamine were from Merck-Schuchardt, Darmstadt; 4-nitroso-N,N-dimethylaniline from Riedel-de-Haen, Seelze-Hannover. 4-Chloronitrosobenzene, 3,4-dichloronitrosobenzene, 4-chlorophenylhydroxylamine, 3,4-dichlorophenylhydroxylamine and 4-N-hydroxyphenetidine were kindly provided by Prof. Dr. W. Lenk, Institut für Pharmakologie und Toxikologie der Universität München.

Phenylhydroxylamine, m. p. 84°, and 4-N-hydroxylamino-acetophenone, m. p. 106-107°, were prepared from the nitro derivatives according to GOLDSCHMIDT[10]. 4-Nitrosoacetophenone was obtained by oxidation of the hydroxylamine with $FeCl_3$; m. p. 109° after recrystallisation from CH_2Cl_2. 4-Nitrosophenetol was prepared by oxidation of phenetidine with persulfate in sulfuric acid (5 min reaction, 0°) according to HIROTA and ITANO[11]. The 4-nitrophenetol impurity was removed by chromatography on Sephadex LH 20 with methanol as eluant, m. p. 35° (58% yield). M. p. and u. v. spectra were identical with the data reported by UEHLEKE[12]. All other analytical grade chemicals were from Merck, Darmstadt, the biochemicals from Boehringer, Mannheim.

HPLC was performed with a chromatograph ALC/GPC 244 equipped with a 254 nm detection and integration assembly (Waters, Milford, Mass.) on μ-Bondapak C_{18} (4 mm ID x 30 cm) with methanol: water containing 20 mM sodium phosphate, pH 7.4 (flow rate 2 ml/min). Retention vol. were: 30% Methanol: 4-Phenetidine (11.2 ml), 4-N-hydroxyphenetidine (9.6 ml), 4-nitrosophenetol (19.8 ml), 4-N,N-dimethylphenylenediamine (8.8 ml), 4-nitroso-N,N-dimethylaniline (14.3 ml). 40% Methanol: Aniline (5.8 ml), phenylhydroxylamine (5.0 ml), nitrosobenzene (14 ml), 4-aminoacetophenone (5.1 ml), 4-N-hydroxylaminoacetophenone (4.3 ml), 4-nitrosoacetophenone (13 ml). 60% Methanol: 4-Chloroaniline (5.6 ml), 4-chlorophenylhydroxylamine (4.6 ml), 3,4-dichloroaniline (7.8 ml), 3,4-dichlorophenylhydroxylamine (6.8 ml).

Isolated metabolites were additionally identified by t. l. c. and u. v. spectra in comparison with authentic specimens. Reduced glutathione, glutathione disulfide and glutathione sulfinic acid were determined as already described[7]. Haemoglobin and ferrihaemoglobin were measured according to KIESE[1].

Rapid changes in extinction of the nitrosoarenes during the reaction with thiols were followed at their maximum absorption between 290 and 450 nm in a DW-2 double wavelength photometer equipped with an Aminco-Morrow stopped-flow accessory and a DASAR digital unit (Aminco, Silver Spring, Md.). Initial velocities were used to determine reaction constants in the reaction of nitrosoarenes with thiols. To determine the order of reaction the components were varied by at least one order of magnitude each.

RESULTS and DISCUSSION

Table 1. Second Order Rate Constants of the Reaction of Nitrosoarenes (ArNO) with Thiols (RSH); (M^{-1} sec^{-1})

v = k x (ArNO) x (Glutathione)		v = k x (Nitrosobenzene) x (RSH)	
4-Nitroso- N,N-dimethylaniline	1.5	ß-Mercaptoethanol	1 300
4-Nitrosophenetol	170	Benzylmercaptan	1 500
Nitrosobenzene	5 000	Glutathione	5 000
4-Cl-Nitrosobenzene	8 500	Cysteine	12 000
3,4-Cl$_2$-Nitrosobenzene	20 000	Cysteamine	12 000
4-Nitrosoacetophenone	80 000		

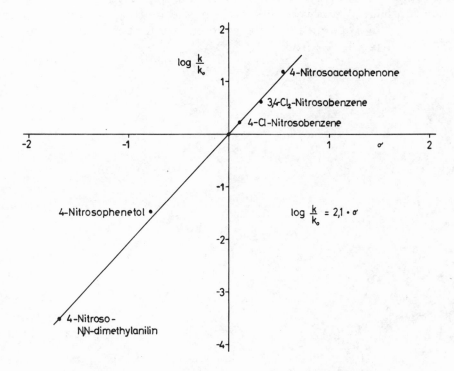

Fig. 1.   Relative Reactivities of Different p-Substituted Nitrosoarenes in the Reaction with Glutathione vs HAMMET
Constants   (pH 7.4, 37°, N$_2$).

All nitrosoarenes tested reacted with thiols in a second order reaction. The rate constants (pH 7.4, 37°, N_2) are given in Table 1. As shown in Fig. 1 the reaction velocities with glutathione increased with increasing HAMMET constant of the substituent.

The kinetics of the decrease in nitrosoarene absorption was biphasic except for 4-nitrosophenetol and 4-nitroso-N,N-dimethylaniline, and the proportion of the rapid reaction dependent on the glutathione concentration. Because the nitroso compound was extractable with ether although the u.v. absorbance in incubates with glutathione had already disappeared, reversible formation of an intermediate had to be considered[7]. Subsequently, the intermediate was either transformed into the sulfinamide (Fig. 2, pathway ①) or thiolytically cleaved by another thiol with formation of the thiol disulfide and the hydroxylamine (pathway ②). The latter reaction resembles to the diazene reduction by glutathione as revealed by the KOSOWER's[13]. As may be anticipated from the scheme in Fig. 2 thiolytic cleavage of the labile intermediate was favoured by increasing the thiol concentration at the expense of sulfinamide isomerisation.

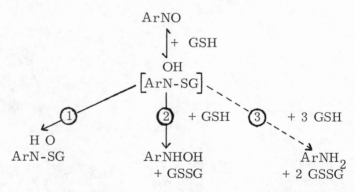

Fig. 2. Pathways Observed During the Reaction of Nitrosoarenes with Glutathione (pH 7.4, 37°, N_2).

Fig. 3 shows the yields of hydroxylamine formation at various glutathione concentrations. The data fit slopes calculated by the function

$$\frac{\text{Hydroxylamine formed}}{\text{Nitrosoarene added}} = \frac{p \times (GSH)}{1 + p \times (GSH)}.$$

In addition, lowering the pH favours sulfinamide isomerisation and decreases phenylhydroxylamine formation.

Furthermore, reduction of 4-nitrosotoluene to the amine giving two molecules of glutathione disulfide was reported by NEUMANN[8, 9]. Similar stoichiometry we observed with 4-nitroso-N,N-dimethylaniline and partly with 4-nitrosophenetol. With these two nitrosoarenes no hydroxylamine was produced. Rather, HPLC analysis revealed formation of short lived adducts which liberated the amine and glutathione disulfide upon treatment with acid. Thus, a third pathway ③ had been proposed by NEUMANN[8, 9] in which the amine might be formed from an intermediate mercaptan. The exact structure of these intermediates remains still to be resolved.

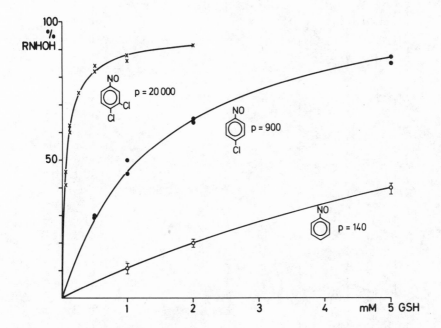

Fig. 3.   Formation of the Hydroxylamine in the Reaction of the Nitrosobenzene (0.05 mM) with GSH (37°C; N_2;pH 7.4; 10 min reaction $\quad p = \dfrac{\% \text{ RNHOH}}{(100 - \% \text{ RNHOH})\,(GSH)}.$

Table 2. Pattern of Reaction Products in the Reaction of Nitrosoarenes with Glutathione

Incubation mixture at pH 7.4, 37°, N$_2$	ArNO mM	GSH mM	Products formed during 10 min reaction					Pathway preferred c.f. Fig.1
			ArNHSOG mM	ArNHOH mM	ArNH$_2$ mM	GSSG mM	Adducts mM	
4-Nitroso- N,N-dimethylaniline	5.0	20.0	0	0	3.97	7.95	1.0	3
+	1.0	2.0	0	0	0.41	0.80	0.08	3
4-Nitrosophenetol	1.0	2.0	0.1	0	0.21	0.44	0.65	3 > 1
Nitrosobenzene	1.0	2.0	0.80	0.20	0	0.20	0	1 > 2
	0.1	2.0	0.08	0.02	0	0.02	0	1 > 2
4-Cl-Nitrosobenzene	0.1	2.0	0.036	0.064	0	0.065	0	2 > 1
3,4-Cl$_2$-Nitrosobenzene	0.1	2.0	0.008	0.092	0	0.095	0	2 ≫ 1
4-Nitrosoacetophenone	0.1	2.0	0	0.1	0	0.1	0	2

Means of at least duplicate measurements.

+ Reaction time extended to 2 h for completion.

Table 2 shows the pattern of reaction products after completion of the reaction between nitrosoarenes and glutathione. It
varied clearly with the ring substituent. Whereas the hydroxylamine was the only reaction product when 4-nitrosoacetophenone
reacted with two equivalents of glutathione (at equimolar ratios
the bis-azoxy-derivative was formed), no hydroxylamine formation was observed with 4-nitrosophenetol and 4-nitroso-N,N-di-
methylaniline. In addition, 4-nitroso-N,N-dimethylaniline
formed small amounts of an adduct which liberated no amine even
after vigorous treatment with mineral acid. Because the quinone-
oxime structure is favoured by the dimethylamino substituent ring
addition may have occurred.

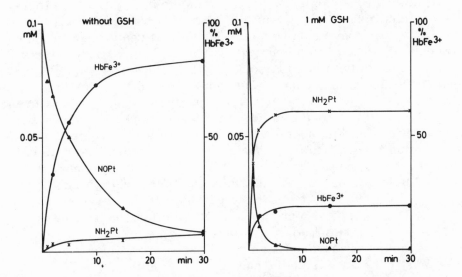

Fig. 4.   Influence of GSH on the Decrease in Nitrosophenetol
          (NOPt); Increase in Phenetidine (NH$_2$Pt) and Ferrihae-
          moglobin During the Reaction of Nitrosophenetol with
          Haemolysate.
          10 g Hb/100 ml of freshly dialysed human haemolysate
          reacted with 0.1 mM nitrosophenetol ± 1 mM GSH in
          presence of   0.02 mM NADPH in a regenerating system
          at pH 7.4, 37° under air (means of 3-4 experiments).

The biological significance of the reaction of nitrosoarenes with thiols can be demonstrated in red cells where ferrihaemoglobin formation and enzymatic reduction of the nitrosoarene to the hydroxylamine and the amine are well known reactions[14]. In human red cells incubated with 1 mM nitrosobenzene about 0.4 mM glutathione sulfinanilide and 0.2 mM haemoglobin-linked cysteine-sulfinanilide were formed[15]. It has been demonstrated that most of the aniline formed originated from glutathione sulfinanilide[15]. From the studies outlined in the previous section it was expected that formation of the amine should not be influenced by glutathione in case of 4-nitrosoacetophenone but should depend largely on glutathione in case of 4-nitrosophenetol. In fact, as shown in Fig. 4 phenetidine formation was negligible in the absence of glutathione but was produced rapidly in its presence. In turn, ferrihaemoglobin formation was quickly terminated in presence of glutathione because 4-nitrosophenetol was quickly removed. Thus, the concentration of glutathione in the red cell may alter significantly ferrihaemoglobinaemia and covalent binding of 4-nitrosoarenes with negative HAMMET constants. In case of 4-nitrosoacetophenone, however, no significant dependence on glutathione in ferrihaemoglobin formation and amine production was observed.

We were interested whether these fast reactions of nitrosoarenes with glutathione occurred also in the liver where the concentration of glutathione is about 10 mM, and N-oxygenation products after ingestion of an amine should be highest. During haemoglobin-free single pass perfusion of rat livers with nitrosobenzene[16] the efflux of glutathione disulfide increased markedly in bile and venous effusate. Furthermore, release of glutathione sulfinanilide in the effusate confirmed the reaction of nitrosobenzene with glutathione within the intact organ. About 1/10 of nitrosobenzene had reacted with glutathione and diminished thereby the glutathione content of the liver. When the glutathione concentration in the liver fell below 1/3, sudden increase in lactate dehydrogenase release indicated membrane damage. Whether these reactions are also more pronounced with nitrosoarenes with negative HAMMET constants is under current investigation.

REFERENCES

1. M. Kiese, Oxidation von Anilin zu Nitrosobenzol im Hunde, Arch. exp. Path. Pharmak. 235: 354 (1959).
2. J.W. Cramer, J.A. Miller, and E.C. Miller, N-hydroxylation: A new metabolic reaction observed in the rat with the carcinogen 2-acetylaminofluorene, J. Biol. Chem. 235: 885 (1960).

3. P. D. Lotlikar, E. C. Miller, J. A. Miller, and A. Margreth, The enzymatic reduction of N-hydroxy derivatives of 2-acetylaminofluorene and related carcinogens by tissue preparations, Cancer Res. 24: 2018 (1964).

4. E. Boyland, D. Manson, and R. Nery, The reaction of phenylhydroxylamine and 2-naphthyl-hydroxylamine with thiols, J. Chem. Soc. 606 (1962).

5. M. Kiese and K. Taeger, The fate of phenylhydroxylamine in human red cells, Arch. Pharmacol. 292: 59 (1976).

6. H.-G. Neumann, M. Metzler, and W. Töpner, Metabolic activation of diethylstilbestrol and aminostilbene-derivatives, Arch. Toxicol. 39: 21 (1977).

7. P. Eyer, Reactions of nitrosobenzene with reduced glutathione, Chem.-Biol. Interactions 24: 227 (1979).

8. B. D. Dölle and H.-G. Neumann, Reaction of aromatic nitroso-compounds with mercaptans, Arch. Pharmacol. 311: R 25 (1980).

9. B. D. Dölle, W. Töpner, and H.-G. Neumann, Reaction of arylnitroso compounds with mercaptans, Xenobiotica in press (1980).

10. C. Goldschmidt, Zur Darstellung von Phenylhydroxylamin, Ber. dtsch. chem. Ges. 29: 2307 (1896).

11. K. Hirota and H. A. Itano, Influence of ring substituents on the binding of nitrosobenzene by ferrohemoglobin, J. Biol. Chem. 253: 3477 (1978).

12. H. Uehleke, N-Hydroxylierung von p-Phenetidin in vivo und durch isolierte Mikrosomen aus Lebern und Nieren: Stimulierung durch Phenobarbital-Vorbehandlung, Arch. Pharmak. 264: 434 (1969).

13. N. S. Kosower, E. M. Kosower, and B. Wertheim, Diamide, a new reagent for the intracellular oxidation of glutathione to the disulfide, Biochem. Biophys. Res. Commun. 37: 593 (1969).

14. M. Kiese, Methemoglobinemia: A comprehensive treatise, CRC Press, Cleveland, Ohio (1974).

15. P. Eyer and E. Lierheimer, Biotransformation of nitroso-benzene in the red cell and the role of glutathione, Xenobiotica in press (1980).

16. P. Eyer, H. Kampffmeyer, H. Maister, and E. Rösch-Oehme, Biotransformation of nitrosobenzene, phenylhydroxylamine, and aniline in the isolated perfused rat liver, Xenobiotica in press (1980).

REACTIVE INTERMEDIATES OF 2-ACETYLAMINOFLUORENE METABOLISM IN VITRO COVALENTLY LABEL SPECIFIC RAT LIVER MICROSOMAL PROTEINS

M.A. Kaderbhai, R.B. Freedman and T.K. Bradshaw*

Biological Laboratory, University of Kent, Canterbury, CT2 7NJ, U.K. and *Shell Toxicology Laboratory (Tunstall), Sittingbourne Research Centre, Sittingbourne ME9 8AG, U.K.

Introduction:

Reactive intermediates in the metabolism of many xenobiotics often react extensively with proteins of the endoplasmic reticulum. This covalent modification may, in some cases, be relevant to mechanisms of toxicity and carcinogenicity. In general, the process has only been characterized in terms of the overall binding of reactive intermediates to total microsomal proteins, but much information about the properties, reactivity and mobility of the intermediates could be derived from a detailed analysis of the modified proteins.

We have used electrophoretic techniques to study the labelling of individual microsomal proteins during the metabolism of 2-acetylaminofluorene in vitro. We find that labelling is concentrated in specific proteins and that the pattern of labelling is altered in microsomes from rats pre-treated with specific inducers.

Results:

After incubation of rat liver microsomes with $(9-^{14}C)-2-$ acetylaminofluorene in the presence of NADPH, followed by extensive washing and extraction of the microsomes, 3-5% of the added label is tightly bound to the membranes. Solubilisation of this label by pronase and trypsin treatment indicates that the labelled derivative is covalently bound to microsomal protein.

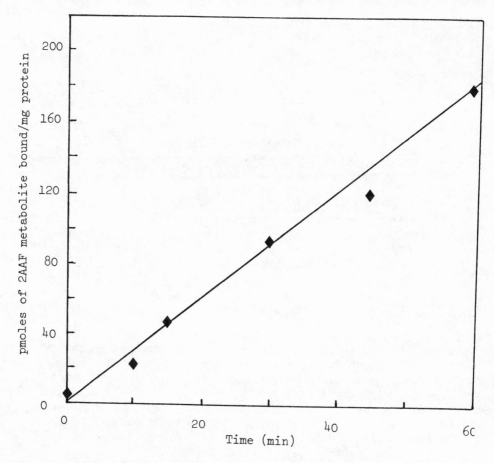

Figure 1. Time course of covalent binding of reactive derivative(s) of (9-^{14}C)-2-acetylaminofluorene to microsomal membrane proteins.

The binding increases linearly with time (Figure 1), is totally dependent on the presence of NADPH (Figure 2) and is inhibited by SKF 525A (1 mM) and by 7,8-benzoflavone (2 mM).

With microsomes from 3-methylcholanthrene-induced rats, binding is increased 8-fold compared to controls, but the increase is only 2-fold with microsomes from phenobarbital-induced rats (Figure 3).

All these findings suggest that the reactive intermediate is generated by a cytochrome P-448-dependent reaction. N-Hydroxylation may be an obligatory step in generation of the reactive intermediate. Binding to microsomes is slightly enhanced by the

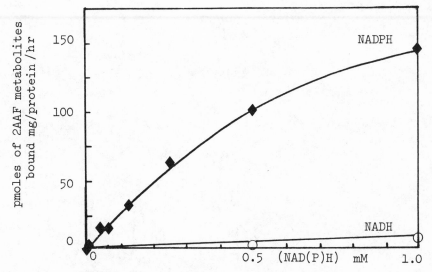

Figure 2.   Dependence on NADPH of 2AAF binding
              to microsomal membrane proteins

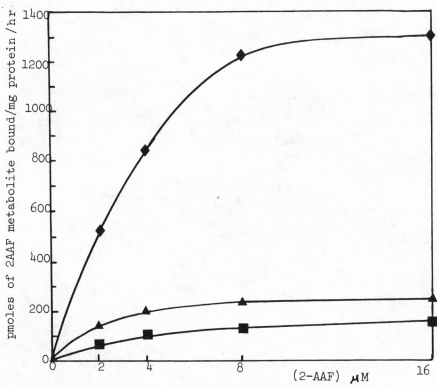

Figure 3.   Binding of radioactive 2AAF metabolites _in vitro_ to
membrane protein of control, 3-MC- and PB-induced microsomes.

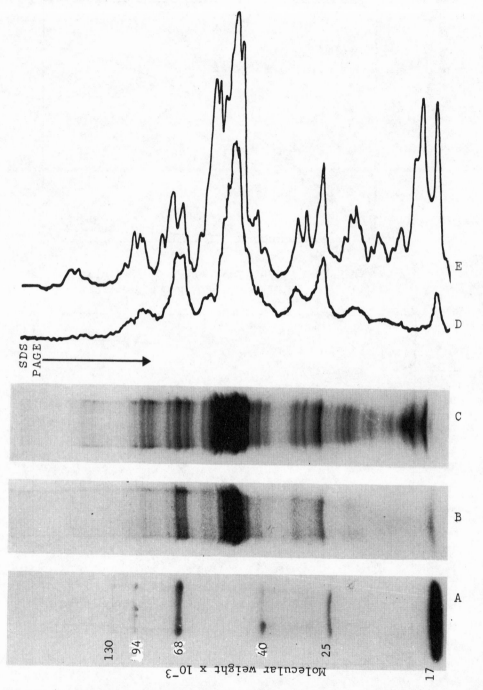

Figure 4. SDS-PAGE of microsomal proteins after incubation with 2AAF
A: Standards. B: Autoradiogram. C: Stained proteins. D,E: Scans of B,C

presence of cytosolic proteins, which are also themselves
slightly labelled. Labelling of microsomal proteins is reduced in
the presence of glutathione or other thiol compounds (10 mM).

SDS-Polyacrylamide gel electrophoresis of microsomal proteins
after incubation in vitro with radiolabelled 2-acetylaminofluorene
shows that the incorporation of label is mainly in a few protein
bands. The molecular weights of these bands are close to those of
cytochrome P-450, cytochrome b5 and their reductases (Figure 4).

Although the extent of labelling of these proteins increases
with time of reaction, the pattern of labelling is not altered
(Figure 5).

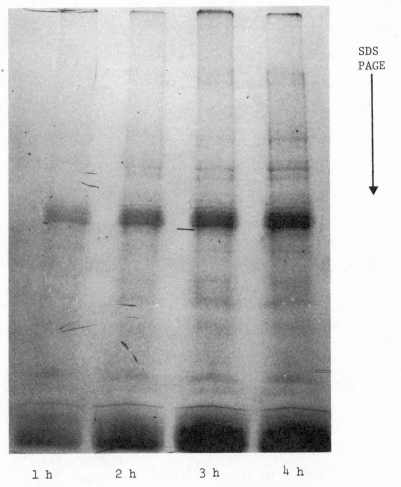

SDS
PAGE

1 h   2 h   3 h   4 h

Figure 5. Time-course of incorporation of label from 2AAF into
specific microsomal proteins. Autoradiograms of SDS-PAGE gels

With microsomes from 3-methylcholanthrene-induced rats, the pattern of labelling is similar to that with control microsomes, but with microsomes from phenobarbital-induced rats only a single band is labelled significantly. This band corresponds in molecular weight to a phenobarbital-induced cytochrome P-450 apoprotein (Figure 6).

With microsomes from rats pretreated with 2-acetylamino-fluorene, the pattern of labelling is similar to that with controls and 3-methylcholanthrene-induced rats, but there is also extensive labelling of a further band (Figure 7). This band has a molecular weight of 48,000 - 50,000 and is specifically induced by 2-acetylaminofluorene pretreatment (Figure 7), and on these grounds we identify it as epoxide hydratase.

The selectivity of labelling of microsomal proteins after incubation with radiolabelled 2-acetylaminofluorene in vitro is confirmed by two-dimensional resolution of the microsomal proteins, using isoelectric focussing in non-ionic detergent followed by SDS/polyacrylamide gel electrophoresis (Figure 8).

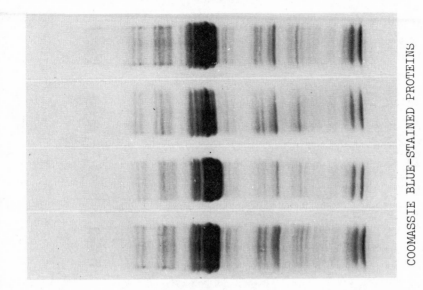

COOMASSIE BLUE-STAINED PROTEINS

Control   PB   3-MC   3-MC
           Day 1   Day 3

SDS PAGE

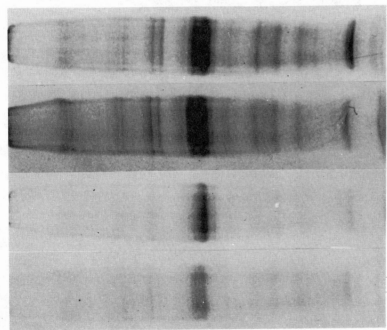

AUTORADIOGRAMS

Control   PB   3-MC   3-MC
           Day 1   Day 3

Figure 6. The effect of pre-treatments of rats with PB and 3-Mc on the pattern of labelling after in vitro incubation with radiolabelled 2-acetylaminofluorene.

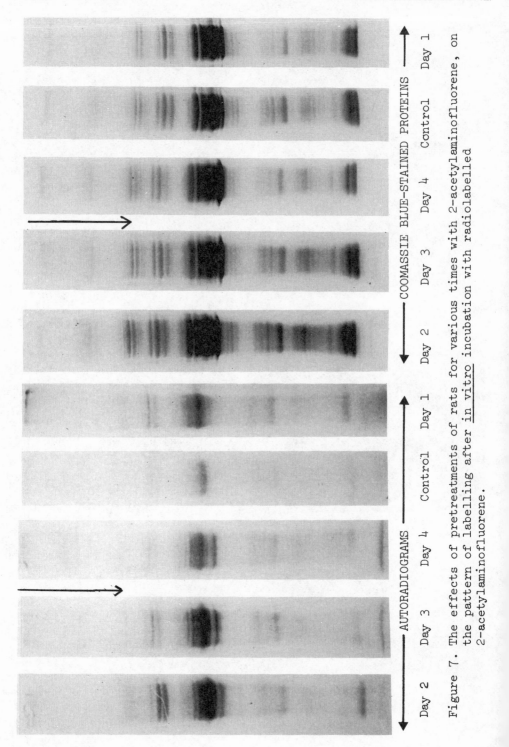

Figure 7. The effects of pretreatments of rats for various times with 2-acetylaminofluorene, on the pattern of labelling after in vitro incubation with radiolabelled 2-acetylaminofluorene.

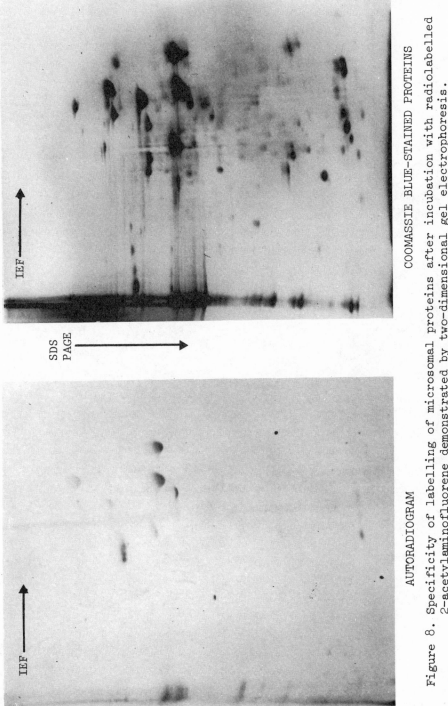

COOMASSIE BLUE-STAINED PROTEINS

AUTORADIOGRAM

Figure 8. Specificity of labelling of microsomal proteins after incubation with radiolabelled 2-acetylaminofluorene demonstrated by two-dimensional gel electrophoresis.

METABOLISM AND MUTAGENIC ACTIVATION OF 2-ACETYLAMINOFLUORENE BY HUMAN LIVER AND LUNG

A R Boobis, M J Brodie, M E McManus, N Staiano[*],
S S Thorgeirsson[*] and D S Davies

Department of Clinical Pharmacology, Royal Postgraduate
Medical School, London, UK and [*]National Cancer
Institute, NIH, Bethesda, USA

INTRODUCTION

2-Acetylaminofluorene (2-AAF) is a potent hepatocarcinogen (e.g. Kreik, 1974). The parent compound is not mutagenic but is activated by N-hydroxylation and subsequent deacetylation or conjugation (Miller, 1970; Schut et al, 1978). In rodents, 2-AAF is N-hydroxylated by a form of cytochrome P-450 inducible by hydrocarbons and associated with benzo(a)pyrene hydroxylase activity (Thorgeirsson et al, 1975).

With microsomal fractions the rate limiting step in the conversion of 2-AAF to a mutagen was N-hydroxylation (Felton et al, 1976). The tumorigenic effects of 2-AAF have also been shown to correlate well with N-hydroxylating activity (Miller et al, 1964). However, it has recently been shown that the Cotton rat, although possessing high N-hydroxylating activity and the ability to activate 2-AAF to a mutagen in vitro, is resistant to the tumorigenic effects of this arylamine (Schut & Thorgeirsson, 1978). In addition, 2-AAF does not appear to be carcinogenic in the Rhesus monkey, despite this species having N-hydroxylating activity comparable to that of the Sprague-Dawley rat (Thorgeirsson et al, 1978).

It is not known whether 2-AAF is carcinogenic in man. We have therefore investigated the ability of microsomal fractions of human liver to metabolise 2-AAF and the capacity of subcellular fractions of human liver and lung to convert 2-AAF and N-hydroxy-2-acetylaminofluorene (N-OHAAF) to mutagens.

METHODS

Human liver samples were obtained by wedge biopsy at laparotomy in patients undergoing biopsy for diagnostic histology. Tissue surplus to histological requirement was made available for these studies with permission of the local Research Ethics Committee. Microsomal fractions were isolated from biopsy samples as previously described (Boobis et al, 1980) and kept frozen at -80°C until required. Human lung samples were from patients undergoing pneumonectomy, usually for bronchogenic carcinoma. Only non-tumerous tissue was used in these studies, the samples being kept at -80°C until preparation of subcellular fractions which were used on the day of preparation.

Aryl hydrocarbon hydroxylase activity (Atlas et al, 1976) and cytochrome P-450 content (Boobis et al, 1980) were assayed as previously described. 2-AAF hydroxylation was determined by the method of Schut & Thorgeirsson (1978) under conditions that were linear with respect to protein and time. Mutagenesis towards S. typhimurium TA98A was determined as described in detail elsewhere (Schut et al, 1978). With human liver samples 100-200 ug microsomal protein were incubated per plate and with human lung samples 1-2.5 mg postmitochondrial protein were incubated per plate. Substrate concentrations were 25 ug 2-AAF and 10 ug N-OHAAF per plate.

Table 1.   2-Acetylaminofluorene metabolism by microsomal fractions of human liver

Activity[a] (pmol product mg^{-1} min^{-1})	Species		
	Man	Sprague-Dawley rat[b]	Rhesus monkey[c]
7-hydroxylation	497 + 56	387 + 17	1290
9-hydroxylation	57 + 10	322 + 13	N.D[d]
5-hydroxylation	39 + 10	186 + 10	113
3-hydroxylation	239 + 39	34 + 2	73
N-hydroxylation	145 + 22	38 + 5	26
AHH activity	30.9 + 5.2	61 + 8	244
Cytochrome P-450 (nmol mg^{-1})	0.51 + 0.05	N.D[d]	0.89

[a] Activities represent mean + SEM for man (n = 9; all with preserved hepatic architecture) and Sprague-Dawley rat (n = 3 pools of microsomes) and mean for Rhesus monkey (n = 2)
[b] Data taken from Schut & Thorgeirsson (1978)
[c] Data taken from Thorgeirsson et al (1978)
[d] ND = not determined

The source and quality of all materials used were as described in the methods papers cited above.

RESULTS

Microsomal fractions of human liver were relatively active in metabolising 2-AAF, total oxidative metabolism being greater than that of male Sprague-Dawley rats (Table 1). Rank order of activities with human liver was 7-hydroxylation > 3-hydroxylation > 9-hydroxylation > 5-hydroxylation. Thus man is much more active at N- and 3-hydroxylation of 2-AAF than either rat or Rhesus monkey but less active at 9- and 5-hydroxylation.

In biopsies with preserved hepatic architecture cigarette smoking was associated with significant increases in AHH and 2-AAF 3-hydroxylase activities (Table 2) but not with any significant change in the other 2-AAF hydroxylases or in total cytochrome P-450 content.

A correlation matrix for 2-AAF hydroxylases, AHH and cytochrome P450 content is shown in Table 3. The correlation between 2-AAF 3-hydroxylase and AHH is significant (r = 0.778) but both enzyme activities were increased in cigarette smokers. When the correlation coefficient was calculated for the non-smokers it was 0.532 (p > 0.1). Also, the correlation between 2-AAF 7-hydroxylase

Table 2. Effect of cigarette smoking on 2-AAF metabolism, AHH activity and cytochrome P-450 content of microsomal fractions of human liver

| Enzyme | Activity (pmol product mg^{-1}min^{-1})[a] | | |
	Non-smokers	Smokers	p[b]
2-AAF			
N-hydroxylase	135 + 33	162 + 25	>0.5
7-hydroxylase	473 + 90	536 + 37	>0.5
9-hydroxylase	68 + 14	39 + 7	>0.1
5-hydroxylase	36 + 16	43 + 5	>0.5
3-hydroxylase	174 + 31	347 + 39	<0.02
AHH	20.4 + 2.2	37.0 + 3.3	<0.01
Cytochrome P-450 (nmol mg^{-1})	0.45 + 0.06	0.54 + 0.06	>0.25

[a]Values are mean + SEM for 5 non-smokers and 3 smokers in whom liver biopsies had preserved hepatic architecture

[b]Comparison between non-smokers and smokers by 2-tailed Student's t test.

and AHH is relatively poor (r = 0.668) but only AHH activity was increased in cigarette smokers. In non-smokers the correlation coefficient increases to 0.770. Of particular note is the lack of a significant correlation between 2-AAF N-hydroxylase and AHH activities (r = 0.554). In non-smokers the correlation coefficient was only 0.452 (p > 0.1).

Human liver microsomal fractions activated both 2-AAF and N-OHAAF into a mutagen (Table 4) whereas human lung post-mitochondrial fractions activated only N-OHAAF into a mutagen. There was considerable inter-individual variation in the ability of human tissue samples to activate 2-AAF to a mutagen. Liver microsomes showed a 20-fold range in the activation of 2-AAF and a six-fold range in the activation of N-OHAAF. The ability of microsomal fractions of human liver to activate 2-AAF to a mutagen did not correlate with N-hydroxylase activity (Table 5) nor with any other enzyme activity measured. Mutagenic activation of 2-AAF and N-OHAAF was unaffected by cigarette smoking. When activation of 2-AAF was compared with activation of N-OHAAF (Fig. 1) a highly significant correlation was found. (r = 0.903, p < 0.001).

Table 3.   Correlation matrix[a] for 2-AAF hydroxylase activities, AHH activity and cytochrome P-450 content of microsomal fractions of human liver

			2-AAF			
	Cyt P-450	AHH	7-OH	9-OH	5-OH	3-OH
AHH	0.790**					
2-AAF 7-OH	0.863***	0.668*				
9-OH	0.455**	0.063	0.721*			
5-OH	0.743**	0.439**	0.814**	0.649*		
3-OH	0.703*	0.778**	0.575**	-0.015	0.534***	
N-OH	0.810*	0.554	0.739**	0.478	0.888***	0.698*

[a]Correlation coefficients (r) are shown for 11 biopsies

*p < 0.05   **p < 0.01   ***p < 0.001

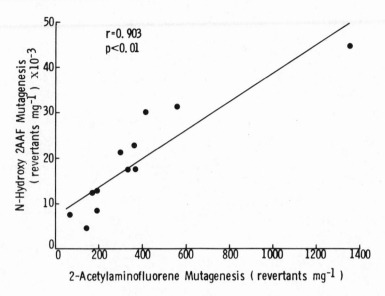

Fig. 1.   Correlation between in vitro activation of 2-AAF and of
          N-OHAAF to mutagens by human liver microsomal fractions
          from 12 subjects.

Table 4.   Mutagenic activation of 2-AAF and N-OHAAF by human liver
           and lung

Mutagen	Reversion rate (revertants mg^{-1}protein plate^{-1})[a]	
	Human liver microsomes[b]	Human lung S9
2-AAF[c]	418 + 127 (67 − 1348)	< 2 (0 − 3.4)
N-OHAAF[d]	20960 + 4300 (7490 − 45,210)	151 + 27 (0 − 239)

[a]Values are mean + SEM (range) of 9 human liver samples and 8
 human lung samples. Reversion rates were corrected for values
 obtained in blank incubations and for amount of protein used.

[b]Microsomal fractions from only those biopsies with preserved
 hepatic architecture were included

[c]25 ug 2-AAF per plate

[d]10 ug N-OHAAF per plate

Table 5. Comparison of reversion rates by 2-AAF and N-OHAAF with
rates of 2-AAF metabolism by human liver

| Enzyme activity | Correlation coefficient (r)[a] | |
	2-AAF	N-OHAAF
7-hydroxylase	0.508	0.562
9-hydroxylase	0.240	0.448
5-hydroxylase	0.143	0.208
3-hydroxylase	0.393	0.207
N-hydroxylase	0.042	0.005
AHH	0.331	0.428
Cytochrome P-450 content	0.178	0.288

[a]None of the correlation coefficients is significant at the 5% level.
n = 12 (for $p = 0.05$ and n = 12, r = 0.576).

DISCUSSION

Microsomal fractions of human liver are active in the oxidative
metabolism of 2-AAF, confirming the work of others (Weisburger et
al, 1964; Dybing et al, 1979). When compared with rat liver there
were important differences in the regio-specific metabolism of 2-
AAF. Human liver is more active at 3- and N-hydroxylation than the
rat but less active at 5- and 9-hydroxylation. Thus, either the
complement of cytochromes P-450 in rat and human liver differ
considerably or the regio-specificity of human and rat liver cytochrome P-450 varies markedly.

Cigarette smoking increases AHH activity of human liver
(Pelkonen et al, 1975; Boobis et al, 1980) and this has been
attributed to the inducing effects of polycyclic aromatic hydrocarbons absorbed from the cigarette smoke. Polycyclic aromatic
hydrocarbons induce AHH activity in many species (Thorgeirsson et
al, 1979) and their inducing effects are also associated with
increases in a number of other enzymes including 2-AAF N-hydroxylase, particularly in rodents (Thorgeirsson et al, 1975) but also
in the Rhesus monkey (Thorgeirsson et al, 1978). However, in
neonatal rabbit liver, and rabbit kidney at all ages, polycyclic
aromatic hydrocarbons selectively induce AHH activity in the
absence of any change in 2-AAF N-hydroxylase activity (Atlas et
al, 1977). From the effects of cigarette smoking on the two
enzyme activities in the present study it appears that this is the
situation that attains in human liver. However, much more work is
required before this can be established.

Correlation analysis of the 2-AAF hydroxylase activities suggests that at least three different forms of cytochrome P-450 are involved in the oxidative metabolism of 2-AAF by human liver. One form appears to be involved primarily in the 5-, 7- and N-hydroxylation of 2-AAF, a second form appears to produce mainly 9-OHAAF and a third form, which is inducible by cigarette smoking, is responsible for 3-hydroxylation of 2-AAF and benzo a pyrene hydroxylation (AHH activity). Further work is in progress to characterise the specificity of the different forms of human liver P-450. In contrast to rodent liver, with human liver there is no correlation between the ability to N-hydroxylate 2-AAF and to activate it to a mutagen. Yet, undoubtedly, N-hydroxylation is the initial step in this process (Miller, 1970) and it has been shown to be rate limiting in several species (Felton et al, 1976). However, with microsomal fractions of human liver, activation of 2-AAF to a mutagen did correlate with activation of N-OHAAF to a mutagen. This latter process is believed to depend upon the activity of a microsomal deacetylase (Schut et al, 1978). Thus, it is tempting to conclude from these data that the deacetylation of N-OHAAF is the rate limiting step in the activation of 2-AAF to a mutagen by microsomal fractions of human liver. This hypothesis is currently under investigation.

The importance of the initial N-hydroxylation of 2-AAF in its conversion to a mutagen is underlined by the results obtained with human lung samples. We had already demonstrated that human lung possesses very low cytochrome P-450 levels and correspondingly low monooxygenase activity (McManus et al, 1980). In the present study it was found that human lung samples did not activate 2-AAF to a mutagen. That this was due to low monooxygenase activity was demonstrated by the fact that the same lung samples could convert N-OHAAF to a mutagen, thus showing that human lung possesses the enzymes necessary for the second step in the activation of 2-AAF.

Thus, both human liver and lung are capable of activating N-OHAAF to a mutagen but only liver is able to activate the parent compound. The implications of these observations for the toxicity of 2-AAF and related compounds in man have yet to be determined.

ACKNOWLEDGEMENTS

We are indebted to Professor J R Belcher, Mr A J Armistead and Mr R Y Frempong of the London Chest Hospital for making human lung samples available to us for these studies. We also acknowledge the excellent cooperation of the Department of Surgery, Hammersmith Hospital in making human liver samples available for use in these studies. We thank Mr D Levitt for his assistance in parts of this work. The expert secretarial assistance of Ms Anne Morgan is greatly appreciated. The work was supported in part by a grant from the Medical Research Council.

REFERENCES

Atlas, S.A., Boobis, A.R., Felton, J.S., Thorgeirsson, S.S. &
    Nebert, D.W., 1977, Ontogenetic expression of polycyclic
    aromatic compound-inducible monooxygenase activities and forms
    of cytochrome P-450 in rabbit. Evidence for temporal control
    and organ specificity of two genetic regulatory systems,
    J. Biol. Chem., 252: 4712
Atlas, S.A., Vesell, E.S. & Nebert, D.W., 1976, Genetic control of
    inter-individual variations in the inducibility of aryl hydro-
    carbon hydroxylase in cultured human lymphocytes, Cancer Res.,
    36: 4619
Boobis, A.R., Brodie, M.J., Kahn, G.C., Fletcher, D.R.,
    Saunders, J.H. & Davies, D.S., 1980, Monooxygenase activity of
    human liver in microsomal fractions of needle biopsy specimens,
    Brit. J. Clin. Pharmacol., 9: 11
Dybing, E., von Bahr, C., Aune, T., Glaumann, H., Levitt, D.S. &
    Thorgeirsson, S.S., 1979, In vitro metabolism and activation of
    carcinogenic aromatic amines by subcellular fractions of human
    liver, Cancer Res., 39: 4206
Felton, J.S., Nebert, D.W. & Thorgeirsson, S.S., 1976, Genetic
    differences in 2-acetylaminofluorene mutagenicity in vitro
    associated with mouse hepatic arylhydrocarbon hydroxylase
    induced by polycyclic aromatic compounds, Mol. Pharmacol., 12: 225
Kreik, E., 1974, Carcinogenesis by aromatic amines, Biochim. Biophys.
    Acta., 355: 177
McManus, M.E., Boobis, A.R., Pacifici, G.M., Frempong, R.Y.,
    Brodie, M.J., Kahn, G.C., Whyte, C. & Davies, D.S., 1980,
    Xenobiotic metabolism by the human lung, Xenobiotica, 26: 481
Miller, J.A., 1970, Carcinogenesis by chemicals; an overview -
    C.H.A. Claves Memorial Lecture, Cancer Res., 30: 559
Miller, E.C., Miller, J.A., Enomoto, M., 1964, The comparative
    carcinogenicities of 2-acetylaminofluorene and its N-hydroxy
    metabolite in mice, hamster and guinea pigs, Cancer Res., 24: 2018
Pelkonen, O., Jouppila, P., Kaltiala, E.H. & Karki, N.T., 1975,
    Aryl hydrocarbon hydroxylase and cytochrome P-450 in human liver:
    fetal development and cigarette smoking, in: "Proc. Eur. Soc.
    Tox. Vol. XVI, Developmental and genetic aspects of drug and
    environmental toxicity," W.A.M. Duncan, D. Julou & M. Kramer,
    eds., Excerpta Medica, Amsterdam
Schut, H.A. & Thorgeirsson, S.S., 1978, In vitro metabolism and
    mutagenic activation of 2-acetylminofluorene by subcellular
    liver fractions from Cotton rats, Cancer Res., 38: 2501
Schut, H.A.J., Wirth, P.J. & Thorgeirsson, S.S., 1978, Mutagenic
    activation of N-hydroxy-2-acetylaminofluorene in the Salmonella
    test system: the role of deacetylation by liver and kidney
    fractions from mouse and rat, Mol. Pharmacol., 14: 682

Thorgeirsson, S.S., Atlas, S.A., Boobis, A.R. & Felton, J.S.,
    1979, Species differences in the substrate specificity of hepatic
    cytochrome P-448 from polycyclic hydrocarbon-treated animals,
    Biochem. Pharmacol., 28: 217
Thorgeirsson, S.S., Felton, J.S. & Nebert, D.W., 1975, Genetic
    differences in the aromatic hydrocarbon-inducible N-hydroxylation
    of 2-acetylaminofluorene and acetaminophen-induced toxicity in
    mice, Mol. Pharmacol., 11: 159
Thorgeirsson, S.S., Sakai, S. & Adamson, R.H., 1978, Induction of
    monooxygenase in Rhesus monkeys by 3-methylcholanthrene:
    metabolism and mutagenic activation of N-2-acetylaminofluorene
    and benzo a pyrene, J. Nat. Cancer Inst., 60: 365
Weisburger, J.H., Grantham, P.H., Vanhorn, E., Steigbigel, N.H.,
    Rall, D.P. & Weisburger, E.K., 1964, Activation and detoxification
    of N-2-fluorenylacetamide in man, Cancer Res., 24: 475

IN VITRO INHIBITION OF LIVER MICROSOMAL ARYLAMIDE N-HYDROXYLASE

C.Razzouk, M.Mercier and M.Roberfroid
Laboratory of Biotoxicology
Université Catholique de Louvain
UCL-73.69 B-1200
Brussels, Belgium

INTRODUCTION

2-acetylaminofluorene (2-AAF) is carcinogenic to various animal species including rat, hamster and mouse (Miller et al, 1960 and Miller et al, 1964). It is also mutagenic in the Ames test in the presence of liver S9, liver microsomes or liver nuclei (Nelson and Thorgeirsson, 1976) and (Sakai et al, 1978). Both carcinogenicity and mutagenicity require cytochrome P-450 dependent N-hydroxylation, the first step in the metabolic activation of this precursor genotoxic agent (Cramer et al, 1960).

Like most mixed function oxidases, the liver microsomal N-hydroxylase is inducible by 3-methylcholanthrene (3-MC) (Lotlikar et al, 1967).

The aim of the present communication is to analyse the in vitro effects of various chemicals on the liver microsomal N-hydroxylase activity from control or 3-MC pretreated rat, hamster and mouse.

MATERIALS and METHODS

Chemicals

All chemicals used were of the purest grade available. 1- and 3-acetylaminofluorene were kindly provided by Dr.H.Gutmann.

Animals and treatments

Male Wistar rats (200-250 g), male golden Syrian hamsters
(100-120 g), male NMRI mice (20-30 g) were obtained from the
"proefdierencentrum, Katholieke Universiteit te Leuven "Belgium".
They were fed semi-synthetid diet (AO3, UAR, Epinay-sur-Orge,
France) and water ad libitum. Food was withdrawn 24h before sacrifice by decapitation.

Pretreatment consisted of two i.p. injections of 3-MC (40 mg/
kg b.w. in corn oil) 48 and 24h prior to sacrifice. Controls
received an equivalent amount of corn oil.

Liver cell fractionation, protein determination and enzymatic
assay

The liver microsomes from control and pretreated rats,
hamsters and mice were prepared by applying the method of de Duve
as described by Amar-Costesec et al. (1974). Protein content was
determined by the method of Lowry et al., (1951), with crystalline
bovine serum albumin as a standard. The activity of the N-hydroxylase was assayed by applying the electron capture gas chromatographic method previously reported to measure the N-hydroxymetabolite
formed after incubation of 2-AAF in the presence of microsomes
(Razzouk et al, 1978). For details concerning the conditions of
the incubations please refer to (Razzouk et al, 1980 a).

RESULTS

Biochemical properties of the liver microsomal N-hydroxylase

Among the three animal species, the mouse has the highest
liver microsomal N-hydroxylase activity, followed by the hamster
and the rat. In term of enzyme affinity the rat appears to be
the leader followed by the hamster and the mouse. Pretreatment
with 3-MC largely induces the N-hydroxylase activity in all three
species : 30 fold in the rat, 12 fold in the hamster and 5 fold
in the mouse. That pretreatment also modifies the K_M of the
enzyme. It reduces it in both the hamster and the mouse liver
whereas it increases it in the rat liver (table 1).

Table 1. Kinetic parameters of liver microsomal 2-AAF N-hydroxylase from control and 3-MC pretreated rat, hamster and
mouse. The K_M and V_{max} are expressed as their computerized values $\pm$ S.D.

Species	Rat		Hamster		Mouse	
Treatment	Control	3MC treated	Control	3MC treated	Control	3MC treated
KM × 10⁻⁶M 2-AAF	0.53 ± 0.02	1 ± 0.01	0.93 ± 0.15	0.23 ± 0.03	1 ± 0.03	0.17 ± 0.04
V. max N-OH-2-AAF/mg Prot./min	13.2 ± 0.4	423 ± 17	139 ± 7.2	1664 ± 56	225 ± 27	1248 ± 54

Computerized values ± s.d.

In vitro inhibition of liver microsomal N-hydroxylase

It has been reported previously (Razzouk et al, 1980 c) that
submicromolar concentrations of 3-MC, 7,8-benzoflavone and
benzo(a)pyrene inhibits the liver microsomal 2-AAF N-hydroxylase
from both control and 3-MC pretreated rat and hamster. That effect
is also shown with liver microsomes from control or induced mouse.
At 0.1 micromolar concentration the inhibition is 60 to 70 % for
microsomes from control mouse, it reaches 80 % after 3-MC pretreatment. Paraoxon (diethyl-p-nitrophenylphosphate) which is an
inhibitor of microsomal deacetylase (Irving, 1966) also acts as
an inhibitor of the N-hydroxylase (Razzouk et al, 1980 c). Such an
effect previously reported for rat and hamster liver microsomes
has also been shown to occur with mouse liver microsomes. In that
case the N-hydroxylase activity decreases progressively with
increasing concentrations of paraoxon. At 5 µM the enzyme activity
(at 0.25 µM concentrations of 2-AAF) is inhibited by 20 %, at
50 µM by 70 % and at 200 µM by 85 %.

Among the various position-isomers of acetylaminofluorene,
the 2-isomer is the only strong carcinogen. The 1- and 3-isomers
are weak carcinogens whereas the 4-isomer is totally inactive
(Morris et al, 1960). Such differences could be due to a selective
affinity of the microsomal N-hydroxylase for the isomer in position 2 since no one of the other isomers appears to be N-hydroxylated to a significant content (Razzouk unpublished results).

As shown in figure 1, 1- and 3-acetylaminofluorene strongly inhibit the 2-AAF N-hydroxylation whereas the 4-isomer does not. The 1- and 3-isomers appear thus to bind to the enzyme which is unable to N-hydroxylate them whereas the 4-isomer does not even bind.

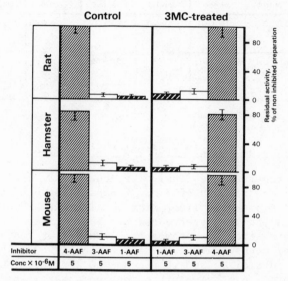

Figure 1. In vitro effect of 1,3 and 4-acetylaminofluorene on the 2-AAF N-hydroxylase of control and 3-MC treated rat, hamster and mouse. The concentration of 2-AAF was 0.25 μM. Each value is the mean of at least 3 different incubations ± S.D.

Added in vitro to the incubation mixture, norharman has been shown to activate the S9-mediated mutagenicity of various chemicals including 2-AAF toward Salmonella typhimurium on the Ames test (Umezawa et al, 1978). Added in vitro to the incubation medium (at 50 μM) norharman but also harman inhibit the liver microsomal N-hydroxylation of 2-AAF (0.25 μM). Liver microsomes from control or 3-MC-treated rat, hamster and mouse are all sensitive to that inhibitory effect which varies from 60 % (rat) to 80 % (hamster and mouse). As shown in figure 2, the effect of harman is dose-dependent.

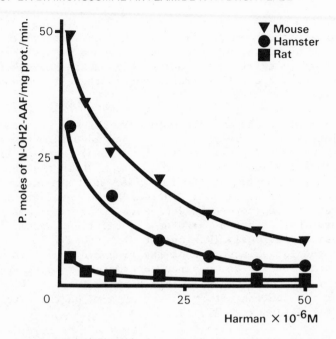

Figure 2. In vitro inhibition of liver microsomal 2-AAF N-hydroxylase by Harman. Dose-effect relationships. The concentration of
2-AAF was 0.25 µM.

DISCUSSION

    Among the three animals studied, the mouse appears to be the
most effective in N-hydroxylating 2-AAF. Although it is a species
less susceptible to its carcinogenic effects (Miller et al, 1964).
There are many possible explanations for such an apparent discrepancy. The low affinity of the murine liver enzyme for 2-AAF as
compared to the rat could be part of it. The absence or low activity of some of the further activating enzymes (sulfotransferase,
deacetylase...) could be another hypothesis.

The presence in mouse liver microsomes of enzymic activiti(es) further catalyzing the metabolism of N-hydroxy metabolite(s) (results not shown) cannot be excluded. Such a metabolic pathway has been shown to exist in the Guinea-pig liver (Razzouk et al, 1980b), a species known to be resistant to the carcinogenicity of 2-AAF (Miller et al, 1964).

Pretreatment of the animals with 3-MC largely induces the N-hydroxylase activity in rat, hamster and mouse. That effect appears to be inversely related to the initial level of enzyme activity since the inducing effect is the most in the rat and the least in the mouse. In the rat liver such a pretreatment has a very peculiar effect since it increases the K_M of the microsomal N-hydroxylase whereas it decreases it in both hamster and mouse liver. That effect of 3-MC on the rat liver enzyme has been discussed previously (Razzouk et al, 1980 a). It could be related to the large induction of the ring hydroxylation which occurs in that species as compared to hamster or mouse (Lotlikar et al, 1967). We have suggested that this could be part of the explanation for the inhibitory effect of 3-MC on the carcinogenicity of 2-AAF (Razzouk et al, 1980 c).

Among the position-isomers of AAF, 1- and 3-acetylaminofluorene which are known to be weak carcinogens when compared to 2-AAF (Morris et al, 1960), strongly inhibit its N-hydroxylation. The isomer in position 4 is completely ineffective, it is also non-carcinogenic. Those data indicate that the spatial configuration of 4-acetylaminofluorene could be significantly different from that of the other isomers. The microsomal enzyme being unable to recognize it as a substrate would not bind it. The 1- and 3-isomers are probably different enough not to be good substrate for the N-hydroxylation reaction but similar enough to bind to the active site of the enzyme.

Norharman is known to activate the S9-mediated mutagenicity of various chemicals including 2-AAF (Umezawa et al, 1978) Such an effect has been called co-mutagenesis. It was thus of interest to study the effect of norharman and its analog harman on the microsomal N-hydroxylation of 2-AAF. Both compounds which are produced by the pyrolysis of tryptophan are strong inhibitors of the arylamide N-hydroxylase in a dose dependent manner. The consequence, if any, of such an effect should be a reduction and not an activation of the enzyme-mediated mutagenicity.

There is thus a discrepancy between the biochemical data and the results in mutagenicity testing. Such a discrepancy could have many explanations : (1) in the mutagenicity test, S9 is used as an activating enzyme system instead of purified microsomes ; (2) the concentrations (μM range) of both 2-AAF and norharman or harman we used are much lower than those used in mutagenicity experiments (mM range) ; (3) the concentration of the enzyme in the Ames test (100 to 200 μl/plate) is also much higher than the concentration of microsomal proteins in our assay (0.02 to 0.1 mg/ml). Furthermore according to Sugimura (these proceedings) the effect of norharman could be of the co-initiation type at the level of the genome and not at all related to the metabolic activation process. Finally it is important to indicate that norharman is more co-mutagenic than harman (Umezawa et al, 1978) and that both compounds also inhibit the activity of aryl hydrocarbon hydroxylase (Levitt et al, 1977) in vitro even though they activate the S9 mediated mutagenicity of benzo(a)pyrene (Umezawa et al, 1978). Co-mutagenicity by norharman and inhibition of the microsomal activation of 2-AAF appear thus to be unrelated phenomena. In conclusion, liver microsomal N-hydroxylases from rat, hamster and mouse appear to be very similar enzymes even though the levels of their activities are different. They are all inducible by 3-MC, they are all inhibited by the same chemicals (3-MC, benzo(a)pyrene, 7,8 benzoflavone, paraoxon, 1- and 3-acetylaminofluorene, harman and norharman), they are all not affected by the presence of 4-acetylaminofluorene. The induction by 3-MC does not modify their susceptibility to the various inhibitors. The only differences between the three species are at the level of the K_M of the enzyme. Such differences in enzyme affinities are however apparent differences due to feed back effects of the ring hydroxylated metabolites on the N-hydroxylating enzyme.

REFERENCES

Amar-Costesec, A., Beaufay, H., Wibo, M., Thines-Sempoux, D., Feytmans, R., Robbi, M. and Berthet, J., 1974, Analytical study of microsomes and isolated subcellular membranes from rat liver. II. Preparation and composition of the microsomal fraction.
J.Cell.Biol., 61:201-212.

Cramer, J.W., Miller, J.A. and Miller, E.C., 1960, N-hydroxy-
        lation : a new metabolic reaction observed in the rat
        with the carcinogen 2-acetylaminofluorene.
        J.Biol.Chem., 235:885-888.
Irving, C.C., 1966, Enzymatic deacetylation of N-hydroxy-2-ace-
        tylaminofluorene by liver microsomes.
        Cancer Res., 26:1390-1396.
Levitt, R.C., Legaverend, C., Nebert, D.W. and Pelkonen, O.,1977,
        Effects of harman and norharman on the mutagenicity and
        binding to DNA of benzo(a)pyrene metabolite in vitro
        and on aryl hydrocarbon hydroxylase induction in cell
        culture.
        Biochem.Biophys.Res.Commun., 79:1167-1175.
Lotlikar, P.D., Enomoto, M., Miller, J.A. and Miller, E.C.,
        1967, Species variations in the N-and ring-hydroxyla-
        tion of 2-acetylaminofluorene and effects of 3-methyl-
        cholanthrene pretreatment.
        Proc.Exp.Bio.Med., 125:341-346.
Lowry, L.H., Rosebrough, M.J., Farr, A.L. and Randall, R.L.,
        1951, Protein measurement with the folin phenol reagent,
        J.Biol.Chem., 193:265-275.
Miller, J.A., Cramer, J.W. and Miller, E.C., 1960, The N- and
        ring-hydroxylation of 2-acetylaminofluorene during
        carcinogenesis in the rat.
        Cancer res., 20:950-962.
Miller, E.C., Miller, J.A. and Enomoto, M., 1964, The compara-
        tive carcinogenicities of 2-acetylaminofluorene and its
        N-hydroxy metabolite in mice, hamster and Guinea-pigs.
        Cancer res., 24:2018-2031.
Morris, H.P., Velat, C.A., Wagner, B.P., Dahlgard, M. and
        Ray, F.E., 1960, Studies of carcinogenicity in the rat
        of derivatives of aromatic amines related to N-2-fluo-
        renylacetamide.
        J.Natl.Cancer Inst., 24:149-180.
Nelson, W.L. and Thorgeirsson, 1976, Structural requirements
        to mutagenic activity of 2-acetylaminofluorene in the
        Salmonella test system.
        Biochem.Biophys.Res.Commun., 71:1201-1206.
Razzouk, C., Agazzi-Léonard, E., Batardy-Grégoire, M.,
        Mercier, M., Poncelet, F. and Roberfroid, M., 1980 a,
        Competitive inhibitory effect of microsomal N-hydroxy-
        lase. A possible explanation for the in vivo inhibition
        of 2-acetylaminofluorene carcinogenicity by 3-methyl-
        cholanthrene.
        Toxicol.Lett., 5:61-67.

Razzouk, C., Evrard, E., Lhoest, G., Roberfroid, M. and
        Mercier, M., 1978, Isothermal gas chromatography with
        wall-coated glass capillary columns. Electron-capture
        detection and a solid injector. II. Application to the
        assay of 2-fluorenylacetamide N-hydroxylase activity
        in a rat-liver microsomal system.
        J.Chromatogr., 161:103-109.
Razzouk, C., Mercier, M. and Roberfroid, M., 1980 b, Characte-
        rization of the Guinea-pig liver microsomal 2-fluoreny-
        lamine and 2-fluorenylacetamide N-hydroxylase,
        Cancer Lett., 9, 123-131.
Razzouk, C., Mercier, M. and Roberfroid, M., 1980 c, Induction,
        activation and inhibition of hamster and rat liver
        microsomal arylamide and arylamine-N-hydroxylase,
        Cancer Res., in press.
Sakai, S., Reinhold, C.E., Wirth, P.J. and Thorgeirsson, S.S.,
        1978, Mechanism of in vitro mutagenic activation and
        covalent binding of N-hydroxy-2-acetylaminofluorene in
        isolated liver cell nuclei from rat and mouse.
        Cancer Res., 38:2058-2067.
Umezawa, K., Shirai, A., Matsushima, T. and Sugimura, T., 1978,
        Comutagenic effect of norharman and harman with 2-ace-
        tylaminofluorene derivatives,
        Proc.Natl.Acad.Sci., USA, 75:928-930.

REACTION SCHEMES FOR THE DEGRADATION OF CYTOCHROME P-450

BY ALLYL-ISO-PROPYLACETAMIDE AND FLUROXENE

Kathryn M. Ivanetich, Melanie R. Ziman and
Jean J. Bradshaw

Department of Medical Biochemistry, University of
Cape Town Medical School, Cape Town, South Africa 7925

The porphyrinogenic compound allyl-iso-propylacetamide (AIA) and the anesthetic agent fluroxene (2,2,2-trifluoroethyl vinyl ether) degrade the heme of hepatic cytochrome P-450 in vivo and in vitro.[1-4] Both of these compounds require prior metabolic activation to reactive species by cytochrome P-450 in order to degrade the heme of cytochrome P-450.[3,4] In order to gain further insight into the mechanism of the degradation of hepatic cytochrome P-450 by AIA and fluroxene, several possible reaction schemes have been evaluated using computer analysis.

EXPERIMENTAL

All experiments were conducted with male Wistar rats (180-220g). Hepatic microsomes were isolated by gel filtration as described by Tangen et al.[5] Other experimental techniques are as described elsewhere.[3,4,6,7]

Reaction schemes were assessed using the mimic program on an Univac model 1180 computer. The mimic program – a digital simulation program – was written by H.E. Petersen and F.J. Samson of the Systems Engineering Group of Wright-Patterson Airforce Base in May 1965. The program was updated in July 1967 by the I.S.D. systems group of Univac, and is commercially available from this firm.

RESULTS AND DISCUSSION

Eight reaction schemes were devised for the degradation of hepatic microsomal cytochrome P-450 hemoproteins by AIA and fluroxene. The cytochrome P-450 dependent conversion of fluroxene to 2,2,2-trifluoroethanol is included in the schemes for fluroxene (see

Figure 1). The reaction schemes for AIA are identical to those for fluroxene except that k_3 (the rate constant for the production of 2,2,2-trifluoroethanol) has been set to zero.

Fig. 1. Reaction schemes for the degradation of cytochrome P-450 hemoproteins by fluroxene[*]

SCHEME 1[+]

$$F + PB450 \underset{k_2}{\overset{k_1}{\rightleftharpoons}} [F\text{-}PB450] \xrightarrow{k_3} TFE + PB450$$
$$\downarrow k_4$$
$$M$$

$$M + PB450 \xrightarrow{k_5} X450$$

$$M + P448 \xrightarrow{k_6} X448$$

SCHEME 2[+]

$$F + PB450 \underset{k_2}{\overset{k_1}{\rightleftharpoons}} [F\text{-}PB450] \xrightarrow{k_3} TFE + PB450$$

$$F + P448 \underset{k_6}{\overset{k_5}{\rightleftharpoons}} [F\text{-}P448] \xrightarrow{k_7} M + P448 \xrightarrow{k_8} X448$$

$$M + PB450 \xrightarrow{k_4} X450$$

SCHEME 3

$$F + PB450 \underset{k_2}{\overset{k_1}{\rightleftharpoons}} [F\text{-}PB450] \xrightarrow{k_3} TFE + PB450$$
$$\downarrow k_4$$
$$X450$$

$$F + P448 \underset{k_6}{\overset{k_5}{\rightleftharpoons}} [F\text{-}P448] \xrightarrow{k_7} X448$$

SCHEME 4

$$F + PB450 \underset{k_2}{\overset{k_1}{\rightleftharpoons}} [F\text{-}PB450] \xrightarrow{k_3} TFE + P450$$
$$k_4 \updownarrow k_5$$
$$[F\text{-}PB450]^{\ddagger} \xrightarrow{k_6} X450$$

$$F + P448 \underset{k_8}{\overset{k_7}{\rightleftharpoons}} [F\text{-}P448] \xrightarrow{k_9} X448$$

SCHEME 5

$$F + PB450 \; \underset{k_2}{\overset{k_1}{\rightleftharpoons}} \; [F\text{-}PB450] \; \overset{k_3}{\longrightarrow} \; TFE + PB450$$

$$F + PB450 \; \underset{k_5}{\overset{k_4}{\rightleftharpoons}} \; [F\text{-}PB450]^{\ddagger} \; \overset{k_6}{\longrightarrow} \; X450$$

$$F + P448 \; \underset{k_8}{\overset{k_7}{\rightleftharpoons}} \; [F\text{-}P448] \; \overset{k_9}{\longrightarrow} \; X448$$

SCHEME 6

$$F + PB450 \; \underset{k_2}{\overset{k_1}{\rightleftharpoons}} \; [F\text{-}PB450] \; \overset{k_3}{\longrightarrow} \; TFE + PB450$$

$$F + PB450 \; \overset{k_4}{\longrightarrow} \; X450$$

$$F + P448 \; \overset{k_5}{\longrightarrow} \; X448$$

*Abbreviations used are: F, fluroxene; PB450, phenobarbital induced form of cytochrome P-450; P448, cytochrome P-448; X450 and X448, degraded forms of PB450 and P448, respectively; TFE, 2,2,2-trifluoroethanol; M, reactive metabolite; $[F\text{-}PB450]^{\ddagger}$, an enzyme-substrate complex differing from $[F\text{-}PB450]$.

+Schemes IA and 2A are modifications of schemes 1 and 2 where M reacts with the $[F\text{-}PB450]$ and $[F\text{-}P448]$ complexes, rather than with free PB450 and P448, in the step of the reaction producing X450 and X448.

The levels of the substrates and different forms of cytochrome P-450 were set as follows; AIA, 5.0 mM; fluroxene, 30.0 mM; the phenobarbital inducible form of cytochrome P-450, 1.96 and 0.75 µM in hepatic microsomes from phenobarbital induced and 3-methylcholanthrene induced rats; cytochrome P-448, 1.4 and 2.0 µM in hepatic microsomes from phenobarbital induced and 3-methylcholanthrene induced rats.

The proposed reaction schemes were assessed by comparison with the following experimental data:

(1) The K_i values for AIA and the K_s values for fluroxene for their interaction with hepatic microsomal cytochrome P-450 in vitro.[7,8]

(2) The K_M and V_{max} values for the degradation of cytochrome P-450 by AIA and fluroxene and for the production of 2,2,2-trifluoroethanol by hepatic microsomes in vitro[4].

(3) The relative extents of degradation of different forms of cyto-
     chrome P-450 by AIA and fluroxene in vivo.[9]

(4) The pseudo first order rate constants (k_{obs}) for the degradation
     of cytochrome P-450 by AIA and fluroxene and for the production
     of 2,2,2-trifluoroethanol from fluroxene by hepatic microsomes
     in vitro.[10]

     In assessing each reaction scheme, the rate constants (e.g.,
k_1, k_2, k_3, etc.) were adjusted to provide the best correlation
between the experimentally determined parameters and values of the
parameters calculated from the model using the mimic program.

     For all schemes for AIA and fluroxene, the rate constants were
adjusted so that good agreement (+10%) was achieved between experimental data and the computer mimicked data for the K_i or K_s value,
the V_{max} values and the relative extents of degradation of different
forms of cytochrome P-450 (Data not shown). The rate constants of
the reaction schemes were also adjusted so that the K_M values for the
degradation of cytochrome P-450 and for the production of 2,2,2-tri-
fluoroethanol were as close to the experimental values as possible,
but good agreement between the experimental and calculated results
was not achievable in all cases (Tables 1 and 2). The parameters
for which the rate constants in the schemes were not adjusted and
which were assessed only in the final versions of each scheme were
the pseudo first order rate constants and the extents of reaction
over 65 min for the degradation of cytochrome P-450 and the production of 2,2,2-trifluoroethanol.

     Reaction schemes 1, 1A, 2 and 2A could not provide first order
kinetics for the degradation of cytochrome P-450 by AIA or fluroxene
or for the production of trifluoroethanol. In all cases, plots of
the concentration of cytochrome P-450 versus time were sigmoid, with
plots of ln (cytochrome P-450) versus time being curved. Reaction
schemes 3 through 6 followed pseudo first order kinetics for the
degradation of hepatic microsomal cytochrome P-450 by AIA and
fluroxene and for the production of 2,2,2-trifluoroethanol. For
AIA and fluroxene, all of these schemes except scheme 6 - for which
K_M values were not calculable - could in large part closely mimic
the experimental results. All of these schemes, however, for both
AIA and fluroxene, gave rise to total extents of degradation of
cytochrome P-450 which were equivalent to the total amounts of
hepatic microsomal cytochrome P-450, a phenomenon which was not
observed experimentally (Tables 1 and 2). One possible explanation
for this discrepancy is that one or more forms of cytochrome P-450
present in hepatic microsomes, but not considered in the model
reaction schemes, are not degraded by AIA or fluroxene and not involved in the metabolism of fluroxene to 2,2,2-trifluoroethanol.

Table 1.   Parameters calculated from the final runs of the
           reaction schemes and the experimentally determined
           parameters for the degradation by AIA of cytochrome
           P-450 in hepatic microsomes from phenobarbital
           induced rats*

Scheme	K_M (XTP450) (mM)	$10^2 k_{obs}$ (XTP450) (min^{-1})	XTP450 (μM/65 min)
1	1.0	N C	3.4
1A	1.0	N.C.	3.4
2	1.0	N.C.	3.4
2A	-	N.C.	3.4
3	1.0	5.2	3.2
4	0.5	4.9	3.1
5	1.0	5.0	3.2
6	N.C.	4.8	3.2
Experi-mental Results	0.44	5.2	2.8

* Abbreviations used are XTP450, degraded total type P-450
  cytochromes; N.C., not calculable.

It is concluded that the reaction schemes for the degradation
of cytochrome P-450 by AIA and fluroxene share several similarities.
Suitable reaction schemes involve the production of reactive species
from AIA and fluroxene by at least two forms of cytochrome P-450.
The reactive species in question appear to degrade the same form of
the enzyme (e.g., phenobarbital induced cytochrome P-450, or 3-methyl-
cholanthrene induced cytochrome P-448) that produced the reactive
species.   It might be anticipated that the same enzyme molecule
that activates fluroxene or AIA is subsequently degraded by the activated metabolite that it produced, before that reactive species
diffuses off of the enzyme.

Reaction schemes involving the production of a reactive species
by one form of cytochrome P-450 and the degradation of another form
of cytochrome P-450 by that reactive species are not consistent with
experimental data.   Reaction schemes in which a bi-molecular reaction
of ferri- or ferrocytochrome P-450 (oxygenated or not) with AIA or

Table 2. Parameters calculated from the final runs of the reaction
          schemes and the experimentally determined parameters for
          the degradation of cytochrome P-450 by fluroxene and for
          the metabolism of fluroxene to 2,2,2-trifluoroethanol.*

Scheme	K_M (X450) (mM)	$10^2 k_{obs}$ (XTP450) (min^{-1})	XTP450 (µM/ 65 min)	K_M(TFE) (mM)	$10^2 k_{obs}$ (TFE) (min^{-1})	TFE (µM/65min)
Phenobarbital Induction						
1	1.0	N.C.	3.4	1.0	N.C.	347
1A	1.0	N.C.	3.4	1.0	N.C.	349
2	0.9	N.C.	3.4	0.9	N.C.	353
2A	–	N.C.	3.4	0.9	N.C.	343
3	1.1	5.0	3.2	1.1	5.5	630
4	0.5	5.3	3.2	0.5	5.5	568
5	1.0	4.4	3.1	1.0	5.0	689
6	N.C.	4.2	3.1	0.9	4.7	742
Experimental Results	0.8	4.3	2.1	0.7	3.7	728

Scheme	K_M (X448) (mM)	$10^2 k_{obs}$ (XTP450) (min^{-1})	XTP450 (µM/ 65 min)	K_M(TFE) (mM)	$10^2 k_{obs}$ (TFE) (min^{-1})	TFE (µM/65min)
3-Methylcholanthrene Induction						
1	1.3	N.C.	2.7	1.3	N.C.	174
1A	1.2	N.C.	2.7	1.2	N.C.	166
2	3.3	N.C.	2.7	1.3	N.C.	195
2A	–	N.C.	2.7	N.C.	N.C.	197
3	3.3	8.0	2.6	1.2	4.3	321
4	3.0	6.5	2.6	1.1	3.6	325
5	3.2	7.1	2.7	1.3	6.0	198
6	N.C.	5.9	2.6	1.3	3.7	327
Experimental Results	3.3	4.9	2.8	1.3	3.4	232

*Abbreviations used are TFE, 2,2,2-trifluoroethanol; X450, degraded
phenobarbital inducible form of cytochrome P-450; X448, degraded
cytochrome P-448; XTP450, degraded total type P-450 cytochromes;
N.C., not calculable.

fluroxene gives rise to the degradation of cytochrome P-450 do not fit the experimental data, since it is not possible to characterize such reaction schemes with K_M values (Scheme 6) (Tables 1 and 2).

Since fluroxene does not produce a type I difference spectrum of appreciable magnitude with cytochrome P-448[7] and since there is no spectral evidence for the binding of AIA to the substrate binding site of cytochrome P-450 [8], it would appear that for both fluroxene and AIA a cytochrome P-450 – substrate complex distinct from that giving rise to a type I difference spectrum (Scheme 5) may result in the degradation of cytochrome P-450.

ACKNOWLEDGEMENTS

This research was supported by grants from the Medical Research Council, the University of Cape Town Staff Research Fund and Computer Fund, and the Nellie Atkinson Bequest.

REFERENCES

1. F. De Matteis, Loss of haem in rat liver caused by the porphyro-
       genic agent 2-allyl-2-isopropylacetamide, Biochem. J. 124:767
       (1971).
2. W. Levin, E. Sernatinger, M. Jacobson and R. Kuntzman, Destruc-
       tion of cytochrome P-450 by secobarbital and other barbitu-
       rates containing allyl groups, Science 176:1341 (1972).
3. K.M. Ivanetich, J.A. Marsh, J.J. Bradshaw and L.S. Kaminsky,
       Fluroxene mediated destruction of cytochrome P-450 in vitro,
       Biochem. Pharmac. 24:1933 (1975).
4. J.A. Marsh, J.J. Bradshaw, S.A. Lucas, L.S. Kaminsky and K.M.
       Ivanetich, Further investigations of the metabolism of
       fluroxene and the degradation of cytochromes P-450 in vitro,
       Biochem. Pharmac. 26:1601 (1977).
5. O. Tangen, J. Jonsson and S. Orrenius, Isolation of rat liver
       microsomes by gel filtration, Anal. Biochem. 54:597 (1973).
6. K.M. Ivanetich, J.J. Bradshaw, J.A. Marsh, G.G. Harrison and
       L.S. Kaminsky, The role of cytochrome P-450 in the toxicity
       of fluroxene anaesthesia in vivo, Biochem. Pharmac. 25:773
       (1976).
7. K.M. Ivanetich, J.J. Bradshaw, J.A. Marsh and L.S. Kaminsky,
       The interaction of hepatic microsomal cytochrome P-450 with
       fluroxene in vitro, Biochem. Pharmac. 25:779 (1976).
8. G.D. Sweeney and J.D. Rothwell, Spectroscopic evidence of inter-
       action between 2-allyl-2-isopropylacetamide and cytochrome
       P-450 of rat liver microsomes, Biochem. Biophys. Res. Commun.
       55:798 (1973).

9. J.J. Bradshaw, M.R. Ziman and K.M. Ivanetich, The degradation
    of different forms of cytochrome P-450 in vivo by fluroxene
    and allyl-iso-propylacetamide, Biochem. Biophys. Res. Commun.
    85:859 (1978).

10. K.M. Ivanetich, J.A. Marsh, J.J. Bradshaw and L.S. Kaminsky,
    Further studies of the fluroxene mediated destruction of
    hepatic microsomal cytochromes P-450 in vitro, in "Microsomes
    and Drug Oxidations," V. Ullrich et al., eds., Pergamon Press,
    Oxford (1977).

SPECIFIC INACTIVATION OF ANILINE HYDROXYLASE BY A REACTIVE INTERMEDIATE FORMED DURING ACRYLAMIDE BIOTRANSFORMATION BY RAT LIVER MICROSOMES

Elsa Ortiz, J. M. Patel, and Kenneth C. Leibman

Department of Pharmacology
University of Florida Medical School
Gainesville, Fla. 32610

Forms of cytochrome P-450 act as "suicide enzymes" when acting upon certain unsaturated substrates. In some cases, green pigments that are adducts of the activated substrate with protoporphyrin IX from the cytochrome heme are formed; the substrate specificity for this phenomenon appears to require merely a terminal double bond with at least one hydrogen atom on the β-carbon atom and a lack of steric hindrance at the allylic position (Ortiz de Montellano and Mico, 1980). Other unsaturated compounds cause conversion of cytochrome P-450 to P-420 without formation of green pigments or loss of heme (Patel et al., 1980). We report here the selective inactivation of a kinetically distinguishable form of aniline hydroxylase by a metabolic product of acrylamide ($CH_2 = CH - CONH_2$), with corresponding loss of heme, but without green pigment formation.

When acrylamide was incubated for 30 min at pH 7.7 with microsomes prepared from lyophilized 9000g supernatant fraction of livers of phenobarbital-pretreated, male, Holtzman rats (Leibman, 1971) in the presence of an NADPH-generating system, cytochrome P-450 (18-30% of different preparations) was lost (Table 1). No loss occurred during incubation in the absence of acrylamide, or when the NADPH-generating system was omitted. ATP and nicotinamide had no effect upon the activation of acrylamide, and were omitted from subsequent experiments. Mg^{++} and EDTA accelerated cytochrome P-450 loss.

Activation of acrylamide was linear with time for 30 min. Substrate saturation occurred at about 4 mM acrylamide. The pH dependence of activation was remarkable; no loss of cytochrome P-450 occurred at pH 7.5 (indeed, the phenomenon was discovered

Table 1. Cofactor Requirements for Formation of Reactive
          Intermediate from Acrylamide

Omission[a]	% Loss of Cytochrome P-450
None	18.6
NADPH-generating system	0.0
ATP	17.1
Nicotinamide	17.1
Mg^{++}	10.0
EDTA	6.3
Acrylamide	0.0

[a]The standard system contained 0.24 mM $NADP^+$, 10 mM
glucose 6-phosphate, 2 units of glucose 6-phosphate
dehydrogenase, 6 mM ATP, 10 mM $MgSO_4$, 1.2 mM EDTA,
8 mM acrylamide, and 10 mg of microsomal protein in
5 ml of 0.2 M Tris, pH 7.7, incubated in air-filled,
stoppered bottles for 30 min at 37°C in the dark.

when a faulty pH meter allowed the preparation of a falsely
labeled buffer), and the optimum was rather narrow (pH 7.7-7.9).
Activity at pH 8.1 was about half that at pH 7.9. The optimal
microsomal protein concentration was 6-12 mg per 5-ml incubation
volume; when greater amounts of microsomes were added, the
percentage of the cytochrome P-450 destroyed was decreased.
NADH did not substitute for NADPH, nor were the two nucleotides
synergistic (Table 2). The activation required oxygen, and was
inhibited by carbon monoxide and by metyrapone, an inhibitor of
many cytochrome P-450-dependent mixed-function oxidase reactions
(Netter et al., 1967; Leibman, 1969). Neither the epoxide
hydrase inhibitor (Oesch et al., 1971) 1,2-epoxy-3,3,3-trichloro-
propane (0.05 or 0.5 mM) nor styrene oxide (5 mM) diminished the
loss of cytochrome P-450 caused by incubation with acrylamide
(data not shown).

Incubation with acrylamide caused similar cytochrome P-450
losses in liver microsomes from phenobarbital-treated and
untreated rats (Table 3). The loss of heme was similar to that
of cytochrome P-450. There was no corresponding loss of NADPH-
cytochrome c reductase activity, and the form of cytochrome

Table 2. Nucleotide, Oxygen, and Cytochrome P-450
          Requirements for Formation of Reactive
          Intermediate from Acrylamide

Conditions[a]	% Loss of Cytochrome P-450
Nucleotides (0.3 mM each)	
None	0.0
NADPH	26.2
NADH	0.0
NADPH + NADH	24.5
Gas phase	
Air	26.2
N_2	0.0
CO/O_2 (4:1, v/v)	0.0
Metyrapone	
0.01 mM	26.2
0.1 mM	12.0
0.5 mM	0.0

[a]Except as noted, the system consisted of 10 mM $MgSO_4$,
1.2 mM EDTA, 8 mM acrylamide, 3 mM nucleotide, and
12 mg of microsomal protein in 5 ml of 0.2 M Tris
buffer (pH 7.7), incubated in air-filled, stoppered
bottles for 30 min at 37°C in the dark.

P-450 lost was apparently not involved in benzphetamine N-
demethylase activity.

Loss of cytochrome P-450 was not accompanied by an increase
in absorbance at 420 nm in the reduced CO difference spectrum.
Furthermore, no green pigment absorbing at 416 nm was extractable
by the method of Unseld and De Matteis (1976) (data not shown).

Two kinetically distinguishable forms of aniline hydroxylase,
one of high affinity (low K_M) and the other of low affinity
(high K_M) have recently been described for hamster liver microsomes (McCoy, 1980). We have found that similar forms of
aniline hydroxylase activity can be demonstrated in liver
microsomes from male, Holtzman rats (Table 4). The high-affinity
form I activity was lost to the same extent after incubation
with acrylamide in preparations from untreated and phenobarbital-
treated rats, whereas no loss of the low-affinity form II
activity occurred in either preparation. The K_M value for
form I was increased 3- to 3.5-fold after acrylamide preincubation

Table 3. Effects of Acrylamide Metabolite on Rat Liver Microsomal Enzyme Activities

Pretreatment of Rats	Preincubation with Acrylamide	Cytochrome P-450[a,b]	Heme[a,c]	NADPH-Cytochrome c Reductase[a,d]	Benzphetamine N-Demethylase[a,e]
		nmol/mg		nmol/min/mg	
None	0	0.95		160	2.6
	+	0.72 (24%)		158 (1%)	2.5 (4%)
Phenobarbital[f]	0	2.4	3.1	245	4.2
	+	1.8 (25%)	2.3 (26%)	234 (4%)	4.1 (2%)

[a] Numbers in parentheses represent percentage loss. Protein was determined by the method of Lowry et al. (1951). Data are from one representative experiment of three.

[b] Determined by the method of Omura and Sato (1964).

[c] Determined by the method of Falk (1964).

[d] Assayed by the method of Williams and Kamin (1962) as modified by Hook et al. (1972).

[e] Assayed by the method of Fouts and Devereux (1972).

[f] 75 mg/kg/day × 3 days.

Table 4. Effect of Acrylamide Metabolite upon Aniline
          Hydroxylase Activities

Pretreatment of Rats	Preincubation with Acrylamide	Aniline Hydroxylase[a]		
		Total (16 mM Aniline)	Form I (0.1 mM Aniline)	Form II (Total − I)
		nmol/min/mg		
None	0	1.5	0.9	0.6
	+	1.2	0.6 (33%)	0.6 (0%)
Phenobarbital[b]	0	2.8	2.0	0.8
	+	2.2	1.4 (30%)	0.8 (0%)

[a]Microsomes were preincubated for 30 min with 0 or 8 mM acrylamide
and appropriate cofactors. The incubation mixtures were then
centrifuged for 1 hr at 105,000g, resuspended, and assayed by
the method of Chhabra et al. (1972), except that 0.1 M Tris
buffer (pH 7.7) was used in 15-min incubations. Data are from
one representative experiment of three.
[b]75 mg/kg/day × 3 days.

in preparations from untreated and phenobarbital-treated rats,
whereas that of form II was unchanged by the acrylamide metabolite
in either preparation (Table 5).

Table 5. Effect of Acrylamide Metabolite on Rat Liver
          Microsomal Aniline Hydroxylase K_M

Preincubation with Acrylamide[a]	Uninduced		Phenobarbital	
	I	II	I	II
	mM			
0	0.025	5.8	0.036	1.70
+	0.077	5.8	0.125	1.70

[a]Conditions were as in Table 4.

The nature of the reactive intermediate formed from acrylamide is as yet unknown. The lack of inhibition by epoxide hydrase inhibitors of cytochrome P-450 destruction does not demonstrate that an epoxide intermediate is not involved; if an epoxide reacted at the site on cytochrome P-450 at which it was formed, it would not be subject to hydration by epoxide hydrase.

Destruction of cytochrome P-450 by incubation of rat liver microsomes with acrylamide had been previously noted (Ivanetich et al., 1978). In that report, however, extensive loss of NADPH-cytochrome c reductase activity was also observed, whereas no such loss was found in our experiments; the reason for this discrepancy is not immediately apparent. It is interesting to note that a related compound, acrolein ($CH_2 = CH - CHO$), causes conversion of cytochrome P-450 to P-420 without loss of heme or formation of green pigment (Patel et al., 1980), and that another related compound, allyl chloride ($CH_2 = CH - CH_2Cl$), causes the formation of green pigment when incubated with liver microsomes (Patel and Leibman, unpublished data). Further studies on the effects of compounds of this series are in progress in our laboratory.

REFERENCES

Chhabra, R. S., Gram, T. E., and Fouts, J. R., 1972, A comparative study of two procedures used in the determination of hepatic microsomal aniline hydroxylation, Toxicol. Appl. Pharmacol., 22:50.

Falk, J. E., 1964, "Porphyrins and Metalloporphyrins," p. 181, Elsevier, Amsterdam.

Fouts, J. R., and Devereux, T. R., 1972, Developmental aspects of hepatic and extrahepatic drug-metabolizing enzyme systems: microsomal enzymes and components in rabbit liver during the first month of life, J. Pharmacol. Exp. Ther., 183:458.

Hook, G. E. R., Bend, J. R., and Fouts, J. R., 1972, Mixed-function oxidases and the alveolar macrophage, Biochem. Pharmacol., 21:3267.

Ivanetich, K. M., Lucas, S., Marsh, J. A., Ziman, M. R., Katz, I. D., and Bradshaw, J. J., 1978, Organic compounds: their interaction with and degradation of hepatic microsomal drug-metabolizing enzymes *in vitro*, Drug Metab. Dispos. 6:218.

Leibman, K. C., 1969, Effects of metyrapone on liver microsomal drug oxidations, Mol. Pharmacol., 5:1.

Leibman, K. C., 1971, Reduction of ketones in liver cytosol, Xenobiotica, 1:97.

Lowry, O. H., Rosebrough, N. J., Farr, A. L., and Randall, R. J., 1951, Protein measurement with the Folin phenol reagent, J. Biol. Chem., 193:265.

McCoy, G. D., 1980, Differential effects of ethanol and other inducers of drug metabolism on the two forms of hamster liver microsomal aniline hydroxylase, Biochem. Pharmacol., 29:685.

Netter, K. J., Jenner, S., and Kajuschke, K., 1967, Über die Wirkung von Metyrapon auf den mikrosomalen Arzneimittelabbau, Naunyn-Schmiedebergs Arch. Pharmakol. Exp. Pathol., 259:1.

Oesch, F., Kaubisch, N., Jerina, D. M., and Daly, J. W., 1971, Hepatic epoxide hydrase: structure-activity relationships for substrates and inhibitors, Biochemistry, 10:4858.

Omura, T., and Sato, R., 1964, The carbon monoxide-binding pigment of liver microsomes, II. Solubilization, purification, and properties, J. Biol. Chem., 239:2379.

Ortiz de Montellano, P. R., and Mico, B. A., 1980, Destruction of cytochrome P-450 by ethylene and other olefins, Mol. Pharmacol., 18:128.

Patel, J. M., Ortiz, E., and Leibman, K. C., 1980, Selective inactivation of rat liver microsomal NADPH-cytochrome c reductase by acrolein, Fed. Proc., 39:865.

Unseld, A., and De Matteis, F., 1976, Isolation and partial characterisation of green pigments produced in rat liver by 2-allyl-2-isopropylacetamide, in: "Porphyrins in Human Diseases," M. Doss, ed., S. Karger, Basel.

Williams, C. H., Jr., and Kamin, H., 1962, Microsomal triphosphopyridine nucleotide cytochrome c reductase of liver, J. Biol. Chem. 237:587.

MICROSOMAL METABOLISM OF ACRYLONITRILE IN LIVER AND BRAIN*

A.E. Ahmed and M.E. Abreu

Department of Pathology and
Department of Pharmacology and Toxicology
University of Texas Medical Branch
Galveston, Texas 77550

INTRODUCTION

Acrylonitrile (VCN) is a widely used monomer in the plastic and acrylic fiber industry. Epidemiological studies indicated a significantly increased incidence of lung and colon cancer among workers exposed to VCN [1]. Chronic exposure of rats to VCN resulted in development of microgliomas of the central nervous system, zymbal gland carcinoma, cell papillomas of the stomach, and squamous cell carcinoma [2,3]. VCN is a mutagen in several Salmonella typhimurium strains in the presence of metabolic activation systems [4]. High doses of VCN produce CNS dysfunction, adrenal necrosis, and congestive lung edema [5,6,7].

Based on the elevation of plasma urinary thiocyanate concentrations of VCN treated animals, metabolic studies suggested that VCN is metabolized to cyanide which is then detoxified to thiocyanate [8,9]. However little is known concerning the mechanism of cyanide liberation from VCN. The work presented herein describes our in vitro studies on the metabolism of VCN to cyanide. The objective is to characterize the metabolizing enzymes and the metabolic pathway in liver, the major site of xenobiotic metabolism, as well as in brain, the major site of VCN toxicity.

MATERIALS AND METHODS

Acrylonitrile (99%) was obtained from Aldrich Chemical Co. All other chemicals, reagents, and cofactors were of reagent grade

---

* Supported by Grant No. ES-01871 from the National Institutes of Health.

and were obtained from commercial sources.

Rats (Male, Sprague-Dawley, Charles River Breeding Co.) weighing 225-275 g were used through these studies. Animals were allowed free access to food and water. For enzyme modulation studies, rats (3 to 4/group) were treated as follows: Phenobarbital was administered in drinking water (1 g/liter water) for 10 days; Aroclor 1254 was injected i.p. in corn oil daily (500 mg/kg) for 5 days prior to sacrifice; $CoCl_2$ was injected s.c. at a dose of 40 mg/kg, 24 and 48 hours before sacrifice; or 3 methylcholanthrene was injected i.p. in corn oil at a dose of 40 mg/kg once daily for 4 days. Control animals received either 0.5 ml corn oil i.p. or were left untreated.

Animals were sacrificed by decapitation, livers were perfused with iced 0.25 M sucrose. Brains were processed without perfusion. In experiments using brain tissue, whole brains were pooled from 15 rats. Subcellular fractionation was carried out by standard differential centrifugation techniques [10]. Fractions were used immediately or stored at -60° up to 2 weeks with no change observed in metabolic activity.

Incubation mixtures contained 50 µmol of potassium phosphate buffer (pH 7.5), 15 µmol $MgCl_2$, 3 µmol NADPH, 3 mg liver microsomal protein and 150 µmol VCN in a final volume of 3 ml contained in tightly capped 30 ml glass serum vials. Reactions were begun by addition of VCN and stopped by immersion of vials in ice. Except where indicated, incubations were carried out at 37° for 30 min in a shaking water bath.

Cyanide content of the incubation mixtures was determined by a silver sulfide electrode as previously described [11]. Protein was determined by the Lowry method [12]. Cytochrome P-450 concentration was determined by the method of Dallner et al. [13].

Kinetic parameters (K_m and V_{max}) of VCN metabolism were calculated by Lineweaver-Burke analysis using a BASIC program [14]. Statistical analysis was accomplished using Student's t test.

RESULTS AND DISCUSSION

In vitro studies were designed to elucidate the reaction mechanism and to identify the enzyme system involved in the biotransformation of VCN to CN^-.

Subcellular localization studies of VCN metabolism to CN^- in liver and brain demonstrate that the enzymes involved in this biotransformation are localized in the microsomal fraction of liver and brain (Table 1). Because of the limited rate of metabolism in brain, hepatic enzymes were used throughout this study to further

TABLE 1. Subcellular Localization of VCN Metabolism to CN in Liver and Brain.

Fraction	$\dfrac{\text{pmoles } CN^-}{\text{mg prot x min}}$	
	Liver	Brain
Nuclear	15.7 ± 3.3	4.76 ± 1.57
Mitochondrial	N.D.	N.D.
Cytosolic	N.D.	N.D.
Microsomal	143.1 ± 15.1	11.78 ± 1.09

N.D. Not detectable.

characterize the metabolism of VCN to cyanide. Maximal activity of the microsomal metabolism of VCN is observed upon the addition of $MgCl_2$ and NADPH to the incubation mixture (Table 2). Molecular oxygen was also required as shown by the decreased rate (21% of control) of VCN metabolism to CN⁻ under anaerobic conditions (Table 2).

TABLE 2 . Co-factor Requirements for Microsomal Metabolism of VCN to CN.

Co-factor	$\dfrac{\text{pmoles } CN^-}{\text{mg prot x min}}$
None	N.D.
NADPH	55.8 ± 12.6
NADPH+ $MgCl_2$	133.3 ± 11.6
NADPH generating systems	88.7 ± 13.9

NADPH was added to give a final concentration of 1 mM and the $MgCl_2$ concentration was 5 mM. The NADPH generating system consisted of isocitrate (3.3 mM), $NADP^+$ (0.33 mM) and isocitrate dehydrogenase (1 unit).

N.D. Not detectable.

Assay conditions for the hepatic microsomal conversion of VCN to cyanide were characterized with respect to protein concentration, time of reaction and the pH and temperature of incubation mixtures. The reaction was linear with time approximately 30 minutes, showed a pH optimum near pH 7.5 and had an optimal incubation temperature around 37°C. CN^- production increased as a function of microsomal protein concentration up to 1 mg/ml. Based on these studies the following conditions were employed for subsequent experiments: a temperature of 37°C, pH of 7.5, 30 min incubation period and a protein concentration of 1 mg/ml.

The possible involvement of the microsomal cytochrome P-450 mixed function oxidase system was further characterized by investigating the effects on VCN metabolism to CN^- of various compounds known to alter the content and substrate characteristics of cytochrome P-450. As can be seen from the data in Table 3, liver microsomes obtained from PB or Aroclor 1254 treated rats metabolized VCN to CN^- at a greater rate than control microsomes. Induction of VCN microsomal metabolism was also observed with 3-MC treatment, however the increase was not as substantial as that obtained with either PB or Aroclor 1254. Decreasing the hepatic P-450 content by treatment with $CoCl_2$ resulted in a parallel decrease in the rate of VCN metabolism.

TABLE 3 . Effect of P-450 Modulation on Microsomal

Metabolism of VCN to CN.

Treatment	$\dfrac{\text{nmoles P-450}}{\text{mg prot}}$ % control	$\dfrac{\text{pmoles } CN^-}{\text{mg prot} \times \text{min}}$ % control
Phenobarbital	201%*	478%*
Aroclor 1254	202%*	414%*
3-Methylcholanthrene	133%	141%
$CoCl_2$	45%*	53%*

Incubations and animal treatments were carried out as described in Materials and Methods.

*Significantly different from control (p<.001).

The kinetics of the metabolism of VCN to CN⁻ was determined using hepatic microsomal preparations from PB or Aroclor treated rats. The apparent Km for VCN of microsomes from PB or Aroclor treated rats was approximately 20% the value for the control preparation (Table 4).

TABLE 4 . Effect of Aroclor and Phenobarbital Treatment on

Kinetic Parameters of Microsomal Metabolism of VCN to CN.

Treatment	Km (mM)	Vmax
Control	190.7 ± 19.7	429.4 ± 36.2
Phenobarbital	54.8 ± 9.5[a]	2192.0 ± 281.6[a]
Aroclor 1254	40.9 ± 4.1[a]	896.8 ± 62.4[a]

Six VCN concentrations (10 to 300 mM) were used with each preparation. Vmax is expressed as pmoles CN mg protein^{-1} minute^{-1}.

[a] Significantly different from control (p<.001).

The effects of inhibitors (SKF 525-A and CO) of the cytochrome P-450 mixed function oxidase system were investigated (Table 5). SKF 525-A markedly inhibited the microsomal metabolism of VCN to CN⁻. Incubation mixtures carried out under 80% CO / 20% O_2 resulted in a substantial decrease in VCN metabolism (Table 5).

TABLE 5 . Effect of SKF 525-A and CO Addition on

Microsomal Metabolism of VCN to CN.

	% Control
Control	100
+ 5 x 10^{-6}M SKF 525-A	45
+ 5 x 10^{-5}M SKF 525-A	36
+ 5 x 10^{-4}M SKF 525-A	28
+ 80% CO / 20% O_2	44

The established relationship between the rate of metabolism of VCN to CN⁻ and treatments known to alter cytochrome P-450 contents or activity strongly support the contention that this compound is metabolized to CN⁻ via cytochrome P-450 dependent systems.

Since the oxidative metabolism of numerous other aliphatic olefins proceeds via epoxide intermediates[15, 16], we proposed that the oxidative metabolism of VCN may also proceed via an epoxide. Such an epoxide intermediate would be highly unstable and could undergo further transformation as shown in Figure 1. One possible transformation, shown in the middle pathway of Figure 1, would be a nonenzymatic rearrangement with subsequent release of CN⁻ and oxidation products. This rearrangement could involve hydride or cyanide transfer followed by hydrolysis.

A second possible transformation is an enzymatic hydration via epoxide hydratase. This reaction would also yield oxidation products and CN⁻. Thus, the effect on VCN metabolism to CN⁻ of 1,1,1-trichloropropene, 2,3-oxide (TCPO), a potent inhibitor of epoxide hydratase [17] was investigated. The data in Table 6 indicate that addition of 10^{-4} or 10^{-3} M TCPO in DMSO decreased CN formation from VCN 26%, while the addition of 10^{-2} M TCPO resulted in an 81% decrease. These results are in agreement with the proposed transformation pathway.

TABLE 6 . Effect of TCPO on Microsomal Metabolism of

VCN to CN.

TCPO Concentration	$\dfrac{\text{pmoles CN}^-}{\text{mg prot} \times \text{min}}$	% Control
0	130.6 ± 7.3	100
1×10^{-4}M	96.5 ± 4.9	74
1×10^{-3}M	97.1 ± 6.6	74
1×10^{-2}M	24.4 ± 1.3	19

TCPO (1,1,1-trichloropropane 2,3-oxide) was dissolved in DMSO. DMSO was present in all incubations at a concentration of 0.3% (v/v).

Alternatively, the proposed epoxide intermediate could be transformed via an interaction with a sulfhydryl compound. Attack of a
nucleophilic sulfur atom on the β-carbon of the proposed epoxide
intermediate could lead to formation of a β-thioethylene cyanohydrin. Rearrangement of this cyanohydrin to an aldehyde could result in CN^- release. Addition of GSH and other sulfhydryl compounds
to the incubation media enhanced the liberation of CN^- from VCN
(Table 7).

TABLE 7 . Effect of Sulfhydryl Compound Addition on

Microsomal Metabolism of VCN to CN.

Addition	% Control
None	100
GSH	226
Cysteine	154
D-penicillamine	801
2-mercaptoethanol	264

Solutions of the sulfhydryl compounds were adjusted to
pH 7.5 prior to addition to incubation flasks. All
sulfhydryl compounds were added to give a final concentration of 10 mM.

The enhancement of CN^- liberation from VCN by a variety of sulfhydryl compounds suggests that this effect is chemically mediated.
Enzymatic oxidation apparently precedes the chemical reaction with
sulfhydryl groups because CN^- formation was not observed when GSH
was incubated with VCN in the absence of microsomal enzymes. The
detection of S-β-hydroxy ethyl N-acetyl cysteine in urine of VCN
treated rats [18] is consistent with the oxidative mechanism proposed
in Figure 1.

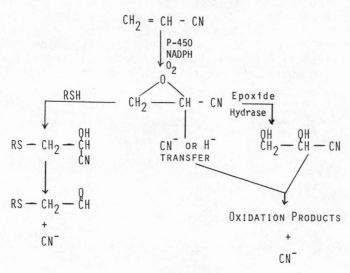

FIGURE 1

In summary, these studies indicate that VCN is metabolized to CN^- via the cytochrome P-450 mixed function oxidase system. An epoxide could be an intermediate product of this oxidative metabolism.

REFERENCES

1.   M. T. O'Berg, Epidemiologic study of workers exposed to acrylonitrile, J. Occup. Med. 22: 245, 1980.

2.   C. Maltoni, A. Ciliberti and V. Dimaio, Carcinogenicity bioassays on rats of acrylonitrile administered by inhalation and by ingestion, Med Lav. 68: 401, 1977.

3.   J. F. Quast, R. M. Enriquez, C. E. Wade, C. C. Humiston, and B. A. Schwetz, Toxicity of drinking water containing toxicity of acrylonitrile in rats, Fed. Regist. 43: 2589, 1978.

4. C. DeMeester, M. Duverger-Van Bogaert, M. Lambotte-Vandepaer, M. Roberfroid, F. Poncelot, and M. Mercier, Liver extract mediated mutagenicity of acrylonitrile, Toxicology 13: 7, 1979.

5. K. Hashimoto and R. Kanai, Studies on the toxicology of acrylonitrile: Metabolism, mode of action and therapy, Ind. Health 3: 3, 1965.

6. M. Abreu and A. E. Ahmed, Studies on the mechanism of acrylonitrile neurotoxicity, Toxicol. Appl. Pharmacol. 48: A54, 1979.

7. S. Szabo, E. S. Reynolds, and K. Kovacs, Animal model of Waterhouse Friederichsen syndrome: Acrylonitrile-induced adrenal apoplexy, Am. J. Pathol. 82: 653, 1976.

8. H. Brieger, F. Rieders, and W. A. Hodes, Acrylonitrile: Spectrophotometric determination, acute toxicity, and mechanism of action. Arch. Int. Hyg. Occup. Med 6: 128, 1952.

9. A. E. Ahmed and K. Patel, Studies on the metabolism of aliphatic nitriles. Toxicol. Appl. Pharmacol. 48: A91, 1979.

10. G. H. Hogeboom, W. C. Schneider and G. E. Palade, Cytochemical studies of mammalian tissues. J. Biol. Chem. 172: 619, 1948.

11. M. E. Abreu and A. E. Ahmed, Metabolism of acrylonitrile to cyanide in vitro studies. Drug Metab. Dispos. (in press) 1980.

12. O. H. Lowry, N. J. Rosebrough, A. L. Farr and R. J. Randall, Protein measurement with the folin phenol reagent. J. Biol. Chem. 193: 265, 1951.

13. G. Dallner, P. Siekevitz and G. E. Palade, Biogenesis of endoplasmic reticulum membranes. II. Synthesis of constitutive microsomal enzymes in developing rat hepatocyte. J. Cell Biol. 30: 97, 1966.

14. W. Cleland, The statistical analysis of enzyme kinetic data. Ad. Enzymol. 29: 1, 1967.

15. E. W. Maynert, R. L. Foreman, and T. Watabe, Epoxides as obligatory intermediates in the metabolism of olefins to glycols. J. Biol. Chem. 245: 5234, 1970.

16. K. C. Leibman and E. Ortiz, Epoxide intermediates in microsomal oxidation of olefins to glycols. J. Pharmacol. Exp. Ther. 173: 242, 1970.

17. F. Oesch and J. Daly, Conversion of naphthalene to trans-naph-
     thalene dihydrodiol: Evidence for the presence of a coupled
     aryl monooxygenase - epoxide hydrase system in hepatic micro-
     somes. Biochem. Biophys. Res. Commun. 46: 1713, 1972.

18. Y. Lin and A. E. Ahmed, Analysis of the urinary metabolites of
     the carcinogen acrylonitrile. Extended Abstr., Annu. Meeting
     Am. Soc. Mass. Spectro., 613, 1979.

TOXICITY AND METABOLISM OF THIOBENZAMIDE

DERIVATIVES IN THE RAT

Robert P. Hanzlik, John R. Cashman, and George J. Traiger

Department of Medicinal Chemistry
University of Kansas
Lawrence, KS 66045

INTRODUCTION

Considerable evidence has been presented to implicate a role
for S-oxidation in the production of tissue injury by a variety of
thiocarbonyl compounds, including arylthioureas (DeMatteis, 1974;
Boyd and Neal, 1976), thioacetamide (Hunter et al., 1977; Porter et
al., 1979), alpha-naphthylisothiocyanate (DeMatteis, 1974; Vyas,
1979), and others. In order to explore the relationship between S-
oxidation and the production of tissue injury in a systematic
fashion, we have investigated the toxicity and metabolism of thiobenzamide and a number of its derivatives, including chemically
synthesized S-oxides, in the rat.

METHODS

Methods for the toxicological studies have been given in references cited in the discussion below.

For the metabolism studies, ^{3}H-thiobenzamide was synthesized in
our laboratory and incubated with rat liver microsomes using standard procedures. Ethanol-quenched aliquots of incubations were
analyzed on silica gel tlc plates having a preadsorbant loading zone.
The ethanol quench solution contained cold standards and 10 ppm
diphenylamine as an antioxidant. The plates were eluted with ethyl
acetate/hexane (63:37) containing 1 ppm diphenylamine.

Alternatively, the enzymatic formation or consumption of thio-
amide-S-oxides could be monitored spectrophotometrically. This
method was verified using the radiometric tlc assay for ^{3}H-thiobenz-

1239

amide oxidation; excellent quantitative agreement was obtained.

RESULTS AND DISCUSSION

Toxicity Studies with Thiobenzamide Derivatives

Thiobenzamide (TB) is a reasonably potent hepatotoxin in rats, producing centrilobular necrosis and increases in plasma glutamic pyruvic transaminase (GPT), plasma bilirubin, and hepatic triglyc- erides, as well as other biochemical changes in the liver (Malvaldi, 1977; Hanzlik et al., 1978; Chieli et al., 1979). The increase in plasma bilirubin is not due to hemolysis, but is associated with a decrease in the ability of the liver both to extract organic anions from plasma and to excrete them into bile (Vyas, 1979). In addition a single dose of TB greatly depresses bile flow (Vyas, 1979). Chronic administration of TB to rats leads to hyperplasia of the bile ducts and extensive morphological alteration of the liver tissue (Malvaldi, 1977).

Para-substituted thiobenzamides produce the same pattern of acute toxic responses as TB, but their potency depends quite strongly on the electronic effect of the substituent, being increased or decreased by substituents which are electron-donating or -withdrawing, respectively (Hanzlik et al., 1978, 1980). At dosages of 0.8 and 2.0 mmole/kg Hammett plots of log [plasma GPT] vs. the substituent parameter σ were strictly linear and parallel, having a slope (Hammett rho value) of -3.4. A similar situation obtained for Hammett plots of log [plasma bilirubin] vs. σ, but the rho value was -1.4 in this case. These results indicate that the biological effects of the thioamides $p\text{-}XC_6H_4CSNH_2$ (where $X = CH_3O$, CH_3, H, Cl, CF_3), are dependent on their chemical properties in a fashion con- sistent with a requirement for S-oxidation(s) for expression of toxicity.

Thiobenzamide-S-oxide (TBSO) is the major metabolite of TB in vitro (see below). When administered to rats it produces the same qualitative pattern of liver injury as TB, but the plasma GPT and bilirubin responses occur earlier and/or are more intense (Hanzlik et al., 1980). Electron donating para substituents enhance the potency of TBSO for elevating plasma GPT (rho = -3.2) but substitu- tion does not affect the plasma bilirubin response (rho = 0). The reason for this divergence is not known, but it is interesting to speculate that the S-oxides directly interfere with bilirubin pro- cessing by the liver, while a second S-oxidation might be required for cytotoxicity and the increase in plasma GPT. Further studies will be required to resolve this issue.

N-Substitution also produced dramatic changes in the toxicity of TB. N,N-Dimethylthiobenzamide produced no overt signs of acute

toxicity in rats at doses up to 2.0 mmole/kg. Another tertiary thiobenzamide derivative, N-(3,4,5-trimethoxybenzoyl)-morpholine (Trithiozine) has been reported to be relatively non-toxic in a variety
of animals and man; it has also been shown to be metabolized, via
its S-oxide, to the corresponding amide (Pifferi et al., 1977). In
contrast to the apparent lack of toxicity with the tertiary thiobenzamides, and to the *hepato*toxicity of the primary thiobenzamides, N-
methylthiobenzamide (NMTB) proved to be a very potent lung toxin in
rats. Like the arylthioureas, NMTB produces pleural effusions and
pulmonary perivascular edema. Relatively few animals survive for
24 hr after a dose of 0.5 mmole/kg. Eight hours after administration
of a 0.3 mmole/kg dose ip to a group of four rats an average (±SE)
of 4.3 ± 0.9 ml of fluid had accumulated in the thoracic cavity,
whereas vehicle-treated controls showed no fluid accumulation.
Furthermore, perivascular edema was clearly apparent upon histological examination of lung sections (Figure 1). We have not
detected significant hepatotoxicity with NMTB at non-lethal doses,
but this question is being explored further through the use of isolated hepatocytes. The extent of analogy in metabolism and toxicity
between NMTB and the arylthioureas is also under active investigation
in our laboratory.

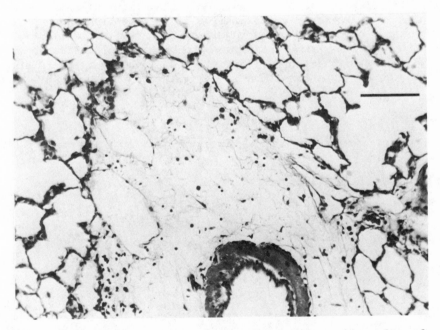

Fig. 1.   Section of rat lung showing extensive perivascular edema
          12 hr after administration of NMTB (0.25 mmole/kg, ip);
          hematoxylin and eosin stain. The marker bar indicates a
          distance of 0.1 mm on the section.

Lastly, we have investigated the toxicity of several 2,6-disub-
stituted thiobenzamides in the rat, including the herbicide Prefix,
which is 2,6-dichlorothiobenzamide. The latter was not hepatotoxic
at doses up to 2.0 mmole/kg. Evidently the lack of toxicity was not
due to the inductive effect of the chlorines, since both 2-chloro-
6-methyl- and 2,6-dimethylthiobenzamide were also non-toxic at that
dose. It is interesting to note that both 2,6- and N,N-disubstitu-
tion forces the thioamide group out of plane and out of conjugation
with the aromatic ring. This sterically-induced electronic effect
alters the chemical properties of the thioamide group, and could
potentially .affect its metabolism as well. However, we have found
that Prefix is efficiently S-oxidized by rat liver microsomes in
vitro. Thus the effect of the substituents is on the chemical and/or
pharmacological properties of the S-oxide, and not on its formation.
The non-toxic tertiary thioamides also undergo enzymatic S-oxidation
as a major route of metabolism (see above).

In Vitro Metabolism of Thiobenzamide

Enzymatic S-oxidation is an important step in the biotransformation of thioacetamide (Ammon et al., 1967; Porter et al., 1979),
ethionamide (Johnston et al., 1967) and trithiozine (Pifferi et al.,
1977), all of which lead to significant quantities of the S-monoxide
or sulfine metabolite. Thiobenzamide is no exception, and rat liver
microsomes S-oxidize ^{3}H-TB extremely efficiently (Table 1). TBSO
formation requires NADPH and oxygen, but is neither inhibited by
SKF-525A or N-octylimidazole, nor induced by phenobarbital pretreatment. However, TBSO formation is partially blocked by phenylthiourea

Table 1. Microsomal Metabolism of Thiobenzamide.[a]

Incubation Conditions	Product Formation	
	TBSO	Benzamide
Complete system	7.44	4.07
Omit cofactor	0.16	0.13
+ SKF-525A (1 mM)	7.55	3.62
+ N-Octylimidazole (1 mM)	7.36	3.37
+ n-Octylamine (1.5 mM)	8.60	4.32
+ Phenylthiourea (2 mM)	5.65	3.25
+ Methimazole (5 mM)	4.56	3.17

[a]Incubations were conducted for 7 min under air
at 33°C with 1 mM ^{3}H-TB and 0.75 mg microsomal
protein/ml. Product formation expressed as
nmole/min/mg protein; average of duplicates.

and methimazole, while it is stimulated by n-octylamine. Based on
these observations the oxidation of TB with rat liver microsomes is
attributable mainly, if not exclusively, to the flavin containing
monooxygenase rather than cytochrome P-450 systems.

In addition to TBSO, a small amount of benzamide is formed but
benzonitrile and p-hydroxythiobenzamide are not observed as metabolites. It is not clear whether the benzamide is formed directly
from TB by chemical decomposition of TBSO, or by further oxidation
of TBSO. Sulfines do break down chemically to amides, but this
process appears to be too slow to account for the benzamide we
observe. On the other hand, the photometric assay clearly shows that
TBSO undergoes a second enzymatic S-oxidation which is slower than
the first, but the products of this oxidation have not yet been identified. In kinetic studies with TB concentrations of 0.1-2.0 mM, and
microsomal protein concentrations of 0.5-2.0 mg/ml, TBSO formation
was linear for up to 15 min, and at longer times the TBSO/benzamide
ratio appears to increase. Lineweaver-Burke plots with various
microsomal preparations yielded apparent K_m and V_{max} values for TBSO
formation in the range of 0.3-1.0 mM and 15-20 nmole/min/mg protein,
respectively.

CONCLUSION

Thiobenzamide and its derivatives produce an interesting spectrum of toxic effects, some of which appear to require metabolic
S-oxidation for expression. The microsomal flavin-containing monooxygenase S-oxidizes all thiobenzamide derivatives we have investigated to their S-monoxide, and in some cases the latter undergoes
a second S-oxidation. Given the dependence of the toxicity of these
compounds on their chemical structure and reactivity, and their
relative amenability to radiolabeling and metabolic studies, they
should prove to be a useful group of model agents for investigating
mechanisms of tissue injury by thiocarbonyl compounds.

ACKNOWLEDGEMENT

Financial support for this research was provided by the NIH
(RR-5606) and the University of Kansas General Research Fund.

REFERENCES

Ammon, R., Berninger, H., Haas, H. J., and Landsberg, I., 1967,
    Thioacetamide-sulfoxide, ein stoffwechselprodukt des thio-
    acetamids, Arzneim. Forsch., 17:521.
Boyd, M. R., and Neal, R. A., 1976, Studies on the mechanism of
    toxicity and of development of tolerance to the pulmonary

toxin, α-naphthylthiourea (ANTU), Drug Metab. Dispos., 4:314.

Chieli, E., Malvaldi, G., and Tongiani, R., 1979, Early biochemical
     liver changes following thiobenzamide poisoning, Toxicology,
     13:101.

DeMatteis, F., 1974, Covalent binding of sulfur to microsomes and
          loss of cytochrome P-450 during oxidative desulfurization
          of several chemicals, Molec. Pharmacol., 10:849.

Hanzlik, R. P., Vyas, K. P., and Traiger, G. J., 1978, Substituent
     effects on the hepatotoxicity of thiobenzamide derivatives
     in the rat, Toxicol. Appl. Pharmacol., 46:685.

Hanzlik, R. P., Cashman, J. R., and Traiger, G. J., 1980, Relative
     hepatotoxicity of substituted thiobenzamides and thiobenz-
     amide-S-oxides in the rat, Toxicol. Appl. Pharmacol., 55:
     (in press).

Hunter, A. L., Holscher, M. A., and Neal, R. A., 1977, Thioacetamide
     induced hepatic necrosis. I. Involvement of the mixed
     function oxidase enzyme system, J. Pharmacol. Exp. Ther.,
     200:439.

Johnston, J. P., Kane, P. O., and Kibby, M. R., 1967, The metabolism
          of ethionamide and its sulfoxide, J. Pharm. Pharmac., 19:1.

Malvaldi, G., 1977, Liver changes following thiobenzamide poisoning,
     Experientia, 33:1200.

Pifferi, G., Ventura, P., Farina, C., and Frigerio, A., 1977, Struc-
     tural elucidation of an S-oxidized metabolite of trithiozine
     in rat and dog, in: "Mass Spectrometry in Drug Metabolism,"
     A. Frigerio and E. L. Ghisalberti, eds., Plenum, New York.

Porter, W. R., Gudzinowicz, M. J., and Neal, R. A., 1979, Thioacet-
     amide-induced hepatic necrosis. II. Pharmacokinetics of
     thioacetamide and thioacetamide-S-oxide in the rat, J. Pharm-
     acol. Exp. Ther., 208:386.

Vyas, K. P., 1979, "Metabolism and Hepatotoxicity of Thiocarbonyl
     Compounds," Ph.D. Thesis, University of Kansas.

THE MICROSOMAL FAD-DEPENDENT MONOOXYGENASE AS AN

ACTIVATING ENZYME: FONOFOS METABOLISM

N. P. Hajjar and E. Hodgson

Interdepartmental Toxicology Program
North Carolina State University
Raleigh, NC 27650 USA.

INTRODUCTION

The mammalian microsomal FAD-dependent monooxygenase (EC 1.14. 13.8), formerly known as an amine oxidase, has been purified to homogeneity and extensively characterized by Ziegler and co-workers. This enzyme has been shown to oxidize sulfur atoms in a wide range of organic compounds (Poulsen et al., 1974, Poulsen and Ziegler, 1979). Recently (Hajjar and Hodgson, 1980a,b) we have shown that thioether sulfur in a number of pesticidal organo-phosphates and carbamates is oxidized to the sulfoxide by this enzyme, a reaction formerly attributed to the microsomal cytochrome P-450-dependent monooxygenase system (Kulkarni and Hodgson, 1980).

The sole product of this reaction appears to be the optically active sulfoxide (Hajjar and Hodsgon, 1980a,b) and further oxida-tion to the sulfone does not occur. Even though the thiono and thiolo sulfur atoms of phosphorothioates are not oxidized by this enzyme, certain phosphonates, such as fonofos (O-ethyl S-phenyl ethylphosphonothioate) and its phenylphosphonate analog, are sub-strates, even though they do not contain a thioether sulfur.

The metabolism of fonofos has been studied extensively. McBain et al. (1971a) showed that fonofos is metabolized in vitro by rat liver microsomes to its oxygen analog and the corresponding thiophosphonic acid. In vivo studies (McBain et al., 1971b) indicate that the oxygen analog is the intermediate in the forma-tion of the terminal residues, O-ethyl ethane phosphonic and -phos-phonothioic acids. Recently investigations have been conducted on the metabolism of the racemic mixture and the chiral isomers of fonofos both in vivo and in vitro (Lee et al., 1978a,b).

Metabolism of the chiral isomers to their oxygen analogs proceeds
via retention of configuration, although inversion occurs to an
extent of 21-28 percent.

The mechanism of oxidative desulfuration of thiono-sulfur
compounds has been reviewed recently (Neal, 1980). Parathion
(0,0-diethyl 0-p-nitrophenyl phosphorothioate), which is not a
substrate for the FAD-dependent monooxygenase (Hajjar and Hodgson,
1980a,b), has been studied extensively and its oxidative desulfuration shown to be dependent on the cytochrome P-450 monooxygenase
system (Neal, 1980).

The present study further defines the metabolism of fonofos
and other phosphonates by the FAD-dependent monooxygenase and
provides evidence for its oxidative desulfuration by this enzyme
with the concomitant production of the oxygen analog, a potent
cholinesterase inhibitor.

MATERIALS AND METHODS

Fonofos and its oxygen analog were obtained from the U.S.
Environmental Protection Agency (Research Triangle Park, NC) and
Stauffer Chemical Company (Mountain View, CA). The oxon was
purified by preparative TLC on Silica gel 60, F_{254} plates developed
in ether:hexane (4:1). (Phenyl-^{14}C)-fonofos, specific activity
17 μCi/μmole, was a gift from the Stauffer Chemical Company.

0-Ethyl, 0-phenyl ethylphosphonothioate and 0-ethyl, 0-phenyl
phenylphosphonothioate were synthesized by reacting 0-ethyl ethyl-
phosphonothioic and 0-ethyl phenylphosphonothioic chloridates with
the appropriate phenol. 0,0-Diethyl S-phenyl phosphorodithioate
and 0-ethyl S-phenyl phenylphosphonodithioate were synthesized by
reacting thiophenol with the corresponding phosphoro- and phosphonothioic chloridates (Sittig, 1977). 0-Ethyl, 0-p-nitrophenyl
ethylphosphonothioate and 0-ethyl 0-p-methoxyphenyl phenylphosphonothioate were synthesized by reacting the sodium salts of
p-nitrophenol and p-methoxyphenol with the corresponding phosphonothioic chloridate, in a reaction similar to that used for EPN
synthesis (Nomeir and Dauterman, 1979). Compounds synthesized
were further purified by preparative TLC and the structures confirmed by NMR spectroscopy.

EPN and 0,0-diethyl S-p-nitrophenyl phosphorodithioate were
provided by Dr. A. A. Nomeir (National Institute of Environmental
Health Sciences, Research Triangle Park NC). Methylphenyl sulfide,
methylphenyl sulfoxide and diphenyl disulfide (DPDS) were purchased
from Fairfield Chemical Company (Blythewood SC).

The rate of oxidation is conveniently measured by following
the substrate-dependent oxidation of NADPH, as previously described

(Poulsen and Ziegler, 1979, Hajjar and Hodgson, 1980a,b). The FAD-dependent monooxygenase, purified to homogeneity from pig liver, was a gift from Dr. D. M. Ziegler of the University of Texas at Austin.

(Phenyl-^{14}C)-fonofos was purified by TLC on analytical Silica Gel G F_{254} plates developed in hexane:chloroform (3:2), resulting in a final radiochemical purity of greater than 99 percent. Various concentrations of ^{14}C-fonofos were incubated for various time periods with the monooxygenase while the substrate-dependent oxidation of NADPH was monitored. The reaction was stopped and the products removed by four extractions, each with 3.0 ml of chloroform. The chloroform extracts were combined, concentrated and subjected to two dimensional TLC on Silica Gel plates. The following pairs of solvent systems (as identified in Table 1) were used: I and II; II and III; III and IV.

Following the incubation of fonofos (15 nmoles) or phenyl fonofos (150 nmoles) with the monooxygenase for 5 and 15 minutes respectively, in the presence or absence of NADPH, aliquots were incubated with purified butyryl cholinesterase (BuChE) for 15 minutes. The residual cholinesterase was determined by the method of Ellman et al. (1961) and the results compared with the inhibition of BuChE by various concentrations of fonofos, fonofos oxon and phenyl fonofos added to 0.25 units of enzyme in 5 µl of acetone. I_{50} values were determined from log-probit curves using data from duplicate experiments. Purified BuChE and butyryl choline were a gift from Dr. A. R. Main of North Carolina State University.

RESULTS

Two dimensional TLC analysis of the products of ^{14}C-fonofos metabolism by the FAD-dependent monooxygenase confirm the identification of the oxygen analog and diphenyl disulfide as the only major metabolites (Table 1 and 2). The metabolites corresponded to authentic standards in all solvent systems in either one or two dimensions. Oxidative desulfuration is NADPH-dependent since the substrate, ^{14}C-fonofos, is quantitatively recovered in its absence. Fonofos oxon is a potent inhibitor of BuChE and is at least 240-fold more potent than fonofos itself, which is 10-fold more potent than its phenyl analog (Table 3). Following the incubation of fonofos and its phenyl analog with the FAD-dependent monooxygenase it is apparent that activation has occurred since the I_{50} values are approximatley 200-fold lower. Under the experimental conditions used the concentrations of fonofos and phenyl fonofos in the aliquots taken for the estimation of BuChE inhibition cause no inhibition without activation. No inhibition is seen in the absence of NADPH. The rate of butyrylcholine hydrolysis in the absence of inhibition is 60.8 µmoles/minute/mg protein under the experimental conditions used.

Table 1.   Thin Layer Chromatography on Fonofos, Phenyl
           Fonofos and their Metabolites

	Rf Values			
	Solvent Systems[a]			
Compound	I	II	III	IV
Fonofos	0.91	0.68	0.67	0.91
Fonofos Oxon	0.40	0.07	0.18	0.57
Phenyl Fonofos	–	0.62	0.60	–
Phenyl Fonofos Oxon[b]	–	0.06	0.13	–
Diphenyl Disulfide	0.93	0.86	0.81	0.91

[a]Solvent systems are as follows: I = ether:hexane (4:1);
II = hexane:chloroform (3:2); III = hexane:acetone (5:1);
IV = hexane:ethylacetate:benzene (1:4:1).

[b]Identified by analogy to fonofos oxon and inhibition of
butyryl cholinesterase.

Table 2.   Metabolism of (Phenyl-^{14}C)-Fonofos by the FAD-
           Dependent Monooxygenase.

Time (mins.)	Fonofos (nmoles)		Metabolites (nmoles)				NADPH Consumed (nmoles)
	Added	Un-reacted	Oxon	DPDS	Water Sol.	Total	
10	130.2	71.6	45.3	12.3	1.0	58.6	58.6
20	157.8	93.1	46.1	15.2	3.4	64.7	79.5
30	52.9	6.2	36.6	5.9	4.2	46.7	68.8

   Although both fonofos and phenyl fonofos are oxidized rapidly
by the FAD-dependent monooxygenase (Table 4) the time course for
the reaction is always curvilinear, regardless of substrate concentration, returning to the endogenous rate of NADPH consumption
before all of the substrate has been consumed. This is unlike the
sulfoxidation reaction which, at saturating substrate concentrations,
is essentially linear until most of the substrate has been oxidized
(Hajjar and Hodgson, 1980). However, as shown in Table 2, the
extent of oxidation, before a return to the endogenous rate of
NADPH consumption, increases with time and decreases at high

Table 3. Activation of Fonofos and Phenyl Fonofos by
          the FAD-Dependent Monooxygenase

Compound	Before Activation	After Activation
Fonofos	$\geq 6.0 \times 10^{-5}$	6.0×10^{-7}
Phenyl Fonofos	$> 6.0 \times 10^{-4}$	$8.5 \times 10^{=6}$
Fonofos Oxon	2.5×10^{-7}	

substrate concentrations. The reaction stoichiometry between NADPH
consumed and substrate metabolized is approximately 1:1 (Table 2).

Structure-activity relationships shown in Table 4 indicate
that fonofos and phenyl fonofos (Compounds 1 and 4) are excellent
substrates for the FAD-dependent monooxygenase. Their O-phenyl
analogs (Compounds 3 and 5), although substrates, are oxidized at
a very slow rate, while fonofos oxon (Compound 2), the putative
product of fonofos oxidation, is not a substrate. Other O-phenyl
phosphonates, such as EPN (Compound 6), its p-methoxyphenyl analog
(Compound 8), and its ethyl phosphonate analog (Compound 7), tend
to have very low activity or, in the case of the ethyl phosphonate
analog, to be inactive.

Neither the phosphorodithioate analog of fonofos (Compound 9)
nor the other phosphorodithioate tested, the S-p-nitrophenyl
analog of parathion (Compound 10), showed any substrate activity.

Methyl phenyl sulfide (Compound 11) is an excellent substrate,
presumably being metabolized to the sulfoxide (Hajjar and Hodgson,
1980), while its diphenyl analog (Compound 12) shows a much reduced
activity. Methyl phenyl sulfoxide (Compound 13), the product of
methyl phenyl sulfoxidation, is not a substrate.

DISCUSSION

The FAD-dependent monooxygenase catalyzes the oxidation of
fonofos and phenyl fonofos. Reaction stoiciometry, product identification and the formation of a potent BuChE inhibitor all indicate
that the reaction is an oxidative desulfuration producing the oxon
as the principal product. Previous studies on fonofos activation
(McBain et al., 1971a,b, Lee et al., 1978a,b) all implicate the
cytochrome P-450-dependent monooxygenase system and propose a
mechanism involving a cyclic phosphooxathiran intermediate similar
to that proposed for parathion (Ptashne and Neal, 1971). Since all
of these studies on fonofos were carried out on microsomal preparations and, since both the FAD-dependent monooxygenase and the
cytochrome P-450-dependent monooxygenase system are NADPH and O_2

Table 4. Structure-Activity Relationships for the Oxidation of Phosphonates and Related Compounds by the FAD-Dependent Monooxygenase

Structure	Name	Activity (nmoles NADPH oxidized/min/nmole enzyme)
1. $(C_2H_5O)(C_2H_5)P(S)S\emptyset^a$	Fonofos	6.45 ± 0.30^b
2. $(C_2H_5O)(C_2H_5)P(O)S\emptyset$	Fonofos oxon	0.00
3. $(C_2H_5O)(C_2H_5)P(S)O\emptyset$	0-Ethyl Ethylphosphonothioate	0.80
4. $(C_2H_5O)(\emptyset)P(S)S\emptyset$	0-Ethyl S-Phenyl Phenylphosphonothioate (Phenyl Fonofos)	10.08 ± 0.21
5. $(C_2H_5O)(\emptyset)P(S)O\emptyset$	0-Ethyl 0-Phenyl Phenylphosphonothioate	0.87
6. $(C_2H_5O)(\emptyset)P(S)O\emptyset NO_2$	EPN	1.32
7. $(C_2H_5O)(C_2H_5)P(S)O\emptyset NO_2$	0-Ethyl 0-p-Nitrophenyl Ethylphos-phonothioate	0.00
8. $(C_2H_5O)(\emptyset)P(S)O\emptyset OCH_3$	0-Ethyl 0-p-Methoxyphenyl Phenylphos-phonothioate	1.00
9. $(C_2H_5O)_2P(S)S\emptyset$	0,0-Diethyl S-Phenyl phosphorodithioate	0.00
10. $(C_2H_5O)_2P(S)S\emptyset NO_2$	0,0-Diethyl S-p-Nitrophenyl phosphoro-dithioate	0.00
11. $CH_3S\emptyset$	Methyl Phenyl Sulfide	27.44 ± 4.82
12. $\emptyset SS\emptyset$	Diphenyl Sulfide	2.28
13. $CH_3S(O)\emptyset$	Methyl Phenyl Sulfoxide	0.00

[a] phenyl ring   [b] mean ± standard deviation (n = 3 or more) or mean of two determinations

dependent, the relative roles of the two systems have not been
defined. It is clear, however, that cytochrome P-450 is involved
in the oxidative desulfuration of phosphorothioates, such as parathion, which are not substrates for the FAD-dependent monooxygenase,
since this reaction has been carried out with purified, reconstituted cytochrome P-450 (Kamataki et al., 1976). It is possible
that phosphonates, such as fonofos and its phenyl analog, are
metabolized by both systems. Lee et al. (1978a) showed that retention of configuration, considered to be consistent with the above
mechanism for cytochrome P-450, was the principal stereochemical
course for the activation of fonofos isomers by rat liver microsomes.
However, it is of interest that inversion occurred to some 21 to 28
percent. They speculated that this inversion was the result of a
second mechanism involving an initial attack on the phosphorus
atom. Further research is necessary to determine whether either
of these mechanisms are related to the oxidation of fonofos by the
FAD-dependent monooxygenase.

The explanation for the curvilinear reaction kinetics is not
immediately apparent. Since fonofos is a racemic mixture of two
chiral isomers, it is possible that one of these is either metabolized at a very slow rate or is an inhibitor of the enzyme. Although
this is supported by the fact that following the incubation of
fonofos for 10-12 minutes, at which time NADPH consumption has
returned to the endogenous rate, addition of more fonofos doe's not,
but addition of more enzyme does, cause an increase in the rate
of oxidation, other explanations are possible. It is noteworthy
that thioether sulfoxidation is essentially linear and very fast.

A second explanation of the enzyme kinetic pattern observed
could be product inhibition or inhibition by a reactive intermediate.
Preincubation of the enzyme with the oxygen analog does not inhibit
enzyme activity. However, it has been shown recently (D. M Ziegler,
personal communication) that a transient intermediate in fonofos
oxidation is, in fact, a non-competitive inhibitor of both fonofos
and methimazole oxidation. It did not appear probable that
elemental sulfur (or the thioperoxy anion, HSO^-) is the inhibitor
in question, since the inhibition was not affected by glutathione.

The phosphorodithioate analog of fonofos is neither a substrate
nor an inhibitor for the FAD-dependent monooxygenase. It is not
fully understood why, with this enzyme, oxidative desulfuration
occurs with phosphonates but not phosphorothioates. The former
lack $p\pi$-$d\pi$ bonding due to the absence of free electrons in the
carbon atom covalently bound to the phosphorus, in contrast to the
oxygen atom in the latter. Thus the phosphorus atom of phosphonates
is less electrophilic than that of phosphorothioates and this may
affect the nucleophilicity of the thionosulfur and thus its
reactivity. Thus changes in the reactivity of both the phosphorus
and thionosulfur atoms are brought about by the presence of the

phosphonate bond. The significance of these changes in the reaction mechanism is, as yet, unknown.

The fonofos metabolites, methyl phenyl sulfide and diphenyl disulfide, found both in vivo and in vitro (Lee et al., 1978a,b), are also good substrates for the FAD-dependent monooxygenase. However, this is a sulfoxidation reaction similar to those found for disulfoton and phorate (Hajjar and Hodgson, 1980a,b).

Although both the cytochrome P-450-dependent monooxygenase system and the FAD-dependent monooxygenase have now been shown to be active in oxidative desulfuration of organophosphorus insecticides much remains to be learned of the reaction mechanisms, the substrate specificities and the relative roles of these two enzymes.

ACKNOWLEDGEMENTS

These investigations were supported, in part, by grants number ES-00044 and ES-07046 from the National Institute of Environmental Health Sciences. Particular thanks are due to Dr. D. M. Ziegler of the University of Texas at Austin, not only for providing the purified FAD-dependent monooxygenase used in these studies, but also for much helpful advice and discussion. We would also like to thank the donors of substrates and enzymes enumerated in Materials and Methods.

REFERENCES

Ellman, G. L., Courtney, K. D. Anders, V. and Featherstone, R. M., 1961, A new and rapid colorimetric determination of acetyl-cholinesterase activity, Biochem. Pharmacol., 7:88.
Hajjar, N. P. and Hodgson, E., 1980, Flavin adenine dinucleotide-dependent monooxygenase: its role in the sulfoxidation of pesticides in mammals, Science, 209:1134.
Hajjar, N. P. and Hodgson, E., 1980, Structure-activity relationships in the sulfoxidation of thioether-containing pesticides by the FAD-dependent monooxygenase from pig liver microsomes, submitted to Biochem. Pharmacol.
Kamataki, T., Lee Lin, M. C. M., Belcher, D. H. and Neal, R. A. Studies of the metabolism of parathion with an apparently homogeneous preparation of rabbit liver cytochrome P-450, Drug Metabol. Disp., 4:180.
Kulkarni, A. P. and Hodgson, E., 1980, Metabolism of insecticides by mixed-function oxidase systems, Pharmac. Therap., 8:379.
Lee, P. W., Allahyari, R. and Fukoto, T. R., 1978, Studies on the chiral isomers of fonofos and fonofos oxon: 2. In vitro metabolism, Pestic. Biochem. Physiol., 8:158.
Lee, P. W., Allahyari, R. and Fukuto, T. R., 1978, Studies on the chiral isomers of fonofos and fonofos oxon: 3. In vivo metabolism, Pestic. Biochem. Physiol., 9:23.

McBain, J. B., Yamamoto, I. and Casida, J. E., 1971, Mechanism of activation and deactivation of Dyfonate (O-ethyl S-phenyl ethylphosphonodithioate) by rat liver microsomes, Life Sci., 10(2):947.

McBain, J. B., Hoffman, L. J., Menn, J. J. and Casida, J. E., 1971, Dyfonate metabolism studies 2. Metabolic pathway of O-ethyl S-phenyl ethylphosphonodithioate in rats, Pestic. Biochem. Physiol., 1:356.

Neal, R. A., 1980, Microsomal metabolism of thiono-sulfur compounds: mechanism and toxicological significance, Rev. Biochem. Toxicol., 2:131.

Nomeir, A. A. and Dauterman, W. C., 1979, Studies on the optical isomers of EPN and EPNO, Pestic. Biochem. Physiol., 10:121.

Poulsen, L. L., Hyslop, R. M. and Ziegler, D. M., 1974, S-Oxidation of thioureylenes catalyzed by a microsomal flavoprotein mixed-function oxidase, Biochem. Pharmacol. 23:3431.

Poulsen, L. L. and Ziegler, D. M., 1979, The liver microsomal FAD-containing monooxygenase. Spectral characterization and kinetic studies, J. Biol. Chem., 254:6449.

Ptashne, K. A. and Neal, R. A., 1971, Oxygen-18 studies on the chemical mechanism of the mixed-function oxidase catalyzed desulfuration and dearylation reaction of parathion, J. Pharmacol. Exp. Therap., 179:380.

Sittig, M., "Pesticides Process Encyclopedia", Noyes Data Corporation, Park Ridge NJ (1977).

SOME PROPERTIES OF SULPHOXIDISING ENZYMES

P.A. Hunt[**], S.C. Mitchell[*] and R.H. Waring[**]

[*] Department of Experimental & Biochemical Pharmacology
St Mary's Hospital Medical School, Paddington, London
[**] Department of Biochemistry, University of Birmingham
Birmingham B15 2TT

INTRODUCTION

Recently an enzyme has been purified from the microsomal fractions of guinea-pig liver, spleen, pancreas, kidney and lungs and also from the serum, which catalysed the conversion of ethionamide, chlorpromazine and thioridazine to their sulphoxides (Prema and Gopinathan, 1972, 1974a, 1976). This sulphoxidase required stoichiometric amounts of NADPH and oxygen, but did not contain any haem-bound iron or cytochrome P-450. Other workers have demonstrated the preference for NADPH over other nicotinamide analogues or flavin compounds (Traficante et al., 1979) and studies with inhibitors have suggested the involvement of a thiol group in the activity. SKF525A, a known potent inhibitor of microsomal oxidases has no effect on sulphoxidation, neither have metal chelating compounds or reducing agents (Gillette and Kamm, 1960; Prema and Gopinathan, 1974a).

Earlier work had shown that in guinea-pig liver microsomes a peroxide-generating system (glucose-glucose oxidase) could not replace the microsomal equivalent for NADPH (Gillette and Kamm, 1960), even though the sulphoxide had previously been shown to be formed from chlorpromazine by hydrogen peroxide and horseradish peroxidase (Cavanaugh, 1957), and suggested that the microsomal sulphoxidase was not a peroxidase. However, a non-enzymic system which could generate superoxide anions (NADH-phenazine methasulphate-O_2) was capable of bringing about sulphoxidation (Prema and Gopinathan, 1974b) and the enzyme superoxide dismutase, which is known to catalyse the conversion of superoxide ions into hydrogen peroxide and molecular oxygen, inhibited sulphoxidation completely in the isolated enzyme preparation. The superoxide anion was therefore put forward as the active form of oxygen in this reaction (Prema and Gopinathan, 1974a).

Nevertheless, evidence also exists for the implication of cyto-chrome P-450 in S-oxidation although Illing (1978) has shown that the binding of thiol-containing compounds to this cytochrome may not be followed by their metabolism. The enzymes in rat liver microsomes which catalyse the sulphoxidation of methylated tetrahydrofurfuryl mercaptan to the sulphoxide and sulphone are both inducible by pre-treatment with either phenobarbital or 3-methylcholanthrene (Fujita and Zira, 1967; Fujita et al., 1973). Similarly, the extremely low rate of chlorpromazine sulphoxide formation by rat intestinal microsomes as compared with those from guinea-pig intestine has been shown to have a direct correlation with the intestinal cytochrome P-450 content, which was low in the rat and high in the guinea-pig (Knoll et al., 1977). In addition, the C- and S-oxidation of perazine by rat liver microsomes were inhibited to the same extent by SKF525A and metyrapone whereas N-oxidation was unaffected (Breyer, 1971).

Peak production of the sulphoxides of several compounds has been shown to occur in guinea-pigs at 5 days post partum both in vivo and in vitro (Brooks and Forrest, 1973; Mitchell and Waring, 1976, 1980; Waring et al., 1978) but this maximum is not coincident with the highest in vitro levels of cytochrome P-450 activity, hydroxylation and demethylation, all of which occur at 3 days after birth (Kuenzig et al., 1974). It is probable that a functional cytochrome P-450 system is necessary for the activity of sulphoxidation enzymes.

It is quite possible that more than one enzyme system is res-ponsible for sulphoxidation and in support of this Ziegler (1978) has shown that many thioureylenes, thiocarbamides and thioamides are substrates for microsomal mixed-function amine oxidases, producing S-oxygenated products. Furthermore, these reactions were NADPH and oxygen dependent but were not prevented by n-octylamine which is known to inhibit cytochrome P-450 dependent oxidations (Prough and Ziegler, 1977). Douch and Buchanan (1979) have recently reported that approximately 20% of the sulphoxidase activity occurring in mouse liver was not associated with microsomes but present in the supernatant fraction and was not inhibited by carbon monoxide.

Because the sulphoxidation systems previously studied appeared to have different cellular locations and different responses to in-hibitors, the guinea-pig liver preparation of Prema and Gopinathan (1974a) was investigated with a variety of substrates to determine the specificities and activities of the various fractions.

MATERIALS AND METHODS

Purification of enzyme

The procedure followed is that outlined by Prema and Gopinathan (1974a). All operations were carried out below 4°C, potassium phosphate buffers being used throughout.

Adult male guinea-pigs (Dunkin-Hartley strain) were stunned and exsanguinated. The excised livers were homogenised with 2 vol of 1.15% (w/v) KCl and then centrifuged at 9000 g for 20 min. Solid $(NH_4)_2SO_4$ was added to the 9000 g supernatant and the proteins precipitating between 40% and 70% saturation were removed by centrifugation, dissolved in a minimal volume of 0.02 M buffer, pH 7.4, and desalted overnight by dialysis against 2 mM buffer, pH 7.4.

The dialysed proteins were adsorbed onto calcium phosphate gel (20 mg/ml; protein/gel ratio of 10/3) at pH 7.4, the supernatant removed by centrifugation and the adsorbed proteins eluted with 0.1 M buffer, pH 6.5. Again, the precipitate obtained between 40% and 70% $(NH_4)_2SO_4$ saturation was collected and dissolved in 0.1 M buffer (pH below 6). A saturated solution of $(NH_4)_2SO_4$ (pH adjusted to 7.5) was added until a saturation of 30% was achieved and the precipitated proteins redissolved in 0.02 M buffer, pH 6.5.

The enzyme was then passed through a column (2.2 cm x 43 cm) of Sephadex G-200 previously equilibrated and then eluted with 0.02 M buffer, pH 6.5. Individual fractions (10, 5, 2 ml) were assayed for protein concentration and enzyme activity. In some instances the protein solution obtained from the 30% $(NH_4)_2SO_4$ fractionation was stirred with previously activated (Peterson and Sober, 1962) DEAE-cellulose for 20 min, centrifuged, the supernatant saved and the DEAE-cellulose eluted with 0.1 M buffer, pH 6.5, which was then centrifuged. Both supernatants were tested for enzyme activity.

Protein determination

The protein content in crude extracts was determined by the biuret method (Gornall et al., 1949) and that in purified preparations by the method of Lowry et al., 1951. Bovine serum albumin fraction V (Sigma Chemical Co.) was used as a standard.

Incubation mixtures

A typical assay mixture contained protein (~ 10 mg crude extract; ~ 250 μg enzyme fractions), 10 mg substrate and 0.02 M phosphate buffer to a final volume of 3 ml and was then incubated for 2 h at

37°C under aerobic conditions in the dark, the reaction being stopped with ethanol. Phenothiazine (B.D.H.), cimetidine (SKF), promazine (Wyeth), chlorpromazine and ethionamide (May & Baker) (Fig. 1) were used as substrates for qualitative experiments, the parent drug and sulphoxide being located by thin-layer chromatography (t.l.c.) as previously described (Harnansingh and Eidus, 1970; Mitchell and Waring, 1976, 1979; Taylor et al., 1978).

Quantitative tests were performed using promazine and chlor-promazine, the sulphoxides being synthesised by similar methods (Mitchell and Waring, 1976). Aliquots ($\sim$ 40 μl) of the incubation mixtures were removed for t.l.c. (silica gel G, 0.4 mm) in cyclo-hexane/dioxan/ethanol/NH$_4$OH (sp.gr. 0.88) (14:4:3:1 by vol.) in a nitrogen atmosphere and the compounds were located with 50% H$_2$SO$_4$ and eluted with methanol. The concentration of sulphoxide present was determined by comparing the UV extinction with that obtained from known amounts of authentic sulphoxide taken through this procedure.

The effect of inhibitors on sulphoxide production was also in-vestigated. The enzyme was preincubated with SKF525A (0.1 mM), superoxide dismutase (E.C. 1.15.1.1) (750 units from bovine blood; Sigma) for 1 h at 37°C, or carbon monoxide which was bubbled through the mixture for 15 min before the addition of substrate. To demon-strate that the reaction was protein-mediated, samples were either boiled with 10 M NaOH or preincubated with the non-specific bacterial alkaline protease, Subtilisin Carlsberg (Sigma) for 1 h, in both cases the pH being returned to 7.4 before the addition of the sub-strate.

RESULTS

The protein elution profile from the gel filtration was nearly identical to that reported by the previous workers (Prema and Gopinathan, 1974a), although two resolvable peaks in sulphoxidation activity were detected (Fig. 2). The use of inhibitors (Table 1) showed that the response of both peaks of activity was the same to treatment with NaOH, protease and SKF525A. However, fraction A was only slightly affected by CO treatment, unlike fraction B, and pre-incubation with superoxide dismutase, which completely inhibited activity in fraction B, only brought about a 25% reduction in activity in fraction A.

The use of DEAE-cellulose showed that the superoxide dismutase-sensitive fraction (B) was adsorbed and could be subsequently eluted with phosphate buffer, whereas fraction A remained in the initial supernatant. The omission of this step in the enzyme purification procedure explains why Prema and Gopinathan (1974a) only obtained one peak of activity whereas two were obtained in this study. The incorporation of this step before gel filtration leads to the elution

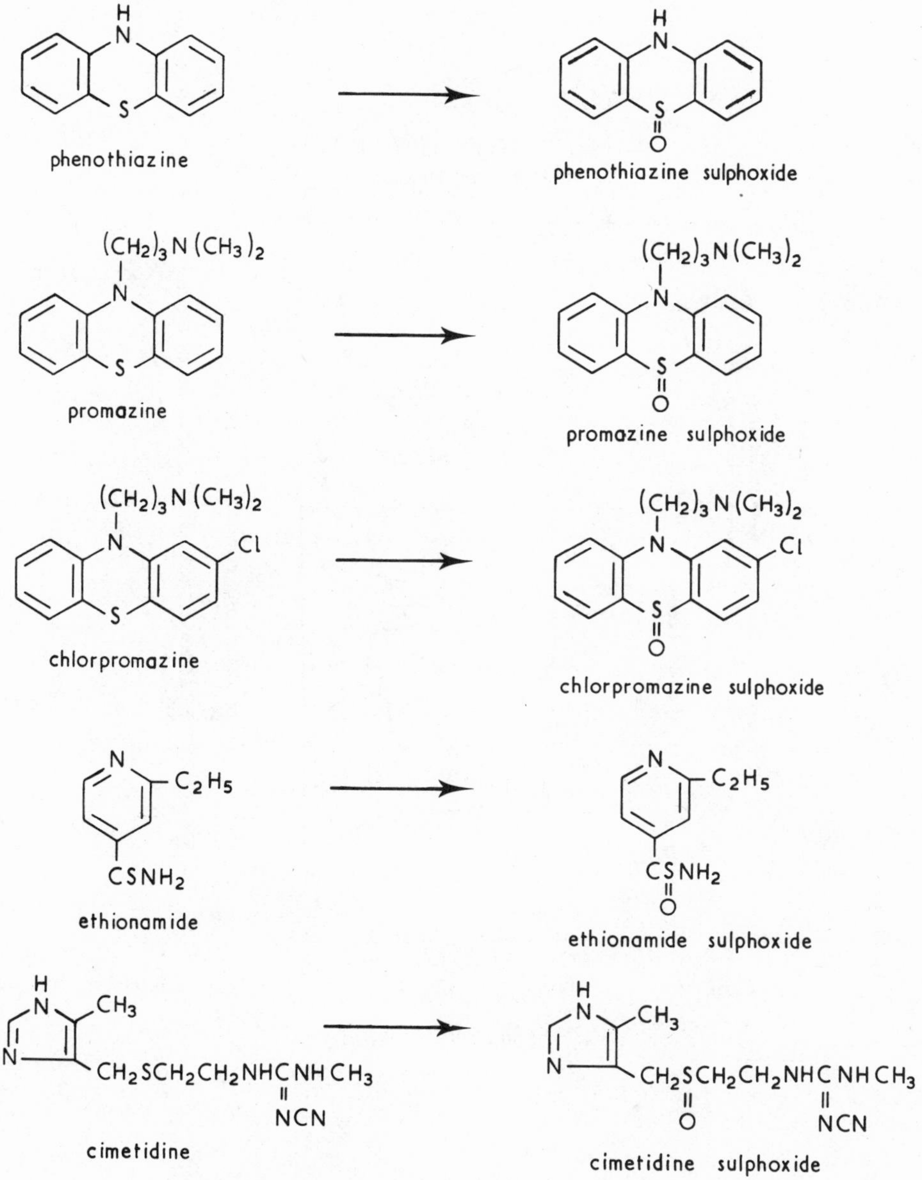

Fig. 1   Structures of compounds tested as substrates for sulphoxidation.

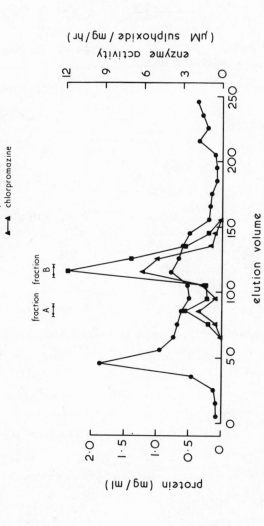

Fig. 2 Sulphoxidising activity of eluted fractions using promazine and chlorpromazine as substrates

Table 1.   Effect of inhibitors on eluted enzyme fractions

Treatment	% inhibition of sulphoxidase activity	
	Fraction A	Fraction B
NaOH	100	100
Protease	100	100
Superoxide dismutase	25	100
Carbon monoxide	2	58
SKF525A	63	64

of only one peak of activity (fraction B) from the column.

Both fractions formed sulphoxides from all the compounds tested; sulphoxidation activity was also found in the initial 9000 g residue, which was incubated with phenothiazine and chlorpromazine as substrates.

DISCUSSION

Although both fractions A and B belong to the same cellular location and have very similar elution volumes, they have markedly different properties. The resistance of enzyme A to inhibition by carbon monoxide suggests that haem proteins are not directly concerned in the sulphoxidations carried out by this fraction. However, the addition of superoxide dismutase has relatively little effect on the activity, so that the superoxide anion seems not to be involved to any great extent. In this case, neither of these postulated mechanisms for sulphoxidation appears convincing. In contrast, enzyme B (Prema and Gopinathan's fraction) is completely inhibited by superoxide dismutase, and equally inhibited by carbon monoxide and SKF525A, so that both the superoxide anion and a haem protein system are required for sulphoxidation to take place.

These fractions may be part of the same microsomal sulphoxidation system which has dissociated during the purification procedure, and where the components still possess some residual activity, or they may represent different sulphoxidation systems with different mechanisms. Both enzymes formed sulphoxides from all the compounds tested i.e. promazine, phenothiazine, chlorpromazine, cimetidine and ethionamide, so that substrates may contain sulphur either in a heterocyclic ring or in a side chain. Sulphoxidation activity was also found in the 9000 g residue, so the multiple enzyme systems involved appear to have not only different pathways but also different locations in the cell.

REFERENCES

Breyer, U. (1971) Biochem. Pharmacol. 20, 3341-3351.
Brookes, L.G. and Forrest, I.S. (1973) Res. Commun. Chem. Pathol. Pharmacol. 5, 741-758.
Cavanaugh, D.J. (1957) Science (N.Y.) 125, 1040-1041.
Douch, P.G.C. and Buchanan, L.L. (1979) Xenobiotica 9, 675-679.
Fujita, T., Susuoki, Z., Kozuka, S. and Oae, S. (1973) J. Biochem. (Tokyo) 74, 723-732
Fujita, T. and Ziro, S. (1967) Biochem. Biophys. Res. Commun. 28, 827-832.
Gillette, J.R. and Kamm, J.J. (1960) J. Pharmacol. Exp. Ther. 130, 262-267.
Gornall, A.B., Bardawill, C.J. and David, M.M. (1949) J. Biol. Chem. 177, 751-766.
Harnansingh, A.H.T. and Eidus, L. (1970) Int. Z. Klin. Pharmacol. Ther. Toxikol. 3, 128-131.
Illing, P.A. (1978) Biochem. Soc. Trans. 6, 89-91.
Knoll, R., Christ, W., Muller-Oerlinghausen, B. and Coper, H. (1977) Naunyn-Schmiedeberg's Arch. Expt. Pathol. Pharmakol. 297, 195-200.
Kuenzig, W., Kamm, J.J., Boublik, M., Jenkins, F. and Burns, J.J. (1974) J. Pharmacol. Exp. Ther. 191, 32-44.
Lowry, O.H., Rosebrough, N.J., Farr, A.L. and Randall, R.J. (1951) J. Biol. Chem. 193, 265-275.
Mitchell, S.C. and Waring, R.H. (1976) Xenobiotica 6, 763-768.
Mitchell, S.C. and Waring, R.H. (1979) Drug Metab. Dispos. 7, 399-403.
Mitchell, S.C. and Waring, R.H. (1980) IRCS Med. Sci. 8, 5.
Peterson, E.A. and Sober, H.A. (1962) Enzymol. 1, 90.
Prema, K. and Gopinathan, K.P. (1972) Proc. Soc. Biol. Chem. (India) 31, 21-22.
Prema, K. and Gopinathan, K.P. (1974a) Biochem. J. 143, 613-624.
Prema, K. and Gopinathan, K.P. (1974b) Biochem. J. 137, 119-121.
Prema, K. and Gopinathan, K.P. (1976) Biochem. Pharmacol. 25, 1299-130
Prough, R.A. and Ziegler, D.M. (1977) Arch. Biochem. Biophys. 180, 363-373.
Taylor, D.C., Cresswell, P.R. and Bartlett, D.C. (1978) Drug Metab. Dispos. 6, 21-30.
Traficante, L.J., Siekierski, J., Sakalis, G. and Gershon, S. (1979) Biochem. Pharmacol. 28, 621-626.
Waring, R.H., Armitage, M. and Mitchell, S.C. (1978) Biochem. Soc. Trans. 6, 91-93.
Ziegler, D.M. (1978) Biochem. Soc. Trans. 6, 94-98.

OPTIMUM S9 CONCENTRATIONS IN BACTERIAL

MUTAGENICITY ASSAYS

R. Forster, M. H. L. Green and A. Priestley

MRC Cell Mutation Unit, University of Sussex
Falmer, Brighton BN1 9QG, England
*Present address: Water Research Centre, Medmenham
                  Marlow, Bucks, England

We have used bacterial mutagenesis to investigate the activation of 2-acetylaminofluorene (AAF) in two assay systems (Forster
et al., 1980).   The Ames test is based in a solid (agar) medium;
the fluctuation test (Green and Muriel, 1976) involves the incubation of the promutagen in multiple replicates in a liquid medium.
We have used Aroclor-induced rat liver S9 and bacterial strain
Salmonella typhimurium TA98 throughout.

The optimum concentration for the detection of AAF is substantially greater in the Ames test than in the fluctuation test (see
Fig. 1).   Why does the S9 optimum differ in the two systems?   and
what factors determine the optimum concentration of S9?   To explain
the divergence in the S9 optima we looked at the differences between
the two tests, which utilise essentially the same functional components.   We demonstrated that over the range of AAF concentrations
used in these experiments, the S9 optimum is not markedly affected
by the concentration of promutagen.   Similarly, the S9 optimum is
not influenced by the inoculum of bacteria.

An experiment was devised to test whether features of the
Ames test such as the presence of soft agar, or the rapid temperature changes are responsible for the different optima in the tests.
A modified incubation mixture was prepared, containing soft agar,
which was dispensed to either Ames test plates or fluctuation test
replicate tubes.   The results (Fig. 2) indicate that the soft agar
and temperature change cause some loss of activity and have a
slight effect on the optimum compared with a conventional fluctuation
test.   However, the same modified mixture produces different optima
when dispensed into the two systems.

1263

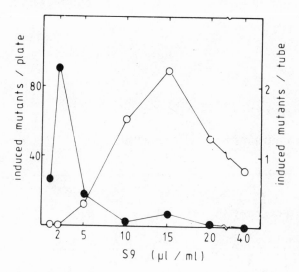

Fig. 1.   Effect of different levels of S9 on activation of AAF in
          the Ames and fluctuation tests.   O — O Ames test,
          0.8 µg/ml AAF in top agar;   ● — ● fluctuation test,
          0.1 µg/ml AAF in incubation mixture.   Levels of S9 are
          µl/ml in top agar or incubation mixture.

     The remaining difference between the two test systems is the
minimal agar bottom layer in the Ames test.   Perhaps diffusion
into the bottom agar is responsible for the different S9 optimum
in the Ames test?

     To simulate Ames test conditions in the fluctuation test, 1 ml
of minimal agar was introduced into each of the replicate tubes
before addition of the incubation mixture.   The S9 optimum of this
modified fluctuation test can be compared with the response in a
conventional fluctuation test (Fig. 3).   The S9 optimum is altered
in the modified test and resembles a characteristic Ames test
response.

The same result is obtained (Fig. 4) when purified microsomes are
used for activation instead of S9-mix.   This indicates that it is
not the diffusion of soluble (cytosolic) constituents of the S9
fraction that is responsible for the difference.

     What is the factor diffusing into the bottom agar which affects
the S9 optimum?   It is not likely to be unmetabolised AAF since the
S9 optimum is unaffected by AAF concentration;   nor is it soluble
(cytosolic) S9 components.   The concentrations of NADP and G-6-P

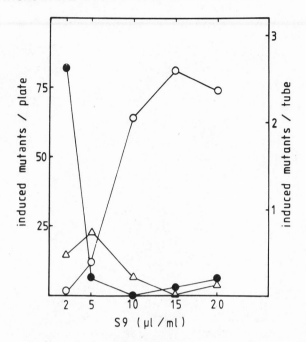

Fig. 2.   Effect of soft agar on mutagenicity of AAF in the Ames
          and fluctuation tests.   ● — ● fluctuation test, no soft
          agar, AAF dose 0.1 µg/ml; Δ — Δ fluctuation test with soft
          agar, AAF dose 0.1 µg/ml= O — O the same mixture dis-
          pensed in an Ames test, AAF dose 0.8 µg/ml.

(NADPH generating system) do not influence the S9 optimum.   It is
possible, however, that AAF metabolites are responsible, and an
attractive explanation for these results is that the S9 optimum is
determined by feedback inhibition of the microsomal enzymes by non-
mutagenic AAF metabolites.   The mechanism for such an effect could
be competitive inhibition of enzyme sites by the non-reactive
metabolites, end-product inhibition or the production of metabolites,
which engage microsomal enzymes in "futile" reactions (viz. Benzo-
a-pyrene quinones: Shen et al, 1979).

We attempted to characterise the diffusing factor(s) by the
addition of a number of agents to a fluctuation test incubate.
The presence of corn oil increases and broadens the S9 optimum
(Fig.5);   this suggests that the agent is lipophilic and is consistent with our model.   The addition of DNA (75 µg/ml) protein

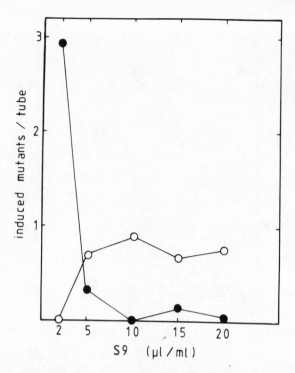

Fig. 3.   Mutagenicity of AAF in a conventional fluctuation test
          (● — ●) and in a fluctuation test with a 1 ml agar
          underlay (O — O).   The respective doses of AAF were
          0.1 μg/ml and 0.8 μg/ml.

(2.5 mg/ml Bovine serum albumin) or glutathione (2.5 mg/ml) and
UDPGA (250 μg/ml) did not affect the S9 optimum.

     To further investigate the possibility that feedback by AAF
metabolites was responsible for the inhibition of S9 activation
we examined the effects of two ring hydroxylated AAF metabolites
(generously provided by Dr. J. A. Miller) on the activation of AAF
in an Ames test (Table 1).   While the 1-OH metabolite had a slight
inhibitory effect on AAF activation, the 7-hydroxy metabolite had
a stimulatory effect on AAF activation.   Similarly, it is reported
by Roberfroid (this symposium) that 1-, 3-, and 5-hydroxy AAF are
strong inhibitors of AAF N-hydroxylase while 7-hydroxy AAF is an
activator.   These results suggest that feedback effects of
relatively unreactive metabolites of AAF may offer a mechanism for

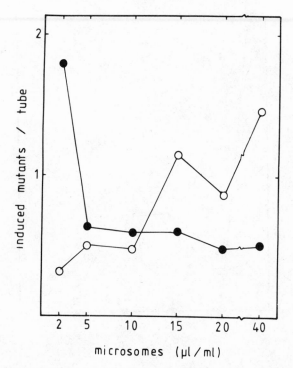

Fig. 4. Mutagenicity of AAF using activation by microsomal
         fraction in a conventional fluctuation test (● — ●) and
         in a fluctuation test with a 1 ml agar underlay (O — O).
         The respective doses of AAF were 0.25 µg/ml and 5.0 µg/ml.

the apparent diffusible inhibition of microsomal activation but if
so the action of the metabolites must be complex.

CONCLUSIONS

1. The optimum S9 concentration for AAF differs markedly in
    bacterial tests based in liquid (fluctuation test) or in agar
    (Ames test).

2. The optimum is not influenced by AAF concentration or bacterial
    inoculum.

3. The same modified incubate dispensed into the two systems pro-
    duces different and characteristic optima.

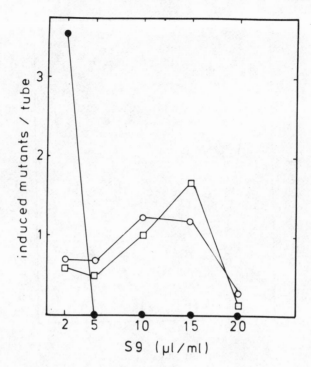

Fig. 5.   Mutagenicity of AAF in a conventional fluctuation test
(● — ●) (AAF dose 0.1 μg/ml) and in fluctuation tests
supplemented with 10% v/v (□ -- □) and 50% v/v (0 — 0)
corn oil (AAF doses 2 μg/ml).

4.   The incorporation of an agar sink for diffusion into fluctuation
     tests activated by S9 or microsomes alone, is effective in
     altering the S9 optimum.

5.   The soluble S9 constituents, AAF, and NADP/G-6-P are not res-
     ponsible for this.

6.   The diffusible factor appears to be lipophilic but is unaffected
     by the presence of DNA, protein or Glutathione and UDPGA.

7.   The microsomal activation of AAF can be modulated by relatively
     unreactive AAF metabolites.   This may offer a mechanism for the
     diffusible inhibition we have observed.

Table 1.   Effect of AAF metabolites on AAF activation in the Ames
           test: Induced revertants per plate

S9 concentration used:

µl S9 fraction per plate	50	25	12.5	5
5 µg AAF	705	561	286	35
5 µg AAF + 100 µg 1.OH AAF	502	213	119	3
5 µg AAF + 100 µg 7.OH AAF	1025	589	269	0

ACKNOWLEDGEMENTS

We acknowledge the assistance of Andrea Hanfmann in performing
one of the experiments.   We are grateful to IRL Press for permission
to reproduce figures used in the text.   This work was in part funded
by the Water Research Centre under a contract from the Department of
the Environment, to whom we are grateful for permission to present
these results.

REFERENCES

Forster, R., Green, M. H. L., and Priestley, A., 1980, Optimal levels
    of S9 fraction in the Ames and fluctuation tests: apparent
    importance of diffusion of metabolites from top agar, Carcino-
    genesis, 1:337.

Green, M. H. L., and Muriel, W. J., 1976, Mutagen testing using Trp+
    reversion in *Escherichia coli*, Mutation Res., 38:3.

Roberfroid, M., 1980, Induction modification and inhibition of
    arylamide N- and C-hydroxylase.   Their roles in the
    mutagenicity of 2-acetyl-aminofluorene, This Symposium.

Shen, A. L., Fahl, W. E., Wrighton, S. A., and Jefcoate, C. R., 1979,
    Inhibition of benzo(a)pyrene and benzo(a)pyrene 7,8, dihydrodiol
    metabolism by benzo(a) pyrene quinones, Cancer Res., 39:4123.

A fuller account of this work appears in Carcinogenesis, 1980 1:337.

EFFECTS OF PLATING EFFICIENCY AND LOWERED CONCENTRATION OF SALTS ON MUTAGENICITY ASSAYS WITH AMES' SALMONELLA STRAINS

Charlotte Witmer, Kathryn Cooper and Joan Kelly

Department of Pharmacology
Thomas Jefferson University
Philadelphia, PA   19107

INTRODUCTION

The classic mutagenicity test of Ames using auxotrophic strains of Salmonella typhimurium which require histidine (Ames et al., 1975) has been demonstrated to detect 90-95% of carcinogens as mutagens when the assay excludes A) those alkyl halides which are metabolized to carcinogens by reductive dehalogenation and B) inorganic carcinogens which may cause infidelity of DNA repair or act by a similar unknown mechanism. Among the inorganic compounds found to be carcinogenic are those of arsenic and chromium (IARC, 1973), cadmium and nickel (IARC, 1976), and iron (Sunderman, 1971). Results of Ames testing of inorganic carcinogens have not been consistent and dose-response relationships are generally lacking.

We have become interested in studies of chromium compounds as carcinogens in the workplace, specifically those compounds supplied to steel mills for use in producing alloys and the waste products which result from such processes. Epidemiological studies of workers in industries with high exposure to these compounds implicate chromium as a lung carcinogen, and laboratory studies indicate that both the metallic and hexavalent forms of chromium can cause cancer in animals when administered in a variety of ways: intraosseously, intramuscularly, intrapeurally, intraperitoneally and subcutaneously (IARC, 1973).

Petrilli and DeFlora (1977) studied the mutagenicity of hexavalent chromium anion using the standard Salmonella typhimurium test of Ames as well as its spot test modification. They reported that the hexavalent chromium anion (CrO_4^{--}) is mutagenic without metabolic activation for strains TA 100, TA 1535, TA 98 and TA 1537; however, Cr(III) cations were not mutagenic with these strains.

1271

Venitt and Levy (1974) demonstrated mutagenic activity for the sodium, potassium and calcium salts of hexavalent chromate for E. coli
strains. Nishioka (1975) published similar reports of potassium
chromate (Cr,VI) mutagenicity using a B. subtilis assay. Recently
Lofroth (1978) reported weak mutagenicity of hexavalent chromium
compounds using strains of Salmonella typhimurium that respond to
frame-shift mutagens; he also detected mutagenic activity with
TA 100 and TA 92. Both of these strains contain the base-pair substitution but with the added pKM101 plasmid factor. His studies
also indicated that an NADPH-dependent microsomal enzyme system
reduced the Cr(VI)-containing anion to a non-mutagenic moiety. In
apparent contradiction to this finding, he found that the Cr(III)
cation is weakly mutagenic (G. Lofroth, personal communication, 1979).
In other studies, Lofroth lowered the concentration of the total
salts in the minimal agar to permit better diffusion of the inorganic ions; he also substituted other divalent anions for SO_4^{--}
to avoid interference with the bacterial transport of CrO_4^{--} anion.
Both techniques increased the sensitivity of the Ames test for
detecting the mutagenic action of CrO_4^{--}.

We have extended the method of Lofroth and concentrated on
studying the effects of altering the concentrations of total salts
and histidine on the following parameters of the standard Ames test:
1) number of spontaneous revertants (background), 2) survival rates
for the auxotrophic strains of bacteria, 3) response to chromium(VI)
as a mutagen, and 4) response to sodium azide as a mutagen. Most of
our studies were carried out with TA 100, which responds to both
frame-shift and base-pair mutagens, and/or TA 1535, which detects
base-pair mutagens only. These became the most useful strains as
a result of their survival and response rates.

MATERIALS AND METHODS

Assays were carried out using the standard plate incorporation
procedure of Ames (Ames et al., 1975). The histidine-requiring
bacterial strains were generously provided by the laboratory of
Bruce Ames.

A brief description of the procedure is as follows: a mixture
of 0.1 ml of the bacterial strain (containing approximately 10^8
bacteria in log phase growth) is mixed with 2.0 ml of molten top
agar (0.6% Difco agar, 0.5% NaCl) to which has been added the suspected mutagen, a small amount of histidine (0.045 mM or 0.25 mM,
as noted) and 0.05 mM biotin (for assays with TA 100 and TA 98).
This top agar mixture is poured onto a (Petri) plate containing
20 ml of solidified minimal agar medium (1.5% Difco agar in Vogel-
Bonner medium (Vogel and Bonner, 1956)). The plates are incubated
in the dark at 37° for 48 hr, unless otherwise specified. As the
only bacteria to survive after the small supply of histidine is

exhausted are the mutants which have regained the capability of
synthesizing histidine, the colonies which appear after the 48 hr
period are counted. Calculation of the mutagenic potential of the
compound tested can be derived directly from the number of mutants
(see Discussion). We varied the minimal agar by using 1/3, 1/2, 2/3
and 3/3 concentrations of the combined salts recommended by Vogel and
Bonner (1956). The complete (3/3) $[S]$ concentrations were: $MgSO_4$,
0.8 mM; citric acid, 10.0 mM; K_2HPO_4, 60 mM; $Na(NH_4)HPO_4$, 13.5 mM.
To obtain the bacteria in the log phase of growth, ten-fold dilutions
of overnight growth of fresh innocula (from frozen cultures) were
reincubated for 4 hr with shaking. These tubes were found to contain approximately 10^8 bacteria/0.1 ml as used in the assay. We
determined the actual number of bacteria plated per assay by colony
counts on Trypticase Soy Agar (TSA) plates, using dilutions calculated to contain 10^3 bacteria. Control samples in which we determined the number of spontaneous mutants/plate were run routinely.
Since the hypothesis underlying the Ames test is that the small
amount of histidine added to each plate limits the number of surviving bacteria, the number of spontaneous mutants arising on each
plate should be constant. To test the validity of this hypothesis,
we plated numbers of bacteria ranging from 10^2 - 10^8/plate on minimal agar and counted the spontaneous revertants. These experiments
were carried out with both (0.045 mM and 0.25 mM) concentrations of
histidine.

Results of these tests (see Tables 1 and 2) led us to use 10^3
bacteria/plate to evaluate survival of the bacterial strains under
conditions of lowered concentrations of salts in the minimal agar,
with both concentrations of histidine. We plated aliquots calculated
to contain 10^3 bacteria on plates containing each of the four concentrations of salts (1/3, 1/2, 2/3 and 3/3 $[S]$) and compared the
numbers of surviving colonies with those surviving when identical
aliquots were plated on TSA plates. The latter plates theoretically
allow 100% survival of all bacteria plated. The ratio of the two
numbers x 100 = % plating efficiency. In the presence of a bacteriotoxic agent, this ratio x 100 = % survival.

The chromium samples tested were 1) chromic chromate (a mixture
of chromic sesquioxide, Cr_2O_3, and CrO_3, 2:1; and 2) a mixture of
iron oxide (Fe_2O_3) with chromic chromate, 1:1. These samples were
supplied by the Bethlehem Steel Company, Bethlehem, PA, for an
evaluation of their mutagenic activity. Analysis of the chromic
chromate, as reported by the Bethlehem Steel Company, was 18.5% Cr
(VI), 50% total chromium.

Colonies were counted either manually or with the use of a
BioTran II Automated Colony Counter (New Brunswick Scientific
Company). To distinguish between survivors and mutants, when we had
plated fewer than 10^4 bacteria/plate, we counted those colonies that
survived 7 transfers onto minimal agar (his⁻) plates, as mutants.

Table 1. Number of Revertants as a Function of the Number of Bacteria Plated and the Concentration of Histidine Strains TA 100 and TA 1535

Bacteria plated [a]	Number of spontaneous revertants 48 hr and (72) hr incubation [b]			
	Histidine 0.25 mM [c]		Histidine 0.045 mM [d]	
	TA 100	TA 1535	TA 100	TA 1535
$4.8 - 5.3 \times 10^7$	153 (179)	33 (53)		
$3.2 - 3.4 \times 10^7$			153 (181)	27 (52)
$4.8 - 5.3 \times 10^6$	89 (121)	14 (39)		
$3.2 - 3.4 \times 10^6$			63 (104)	8 (25)
$4.8 - 5.3 \times 10^5$	23 (136)	11 (37)		
$3.2 - 3.4 \times 10^5$			43 (73)	10 (28)
$4.8 - 5.3 \times 10^4$	52 (122)	0 (35)		
$3.2 - 3.4 \times 10^4$			3 (58)	4 (28)

[a] These numbers were determined by colony counts on TSA plates.
[b] All plates were counted after 48 and 72 hr of incubation at 37°. Numbers in parentheses represent readings at 72 hr.
[c] Histidine concentration recommended by Franz and Malling (1975).
[d] Histidine concentration recommended by Ames et al. (1975).

RESULTS AND DISCUSSION

Table 1 shows that for strains TA 100 and TA 1535 the actual numbers of spontaneous revertants that appeared after 48 hr of incubation varied as a function of the number of bacteria plated rather than appearing to be constant. However, if revertants were scored again at the 72 hr period, the numbers increased so as to approach a constant value. Specifically, the revertants seen with 0.25 mM histidine closely approach constant values, except for the plates with the highest numbers of bacteria added. At the lower histidine concentration, the growth rate of both strains was apparently slower and a constant value for TA 100 was not approached in the incubation times studied. The pattern for TA 1535 (at 0.045 mM histidine) was the same as for both strains at 0.25 mM histidine with a constant number of revertants appearing with time except for the plates with 3.4×10^7 bacteria. Dilutions of the bacteria were made in phosphate

Table 2. Number of Survivors as a Function of the Number of
Bacteria Plated and the Concentration of Histidine
Strains TA 100 and TA 1535

Bacteria plated	Number of survivors 48, (72) and hr incubations [a]			
	Histidine 0.25 mM		Histidine 0.045 mM	
	TA 100	TA 1535	TA 100	TA 1535
2.9×10^3	504 (504) [504]			
1.8×10^3		1,038 (1,107)		
5.8×10^2	402 (402)			
$3.2 - 3.4 \times 10^2$			4 (205) [337]	99 (305) [394]

[a]Plates were counted after 48, 72 and 96 hr, as noted. Numbers
in parentheses represent readings at 72 hr, those in brackets –
96 hr. All other conditions were the same as described in
Table 1.

buffer (100 mM phosphate, 50 mM KCl, pH 7.4) as is routinely used
for colony counts, so as to keep the bacterial population constant,
thus the highest numbers of bacteria were plated with a small additional amount of histidine. However, even those plates containing
the highest numbers of bacteria had increased revertants at the 72
hr period, indicating that a 72 hr period of incubation may provide
a more accurate measure of both the mutagenicity and background
values than the 48 hr period.

Two common methods of evaluating the data from mutagenicity
tests are: 1) expressing the results as "x background" and 2) using
the one-tailed Student's t test to determine the significance of
the number of revertants caused by a suspected mutagen. In both
cases, the background numbers of spontaneous revertants are of
paramount importance. It can be seen that slight differences in
bacteria can greatly influence these results, e.g. if a compound is
a borderline mutagen and 0.5×10^7 bacteria are plated as compared
to 0.5×10^8. The extent to which the calculations are affected by
incubation time cannot be determined until the rate of growth of the
revertants induced by exposure to a mutagen is compared with the
rate of growth of the spontaneous mutants.

In the Ames test, lack of a background lawn is a critical observation that indicates a bacteriotoxic effect by the test substance which may obscure mutagenic effects. There is a significantly thicker lawn on the 0.25 mM histidine plates than on those containing 0.045 mM histidine. The thicker lawn allows a more accurate detection of any changes in the background covering. At lower concentrations of bacteria plated, with both concentrations of histidine, auxotrophs survive and form isolated colonies as opposed to lawn formation (see Table 2). These colonies do not survive transfer onto minimal agar (his⁻) plates.

Table 2 shows the large numbers of isolated colonies appearing after a 48 hr incubation when <10^4 bacteria were plated on the his⁺ minimal agar with conditions identical to those described for Table 1. The numbers of colonies represented a high percentage of the bacteria plated; these colonies were not revertants but surviving auxotrophs as shown by their lack of survival on his⁻ minimal agar plates. The histidine thus did not limit the growth of these numbers of bacteria but allowed them to reach colony size. This is in contrast to the limiting effect on growth of larger numbers of bacteria plated, where growth is limited to background lawn formation. The bacteria on 0.25 mM histidine plates characteristically reached the maximum number of colonies earlier, although those on 0.045 mM histidine approached a higher plating efficiency with time (72 and/or 96 hr), indicating that histidine was available after the 48 hr period. The actual lower plating efficiency at 0.25 mM histidine may represent clumping of the bacteria, as other studies in our laboratory (data not shown) indicate that colony distribution is frequently poor on plates with this histidine concentration.

The relatively high plating efficiency of the strains tested with innocula of 10^3 bacteria on his⁺ minimal agar plates indicates that toxic effects can be assessed under these conditions, i.e. by scoring surviving colonies on addition of the presumptive toxin to the bacteria plated on the his⁺ minimal agar plates. We evaluated both histidine concentrations for such assays and used this system to determine whether lowered concentrations of salts caused toxic effects.

Our studies showed that most strains tested survived poorly with lowered salts available and that apparent survival was dependent on both the histidine and salts concentrations. The increased numbers of bacteria detected after 72 hr of incubation with the lower histidine concentration reinforced the findings of Table 2, as did the better growth rate with 0.25 mM histidine with all strains.

Figure 1 shows the apparent low survival rates of all strains tested except for TA 1535 on the 1/3 (S) plates after 48 hr of incubation with 0.045 mM histidine. At this 48 hr period, only TA 1535 survived these conditions sufficiently well to be valuable

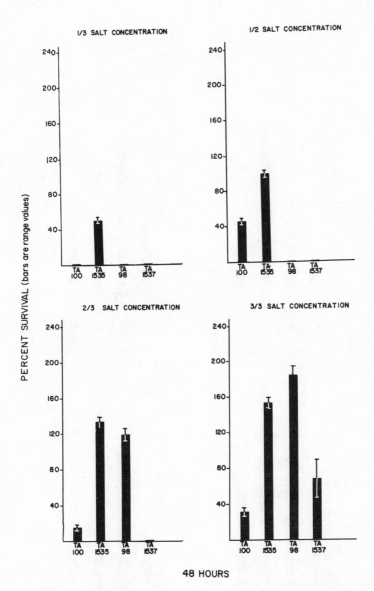

Fig. 1. Survival rates of Salmonella strains after 48 hr incubation with several [S] concentrations (see Methods). Histidine concentration was 0.045 mM. A concentration of $\sim 10^3$ bacteria was added to each plate. The % survival was determined as described in Methods.

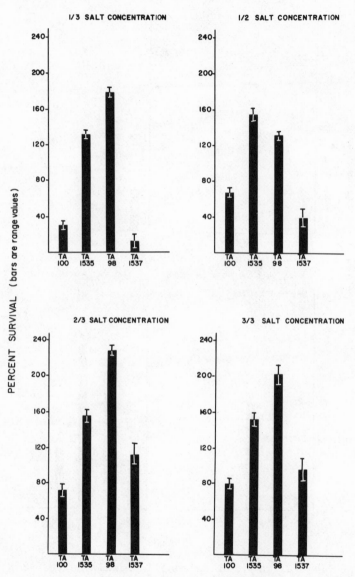

Fig. 2 Survival rates of Salmonella strains after 72 hr
        incubation with several [S] concentrations.
        These colony counts are from the same plates as
        those in Fig. 1 (scored at 48 hr) after an addi-
        tional 24 hr of incubation.

MUTAGENICITY ASSAYS WITH AMES' SALMONELLA STRAINS

in the actual mutagenicity assay. The low response of TA 100 (48 and 72 hr incubations) at all [S] values at the lower histidine concentration also confirmed the findings of Table 2.

Figure 2 shows the surviving colonies when the plates were reincubated and recounted at 72 hr. The increased numbers of colonies (at 0.045 mM histidine) with all salts concentrations indicated that the bacteria grew slowly with the low histidine available. However, the % survival of TA 1535 was high at all [S] concentrations, suggesting that this strain would be useful for detection of inorganic compounds as mutagens with the lowered [S] plates to detect mutagens, unless the revertants are scored after 72 hr of incubation. After 72 hr of incubation, only TA 1537 at 1/3 [S] had a poor survival rate.

Figure 3 shows the apparent % survival of 5 strains in a typical series of experiments with the his$^+$ (0.025 mM) minimal agar plates containing varied concentrations of salts. Although growth of all strains tested was relatively poor at the 1/3 [S] level, all strains exhibited at least 80% survival on 1/2 [S] and above. The increase in numbers of colonies detected after 24 hr additional incubation was not appreciable (data not shown) with the exception of TA 100 on 1/3 [S] which showed a 40% survival rate at the 72 hr period. TA 100 was seen to grow somewhat more slowly than TA 1535, except on the 3/3 [S] plates, with 0.25 mM histidine. All strains grown with 0.25 mM histidine at the four concentrations of salts appeared to be sufficiently responsive for use in mutagenicity and/or toxicity evaluation, with the exception of TA 1537 on 1/3 [S] plates.

Figure 4 shows an example of the use of lowered concentrations of salts in detecting mutagenicity of Cr (VI) compounds. These are the results of a series of tests with a mixture of "chromic chromate" and iron oxide (1:1) (see Methods). When tested at 3/3 [S] (the standard concentration of the Ames test), the sample was not detected as a mutagen. It was also only slightly toxic, with 80% surviving colonies. As the concentrations of salts were lowered, the sample is seen to be more toxic and to be more readily detected as a mutagen. At 1/3 [S], the background number of revertants was low (5) because of the low survival of the strain in the presence of the sample, but the number of mutants increased to 22 x background, indicating a very significant mutagenic effect. It appears that lowering the concentration of salts in the minimal agar allowed the sample to diffuse more readily through the agar (and perhaps be more efficiently transported by the bacteria) to react with DNA at a detectable mutagenic level. The addition of higher amounts of sample (∼ 1.0 mg) proved very toxic to the bacteria so that mutagenic activity could not be calculated.

Sodium azide is a mutagen used as a positive control for studies of mutagens that do not require metabolic activation. Its action is

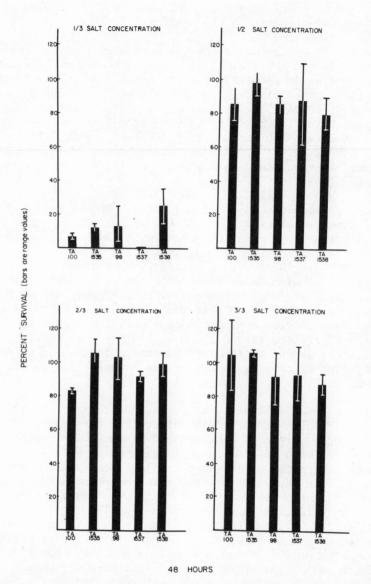

Fig. 3.   Survival rates of Salmonella strains after 48 hr
incubation with several [S] concentrations (see
Methods).   Histidine concentration was 0.25 mM.
All other conditions were the same as those for
Fig. 1.

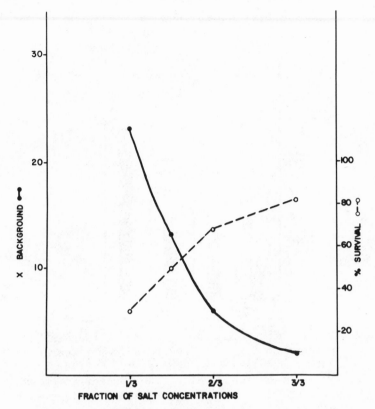

Fig. 4.   Mutagenic response and survival rates of TA 1535
with the addition of 0.5 mg of a 1:1 mixture of
"chromic chromate" and iron oxide (Fe_2O_3). Histidine concentration was 0.25 mM. Revertant
colonies were scored after 48 hr of incubation.
Background values were: 1/3 [S] – 5; 1/2 [S] – 4;
2/3 [S] – 6; 3/3 [S] – 17. Fe_2O_3 at this concentration (0.025 mg/plate) did not elevate backgrounds.
All plates contained approximately 0.5 x 10^8
bacteria. (See Methods for analysis of "chromic
chromate".)

positive for base-pair detecting strains of Salmonella although the
mechanism of the mutagenic activity is not known. TA 100 and
TA 1535 were found to respond almost identically to this compound
(data not shown); we therefore tested TA 1535 and TA 100 with all
four [S] concentrations and at both histidine concentrations.

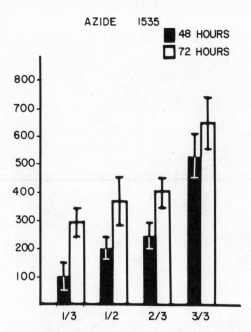

Fig. 5. Numbers of revertants of TA 1535 (ordinate) on
exposure to 0.5 ug of sodium azide at four concentrations of salts (abscissa) at 48 and (72)
hour incubations. Histidine concentration was
0.045 mM. Numbers of revertants are given as revertant colonies minus background colonies. All
plates were innoculated with approximately 0.5 x
10^8 bacteria. Background values were: 1/3 [S] -
7, (20); 1/2 [S] - 11, (24; 2/3 [S] - 13, (26);
3/3 [S] - 20, (36). Values shown in parentheses
are for 72 hr incubations.

      Our results with this compound indicate that decreased concen-
trations of salts resulted in decreased response to azide as a
mutagen (Figure 5). The figure shows that there is some advantage
in scoring the revertant colonies at the 72 hr period, although

maximum sensitivity of the assay is seen with the complete (3/3) concentrations of salts. The azide ion is smaller than the Cr(VI)-containing anions, and is apparently not limited in diffusion by the salts in the agar as is the CrO_4^{--} ion (and possibly other inorganic ions). Similar results were obtained with azide as a mutagen using the higher (0.25 mM) concentration of histidine, i.e. the sensitivity of the assay was decreased by lowering the concentrations of the salts.

An additional finding was that in many studies with low survival rates of the strain plated, we detected "pin-point" size colonies which remained abnormally small despite increased incubation time. Plating of randomly selected colonies on his⁻ minimal agar showed them to be surviving auxotrophs, rather than mutants. Subsequent studies in which we added mutagens to the system indicated that in all cases in which "pin-points" were detected, the suspected mutagen was toxic to the bacteria. When "chromic chromate" was tested in our laboratory with the spot test modification of the standard Ames test, the center of the plate was bare, indicating complete killing of the bacteria and this bare circle was surrounded by an area of hundreds of "pin-points", which we mistakenly counted as revertants (data not shown). These colonies were then transferred to his⁻ minimal agar and detected as survivals. An awareness of these deviants seems a necessary part of testing for toxic inorganic or organic compounds.

Our findings suggest that some inorganic compounds can be more accurately detected as mutagens using modified conditions of the standard Ames test, specifically by studying the effects of lowered concentrations of salts as well as varied histidine concentrations. Another improvement would be to score the revertant colonies after 72 hr of incubation to determine whether they have not reached maximum numbers at the 48 hr period. The importance of accurate colony counts when plating the bacteria from suspensions of phosphate buffer or from broth seems evident, as the numbers of spontaneous revertants and the number of surviving bacteria greatly influence the calculations of mutagenic action. Not all inorganic compounds may exhibit this pattern seen for detection of hexavalent chromium, but it is evident that the calculation of mutagenic activity must be based on numbers of surviving bacteria when the toxic agent kills a high percentage of the bacteria.

ACKNOWLEDGEMENTS

This project was supported by the American Iron and Steel Institute and by the Environmental Protection Agency (Grant #R806279).

REFERENCES

Ames, B.N., Durston, W.E., Yamasaki, E., and Lee, F.D., 1973, Car-
    cinogens are mutagens: a simple test system combining liver
    homogenates for activation and bacteria for detection, Acad.
    Sci., U.S.A., 70: 2281-2285.
Ames, B.N., McCann, J., and Yamaski, E., 1975, Methods for detecting
    carcinogens and mutagens with the Salmonella/mammalian-
    microsome mutagenicity test, Mutat. Res., 31: 347-364.
Franz, C.N., and Malling, H.V., 1975, The quantitative microsomal
    mutagenesis assay method, Mutat. Res., 31:365-380.
IARC, 1973, Some inorganic and organometallic compounds, Vol. 2,
    pp. 100-125, in:"IARC Monographs on the Evaluation of Car-
    cinogenic Risk of Chemicals to Man," International Agency
    for Research on Cancer, Lyon.
IARC, 1976, Cadmium, nickel, some epoxides, miscellaneous industrial
    chemicals and general considerations on volatile anaes-
    thetics, Vol 11, pp. 75-112, in: "IARC Monographs on the
    Evaluation of Carcinogenic Risk of Chemicals to Man," Inter-
    national Agency for Research on Cancer, Lyon.
Löfroth, G., 1978, The mutagenicity of hexavalent chromium is de-
    creased by microsomal metabolism, Naturwissenschaften, 65:
    207-208.
Nishioka, H., 1975, Mutagenic activities of metal compounds in
    bacteria, Mutat. Res., 31: 185.
Petrilli, F.L., and DeFlora, S., 1977, Toxicity and mutagenicity of
    hexavalent chromium on Salmonella typhimurium, Appl. Environ.
    Microbiol., 33: 805-809.
Sunderman, F.W., Jr., 1971, Metal carcinogenesis in experimental
    animals, Food Cosmet. Toxicol., 9: 105.
Venitt, S., and Levy, L.S., 1974, Mutagenicity of chromates in
    bacteria and its relevance to chromate carcinogenesis,
    Nature, 250: 493.
Vogel, H.J., and Bonner, D.M., 1956, Acetylornithinase of Escherichia
    coli: partial purification and some properties, J. Biol.
    Chem., 218: 97-106.

METABOLIC ACTIVATION OF TRIAZENE CYTOSTATICS

G.F. Kolar and M. Wildschütte

Institute for Toxicology and Chemotherapy
German Cancer Research Centre
Heidelberg, F.R.G.

INTRODUCTION

The mechanism by which dimethyltriazene cytostatics
inhibit tumours is not clearly understood, but the
available evidence is consistent with the formation of
reactive intermediates by oxidative transformation at
the terminal dimethylamino group of the triazene side-
chain. Audette et al. (1973) have proposed that the
antineoplastic action of triazenes depended on the enzymic formation of the corresponding monomethyl analogues which had been shown to be direct alkylating agents
both in vitro (Preussmann and von Hodenberg, 1970) and
in vivo (Margison et al., 1979). The authors therefore
concluded that the tumour growth was inhibited mainly by
methylation of crucial nucleophilic sites in cellular
biopolymers. However, the putative active metabolites
- the monomethyltriazenes - proved to be less effective
antitumour agents than the corresponding dimethyltriazene compounds (Vaughan and Stevens, 1978). Moreover,
the finding that the triazenes exhibited a much wider

antineoplastic activity range than that encountered in
the classical alkylating agents, also was not consistent
with this interpretation.

Previous work from our laboratory (Kolar and Schlesiger, 1976) showed that the parent 3,3-dimethyl-1-phenyl-
triazene was rapidly metabolized in rats, and that the
para and ortho ring positions and the terminal dimethylamino group of the triazene side-chain became enzymically oxidized. The triazene was metabolized in high yield
to degraded products (hydroxyanilines) but the proportion
of metabolites that could be cleaved into arenediazonium
cations with cold acid, a reaction characteristic of aromatic diazoamino compounds, was only about 1%.

We selected 1-(2,4,6-trichlorophenyl)-3,3-dimethyl-
triazene and 5-(3,3-dimethyl-1-triazeno)imidazole-4-carb-
oxamide (DIC, DTIC, NSC-45388) for metabolic studies
that were performed in adult male Sprague-Dawley rats.
The ring-chlorinated triazene derivative appeared to be
not only a well-suited model compound, but it also exhibited a pronounced tumour-inhibiting activity (IST
against TLX5 murine lymphoma 79% and therapeutic index
against ADJ/PC6 murine plasma cell tumour 25; T.A. Connors and G.F. Kolar, unpublished results). The substitution of all three ortho and para positions with chlorine
not only prevented ring hydroxylation in vivo but also
stabilized the 2N/3N bond against hydrolytic cleavage.
It was conceivable that, in a compound resisting annular
hydroxylation in vivo, one could well expect enhanced
enzymic transformation at the terminal dimethylamino group
of the side-chain, and possibly also extensive covalent
binding of the modified triazene to a water-soluble moi-

ety. The clinically useful 5-(3,3-dimethyl-1-triazeno)-
imidazole-4-carboxamide is an effective single agent for
the treatment of malignant melanoma (Comis, 1976) and,
in combination with adriamycin, for the control of dis-
seminated soft-tissue sarcoma (Gottlieb et al., 1976).
The mechanisms of its tumour-inhibiting action is thought
to depend on the formation of reactive intermediates by
enzymic transformation, but the relative importance of
the individual metabolites, or the metabolic sequences,
are unknown.

METABOLISM OF 1-(2,4,6-TRICHLOROPHENYL)-3,3-DIMETHYLTRI-
AZENE

Following subcutaneous administration of the title
compound (Fig. 1) a single urinary metabolite with an
intact diazoamino structure was detected by thin-layer
chromatography. The trichlorobenzenediazonium cation,
released by acid catalyzed cleavage of the 2N/3N bond,
was coupled with N-ethyl-1-naphthylamine (NEN reagent)
to yield 4-(2,4,6-trichlorobenzeneazo)-N-ethyl-1-naph-
thylamine, a derivative which was properly identified.
In addition, the chromatograms contained a high moving

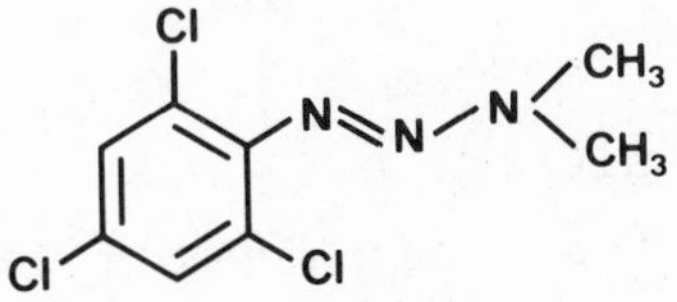

Fig. 1   Structure of 1-(2,4,6-trichloro-
         phenyl)-3,3-dimethyltriazene

spot of 2,4,6-trichloroaniline that arose by breakdown
of the metabolite during handling. The metabolite was
purified from filtered urines by anion and cation exchange chromatography followed by repeated gel filtration
on Sephadex G25. The freeze-dried compound was obtained
as a faintly yellow solid (in about 95% purity) which
appeared as a single spot (R_F 0.32-0.34 in n-butanol:
ethanol:conc. ammonia/5:3:2 v/v/v) on thin-layer chromatograms (Kolar and Carubelli, 1979a). The structure of
the metabolite was elucidated by chemical degradation
and by a combination of spectrophotometric methods: Glucuronic acid, which was liberated by acid hydrolysis and
detected with an alkaline silver nitrate reagent, co-
chromatographed on thin-layers with an authentic sample.
The presence of glucuronic acid in the hydrolysates was
confirmed by spectral analysis of the chromogen (λ max
530 nm) obtained in the carbazole reaction. The combined
analytical data (i.e. the ratio of the developed 4-(2,4,6-
trichlorobenzene)azo dye to the glucuronic acid contents,
determined by the carbazole reaction) indicated the presence of one glucuronosyl residue per molecule. The mode
of linkage between the uronic acid and the dimethyltriazene side-chain was then established. The lability of
the metabolite pointed to the conjugation through an O-
glycosidic bond between glucuronic acid and an 3N-hydroxy-
methyl group, introduced by metabolic oxidation of one
of the 3N-methyl groups of the triazene side-chain. In
addition, the release of formaldehyde by mild acid hydrolysis (0.67 M HCl at 60°C for 30 min), detected as 3,5-
diacetyl-1,4-dihydrolutidine (λ max 415 nm) in the Nash
reaction, provided important support for this structure.

Conclusive evidence for the presence of the 3N-hy-

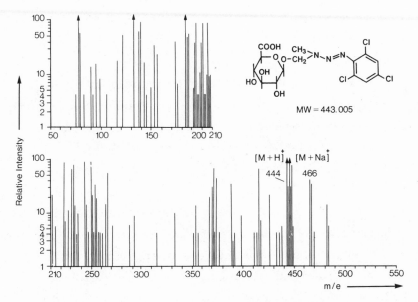

Fig. 2   FD Mass spectrum and structure of
         /1-methyl-3-(2,4,6-trichlorophenyl)-
         2-triazeno/methyl β-D-glucopyrano-
         side uronic acid

droxymethyl glucuronide structure was obtained from NMR
spectroscopy in D_2O which showed the following diagnosti-
cally important signals: (1) a singlet at 7.66 ppm /2H/
of the ring protons; (2) a doublet of doublets at 5.36-
5.52 ppm /2H/ of the 3N-methylene group; and (3) a singlet at 3.33 ppm /3H/ of the N-methyl group. The final
confirmation of the 3N-hydroxymethyl-O-glucuronide structure came from field desorption mass spectrometry
(Fig. 2), which established the molecular ion of the metabolite at m/e 443.005, in agreement with the proposed
structure (Schulten and Lehmann, 1980).

METABOLISM OF 5-(3,3-DIMETHYL-1-TRIAZENO)IMIDAZOLE-4-
CARBOXAMIDE

Thin-layer chromatographic examination of urinary
excretion products from rats that had been dosed with
5-(3,3-dimethyl-1-triazeno)imidazole-4-carboxamide
(Fig. 3) confirmed the presence of both previously identified metabolites, 5-aminoimidazole-4-carboxamide (Householder and Loo, 1969) and 2-azahypoxanthine (Householder
and Loo, 1971). However, besides the injected drug, the
chromatograms also showed the presence of another spot at
a lower R_F, discernible after spraying the developed dried
plates with acidified ethanolic N-ethyl-1-naphthylamine,
followed by UV irradiation (366 nm). This reaction indicates the presence of an intact diazoamino grouping in the
molecule, and this fact suggested to us that the new metabolite might well be important to the biological fate
and mode of action of the parent carcinostatic drug.

The elucidation of structure of this labile compound
required an indirect approach after our attempts to effect isolation of the metabolite by ion exchange or chromatographic methods failed. Similar difficulty has appar-

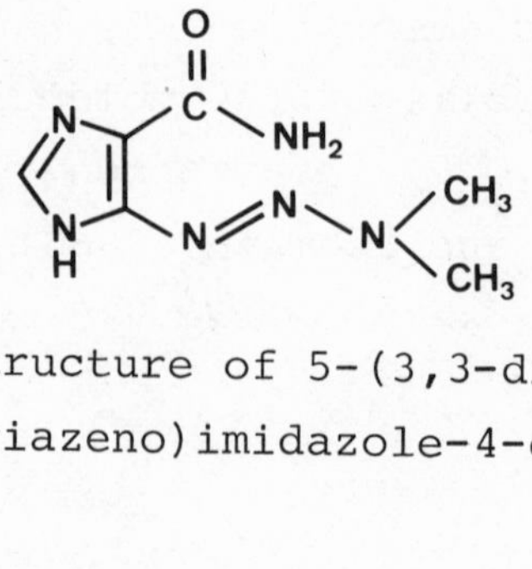

Fig. 3   Structure of 5-(3,3-dimethyl-1-
triazeno)imidazole-4-carboxamide

ently been experienced by Loo et al. (1968) who observed
the excretion of a labile DIC metabolite of low R_F in
dogs.

The light-catalyzed release of 5-diazoimidazole-4-
carboxamide, identified as 5-/(4-ethylamino-1-naphthyl)-
azo/imidazole-4-carboxamide, from the metabolite confirmed
not only the presence of an intact diazoamino group in
the molecule but also indicated the absence of any polar
function (e.g. hydroxyl) in the imidazole moiety. Con-
trary to our previous experience with metabolic transformation of 1-(2,4,6-trichlorophenyl)-3,3-dimethyltri-
azene, the metabolite was not retained by an anionic ex-
changer indicating freedom from covalent binding of the
drug to an anionic cellular constituent. However, the
lower chromatographic mobility of the metabolite was consistent with increased polarity at the dimethylamino terminal of the side-chain. Since demethylation of the drug
to 5-(3-methyl-1-triazeno)imidazole-4-carboxamide could
be excluded from chromatographic evidence, enzymic oxidation of DIC to its 3N-methylol derivative seemed to be
the most likely metabolic transformation.

The identification of the metabolite could be effected through the availability of 5-(3-hydroxymethyl-
3-methyl-1-triazeno)imidazole-4-carboxamide (Fig. 5) which
we synthesized by condensing 5-(3-methyl-1-triazeno)imid-
azole-4-carboxamide with excess of formaldehyde in anhydrous methanol (Kolar et al., 1980). The structure of
the synthetic methylol was established by NMR spectroscopy
(Fig. 4) and confirmed by FD mass spectrometric molecular weight determination (m/e 198). Because the synthetic compound showed appreciable stability in dry methanol, the excreted metabolite (and DIC) were extracted

from freshly collected freeze-dried urines and purified
by fractional precipitation with petroleum ether. The
enriched fractions yielded reproducible separations and
the metabolite was identified as 5-(3-hydroxymethyl-3-
methyl-1-triazeno)imidazole-4-carboxamide by co-chromato-
graphy in two solvent systems. An additional proof for
the 3N-methylol structure of the metabolite (Fig. 5) was

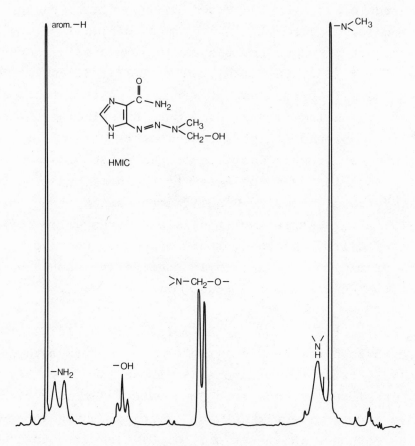

Fig. 4   90 MHz ^{1}H-NMR Spectrum of 5-(3-hydroxy-
         methyl-3-methyl-1-triazeno)imidazole-
         4-carboxamide (doublet of >N-CH$_2$-O- pro-
         tons at δ = 5.10 ppm, J = 7 Hz)

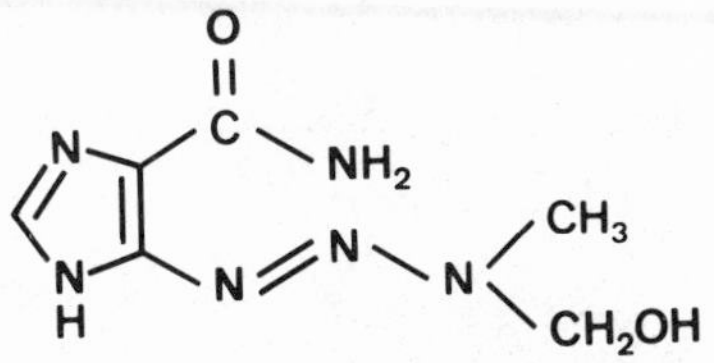

Fig. 5 Structure of 5-(3-hydroxymethyl-3-methyl-
1-triazeno)imidazole-4-carboxamide

obtained from its methylating capacity for nucleophilic
substrates: Thin-layer chromatograms of stored methanolic
extracts from freeze-dried urines (1-2 days at 4°C in the
dark) showed the formation of an artefactual NEN/UV-posi-
tive compound having a higher mobility than DIC which we
identified as 5-(3,3-dimethyl-1-triazeno)-1-methylimid-
azole-4-carboxamide. The methylating capacity of 5-(3-
hydroxymethyl-3-methyl-1-triazeno)imidazole-4-carboxamide
was confirmed by its addition to a ten-fold excess of DIC
in methanol which resulted in the formation of the identical ring-methylated DIC.

CONCLUDING REMARKS

The molecular structure of $/$1-methyl-3-(2,4,6-tri-
chlorophenyl)-2-triazeno$/$methyl β-D-glucopyranoside uronic
acid (Kolar and Carubelli, 1979b) established for the
sole urinary metabolite of 1-(2,4,6-trichlorophenyl)-3,3-
dimethyltriazene supplied the first unequivocal proof for
the formation of triazene 3N-hydroxymethyl intermediates
in living systems and for their stabilization by covalent

binding to glucuronic acid. Since enzymic hydrolysis of
the conjugate, followed by a loss of formaldehyde, leads
to the formation of the direct methylating 1-(2,4,6-tri-
chlorophenyl)-3-methyltriazene as shown by the high mutagenic activity of the metabolite in <u>Salmonella</u> strains
(unpublished results), it is conceivable that the conjugated metabolite acts as one of the carrier forms of
the biologically active 1-(2,4,6-trichlorophenyl)-3,3-
dimethyltriazene.

On the other hand, biotransformation of 5-(3,3-di-
methyl-1-triazeno)imidazole-4-carboxamide did not yield
any conjugated products and the only metabolite with intact diazoamino structure detected on thin-layer chromatograms was identified as 5-(3-hydroxymethyl-3-methyl-
1-triazeno)imidazole-4-carboxamide. Therefore it appears
that enzymic oxidation of 3N-methyl to 3N-methylol sufficiently increases the solubility of the drug so that
conjugation with water-soluble endogenous compounds does
not occur. Since the tumour inhibiting activity of hexa-
methylmelamine has been attributed to the introduction
of N-methylol groups (Rutty et al., 1978), it is plausible that the formation of 3N-methylols in triazene cytostatics may be an important contributory factor to their
broad spectrum of antineoplastic activity.

REFERENCES

Audette, R.C.S., Connors, T.A., Mandel, H.G., Merai, K.
    and Ross, W.C.J., 1973, Studies on the mechanism
    of action of the tumour inhibitory triazenes, <u>Bio-
    chem. Pharmacol.</u>, 22:1855.

Comis, R.L., 1976, DTIC (NSC-45388) in malignant melano-
     ma: A perspective, Cancer Treat. Rep., 60:165.

Gottlieb, J.A., Benjamin, R.S., Baker, L.H. et al., 1976,
     Role of DTIC (NSC-45388) in the chemotherapy of sar-
     coma, Cancer Treat. Rep., 60:199.

Householder, G.E. and Loo, T.L., 1969, Elevated urinary
     excretion of 4-aminoimidazole-5-carboxamide in pa-
     tients after intravenous injection of 4-(3,3-dimethyl-
     1-triazeno)imidazole-5-carboxamide, Life Sci., 8:533.

Householder, G.E. and Loo, T.L., 1971, Disposition of 5-
     (3,3-dimethyl-1-triazeno)imidazole-4-carboxamide,
     a new antitumor agent, J. Pharmacol. Exp. Ther.,
     179:386.

Kolar, G.F. and Schlesiger, J., 1976, Urinary metabolites
     of 3,3-dimethyl-1-phenyltriazene, Chem.-Biol. Inter-
     act., 14:301.

Kolar, G.F. and Carubelli, R., 1979a, Urinary metabolite
     of 1-(2,4,6-trichlorophenyl)-3,3-dimethyltriazene
     with an intact diazoamino structure, Cancer Letters,
     7:209.

Kolar, G.F. and Carubelli, R., 1979b, /1-Methyl-3-(2,4,6-
     trichlorophenyl)-2-triazeno/methyl β-D-glucopyrano-
     side uronic acid is a novel metabolite of 1-(2,4,6-
     trichlorophenyl)-3,3-dimethyltriazene, in: "Glyco-
     conjugates", Proceedings of the Fifth International
     Symposium, Kiel, Federal Republic of Germany,
     R. Schauer, ed., Thieme, Stuttgart.

Kolar, G.F., Maurer, M. and Wildschütte, M., 1980,
     5-(3-Hydroxymethyl-3-methyl-1-triazeno)imidazole-
     4-carboxamide is a metabolite of 5-(3,3-dimethyl-
     1-triazeno)imidazole-4-carboxamide (DIC, DTIC,
     NSC-45388), Cancer Letters, 10:235.

Loo, T.L., Luce, J.K., Jardine, J.H. and Frei, E., III.,
    1968, Pharmacologic studies of the antitumor agent
    5-(dimethyl-triazeno)imidazole-4-carboxamide,
    Cancer Res., 28:2448.
Margison, G.P., Likhachev, A.J. and Kolar, G.F., 1979,
    In vivo alkylation of foetal, maternal and normal
    rat tissue nucleic acids by 3-methyl-1-phenyltri-
    azene, Chem.-Biol. Interact, 25:345.
Preussmann, P. and von Hodenberg, A., 1970, Mechanism of
    carcinogenesis with 1-aryl-3,3-dialkyltriazenes II.
    In vitro alkylation of guanosine, RNA and DNA with
    aryl-monoalkyltriazenes to form 7-alkylguanine,
    Biochem. Pharmacol., 19:1505.
Rutty, C.J., Connors, T.A., Ngyen-Hoang-Nam, Do-Cao-Thang
    and Hoellinger, H., 1978, In vivo studies with hexa-
    methylmelamine, Europ. J. Cancer, 14:713.
Schulten, H.-R. and Lehmann, W.D., 1980, Field desorption
    mass spectrometry, TIBS, 5:142
Vaughan, K. and Stevens, M.F.G., 1978, Monoalkyltriazenes,
    Chem. Soc. Rev., 7:377.

TWO BIOLOGICAL REACTIVE INTERMEDIATES IN AMINOAZO-DYE CARCINOGENESIS AND THEIR GLUTATHIONE ADDUCTS

S.K.S. Srai[1], B. Ketterer[1], and F.F. Kadlubar[2]

[1]The Middlesex Hospital Medical School
London, W1P 7PN, U.K., and [2]National Center for
Toxicological Research, Jefferson, AR 72079
U.S.A.

Two reactive intermediates have now been shown to occur during the course of the metabolism of N,N-dimethyl-4-aminoazobenzene (DAB) or N-methyl-4-aminoazobenzene (MAB). They have proved too labile to be isolated per se, but their existence is recognized by their ability to form adducts which are sufficiently stable to be isolated and characterized. These two reactive intermediates have been identified as N-sulphonoxy-N-methyl-4-aminoazobenzene (N-sulphonoxy-MAB) (Fig. 1, I) and N-methylol-4-aminoazo-benzene (N-methylol-AB) (Fig. 2, II) respectively.

In this paper the evidence for the existence of these biological reactive intermediates is described and the role of their glutathione (GSH) adducts in the detoxification of aminoazo-dye carcinogens is discussed.

A.   N-Sulphonoxy-MAB

The identification of this intermediate has depended to a large extent on comparisons of DAB or MAB with the arylamide carcinogen N-acetyl-2-aminofluorene (AAF). An important discovery was that an essential step in the metabolism of AAF to a biological reactive form was N-hydroxylation (Cramer et al., 1960). While N-hydroxy-AAF was not itself reactive, several synthetic esters such as N-acetoxy-AAF, N-benzoyloxy-AAF and N-sulphonoxy-AAF were. Subsequently N-sulphonoxy-AAF was shown to be a biological reactive intermediate in the liver of the male rat (Lotlikar et al., 1966; DeBaun et al., 1968; King and

Phillips, 1968; Maher et al., 1968; and DeBaun et al., 1970).

That similar metabolic transformations result in the activation of MAB to N-sulphonoxy-MAB is given indirect support by the finding that the synthetic ester N-benzoyloxy-MAB (Poirier et al., 1967) gives adducts identical to those formed in vivo after the administration of DAB or MAB (Scribner et al., 1965; Poirier et al., 1967; Lin et al., 1968; Lin et al., 1969; Lin et al., 1975a and 1975b; Beland et al., 1980; Tarpley et al., 1980), and direct support by the demonstration that the same adducts are also generated when N-hydroxy-MAB, a sulphotransferase system and the appropriate nucleophile are incubated in vivo (Kadlubar et al., 1976a and 1976b).

N-Sulphonoxy-MAB forms adducts with methionyl, tyrosyl and cysteinyl residues in proteins and with guanine in nucleic acids (for references see above). Reactions of this type which involve informational macromolecules may be responsible for the cytotoxicity and carcinogenicity of DAB. For instance MAB forms both an N-(deoxyguanosin-8-yl)-MAB derivative and a 3-(deoxyguanosin-N^2-yl)-MAB derivative with DNA in vivo (Beland et al., 1980), one or both of which may be important in the initiation of carcinogenesis.

GSH which is present in hepatocytes in concentrations up to 10 mM should also react with N-sulphonoxy-MAB. This proves to be so and the extent to which GSH might detoxify N-sulphonoxy-MAB has been examined.

If N-benzoyloxy-MAB is reacted with GSH three adducts result, namely 3-, 2'-, and 4'-(glutathion-S-yl)-MAB (3-, 2'-, and 4'-GS-MAB). 3-GS-MAB (Fig. 1, III) is the major product accounting for 90% of the total GSH adduct. It can be isolated from bile (Ketterer et al., 1979) and is the GSH adduct produced when N-sulphonoxy-MAB is generated in vitro in the presence of GSH (Kadlubar et al., 1980) (see Fig. 1).

Interesting observations emerge from in vitro experiments. For example in the presence of 10 mM GSH (a physiological level) only 6-16% of the N-sulphonoxy-MAB generated is trapped as 3-GS-MAB.

Also when DNA and GSH are both present in the same in vitro system, GSH is not as effective in competing with DNA for N-sulphonoxy-MAB as might be anticipated. For

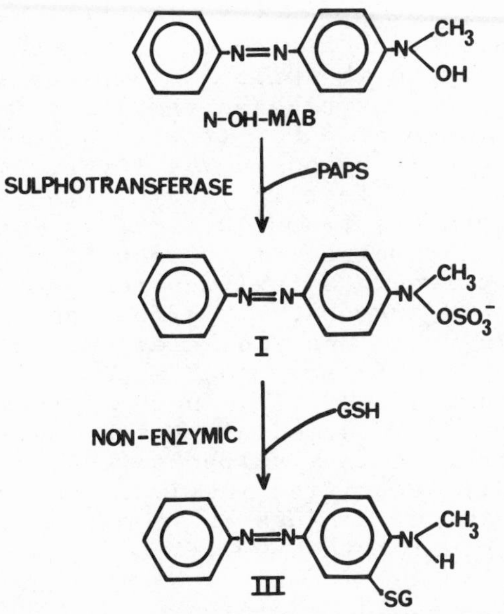

Fig. 1.   Metabolic formation of 3-(glutathion-S-yl)-MAB
          from N-hydroxy-MAB

          N-hydroxy-MAB which results from the microsomal
          N-hydroxylation of MAB (Kadlubar et al., 1976a)
          is sulphonated by a cytosolic sulphotransferase
          which requires 3'-phosphoadenosine-5'-phospho-
          sulphate (PAPS) as a co-substrate to give as a
          product the biological reactive intermediate
          N-sulphonoxy-N-methyl-4-aminoazobenzene (N-
          sulphonoxy-MAB) - (I). This intermediate then
          reacts with GSH to give 3-(glutathion-S-yl)-N-
          methyl-4-aminoazobenzene (3-GS-MAB) - (III)

example 10 mM GSH reduces the amount of N-sulphonoxy-MAB
trapped by 1 mg per ml DNA by only 30%. Furthermore
these results are not affected by the presence of
glutathione-S-transferases or glutathione-S-transferase
inhibitors indicating that N-sulphonoxy-MAB is not a
substrate for these enzymes.

     On the basis of these results one might suppose that
a contributing factor to the carcinogenicity of DAB and
MAB may be the fact that GSH is not sufficiently effective
as a trapping agent for N-sulphonoxy-MAB (Kadlubar et al.,

1980). In order to pursue this idea further, preliminary experiments have been carried out in which the amount of 3-GS-MAB excreted in the bile is compared with the stable macromolecule-bound carcinogen remaining in the liver after a single dose of DAB. It was found that while 0.05% of the total dose is bound in the liver, 0.2% is excreted as 3-GS-MAB in the bile (Srai, Waynforth and Ketterer, 1980). GSH therefore traps 40 times as much N-sulphonoxy-MAB as do the macromolecular nucleophiles and is therefore much more effective in detoxification than the in vitro experiments would suggest. This does not necessarily mean that results in vitro bear no relationship to results in vivo, it may simply show that a potential seen in vitro is exploited more effectively in the intact cell. For example a reactive intermediate which is trapped by GSH in vitro better than is N-sulphonoxy-MAB may be very much more effectively detoxified in vivo. In such a case the ratio of glutathione conjugates to macromolecular adducts would be much greater than 40.

B.   N-Methylol-4-Aminoazobenzene

During the course of the isolation of 3-GS-MAB another aminoazo-dye glutathione conjugate was isolated and found on treatment with N-HCl to decompose to 4-aminoazobenzene (AB).

In its acid lability it was reminiscent of an aminoazo-dye glutathione adduct prepared by Mueller and Miller (1953) during their early study of microsomal oxidative N-demethylation. Using 3-methyl-MAB as their substrate these authors showed that microsomal N-demethylation was associated with the release of the N-desmethyl compound, 3-methyl-AB and a stoichiometric amount of formaldehyde (CH_2O). If however GSH was present in the incubation mixture a polar dye was produced which, when treated with acid, decomposed releasing 3-methyl-AB. A similar polar dye was synthesized by the condensation of 3-methyl-AB, CH_2O and GSH. Mueller and Miller concluded that in the in vitro experiments GSH had trapped an N-methylol intermediate which was formed during microsomal oxidative N-demethylation.

Direct evidence for the proposed N-methylol intermediate comes from the discovery of compounds with N-methyl groups which are metabolized to N-methylol derivatives sufficiently stable to be isolated and identified. Among these are some N-alkylamides (Keberle et al., 1963; McMahon and Sullivan, 1965)

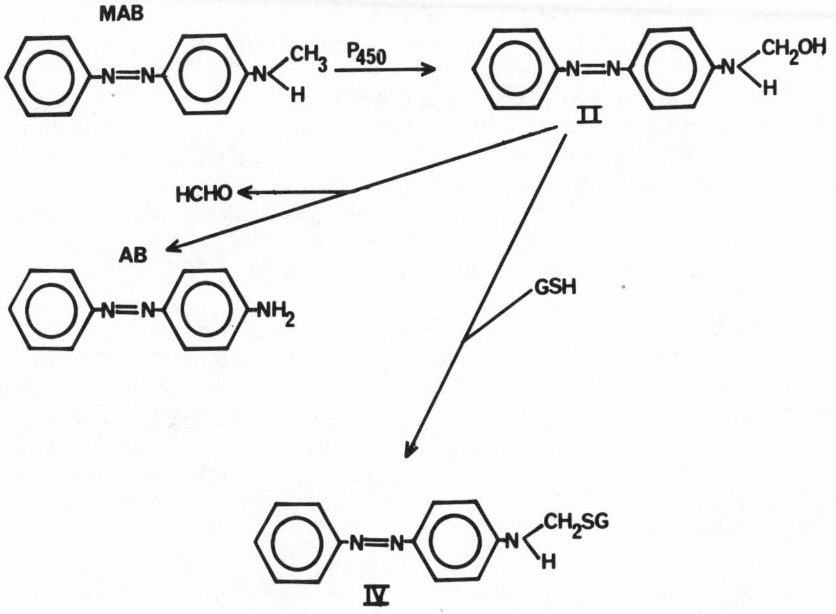

Fig. 2. Metabolic formation of glutathion-S-methylene-
(4-aminoazobenzene) from MAB

Cytochrome P450 oxidations of the N-methyl group
of MAB gives N-methylol-4-aminoazobenzene (N-
methylol-AB) (II). In the absence of GSH this
decomposes to formaldehyde and 4-aminoazobenzene
(Miller and Miller, 1953). In the presence of
GSH it is also trapped as glutathion-S-methylene-
(4-aminoazobenzene) (GS-CH$_2$-AB) (IV).

methylcarbazole (Gorrod and Temple, 1976), triazenes and
methylmelamines (Gescher et al., 1979).

In the present work it seemed likely that we had
isolated a glutathione adduct similar to that proposed
by Mueller and Miller and consequently we reacted AB, GSH,
and an excess of CH$_2$O to give the Mannich base glutathion-
S-methylene-(4-aminoazobenzene) (GS-CH$_2$-AB) (Fig. 2, IV).
The structure of this compound was established by
13C-NMR which gave a spectrum identical in every way with
the combined spectra of MAB and GSH except for downfield
shifts of the resonances of both the N-methyl carbon of
the aminoazo-dye and the Cβ of the cysteinyl residue

which are consistent with the bonding of each of them
to a substituted sulphur atom (Ketterer et al., 1980).

The original biliary conjugate and a biosynthetic
product resulting from the incubation of MAB, microsomes,
NADPH and GSH in vitro are identical with synthetic
GS-CH$_2$-AB in chromatographic properties (TLC and HPLC)
and visible and u-v spectra. On treatment with acid
they all give stoichiometrically equivalent amounts of
AB, CH$_2$O and GSH (Ketterer et al., 1980).

The link between GS-CH$_2$-AB formation and the microsomal oxidation of the N-methyl group of MAB is confirmed
by the use of 2-[(2,4-dichloro-6-phenyl)phenoxy]ethylamine
(DPEA) an inhibitor of microsomal N-demethylation
(McMahon, 1962). When DPEA is added to in vitro systems
containing MAB, microsomes, NADPH and GSH neither N-demethylation nor GS-CH$_2$-AB formation take place (Ketterer
et al., 1980).

Thus microsomal oxidation of the N-methyl groups of
MAB gives rise to N-methylol-AB which, though labile,
readily forms an adduct with GSH namely GS-CH$_2$-AB
(Fig. 2).

Unlike 3-GS-MAB which is stable and a minor compound
accounting for only 0.3% of the biliary metabolite of
DAB, GS-CH$_2$-AB is unstable and a major component
accounting for 20% of the biliary metabolites.

In addition to GS-CH$_2$-AB, DAB might be expected to
give rise to GS-CH$_2$-MAB as a result of the oxidation of
the first N-methyl group. Such a compound is not found
in bile nor is its decomposition product MAB. It would
appear that oxidation of the first N-methyl group of DAB
results solely in N-demethylation while oxidation of the
remaining secondary amine gives rise to GS-CH$_2$-AB.
These two oxidations followed by reaction with GSH are
very rapid since GS-CH$_2$-AB is the first metabolite to
appear in bile (Srai, Waynforth and Ketterer, 1980).

CONCLUSIONS

Evidence for the existence of two biological reactive
intermediates in DAB and MAB metabolism, N-sulphonoxy-MAB
and N-methylol-AB, has been given.

These two intermediates appear to have quite
different roles. N-sulphonoxy-MAB (or its analogue
N-benzoyloxy-MAB) behaves as a strong electrophile, is

linked with carcinogenesis and mutagenesis (Poirier et al., 1967; Maher et al., 1968; Kadlubar et al., 1976b) and forms adducts with DNA (Beland et al., 1980; Tarpley et al., 1980). In order to react with DNA and other macro-molecules, N-sulphonoxy-MAB must escape reaction with small molecular nucleophiles within the cell, in particular GSH and some evidence has been presented that GSH may have limitations as a detoxifying agent for this intermediate (Kadlubar et al., 1980).

N-methylol-AB is clearly important in the detoxifi-cation of DAB since its GSH adduct, GS-CH$_2$-AB accounts for one fifth of the total biliary excretion of DAB. The optimal excretion of DAB metabolites should therefore depend on the maintenance of hepatic GSH levels and indeed Levine and Finkelstein (1978) have shown that the depletion of hepatic GSH levels can suppress total levels of biliary metabolites of DAB. We have shown in our own experiments involving GSH depletion that this is due, in large part, to the selective inhibition of the formation and excretion of GS-CH$_2$-AB (Srai, Waynforth and Ketterer, 1980).

In addition to its adduct with GSH, N-methylol-AB has the potential to form adducts with thiols, phenols and amines in proteins and amines in nucleic acids (Walker, 1944). AB and CH$_2$O have been condensed with nucleic acids and cytosine derivatives: acid labile adducts were formed and the structure of the cytosine adduct was shown to be a triazene derivative into which two molecules of CH$_2$O had been incorporated (Roberts and Warwick, 1963 and 1966). It seems unlikely that this cytosine adduct would form in vivo during DAB carcino-genesis, but other adducts may well occur. Therefore the extent to which in vitro generated N-methylol-AB forms adducts with macromolecules clearly requires examination, a task which is likely to be hampered by their instability. If adducts form with these macromolecules it is important to establish what role, if any, they have in carcinogenesis.

ACKNOWLEDGEMENTS

This research is supported by the Cancer Research Campaign, U.K., and the Food and Drug Administration Department of Health and Human Services, U.S.A.

REFERENCES

Beland, F.A., Tullis, D.L., Kadlubar, F.F., Straub, K.M.
    and Evans, F.E., 1980, Characterization of DNA
    adducts of the carcinogen N-methyl-4-aminoazobenzene
    in vitro and in vivo, Chem.-Biol. Interact., 31:1.
Cramer, J.W., Miller, J.A., and Miller, E.C., 1960,
    N-Hydroxylation: A new metabolic reaction observed
    in the rat with the carcinogen 2-acetylaminofluorene,
    J. Biol. Chem., 235: 885.
DeBaun, J.R., Rowley, J.Y., Miller, E.C., and Miller,
    J.A., 1968, Sulfotransferase activation of N-hydroxy-
    2-acetylaminofluorene in rodent livers susceptible
    and resistant to this carcinogen, Proc. Soc. Exptl.
    Biol. Med., 129:268.
DeBaun, J.R., Miller, E.C., and Miller, J.A., 1970,
    N-Hydroxy-2-acetylaminofluorene sulfotransferase:
    Its probable role in carcinogenesis and in protein-
    (methion-S-yl) binding in rat liver, Cancer Res.,
    30: 577.
Gescher, A., Hickman, J.A., and Stevens, M.F.G., 1979,
    Oxidative metabolism of some N-methyl containing
    xenobiotics can lead to stable progenitors of
    formaldehyde, Biochem. Pharmac., 28: 3235.
Gorrod, J.W., and Temple, D.J., 1976, The formation of
    an N-hydroxymethyl intermediate in the N-demethyla-
    tion of N-methylcarbazole in vivo and in vitro,
    Xenobiotica, 6: 265.
Grasse, F.R., and James, S.P., 1972, Biosynthesis of
    sulphated hydroxypentylmercapturic acids and the
    metabolism of 1-Bromopentane in the rat,
    Xenobiotica, 2: 117.
James, S.P., and Waring, R.H., 1971, The metabolism of
    alicyclic ketones in the rabbit and rat,
    Xenobiotica, 1:573.
Kadlubar, F.F., Miller, J.A., and Miller, E.C., 1976a,
    Microsomal N-oxidation of the hepatocarcinogen
    N-methyl-4-aminoazobenzene and the reactivity of
    N-hydroxy-N-methyl-4-aminoazobenzene, Cancer Res.,
    36: 1196.
Kadlubar, F.F., Miller, J.A., and Miller, E.C., 1976b,
    Hepatic metabolism of N-hydroxy-N-methyl-4-aminoazo-
    benzene and other N-hydroxy arylamines to reactive
    sulfuric acid esters, Cancer Res., 36: 2350.

Kadlubar, F.F., Ketterer, B., Flammang, T.J., and
    Christodoulides, L., 1980, Formation of 3-(glutathion-
    S-yl)-N-methyl-4-aminoazobenzene and inhibition of
    aminoazo dye-nucleic acid binding in vitro by
    reaction of glutathione with metabolically-generated
    N-methyl-4-aminoazobenzene-N-sulfate,
    Chem.-Biol. Interact., 31: 265.
Keberle, H., Riess, W., Schmid, K., and Hoffmann, K.,
    1963, Ueber den mechanismus der biologischen N-
    desalkylierung, Arch. Int. Pharmacodyn., 142: 125
Ketterer, B., Kadlubar, F., Flammang, T., Carne, T., and
    Enderby, G., 1979, Glutathione adducts of N-methyl-4-
    aminoazobenzene formed in vivo and by reaction of N-
    benzoyloxy-N-methyl-4-aminoazobenzene with
    glutathione, Chem.-Biol. Interact., 25: 7.
Ketterer, B., Kadlubar, F., Srai, S.K.S., Tullis, D.,
    Evans, F., and Unruh, L., 1980, Unpublished
    Information.
King, C.M., and Phillips, B., 1968, Enzyme-catalyzed
    reactions of the carcinogen N-hydroxy-2-fluoroenyl-
    acetamide with nucleic acid, Science, 159: 1351.
Levine, W.G., and Finkelstein, T.T., 1978, Biliary
    excretion of N,N-dimethyl-4-aminoazobenzene (DAB) in
    the rat. Effects of pretreatment with inducers and
    inhibitors of the mixed function oxidase system and
    with agents that deplete liver glutathione.
    Drug Metabolism and Disposition, 6: 265.
Lin, J-K., Miller, J.A., and Miller, E.C., 1968, Studies
    on structures of polar dyes derived from the liver
    proteins of rats fed N-methyl-4-aminoazobenzene.
    II. Identity of synthetic 3-(homocystein-S-yl)-N-
    methyl-4-aminoazobenzene with the major polar dye
    P2b, Biochemistry, 7: 1889.
Lin, J-K., Miller, J.A., and Miller, E.C., 1969, Studies
    on structures of polar dyes derived from the liver
    proteins of rats fed N-methyl-4-aminoazobenzene.
    III. Tyrosine and homocysteine sulfoxide polar dyes,
    Biochemistry, 8: 1573.
Lin, J-K., Schmall, B., Sharpe, I.D., Mivra, I., Miller,
    J.A., and Miller, E.C., 1975a, N-Substitution of
    carbon 8 in Guanosine and Deoxyguanosine by the
    carcinogen N-Benzoyloxy-N-methyl-4-aminoazobenzene
    in vitro, Cancer Res., 35: 832.
Lin, J-K., Miller, J.A., Miller, E.C., 1975b, Structures
    of hepatic nucleic acid-bound dyes in rats given
    the carcinogen N-methyl-4-aminoazobenzene, Cancer
    Res., 35: 844.

Lotlikar, P.D., Scribner, J.D., Miller, J.A., and Miller, E.C., 1966, Reaction of esters of aromatic N-hydroxy amines and amides with methionine in vitro: a model for in vivo binding of amine carcinogens to proteins, Life Sciences, 5: 1263.

McMahon, R.E., 1962, The competitive inhibition of the N-demethylation of butynamine by 2,4-dichloro-6-phenylphenoxyethylamine (DPEA), J. Pharmacol. Exptl. Therap., 138: 382.

McMahon, R.E., and Sullivan, H.R., 1965, The metabolism of the herbicide diphenamid in rats, Biochem. Pharmac., 14: 1085.

Maher, V.H., Miller, E.C., Miller, J.A., and Szybalski, W., 1968, Mutations and decreases in density of transforming DNA produced by derivatives of the carcinogens 2-acetyl-aminofluorene and N-methyl-4-aminoazobenzene, Mol. Pharmac., 4: 411.

Miller, J.A., and Miller, E.C., 1953, The carcinogenic aminoazo dyes, Adv. Can. Res., 1: 339.

Mueller, G.C., and Miller, J.A., 1953, The metabolism of methylated aminoazo dyes. II. Oxidative demethylation by rat liver homogenates, J. Biol. Chem., 202: 579.

Poirier, L.A., Miller, J.A., Miller, E.C., and Sato, K., 1967, N-Benzoyloxy-N-methyl-4-aminoazobenzene: Its carcinogenic activity in the rat and its reactions with proteins and nucleic acids and their constituents in vitro, Cancer Res., 27: 1600.

Roberts, J.J., and Warwick, G.P., 1963, Azo-dye liver carcinogenesis: Reaction of hydroxymethylaminoazo-benzene with nucleic acids in vitro, Nature, 197: 87.

Roberts, J.J., and Warwick, G.P., 1966, Azo-dye carcinogenesis. The reactions of 4-hydroxymethyl-aminoazobenzene with cytosine derivatives. Int. J. Cancer, 1: 107.

Scribner, J.D., Miller, J.A., and Miller, E.C., 1965, 3-Methylmercapto-N-methyl-4-aminoazobenzene: An alkaline-degradation product of a labile protein-bound dye in the livers of rats fed N,N-dimethyl-4-aminoazobenzene, Biochem. Biophys. Res. Commun., 20: 560.

Srai, S.K.S., Waynforth, B., and Ketterer, B., 1980, Unpublished information.

Tarpley, W.G., Miller, J.A., and Miller, E.C., 1980, Adducts from the reaction of N-Benzoyloxy-N-methyl-4-aminoazobenzene with deoxyguanosine or DNA in vitro and from hepatic DNA of mice treated with N-methyl- or N,N-dimethylaminoazobenzene, Cancer Res., 40: 2493.

Walker, J.F., 1944, "Formaldehyde", Reinhold, New York.

NUDE MOUSE IN STUDIES ON CHEMICAL CARCINOGENESIS

M.Laaksonen, R. Mäntyjärvi and O. Hänninen

Departments of Clinical Microbiology and Physiology

University of Kuopio, SF-70101 Kuopio 10, Finland

INTRODUCTION

Increased awareness of the significance of chemical carcinogenesis has stimulated the development of several types of tests
for screening of carcinogenic activity. Compared to other techniques, the use of mammalian cells as targets in the tests offers
the advantage of a closer resemblance to the in vivo situation.
Carcinogenicity tests based on mammalian cell transformation models
have been developed using either cell lines or low passage cultures
of fibroblasts as targets (see Rowan, 1978). The possibility to
assess the tumorigenic potential of the transformed cells is one of
the requirements of transformation tests. However, the transplanted
cells may be rejected even in a syngeneic host if transformation is
associated with expression of new cell surface antigens. Immunosuppression of the recipients can be used to overcome this problem,
but another solution was offered by the introduction of nude mice
strains (Nomura et al., 1977). The immunological defect of these
mice allows transplantation of even xenogeneic tumor cells, and
transformed cells derived from fibroblast cultures of nude mouse
tissues should have a good opportunity to grow in nude mice if the
cells are tumorigenic.

One of the factors affecting cell transformation is the metabolic activation of carcinogens. The levels of metabolizing enzymes
vary e.g. between mouse strains as shown by Nebert (1978), and the
variation may also be apparent at the cell culture level. Adaptation of the nude mouse model for carcinogenicity tests precludes
that the mouse strains used have been characterized for their capability to metabolize foreign compounds. Earlier studies suggest
that there may be differences between normal and nude mice in some

of the enzymes involved in liver and kidney tissues (Freudenthal et al., 1976; Litterst et al., 1978). To our knowledge the skin of nude mice has not been analyzed for enzymes metabolizing foreign compounds.

The aim of the present study was to establish a model for carcinogenicity tests based on the use of nude mice. In this report we describe preliminary experiments to set up a transformation focus assay in nude mouse skin fibroblast cultures. To clarify the capability of nude mouse to metabolize procarcinogens we have measured both hydroxylation and conjugation reactions in nude and in the background NMRI mice.

MATERIALS AND METHODS

Animals

Adult female NMRI homozygous normal (+/+) and homozygous nude (nu/nu) mice were used. The nude strain was obtained from Dr. O. Mäkelä, University of Helsinki. The animals used in the present experiments were derived from the matings of nu/nu males with normal outbred NMRI thus obtaining nu/+ heterozygotes. Such nu/+ were used for inter-crosses or for mating nu/nu males to produce the nu/nu females used in the experiments. Mice were produced in the Laboratory Animal Center of the University of Kuopio. They were fed on commercial standard pellets (Hankkija LTD, Finland) ad libitum.

In the induction experiments 3-methylcholantrene (Sigma Chemicals, Saint Louis, Mo) was given intraperitoneally (25 mg/kg or 80 mg/kg) in olive oil once 48 hours prior to sacrifice. Control animals received equivalent volumes of olive oil.

Tissue Preparation and Enzyme Assays

Animals were killed by cervical dislocation. The liver and kidneys were quickly removed into ice-cold 0.25 M sucrose and weighed. Microsomes were isolated as described earlier and finally suspended in ice-cold 0.25 M sucrose to give about 30 mg protein/ml (Hänninen, 1968).

For obtaining skin specimens the dorsal area of killed animals was shaved and the remaining hair was removed using a scalpel. The skin was immediately immersed in ice-cold 0.25 M sucrose containing 10 mM potassium phosphate buffer, pH 7.4. Subcutaneous fat tissue was removed and the remaining tissue was homogenized in 5 volumes of the same medium using an Ultra Turrax homogenizer at full speed

for 5 x 15 sec in an ice bath. The homogenate was then filtered
through cheese cloth and the homogenate was used as such for determination of enzyme activities (Schmassmann et al., 1976).

Protein concentrations were determined by method of Lowry
et al. (1951).

Cytochrome P-450 concentration was measured from ditionite
difference spectra in a Cary spectrofotometer (model 118) as described by Omura and Sato (1964) by using the extinction coefficient
of $91 \times nM^{-1} \times cm^{-1}$.

NADPH Cytochrome c reductase (EC 1.6.2.4) activity was determined by following the reduction of cytochrome c (Sigma Chemicals, Saint Louis, Mo) at 550 nm in Perkin-Elmer 402 spectrophotometer at 38° C (Phillips and Langdon, 1962).

The deethylation of 7-ethoxycoumarin was determined by the
direct fluorometric method as modified by Aitio (1978). The amount
of 7-hydroxycoumarin produced was measured by following the fluorescence (excitation wavelength 390 nm, emission 440) with a
Perkin-Elmer MPF 2 A spectrofluorometer. 7-Ethoxycoumarin and
7-hydroxycoumarin were kindly donated by Dr. H. Vainio, Institute
of Occupational Health, Helsinki.

Aryl hydrocarbon hydroxylase (EC 1.14.14.2) was measured with
3,4-benzpyrene (Sigma Chemicals, Saint Louis, Mo) as substrate as
modified by Nebert and Gelboin (1968). The hydroxylated metabolites
of 3,4-benzpyrene were determined in alkaline solution with a
Perkin-Elmer spectrofluorometer MPF 2 A (excitation wavelength
398 nm and emission 522 nm). The activity was calculated as 3-
hydroxy-benzpyrene fluorescence equivalents.

Native UDPglucuronosyltransferase (EC 2.4.1.17) activity was
determined by the method as modified by Hänninen (1968) by using
p-nitrophenol as aglycone.

Cell Cultures

Primary fibroblast cultures were prepared by trypsinization
from minced skin preparations of newborn nude mice. Cells detached
by 0.25 % trypsin (Merck) in 0.01 M phosphate-buffered 0.15 M NaCl
(PBS) at 37° C were washed and resuspended in culture medium consisting of Eagle's Minimum Essential Medium (MEM; Gibco, Grand
Island, N.Y.) supplemented with 10 % fetal calf serum and penicillin
(200 IU/ml) and streptomycin (200 µg/ml). Primary cultures were
grown in 25 cm^2 plastic flasks, secondary cultures were prepared
by passaging the cells in plastic Petri dishes, diameter 30 mm
(Flow laboratories, Irvine, Scotland). Cell cultures were incubated

at 37° C in atmosphere of 5 % CO_2 in air.

Chemical Transformation

Test chemicals 3-methylcholantrene (3-MC) and 1,2-benzan-
thracene were purchased from Sigma Chemicals, Saint Louis, Mo and
3,4-benzpyrene from Fluka AG, Buchs, Switzerland. Stock solutions
of the chemicals were made in dimethylsulfoxide (DMSO, Merck) at
concentrations of 1 mg/ml or 10 mg/ml. Secondary fibroblast cultures were treated with chemicals for variable times or according
to the two stage schedule described by Casto et al. (1977). In the
latter treatment the cells were exposed for three days to an appropriate concentration of a chemical and for another three days to
the same chemical at half of the preceding concentration. After
the six days the chemical-containing medium was replaced by fresh
medium, and the incubation was continued for a total of 25 days
with weekly feeding of the cells. At the end of the incubation the
cells were fixed in methanol, stained with Giemsa stain (Merck) and
the transformed foci were counted. The proportion of cells surviving the chemical treatment was determined by counting cells in
dishes seeded with 10^3 cells which were treated as described above
and stained five days after the removal of the chemical.

RESULTS

Tissue Metabolism of Xenobiotic Chemicals

The basal levels of some of the hydroxylation and conjugation
enzymes in female nude and normal NMRI mice are shown in Table 1.
The activity of 7-ethoxycoumarin O-deethylase was significantly
higher in the skin of nude mice. The same was probably true for
NADPH-cytochrome c reductase which was not detactable in the skin
of normal mice under the conditions used. UDPGlucuronosyltransferase could not be measured in skin preparations of normal mice
due to technical difficulties in the enzyme determination. Skin
homogenates of nude mice contained colored compounds which hampered
the measurement of cytochrome P-450. In liver and kidney samples
the only difference between the two mouse groups was the significantly higher level of 7-ethoxycoumarin O-deethylase in the kidney
of nude mice.

A more direct way to assess the capacity of tissues to process
polycyclic aromatic hydrocarbon procarcinogens is to measure 3,4-
benzpyrene hydroxylase. The level of this enzyme complex was significantly higher in the liver of nude mice (Table 2). The enzyme
was not detectable in the skin of either mouse group (Table 2).

Table 1.   Enzymes metabolizing xenobiotic chemicals in tissues
           of normal (NMRI) (+/+) and athymic (nu/nu) female mice
           (n=5, mean ± SD).

Tissues and Enzymes	Mouse Group	
	Normal (+/+)	Nude (nu/nu)
SKIN HOMOGENATES		
7-Ethoxycoumarin 0-deethylase[a]	1.00 ± 0.1	2.8 ± 0.9[d]
NADPH Cytochrome c reductase[b]	BDL	9.0 ± 2.0
UDPGlucuronosyltransferase[b]	NT	0.15 ± 0.025
LIVER MICROSOMES		
Cytochrome P-450[c]	0.789 ± 0.355	0.761 ± 0.295
7-Ethoxycoumarin 0-deethylase[b]	2.25 ± 0.09	2.21 ± 0.43
NADPH Cytochrome c reductase[b]	158.5 ± 32.2	157.3 ± 31.42
UDPGlucuronosyltransferase[b]	0.82 ± 0.11	0.75 ± 0.04
KIDNEY MICROSOMES		
Cytochrome P-450[c]	0.26 ± 0.10	0.18 ± 0.01
7-Ethoxycoumarin 0-deethylase[b]	0.028 ± 0.009	0.056 ± 0.013[d]
NADPH Cytochrome c reductase[b]	46.9 ± 8.4	49.1 ± 13.8
UDPGlucuronosyltransferase[b]	0.37 ± 0.05	0.50 ± 0.13

[a] pmoles product x min^{-1} x mg $protein^{-1}$

[b] nmoles product x min^{-1} x mg $protein^{-1}$

[c] nmoles cytochrome x mg $protein^{-1}$

[d] a significant difference between nude and normal mice
(0.01>p>0.001; t-test)
BDL below the detection limit of the method used
NT not tested due to technical difficulties in enzyme determination

Table 2.   Effect of 3-methylcholantrene pretreatment on 3,4-
benzpyrene hydroxylase and UDPGlucuronosyltransferase
activities in normal (+/+) and nude (nu/nu) NMRI mice
(n=5, mean±SD).

Tissues and enzymes	Mouse group			
	Normal (+/+)		Nude (nu/nu)	
	Control	Induced	Control	Induced
SKIN HOMOGENATE[d]				
NADPH Cytochrome c reductase[a]	BDL	BDL	9.0±2.0	17.4±11.3
3,4-Benzpyrene hydroxylase[b]	BDL	BDL	BDL	4.3±2.8
7-Ethoxycoumarin O-deethylase[b]	1.00±0.1	0.9±0.1	2.8±0.9	5.2±2.6[c]
UDPGlucuronosyl-transferase[a]	NT	NT	0.15±0.025	0.20±0.03
LIVER[e]				
3,4-Benzpyrene hydroxylase[a]	0.200±0.08	0.400±0.03	0.370±0.04[c]	0.550±0.04
UDPGlucuronosyl-transferase[a]	0.800±0.09	0.76±0.04	0.920±0.28	0.900±0.11

[a] nmoles product x min^{-1} x mg protein^{-1}
[b] pmoles product x min-1 x mg protein^{-1}
[c] a significant difference to the corresponding value of normal mice
($0.01 > p > 0.001$; t-test)
[d] 3-methylcholantrene administration 25 mg/kg 48 hrs before the
determination of enzyme activities
[e] 3-methylcholantrene administration 80 mg/kg 48 hrs before the
determination of enzyme activities
BDL below the detection limit of the method used
NT not tested due to technical difficulties in enzyme determination

Table 2 also shows the inducibility by 3-MC of the enzymes metabolizing xenobiotic chemicals. The hepatic 3,4-benzpyrene hydroxylase was induced in 48 hours by a single intraperitoneal administration of 3-MC in both mouse groups. On the other hand, the same treatment had no effect on UDPglucuronosyltransferase in 48 hours. Enzyme inducibility in the skin of nude mice was evident already with a smaller amount of 3-MC. The difference between normal and nude mice in monooxygenases was striking. In normal mice the enzymes were not induced (7-ethoxycoumarin O-deethylase) or stayed below the level of detection within the time period following the 3-MC administration. In nude mice there was an increase in all monooxygenase activities measured. The increase was significant for 7-ethoxycoumarin O-deethylase and from undetectable to 4.3 pmoles $\times$ min^{-1} $\times$ mg protein^{-1} for 3,4 benzpyrene hydroxylase. In contrast to the monooxygenase activities there was little change in the UDPglucuronosyltransferase level in the nude mouse skin within 48 hours after exposure to the inducer.

Cell Transformation

A standard carcinogen, 3-MC, was used in the first series of experiments. This carcinogen induced formation of typical multilayered foci of randomly oriented fibroblast-like cells. The periphery of these foci was irregular. The transformed foci appeared 14-20 days after chemical treatment and they were about 2-5 mm in diameter at the end of the experiment. In DMSO control cultures no growth was observed resembling the foci of chemically transformed cells. Exposures longer than 6 days and concentrations higher than 4 µg/ml of culture medium were found to be too toxic for cells. In the following experiments and exposure for 6 days to two concentrations of a carcinogen as described by Casto et al. (1977) was used. The transformation frequency correlated to the concentration of 3-MC but was also dependent on the number of cells plated per dish. An example of the experiments on this aspect is shown in Table 3. The highest number of foci was found in dishes seeded with 2.5 $\times$ 10^4 cells and treated with 2 µg/ml of 3-MC. Table 3 also shows that at the 3-MC concentrations used the surviving cell fraction was about 0.4.

Foci of transformed cells were induced in fibroblast cultures also by another carcinogen, 3,4-benzpyrene and by a weak carcinogen, 1,2-benzanthracene. However, a higher concentration of the latter chemical than of 3-MC was required for transformation. Foci induced by 1,2-benzanthracene resembled epithelial cells and they had less tendency to pile up (data not shown).

Table 3. Effect of cell number on the expression of transformed
          cell foci in nude (nu/nu) NMRI mouse fibroblasts by 3-
          methylcholanthrene (3-MC)

3-MC (µg/ml)	Surviving fraction	Cells seeded/30 mm dish		
		1×10^4	2.5×10^4	5×10^4
2	0.40	6/4[a]	15/5	10/5
1	0.45	0/8	12/8	7/8
Controls				
DMSO	0.79	0/8	0/8	0/8
Medium	1.0	0/8	0/8	0/8

[a] number of foci per number of dishes

DISCUSSION

It would be advantageous if the target cells in a transformation assay were themselves able to activate carcinogenic chemicals.
Therefore we began the present study with experiments characterizing
the capability of our nude mouse strain to metabolize foreign compounds. We did not find any general difference between nude and
normal mouse in hepatic or renal levels of enzymes of the monooxygenase complex. The exceptions were the increased levels of 7-
ethoxycoumarin O-deethylase in the kidney of the nude mouse and,
more remarkably, of the aryl hydrocarbon hydroxylase in the liver
of the nude mouse. These results are in agreement with an earlier
study on nude mouse with a different background (Freudenthal et al.,
1976; Litterst et al., 1978).

Since our fibroblast cultures were derived from skin preparations, we were especially interested in the presence of enzymes
metabolizing xenobiotics in the skin of nude mouse. With the
exception of aryl hydrocarbon hydroxylase, activities of all enzymes tested were found in nude mouse skin preparations. On the
other hand, only 7-ethoxycoumarin O-deethylase was detectable in the
skin of normal mouse. The negative result may be partly explained
by the structure and strength of normal skin affecting the preparation of skin homogenates.

The induction of the monooxygenase enzymes and associated
cytochromes is under the control of the Ah locus, and the control

extends also to extrahepatic tissues (Nebert, 1978). Under the experimental conditions used the hepatic 3,4-benzpyrene hydroxylase increased in both mouse groups after a single intraperitoneal administration of 3-methylcholantrene, but the induction was weaker than in mice with an Ah allele making them genetically responsive (Nebert, 1978). Despite that, 3-methylcholantrene also caused a detectable, if not in all cases significant increase in the monooxygenase enzymes of nude mouse skin. According to earlier reports nude mice are more resistant to the induction of skin tumors by chemical carcinogens than their normal counterparts (Holland et al., 1977; Schjerven et al., 1974). Our results show that this is not due to the lack or uninducibility of the monooxygenase enzyme complex in the nude mouse skin. It may be of significance that no corresponding increase was found in UDPglucuronosyltransferase, an enzyme which could decrease the amount of active forms of carcinogens produced by the monooxygenase enzyme complex. More detailed studies are required to confirm the results of our experiments.

Secondary cultures of fibroblasts prepared from the skin of newborn nude mice proved suitable targets for transformation by carcinogenic chemicals. This was an expected result comparable to those obtained by using fibroblast targets from other species (Casto et al., 1977). The use of secondary fibroblasts has the disadvantage of requiring new batches of primary cells quite frequently. This could be avoided either by using a continuous cell line similar to 3T3 cells (Kakunaga, 1975), or by cryopreservation of primary fibroblasts. The latter alternative has been succesfully adapted for hamster fibroblasts (Pienta et al., 1977). In our hands, cryopreserved fibroblasts were less sensitive to transformation by 3-methylcholantrene than the same batch of cells maintained in culture (results not shown). Refinement of the cryopreservation technique may, however, improve the survival of transformation sensitive cells.

The concentration of chemicals, the length of the exposure and the initial cell density affected transformation in the nude mouse skin fibroblast assay in the same way as reported previously of other transformation assays (Casto et al., 1977).

Conversion of procarcinogens to activated forms can be achieved in vitro by using microsomal preparations e.g. from liver cells. This treatment is a prerequisite for bacterial mutagenicity tests (Ames et al., 1975). In transformation assays using mammalian cells as targets the activation may also be carried out by the enzyme machinery of the target cells. We have not yet analyzed cultured nude mouse skin fibroblasts for enzymes of the monooxygenase complex. However, the in vivo results, the induction of transformation by 3-methylcholanthrene and earlier analyses of other types of cultured cells (Venkatesan et al., 1979) suggest that nude mouse fibroblasts are able to activate procarcinogens. An enhanced activation can probably be expected by using fibroblasts from nude mice crossbred with a background strain of mice with a highly responsive Ah gene allele.

We are currently growing out clones of transformed cell to test their tumorigenicity in nude mice.

SUMMARY

Enzymes of the monooxygenase complex were found in the skin of a nude mouse strain, and the activities increased after an intraperitoneal administration of 3-methylcholantrene. No comparable change was observed in UDPglucuronosyltransferase, an enzyme of the conjugation system. Fibroblast cultures prepared from nude mouse skin proved sensitive for transformation by carcinogenic chemicals. Nude mouse accepts transplants e.g. transformed cells more easily than conventional laboratory animals. The nude mouse model may therefore offer a useful alternative to other transformation assays using mammalian cells as targets.

ACKNOWLEDGEMENTS

This study was supported by the Finnish Foundation for Cancer Research and by the Sigrid Juselius Foudation and by the National Science Council, Finland.

REFERENCES

Aitio, A., 1978, A simple and sensitive assay of 7-ethoxycoumarin deethylation, Anal. Biochem., 85:488.
Ames, B. N., McCann, J., and Yamasaki, E., 1975, Methods for detecting carcinogens and mutagens with the Salmonella/ mammalian-microsome mutagenicity test, Mutation Research 31:347.
Casto, B. C., Janosko, N., and DiPaolo, J. A., 1977, Development of a focus assay model for transformation of hamster cells in vitro by chemical carcinogens, Cancer Research, 37:3508.
Freudenthal, R. I., Leber, A. P., Emmerling, D. C., Kerchner, G. A., and Ovejera, A. A., 1976, Comparison of the drug metabolizing enzymes in the liver and kidneys from homozygous nude swiss, heterozygous normal swiss, homozygous normal swiss and DBA/2 mice, Res. Comm. Chem. Path. Pharmacol., 15:267.

Holland, J.M., Perkins, E.H., and Gipson, L.C., 1977, Resistance of germfree athymic nude mice to two-stage epidermal carcinogenesis, Proc. Am. Assoc. Cancer Res., 18:10.

Hänninen, O., 1968, On the metabolic regulation in the glucuronic acid pathway in the rat tissues, Ann. Acad. Sci. Fenn. A$_2$ No 142:1.

Kakunaga, T., 1975, The role of cell division in the malignant transformation of mouse cells treated with 3-methylcholantrene, Cancer Research, 35:1637.

Litterst, C. L., Sikic, B. I., Mimnaugh, E. G., Guarino, A. M., and Gram, T. E., 1978, In vitro drug metabolism in male and female athymic, nude mice, Life Sci., 22:1723.

Lowry, O. H., Rosebrough, A. L., Fare, A. D., and Randall, R. J., 1951, Protein measurement with the Folin phenol reagent, J. Biol. Chem., 193:265.

Nebert, D. W., 1978, Genetic control of carcinogen metabolism leading to individual differences in cancer risks, Biochemie, 60:1019.

Nebert, D. W., and Gelboin, H.V., 1968, Substrate inducible microsomal arylhydroxylase in mammalian cell culture 1 Assay and properties of induced enzyme, J. Biol. Chem., 243:6242.

Nomura, T., Ohsawa, N., Tamaoki, N., and Fugiwara, K., 1977, "Proceedings of the Second International Workshop on Nude Mice", Gustav Fischer Verlag, Stuttgart, New York.

Omura, T., and Sato, R., 1964, The carbon monoxide binding pigment of liver microsomes, J. Biol. Chem., 239:2370.

Phillips, A. H., and Langdon, R. G., 1962, Hepatic triphosphopyridine nucleotide cytochrome c reductase: Isolation, characterization and kinetic studies, J. Biol. Chem., 237:2652.

Pienta, R. J., Poiley, J. A., and Lebherz III, W. B., 1977, Morphological transformation of early passage golden Syrian hamster embryo cells derived from cryopreserved primary cultures as a reliable in vitro bioassay for identifying diverse carcinogens. Int. J. Cancer, 19:642.

Rowan, A., 1978, Carcinogen testing, World Medical Journal, 25:3.

Schjerven, L., Elgjo, K., Iversen, O. H., and Palbo, V., 1974, Skin carcinogenesis in nude mice, in "Proceedings of the First International Workshop on Nude Mice", J. Rygaard and C. O. Polvsen, eds., Gustav Fischer Verlag, Stuttgart.

Schmassmann, H. U., Glatt, H. R., and Oech, F., 1976, A rapid assay for epoxide hydratase activity with benzo(a)pyrene 4,5-(K-region-) oxide as substrate, Anal. Biochem., 74:94.

Venkatesan, N., Alfred, L. J., Torralba, G., and Benedict, W. F., 1979, Enhancement of aryl hydrocarbon hydroxylase activity in cultured rodent cells by X-irradiation, Life Sci., 24:797.

CONVERSION OF LIVER HAEM INTO N-SUBSTITUTED PORPHYRINS OR GREEN PIGMENTS
Evidence for two distinct classes of products

Francesco De Matteis, Anthony H. Gibbs and
Anne P. Unseld,

M.R.C. Toxicology Unit, M.R.C. Laboratories,
Woodmansterne Road, Carshalton, Surrey SM5 4EF,U.K.

INTRODUCTION

The ability of unsaturated drugs, like 2-allyl-2-isopro-
pylacetamide (AIA) or allylbarbiturates to convert the
prosthetic group of liver cytochrome P-450 into modified porphyrins (or green pigments) has been known for some time. The
original observations (reviewed by Ortiz de Montellano et al.,
1978a and by De Matteis, 1978) indicated that at least one
unsaturated side chain was required for a drug to be active
and also that metabolic activation of the allylic double bond
was essential before the prosthetic group of the cytochrome
could be attacked. These findings have since been extended
to unsaturated compounds containing the acetylenic grouping,
including ethynyl-substituted steroids and acetylene gas
itself (White and Muller-Eberhard, 1977; White, 1978). More
recently Ortiz de Montellano and coworkers have obtained
evidence that a reactive metabolite of the effective drugs
(of either the olefinic or acetylenic group) become covalently
bound onto the porphyrin nucleus of cytochrome P-450 (Ortiz
de Montellano et al., 1978b; 1979a and b); and work from our
own laboratory has shown that the various modified porphyrins
obtained from the action of these unsaturated drugs on haem
are all N-monosubstituted (De Matteis and Cantoni, 1979; De
Matteis et al., 1980a). It can therefore by concluded that
these unsaturated drugs are metabolized by cytochrome P-450 into reactive derivatives which alkylate one of the pyrrole nitrogens of its porphyrin prosthetic group and therefore inhibit
the haemoprotein by a suicidal type of inactivation. The
detailed mechanism of the reaction leading to N-alkylation and
the nature of the reactive metabolite are still unclear.

Another N-monosubstituted porphyrin has recently been isolated from the liver of animals treated with certain drugs which cause experimental hepatic porphyria, for example 3,5--diethoxycarbonyl-1,4-dihydrocollidine, for short DDC (Tephly et al., 1979). In this case the drug responsible for increased pigment production in the liver cannot be demonstrated bound (at least not its pyridine ring) onto the modified porphyrin (De Matteis et al., 1980b) and the pyrrole nitrogen bears instead a much smaller substituent, the methyl grouping (De Matteis et al., 1980c). This second pigment is capable of producing a marked inhibition of the last enzyme of the pathway of haem biosynthesis, protohaem ferro-lyase (E C 4.99.1.1) both in vivo and in vitro (De Matteis et al., 1980d). Because of this, drugs which stimulate the formation in the liver of this second type of green pigment (henceforth referred to as "inhibitory" pigment) will cause marked accumulation of protoporphyrin (the substrate of the inhibited enzyme) and the biochemical picture of hepatic protoporphyria.

The most significant differences between the two classes of pigments (those arising from the action of unsaturated drugs on cytochrome P-450 on the one hand and the pigment with inhibitory activity on protohaem ferro-lyase, on the other) are listed in Table 1. In this paper the evidence that both classes of pigments are N-monosubstituted porphyrins will first be summarized. Then the pathways leading to their formation will be briefly discussed. In a theoretical paper Schoental (1976) suggested that N-alkylated porphyrins might be produced in vivo after treatment with alkylating hepatotoxins, for example carcinogens.

EXPERIMENTAL

Male Porton (Wistar derived) rats were given phenobarbitone sodium in their drinking water (1 mg/ml) for 6 full days, then returned to plain drinking water and starved overnight before treatment with unsaturated compounds: 2-allyl-2-isopropylacetamide was given subcutaneously (400 mg/kg) and ethylene was administered by inhalation as a 5% (by vol) mixture in air. In the labelling experiment [^{59}Fe]-ferric chloride (1 μCi/rat) was injected intraperitoneally 17 h before injection of 2-allyl-2-isopropylacetamide. The isotope had a specific activity of 8.5 mCi/mg Fe (at the time of issue).

The microsomal fractions of the livers from animals injected with [^{59}Fe] were prepared from the post-mitochondrial supernatant of their liver homogenates by the $CaCl_2$ method of Kamath and Rubin (1972), in order to remove most of the non-haem iron (Montgomery et al., 1974). The calcium-sedimented

Table 1. Main differences between the N-substituted porphyrins
          produced by treatment with unsaturated drugs and that
          isolated after 3,5-diethoxycarbonyl-1,4-dihydrocol-
          lidine (DDC).

Class	Drug responsible	Source of the green pigment	Substituent at the pyrrole nitrogen	Inhibition of liver protohaem ferro-lyase
A	Unsaturated drugs containing allyl or ethynyl sub- stituents	Cytochrome P-450 haem	Mono- oxygenated derivative of the drug	None
B	Drug responsible for hepatic protoporphyria, for example DDC or griseofulvin	Probably hepatic haem, but pool unknown	Methyl grouping	Marked

microsomal fractions were then resuspended in 1.15% KCl and
re-centrifuged at 105,000g for 1 h to remove excess calcium.
The washed microsomal pellets were extracted with 10% acetic
acid in acetone (2.5 ml/g wet liver, twice) and the pooled
extracts (equivalent to 6.8 g liver) were applied to a 20 cm
x 2.6 cm column of 3-dibutylamino-2-hydroxypropyl-Sephadex
LH 20.   The modified Sephadex (acetate form) was prepared
according to Alme and Nyström (1971), was allowed to swell in
5% acetic acid in acetone, in which it was then packed.
Elution was by gravity with 5% acetic acid in acetone (250 ml)
followed by 10% acetic acid in acetone. By this technique an
iron-containing green pigment could be separated from haem
(see later) both being obtained as the free carboxylic acid.

     Separation of the green pigments from haem was also
obtained by t.l.c. of a chloroform extract after methylation
of the whole liver or of microsomal fractions with methanol
containing 5% (v/v) H_2SO_4 at room temperature for 16 h in the
dark (Unseld and De Mätteis, 1976).   This technique, which
yielded the green pigments as the metal-free porphyrin methyl
esters, was also used for their further purification.
Alternatively the pigments were purified by chromatography on
Sephadex LH 20 (De Matteis et al., 1980b), followed by methylation by reaction with BF_3 in methanol (Smith and Francis,
1979).   Determination of the porphyrin electronic spectra,
titration with trifluoroacetic acid and formation of the zinc

Table 2. Bathochromic shifts from the absorption maxima of the respective parent porphyrins seen in the neutral spectrum of several green pigments extracted from the livers of treated animals (A) or of authentic N-alkylated porphyrins (B)

Porphyrin	Bathochromic shifts(nm)						References
	Soret	IV	III	II	Ia	I	
A) Green pigments produced by:							
2-Allyl-2-isopropylacetamide	10	7	3	18	20	21	De Matteis & Cantoni (1979)
Ethylene	11	8	5	19.5	21	22	Present work
3,5-Diethoxycarbonyl-1,4-dihydrocollidine	12	8	5	20	22	22	De Matteis & Gibbs (1980)
B) Synthetic porphyrins							
N-Methyl octaethylporphyrin	12	9.5	5	19.5	22	22	De Matteis & Cantoni (1979)
N-Methyl mesoporphyrin	11.5	7	3.5	18.5	21	20.5	Present work
N-Methyl protoporphyrin[a]	12	9	6	19	22	23	De Matteis et al.(1980c)
N-Propyl mesoporphyrin	12.5	8	6	19.5	21	20.5	Present work
N-Propyl protoporphyrin	12.5	9.5	7	20	18.5	21.5	Present work

All porphyrins listed here exhibited an aetio-type spectrum, were examined dissolved in chloroform and, with the exception of N-methyloctaethylporphyrin, were methyl esters. The bathochromic shifts of the various bands were calculated from the absorption maxima of protoporphyrin IX (Falk, 1964) in the case of both the green pigments and the N-alkylated protoporphyrins; or from those of mesoporphyrin IX (Falk, 1964) or of octaethylporphyrin (De Matteis & Cantoni, 1979) for the other modified porphyrins, as appropriate.

[a]Average of values reported for two different chromatographic fractions.

complex derivatives was as described (De Matteis and Cantoni, 1979). Haem was crystallized by the method of Labbe and Nishida (1957) and N-alkylated porphyrins were synthesized by a modification of the method of McEwen (1946) as follows. Protoporphyrin IX or mesoporphyrin IX (2.8 - 3 mg of the methyl esters) were heated with methyl iodide (1 ml) for 1 day at 95°C or with propyl iodide (1 ml) for 4 days at 105°C. The N-mono-alkylated products were purified by two t.l.c. runs in $CHCl_3$/kerosene/methanol (20:5:3, by vol), eluted from the silica with 2.5% (by vol) conc. HCl in methanol and transferred to $CHCl_3$ by addition of dilute ammonia.

EVIDENCE THAT GREEN PIGMENTS OF BOTH CLASSES ARE N-MONOALKY-LATED PORPHYRINS

The following three lines of evidence have so far been obtained.

A) Bathochromic shifts of the porphyrin electronic spectra.

When the green pigments produced by treatment with un-saturated drugs were extracted from the liver of treated animals and purified as the methyl esters (a procedure which involved exposure to strong acid), they exhibited an electronic spectrum of the aetio-type, typical of a porphyrin free base; however all absorption maxima were shifted towards longer wave-lengths, as compared with those of protoporphyrin, the native porphyrin of cytochrome P-450 (McDonagh et al., 1976; Unseld and De Matteis, 1976; Ortiz de Montellano et al., 1979a; De Matteis et al., 1980a). Similar spectra have been obtained with the "inhibitory" green pigment produced by DDC treatment (Tephly et al. 1979, De Matteis and Gibbs, 1980; and see Fig.2). The bathochromic shifts (from protoporphyrin) have been measured for each of the green pigments and found to be very similar to those which authentic N-alkylated porphyrins exhi-bited from their respective parent porphyrins (De Matteis and Cantoni, 1979; De Matteis and Gibbs, 1980, and see Table 2). Similar bathochromic shifts were found for all N-substituted porphyrins examined regardless of the type of porphyrin which was substituted and of size of the substituent at the pyrrole nitrogen.

Even though the bathochromic shifts of the neutral spec-trum were very similar for the two classes of naturally occur-ring pigments (compare in Table 2 the pigments produced by 2-allyl-isopropylacetamide and ethylene with that produced by DDC), significant differences were found between the two classes in the bathochromic shift of the dication spectrum (see Table 3). This may reflect the size of the substituent

Table 3. Absorption spectrum of the dication derivatives of protoporphyrin and mesoporphyrin and bathochromic shifts from their absorption maxima seen in the dication spectrum of various porphyrins obtained by modification of these two parent porphyrins.

Porphyrin	Absorption maxima or bathochromic shifts(nm)			References
	Soret	II	I	
Absorption maxima of:				
A) Protoporphyrin IX	411	556	600	Present work
B) Mesoporphyrin IX	405	549.5	592	Present work
Bathochromic shifts seen in:				
2-Allyl-2-isopropylacetamide-green pigment (from A)	9	11	13	Present work
Ethylene-green pigment (from A)	7	9.5	10	De Matteis et al.(1980d)
3,5-Diethoxycarbonyl-1,4-dihydro-collidine-green pigment (from A)	1	4	4	De Matteis & Gibbs (1980)
N-Methyl protoporphyrin IX (from A)	1	4	4	De Matteis et al.(1980c)
N-Methyl mesoporphyrin IX (from B)	0.5	2.5	4	Present work
N-Propyl protoporphyrin IX (from A)	7.5	9	10	Present work
N-Propyl mesoporphyrin IX (from B)	6.5	9.5	8.5	Present work

All porphyrins listed here were methyl esters and the spectra of their dication derivatives were determined in chloroform.

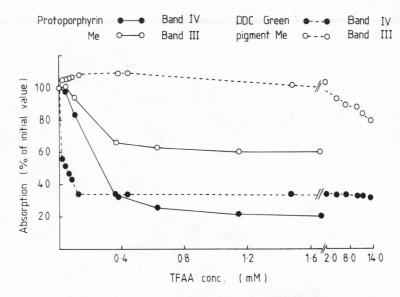

Fig.1. Titration of green pigment produced by 3,5-diethoxy-
carbonyl-1,4-dihydrocollidine (DDC) and of protoporphyrin IX, both methyl esters (Me), with trifluoroacetic acid. The absorption at the wavelengths of
band IV and band III maxima observed at different acid
concentrations are expressed as a percentage of the
initial value of the free base.

at the pyrrole nitrogen which is known to be smaller (only one
carbon unit) in the "inhibitory" pigment (De Matteis et al.,
1980c). Some support for this interpretation is provided by
the finding (Table 3) that when the size of the N-substituent
of either mesoporphyrin or protoporphyrin was increased from
a one to a three carbon unit, the dication spectrum of the
N-methylated and N-propylated porphyrins thus obtained
resembled that of the "inhibitory" pigment and of the green
pigments produced by unsaturated compounds, respectively.

B) Titration with acid

    Additional evidence that the green pigments are N-mono-
substituted porphyrins was provided by titration experiments

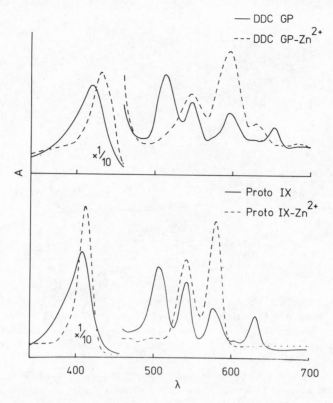

Fig.2. Absorption spectrum of the green pigment produced by
treatment with 3,5-diethoxycarbonyl-1,4-dihydro-
collidine (DDC) and of protoporphyrin. The spectrum
of the porphyrin free bases (--) and of their Zn^{2+}
chelates (- -) were determined in chloroform, using
the porphyrin methyl esters.

with trifluoroacetic acid. On conversion of the porphyrin
free base to the monocation band IV of the neutral spectrum
decreased, whereas band III decreased when the monocation was
converted into the dication. With protoporphyrin bands IV and
III decreased together (Fig.1), as the dication was formed
directly and an intermediary monocation was difficult to demonstrate: this is because the pKa values of the two conjugated acids are close to one another (Neuberger and Scott,
1952). In contrast, with both classes of green pigments and
with authentic N-alkylated porphyrins (De Matteis and Cantoni,

1979; De Matteis and Gibbs, 1980) band IV decreased rapidly at
very low concentrations of acid, but band III decreased slowly
and only when relatively large amounts of acid were added (see
in Fig.1 the titration curve obtained with the DDC green pigment). We can therefore conclude that, like N-methylated porphyrins (Neuberger and Scott, 1952; Jackson and Dearden, 1973)
the various green pigments are more basic than their parent
porphyrin and the pKa values of their conjugated acids are sufficiently apart to allow the existence of the monocation even
after addition of a strong acid.

C) Spectral characteristics of the zinc complex

The green pigments closely resembled authentic N-alkylated porphyrins also in the very rapid rate of chelation of
zinc in vitro (De Matteis and Cantoni, 1979) and in the
spectral characteristics of the resulting zinc complex. The
additional minor band approximately 200 nm towards the red
from the Soret maximum(shown in Fig.2 for the zinc chelate of
the DDC green pigment),has been reported by Ortiz de
Montellano et al. (1978) for the AIA green pigment, and was
also exhibited by authentic N-monoalkylated porphyrins
(De Matteis and Gibbs, 1980). Therefore by all the criteria
listed above under A), B) and C) the green pigments behaved
as authentic N-alkylated porphyrins. They differed in all
these respects from porphyrins substituted at their meso carbon positions or bearing electron-withdrawing substituents
at the β positions of the pyrrole rings (De Matteis et al.,
1980a).

MECHANISM OF PRODUCTION OF GREEN PIGMENTS BY DRUGS

A clear distinction must be made between the two classes
of green pigments which have been considered here. With pigments arising from the metabolism of unsaturated drugs by
cytochrome P-450, the work of Ortiz de Montellano and colla-
borators has clearly indicated that the drugs themselves,
probably after undergoing metabolic activation, become covalently bound onto the porphyrin nucleus. Ortiz de Montellano
et al.(1979c) have also shown that preformed epoxides added
in vitro do not cause destruction of cytochrome P-450 and have
suggested the importance of other structural features of the
postulated reactive metabolites (for example the presence of
an "activated" carbonyl grouping). However, the simple unsaturated compounds acetylene and ethylene have also been
shown to be active (White, 1978; De Matteis et al., 1980c
and d; Ortiz de Montellano and Mico, 1980) demonstrating that
an unsaturated bond between two carbons is the only structural
requirement for the effect. In addition, evidence has been

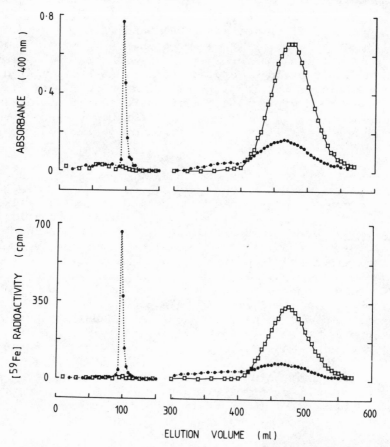

Fig.3.   Chromatography on 3-dibutylamino-2-hydroxypropyl
         Sephadex LH 20 (acetate form) of microsomal extracts
         obtained from control (□———□) or 2-allyl-2-isopro-
         pylacetamide-treated (● ··· ●) rats. The [^{59}Fe]
         radioactivity refers to total counts recovered in
         3 ml fractions.

obtained (De Matteis et al. 1980c) for the presence of a free
hydroxyl group in the pigment produced by treatment with
ethylene, and this could arise from the reaction of an intermediary epoxide (or of a closely related species) with a
pyrrole nitrogen. Perhaps there is an iron-bound intermediate
(a nascent epoxide?) which will facilitate and direct reaction
of the electrophilic carbon onto the pyrrole nitrogen
(De Matteis, 1980 and see Fig.5).

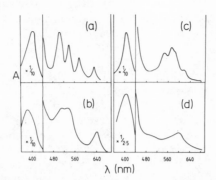

Fig.4. Absorption spectrum of the haem fraction (b) and of
       the iron-containing green pigment fraction (d)
       obtained by column chromatography of a microsomal
       extract from rats treated with 2-allyl-2-isopropyl
       acetamide. The spectrum of the corresponding iron-
       -free green pigment (c) and of protoporphyrin (a)
       are also shown. All spectra were obtained in 5%
       acetic acid in acetone, using the pigment free car-
       boxylic acids.

       The following evidence indicates that an iron-containing
green pigment can be detected in the liver of animals given an
unsaturated drug, provided that precautions are taken to avoid
strong acidic conditions. When the liver haem was prelabelled
with [59Fe] and the animals were then given a dose of 2-allyl-
-2-isopropylacetamide and killed 1 h later, a marked loss of
radioactivity was observed from the haem crystallized form
the microsomal fraction (3,500 dpm/g liver in controls and
1,470 dpm/g liver in treated animals, respectively; averages
of two observations). The microsomal extract from treated
animals showed, when chromatographed on 3-dibutylamino-2-
-hydroxypropyl Sephadex LH 20, two distinct bands (Fig.3): a
first fraction which was green and was rapidly eluted with 5%

acetic acid in acetone; and a second band which was brown and
was eluted more slowly by 10% acetic acid in acetone. The
extract of control microsomes contained only the slow-moving
brown fraction. Both bands were radioactive and showed a
strong absorption at 400 nm. Treatment with 2-allyl-2-iso-
propylacetamide caused a loss of absorbance at 400 nm and also
of [59Fe] radioactivity from the slow-moving band, with a
corresponding gain in both absorbance and radioactivity in the
green pigment fraction (Fig.3). Similar findings were obtained
when microsomes were incubated with the drug in vitro in presence of a NADPH generating system. It will be noticed from
Fig.3 that the chromatographic profile of the extract from
treated rats also showed an increase in the 400 nm absorbance
immediately preceding (as a shoulder) the haem peak. This was
accompanied by a significant increase in the [59Fe] radioactivity (as compared with the corresponding fractions from control
microsomes) and was consistently seen whether the green pigments were produced in vivo or in vitro. These findings may
indicate the presence of more than one iron-containing green
pigments.

The most intensely coloured fractions obtained from column
chromatography were scanned for their visible spectra. The
brown band, whether obtained from microsomal extracts of control or treated animals, exhibited a spectrum identical to
that of authentic haem (Fig.4b). The Soret maximum (in either
5% or 10% acetic acid in acetone) was at 385 nm, with minor
absorbance peaks, at 506, 539 and 640 nm. It was conclu-
sively identified as haem by addition of carrier haem to the
column fractions, followed by crystallization and counting of
the crystallized haem. The main green pigment fraction from
column chromatography showed (in 5% acetic acid in acetone) a
Soret maximum at 406 nm, with a minor peak at about 600 nm
(Fig.4d): this spectrum is unlike that of either haem or protoporphyrin and is also different from that exhibited in the
same solvent by the corresponding metal-free pigment (Fig.4c).
The latter was obtained from the liver of rats treated with
2-allyl-2-isopropylacetamide by methylation in strong acid and
subsequent hydrolysis of the methyl ester. These observations
confirm and extend previous findings (Unseld, 1976; De Matteis
and Unseld, 1976), in line with the suggestion that the green
pigments produced by treatment with unsaturated drugs exist in
the liver as iron-containing compound(s) which can subsequently
generate, on exposure to strong acids, the corresponding metal-
-free porphyrins. The exact structure of the iron-containing
column pigment is not yet known. It is possible that, as
illustrated in Fig.5, in the iron-containing column pigment
one of the pyrrole nitrogens has already become alkylated and
that the effect of strong acid is that of facilitating the

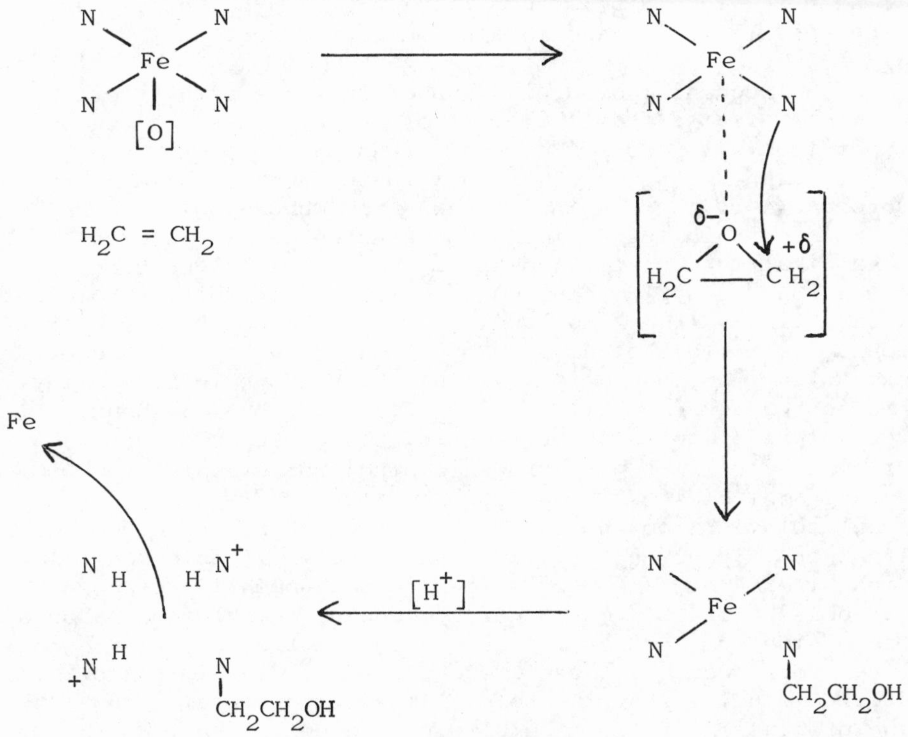

Fig.5. Possible reactions involved in the metabolic activation of an olefin (ethylene) by cytochrome P-450 and
in the consequent alkylation of one of its pyrrole
nitrogens. The following sequence of events is
schematically represented (proceeding from the top
of the figure in a clockwise direction). 1) Ethylene
is bound in the hydrophobic pocket of the cytochrome
close to the iron-bound oxygen (oxene?); 2) a transient and reactive iron-bound intermediate is formed:
this may either dissociate and form a proper epoxide
or alkylate one of the pyrrole nitrogens; 3) the iron
chelate of N-hydroxyethylprotoporphyrin may then be
formed and 4) this will then loose iron on exposure
to strong acids to generate the corresponding
porphyrin.

elimination of the metal ion to generate the corresponding
iron-free porphyrins. It must be emphasized that even though
epoxides are known to arise from the metabolism of olefins by
cytochrome P-450, there is no direct evidence as yet for a role
of either epoxides or of closely related species in the alkylation of the haem moiety of this cytochrome. The scheme

illustrated in Fig.5 is therefore in part hypothetical and
other possibilities should also be considered.

The mechanism of production of the "inhibitory" green
pigment isolated from the liver of animals given DDC is not
understood. In contrast to the green pigments produced by
unsaturated drugs, which all originate from the haem of cytochrome P-450, it is not yet known from which pool of hepatic
haem the "inhibitory" pigment originates. Also, whereas it is
clear that the unsaturated drugs become themselves bound onto
the porphyrin system, in the case of the "inhibitory" pigment
radioactivity from ring-labelled DDC could not be demonstrated
bound (De Matteis et al., 1980b). The "inhibitory" pigment
extracted from porphyric liver has been shown to be N-methyl
protoporphyrin (De Matteis et al., 1980c), but the source of
the methyl group is not yet known. It could conceivably
originate from the porphyrogenic drug, for example from DDC.
A similar inhibitor which close spectral characteristics has
been isolated from the liver of mice which had received no
treatment (Tephly et al., 1979 and 1980). If this control
"inhibitory" pigment proves to be also N-methyl protoporphyrin,
the possibility of some other endogenous methyl donor should
be considered, a donor that is independent of a porphyrogenic
drug. More work is therefore required to clarify the origin
of the methyl group, the role of the porphyrogenic drugs in
the production of the "inhibitory" pigment as well as the detailed mechanism leading to N-methylation of the protoporphyrin molecule.

REFERENCES

Almé, B. and Nyström, E., 1971, Preparation of lipophilic
     anion exchangers from chlorohydroxypropylated Sephadex
     and cellulose, J.Chromat., 59:45.
De Matteis, F., 1978, Loss of microsomal components in drug-
     induced liver damage, in cholestasis and after adminis-
     tration of chemicals which stimulate heme catabolism,
     Pharmac.Ther. A.2:693.
De Matteis, F., 1980, in: Discussion of the paper by De
     Matteis et al., 1980a, p.132 and 134.
De Matteis, F. and Cantoni, L., 1979, Alteration of the por-
     phyrin nucleus of cytochrome P-450 caused in the liver
     by treatment with allyl-containing drugs. Is the modi-
     fied porphyrin N-substituted?, Biochem.J., 183:99.
De Matteis, F. and Gibbs, A.H., 1980, Drug-induced conversion
     of liver haem into modified porphyrins. Evidence for
     two classes of products, Biochem.J., 187:285.
De Matteis,F.and Unseld,A.,1976, Increased haem degradation
     caused by foreign chemicals, Biochem.Soc.Trans., 4:205.

De Matteis, F., Gibbs, A.H., Cantoni, L. and Francis, J., 1980a, Substrate-dependent irreversible inactivation of cytochrome P-450: conversion of its haem moiety into modified porphyrins, in: Ciba Foundation Symposium 76, p.119, Excerpta Medica, Amsterdam.

De Matteis, F., Gibbs, A.H. and Tephly, T.R., 1980b, Inhibition of protohaem ferro-lyase in experimental porphyria. Isolation and partial characterization of a modified porphyrin inhibitor, Biochem.J., 188:145.

De Matteis, F., Gibbs, A.H., Jackson, A.H. and Weerasinghe, S. 1980c, Conversion of liver haem into N-substituted porphyrins or green pigments. Nature of the substituent at the pyrrole nitrogen atom, FEBS Lett., 119:109.

De Matteis, F., Gibbs, A.H. and Smith, A.G., 1980d, Inhibition of protohaem ferro-lyase by N-substituted porphyrins. Structural requirements for the inhibitory effect, Biochem.J., 189:645.

Falk, J.E., 1964, Porphyrins and metalloporphyrins, B.B.A. Library, vol 2, Elsevier, Amsterdam.

Jackson, A.H. and Dearden, G.R., 1973, N-Methylporphyrins, Ann.N.Y.Acad.Sci., 206:151.

Kamath, S.A. and Rubin, E., 1972, Interaction of calcium with microsomes: a modified method for the rapid isolation of rat liver microsomes, Biochem.Biophys.Res.Commun., 49:52.

Labbe, R.F. and Nishida, G., 1957, A new method of hemin isolation, Biochim.Biophys.Acta, 26:437.

McDonagh, A.F., Pospisil, R., Meyer, U.A., 1976, Degradation of hepatic heme to porphyrins and oxophlorins in rats treated with 2-allyl-2-isopropylacetamide, Biochem.Soc. Trans., 4:297.

McEwen, W.K., 1946, Steric deformation. The synthesis of N-methyl etioporphyrin I, J.Am.Chem.Soc., 68:711.

Montgomery, M.R., Clark, C. and Holtzman, J.L., 1974, Iron species of hepatic microsomes from control and pheno-barbital-treated rats, Arch.Biochem.Biophys., 160:113.

Neuberger, A. and Scott, J.J., 1952, The basicities of the nitrogen atoms in the porphyrin nucleus; their dependence on some substituents of the tetrapyrrolic ring, Proc.R.Soc.Lond.A Math.Phys.Sci.; 213:307.

Ortiz de Montellano, P.R. and Mico, B.A., 1980, Destruction of cytochrome P-450 by ethylene and other olefins, Mol. Pharmacol., 18:128.

Ortiz de Montellano, P.R., Mico, B.A., Yost, G.S. and Correia, M.A., 1978a, Suicidal inactivation of cytochrome P-450: covalent binding of allylisopropylacetamide to the heme prosthetic group, in: "Enzyme Activated Irreversible Inhibitors", p.337, M.Seiler, ed., Elsevier/North-Holland, Amsterdam.

Ortiz de Montellano, P.R., Mico, B.A., Yost, G.S., 1978b,
    Suicidal inactivation of cytochrome P-450. Formation
    of a heme-substrate covalent adduct, Biochem.Biophys.
    Res.Commun., 83:132.
Ortiz de Montellano, P.R., Kunze, K.L., Yost, G.S. and Mico,
    B.A., 1979a, Self-catalyzed destruction of cytochrome
    P-450: covalent binding of ethynyl sterols to prosthetic
    heme, Proc.Natl.Acad.Sci.USA, 76:746.
Ortiz de Montellano, P.R., Kunze, K.L., Mico, B.A., Yost, G.
    S. and Beilan, M., 1979b, Inactivation of cytochrome
    P-450 by suicide substrates. Partial structure of the
    resulting prosthetic heme adducts, in: Fourth Interna-
    tional Symposium on Microsomes and Drug Oxidations.
    Ann Arbor, Michigan, p.108 (abstr).
Ortiz de Montellano, P.R., Yost, G.S., Mico, B.A., Dinizo,
    S.E., Correia, M.A. and Kumbara, H., 1979c, Destruction
    of cytochrome P-450 by 2-isopropyl-4-pentenamide and
    methyl 2-isopropyl-4-pentenoate: mass spectrometric
    characterization of prosthetic heme adducts and non-
    participation of epoxide metabolites, Arch.Biochem.
    Biophys, 197:524.
Schoental, R., 1976, Alkylation of coenzymes and the acute
    effects of alkylating hepatotoxins, FEBS Lett., 61:111.
Smith, A.G. and Francis, J.E., 1979, Decarboxylation of por-
    phyrinogens by rat liver uroporphyrinogen decarboxylase,
    Biochem.J., 183:455.
Tephly, T.R., Gibbs, A.H. and De Matteis, F., 1979, Studies
    on the mechanism of experimental porphyria produced by
    3,5-diethoxycarbonyl-1,4-dihydrocollidine, Biochem.J.
    180:241.
Tephly, T.R., Gibbs, A.H., Ingall, G. and De Matteis, F.,
    1980, Studies on the mechanism of experimental porphyria
    and ferrochelatase inhibition produced by 3,5-diethoxy-
    carbonyl-1,4-dihydrocollidine, Int.J.Biochem., in press.
Unseld, A., 1976, A study of the degradation of haem and the
    production of green pigments by 2-allyl-2-isopropyl-
    acetamide in rat liver, Ph.D.Thesis, University of London.
Unseld, A. and De Matteis, F., 1976, Isolation and partial
    characterization of green pigments produced in rat liver
    by 2-allyl-2-isopropylacetamide, in: "Porphyrins in
    Human Disease", p.71, M. Doss, ed., Karger, Basel.
White, I.N.H., 1978, Metabolic activation of acetylenic sub-
    stituents to derivatives in the rat causing the loss of
    hepatic cytochrome P-450 and haem, Biochem.J., 174:853.
White, I.N.H. and Muller-Eberhard, U., 1977, Decreased liver
    cytochrome P-450 in rats caused by norethindrone or
    ethynyloestradiol, Biochem.J. 166:57.

INTERACTION OF REACTIVE INTERMEDIATES WITH DNA

P.N. Magee, C.K. Chu, C.T. Gombar,
D.E. Jensen and L.G. Parchman
Fels Research Institute
Temple University School of Medicine
Philadelphia, PA 19140, U.S.A.

Although conclusive proof is still lacking, there is a wide-spread and growing belief that the initiation of carcinogenic processes by chemicals may result from alterations of cellular DNA (Miller, 1978). There is also substantial evidence that the activated forms of many chemical carcinogens react covalently with cellular DNA in vivo (Brookes, 1966; Sarma et al, 1975; Pegg, 1977; Miller, 1978; Lawley, 1980) and it is obviously possible that these carcinogen-DNA interactions may induce the critical DNA alterations postulated for initiation of carcinogenesis. There is also a growing body of evidence that not all chemical carcinogens act in this manner, and the concepts of genotoxic and epigenetic or non-genotoxic carcinogens (Brookes et al, 1973) have been developed by several authors notably Williams (1979). The broad category of non-genotoxic carcinogens is taken to include agents that appear to act by such mechanisms as chronic tissue injury, immunosuppression, solid state effects, hormonal imbalance, cocarcinogenicity and promotion (Williams, 1980).

A comprehensive literature survey of in vivo covalent binding of organic chemicals to DNA has been recently provided by Lutz (1979) who has proposed that the extent of binding may be a quantitative indicator in the process of chemical carcinogenesis (See also the contribution by W.K. Lutz to the Proceedings of this Symposium). He has suggested that covalent binding may be expressed in quantitative terms as an index, the Covalent Binding Index (CBI) which relates damage to DNA as a function of dose, and is defined as follows:

$$CBI = \frac{\text{Micromole chemical bound per mole nucleotides}}{\text{Millimole chemical administered per kg animal}}$$

Over the wide range of chemicals included in his survey Lutz found a

Table 1.   Liver DNA Labeling by Radioactive
Chemicals In Vivo—Compounds Studied

N-Nitrosodimethylamine	Potent Carcinogen	Rat
N-Nitrosoproline	Non-Carcinogen	Rat
Trichlorethylene	Disputed Carcinogen	Mouse
N-Nitrosocimetidine	No Evidence Available	Mutagen

reasonable correlation between potency of the carcinogen and its CBI.
He proposes measurements of covalent binding, usually in rat liver,
as a screen for carcinogenic potency of chemicals. An obvious draw-
back of this proposal is the requirement for a radioactively labeled
material for the assay, since, with few exceptions, covalent binding
can only be detected and measured in this way. However Lutz points
out that most newly synthesized chemicals for which any widespread use
is anticipated are prepared with a radio label for pharmacokinetic and
toxicity studies and the same applies to many other compounds already
introduced into the environment.

In this presentation some recent studies at the Fels Research
Institute on covalent binding of some selected chemicals to rat liver
DNA will be described and the results related to the work of Lutz.
The chemicals studied are shown in Table 1.

Table 2.   Procedure for DNA Isolation

1) Homogenize Liver in 0.15M NaCl-0.04M EDTA (pH 7.8)

2) Pellet Nuclei at 3000 RPM

3) Phenol Extract Nuclei with Modified Kirby Procedure

4) Treat with RNase

5) Remove Glycogen with 2.5M Phosphate-Methoxyethanol

6) Precipitate DNA with CTAB

7) Extract Lipid with Organic Solvents

8) Centrifuge DNA through Cesium Chloride

9) Dialyze and Lyophilyze DNA Fraction

N-nitrosodimethylamine was included as a positive control, since considerable information is available on its covalent binding to DNA in vivo and on its carcinogenic potency.

In all studies of covalent binding of chemicals to cellular constituents the methodology for extraction and purification of the labeled macromolecular materials is crucial. The procedure used for DNA isolation in most of our work is outlined in Table 2.

NITROSOPROLINE

The number of N-nitroso compounds that are known to be carcino- genic is large and constantly increasing as more compounds are tested (Magee & Barnes, 1967; Druckrey et al, 1967; Magee et al, 1976). Until recently N-nitrosodiphenylamine was thought not to be carcino- genic but it has now been shown to induce liver tumors in rats (Cardy, et al, 1979). In contrast, several published reports have indicated that N-nitrosoproline is not carcinogenic and this finding has recently been confirmed by Mirvish and his colleagues (1980) who fed 39 rats very high dietary levels of the compound throughout their life span and found no indication of any excess tumor incidence in the control over the treated animals. It is there- fore of interest to study the metabolism of nitrosoproline because of its lack of carcinogenicity and to compare it with that of the carcin- ogenic compounds in the hope of increasing knowledge of the mechanisms of action of this group of powerful carcinogens. From another point of view such metabolic studies are important because proline, which is readily nitrosated (Mirvish et al, 1973) may prove to be a valuable test material for study of the rates of endogenous nitrosation in ex- perimental animals and in human beings. Such a test would involve administration of a nitrosatable amine and detection and quantitation of the nitrosated product or a metabolite in the urine or other avail- able body fluid. There is considerable current interest in the possi- ble role of endogenously formed nitrosamines in the causation of human cancer but no quantitative information on the extent to which these nitrosation reactions may occur in man is available.

Proline has been shown to be nitrosated in vivo in the rat after simultaneous oral administration with sodium nitrite (Braunberg & Dailey, 1973). In these experiments [^{14}C]nitrosoproline, prepared from uniformly labeled L-[^{14}C]-proline, was administered by stomach tube to rats, together with sodium nitrite, and the presence of the nitrosated derivative was demonstrated in the stomach contents 15 min. after the treatment. The authors suggested that the methods used might well serve as a model system to determine the possibility of in vivo formation of known carcinogenic nitrosamines from the parent amines and sodium nitrite. The same authors (Dailey et al, 1975) sub- sequently showed that uniformly labeled [^{14}C]nitrosoproline was ab- sorbed from the gastrointestinal tracts of fed and fasted rats after oral intubation. About 99% of the administered ^{14}C-radioactivity was absorbed by 24 h after administration and about 98% of the activity was excreted in the urine, with 1-2% appearing as expired $^{14}CO_2$. These findings strongly support the idea that proline may be a nitrosatable amine that could be safely given to human beings for the purpose of evaluating endogenous nitrosation reactions.

Experiments have been carried out at the Fels Research Institute, with Ms. Cecilia Chu, to investigate further the metabolism of N-nitrosoproline with emphasis on the evaluation of its safety for use as an indicator of endogenous human nitrosation. The radioactive compound has been prepared from two samples of ^{14}C]proline, one universally labeled as in the work of Braunberg and Dailey, and the other labeled only in the carboxyl group. These compounds were administered to rats by stomach tube and the proportions of the injected dose appearing as expired $^{14}CO_2$ and in the urine were determined at 24 h with the results shown in Table 3.

The findings with [U-^{14}C]nitrosoproline are in good agreement with those of Dailey et al (1975) and confirm that there is little metabolic breakdown of the compound with excretion of a very high proportion of the injected dose in the urine. The very small fraction of the ^{14}C-carboxyl labeled compound metabolized to $^{14}CO_2$ at both dose levels is important because it indicates that only a very small fraction of the nitrosoproline, if any, is converted to N-nitrosopyrrolidine which is a known carcinogen. The urine of these animals was analysed by a conventional amino acid analyser and all of the radioactivity present was found to elute with added N-nitrosoproline, again confirming the results of Dailey et al (1975). In addition, no detectable radioactivity was found to elute with proline, indicating that no significant metabolic denitrosation had occurred. This conclusion is supported by the absence of detectable radioactivity in total liver proteins prepared from treated rats, since some of any ^{14}C]proline formed would have been expected to be incorporated by normal biosynthetic pathways.

The possibility of covalent binding of radioactivity from [U-^{14}C]

Table 3. Metabolic Fate of ^{14}C-Nitrosoproline

	Dose (mg/kg)	No. of Rats	$^{14}C-CO_2$ At 24 h (% of Dose)	Nitrosoproline In Urine (% of Dose)
[Carboxyl-^{14}C]Nitrosoproline	1	3	0.66 ± 0.19	100.6 ± 1.45
[Carboxyl-^{14}C]Nitrosoproline	10	3	0.45 ± 0.01	92.9 ± 7.1
[[14]C-(U)]Nitrosoproline	1	4	3.07 ± 0.7	95.7 ± 8.6

Table 4.   Extraction of DNA from Liver of Rat
8 h After N-Nitrosoproline

Dose..... 1.74 mg/kg,   75 uCi per rat,   by Gavage

	Radioactivity	% of Injected Dose
Homogenate	2.6×10^5 CPM	0.168
Supernatant	2.5×10^5 CPM	0.14
Nuclear Pellet	3.48×10^4 CPM	0.022
DNA from the Treated Rat (5.4 mg)	29 CPM	
DNA from Calf-Thymus (5.5 mg)	32 CPM	

nitrosoproline to rat liver DNA was then examined. A rat was given
[U-^{14}C]nitrosoproline, 1.74 mg/kg body wt., total radioactivity 75µCi,
by stomach tube and was killed 8 h later. DNA was extracted from the
liver and its radioactivity compared with that of the same weight of
calf thymus DNA. A flow-sheet of this experiment is shown in Table 4.

As shown in the table, the radioactivity measurement of the DNA
from the liver of the treated rat was the same as that from the same
weight of non-radioactive calf thymus DNA, indicating that there had
been no covalent binding of [^{14}C]nitrosoproline to rat liver DNA. In
view of the very high proportion of the dose that was excreted in the
urine and the very small expiration of $^{14}CO_2$ this finding of zero covalent binding to DNA may seem unsurprising. However, as with many
chemical carcinogens, the fraction of the administered dose of the
nitrosamines that is bound to DNA has always been found to be very
small but it may still be consistent with a carcinogenic effect.

It is concluded that the present findings are consistent with
the non-carcinogenicity of N-nitrosoproline and that they support the
possible use of proline as a nitrosatable secondary amine in studies
of rates of endogenous nitrosation in man.

TRICHLOROETHYLENE

Trichloroethylene was tested in the Bioassay Program of the
National Cancer Institute in 1976 and reported to induce liver tumors
in B6C3Fl mice. Subsequently Henschler and his colleagues (Henschler
et al, 1977) questioned the validity of this bioassay on the grounds
that analysis of a sample of the industrial grade material used showed

the presence of considerable amounts of epichlorohydrin and 1,2-epoxi-
butane, added as stabilizers. Since these epoxides are highly mutagenic in the Ames test they clearly may have played a role in the
observed carcinogenic effect since the dose levels of trichloroethylene used were high.

Based on the possible similarity with the activation of vinyl
chloride, it has been suggested that trichloroethylene might be
metabolized to give the electrophilic intermediate trichloroethylene
epoxide which could then interact with cellular DNA and other macromolecules (Van Duuren, 1975; Uehleke et al, 1977). Evidence supporting this proposal has been provided by Banerjee and Van Duuren (1978)
who reported covalent binding of radioactivity to microsomal proteins
and to exogenous DNA after incubation of [^{14}C]trichloroethylene with
microsomes from B6C3F1 mice in vitro and similar results were obtained
by Laib et al (1979) using rat liver microsomes and exogenous RNA.
The latter workers showed that incubation with [^{14}C]vinyl chloride
under the same conditions resulted in much more extensive binding than
was obtained with [^{14}C]trichloroethylene. Both groups noted that binding to proteins was considerably greater than to either nucleic acid.

Consideration of the chemistry and biochemistry of trichloroethylene oxide led Bonse et al (1976) to conclude that the
conditions within the hepatic and other cells might favor the spontaneous rearrangement of the oxirane to form trichloroacetaldehyde which
is comparatively non-reactive and could thus be regarded as a detoxification reaction. They suggest that trichloroethylene is epoxidized
by mixed function oxidases and that the epoxide rearranges immediately
while still bound to the enzyme and that this mechanism could explain
the relatively weak mutagenic action and doubtful carcinogenicity of
trichloroethylene.

In view of the somewhat equivocal nature of the evidence for the
carcinogenicity of trichloroethylene it was of interest to investigate
the covalent binding of radioactivity from [^{14}C]trichloroethylene to
liver DNA in vivo. Experiments to this end, in collaboration with Dr.
Gerald Parchman, will now be described.

[^{14}C]Trichloroethylene, 3.4 mCi per m mole was kindly provided
by Dr. P. Watanabe of Dow Chemical Corporation, Midland, Michigan and
was administered, in corn oil, to male B6C3F1 mice by intraperitoneal
injection. This strain was chosen because it was the one used in the
National Cancer Institute Bioassay. The animals were killed 6 h after
the treatment and liver DNA, RNA and protein were extracted and covalently bound radioactivity was determined. The results are shown
in Table 5 in which covalent binding to rat liver after administration
of [^{14}C]dimethylnitrosamine is included for comparison. As indicated
earlier, dimethylnitrosamine was taken as a positive control.

It is clear that, in contrast to the findings described above

Table 5. Incorporation of ^{14}C-Dimethylnitrosamine
(DMN) and ^{14}C-Trichloroethylene (TCE)
Metabolites into Rodent Liver Macromolecules

Treatment	Dose		Binding to Macromolecules			
	(mg/kg)	(umole/kg)	Protein (cpm/mg)	RNA (cpm/mg)	DNA (cpm/mg)	CBI*
^{14}C-DMN (42.8 mCi/mmole)	0.05	0.68	247	571	394	2045
^{14}C-TCE (3.4 mCi/mmole)	94	718	10,770	367	83	5.1

$$*CBI = \frac{u\ mole\ carcinogen}{mole\ DNA\ nucleotide} \div \frac{m\ mole\ carcinogen}{kg\ B.W.}$$

Male Sprague-Dawley Rat (250 g). I.P. injection

15 Male B6C3F1 Mice (20-22 g). I.P. injection

with [^{14}C]nitrosoproline, the covalent binding to mouse liver in vivo,
although very low, is not zero. Calculation of the Covalent Binding
Index gave a figure of 5, which is very much less than the calculated
CBI for dimethylnitrosamine in rat liver of 2045. If extent of covalent binding does provide an index of carcinogenic potency these findings clearly support the view that trichloroethylene is only very
weakly carcinogenic, if at all. They are also consistent with the
views of Henschler and his colleagues (1978) on the nature of the
metabolic intermediates of trichloroethylene.

NITROSOCIMETIDINE

Cimetidine (Tagamet, N-cyano-N'-methyl'N"-[2-[[(5-methyl-1H-
imidazol-4-yl)methyl]thio]ethyl]guanidine) is a drug that is used extensively in the treatment of human gastrointestinal disorders, including peptic ulcer (Finkelstein, 1978). Recently Elder and his
colleagues (Elder et al, 1979) reported the occurrence of gastric
cancer in three patients who had received treatment with cimetidine
and suggested a possible causal relationship with the ingestion of the
drug. The same authors drew attention to the chemical structure of
cimetidine and suggested that its nitrosated derivative (Fig. 1) might
be carcinogenic because of its structural resemblance to N-methyl-N'-
nitro-N-nitrosoguanidine (Fig. 1) which is a powerful carcinogen for
the stomach of rodents and dogs. Since nitrosation of various amines
has been shown to occur in the stomachs of experimental animals after
their administration simultaneously with nitrites (Mirvish, 1977) it
was suggested that a similar nitrosation of cimetidine might occur in

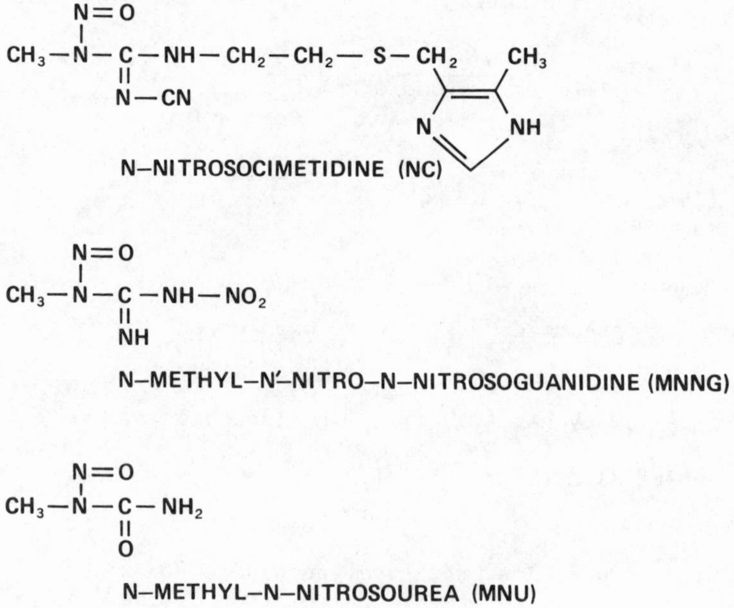

N-NITROSOCIMETIDINE (NC)

N-METHYL-N'-NITRO-N-NITROSOGUANIDINE (MNNG)

N-METHYL-N-NITROSOUREA (MNU)

Figure 1.

human subjects receiving the drug, since nitrites are present in human saliva (Tannenbaum et al, 1974). The proposal by Elder and his colleagues that the human stomach cancers reported by them might have been related to cimetidine was criticised by several authors (Roe, 1979; Ruddell, 1979; Hill, 1979) because, among other reasons, none of these patients had been exposed for more than 6-12 months to the drug. However, N-nitrosocimetidine has been prepared (Bavin et al, 1980) and shown to be a mutagen in the Ames test as well as causing DNA damage as indicated by preferential killing of repair deficient strains of E.Coli (Pool et al, 1979) and by the induction of DNA strand breaks in human lymphoblastoid cell lines (Henderson & Basilio, 1980) and in mouse epithelial cells in culture (Schwartz et al, 1980). Cimetidine itself did not have these effects on DNA. Nitrosocimetidine has also been shown to act as a methylating agent at pH 7 towards 3, 4-dichlorobenzenethiol with an activity comparable to that of N-methyl-N'-nitro-N-nitrosoguanidine (Foster et al, 1980).

We have investigated the capacity of N-nitrosocimetidine to alkylate DNA in vitro and in the organs of rats after its administration

Table 6. Methylation of DNA In Vitro by Methylnitrosourea (MNU),
          Methylnitronitrosoguanidine (MNNG), Nitrosocimetidine
          (NC) and Cimetidine (C)

Agent (10mM)	Cysteine (20mM)	7-meG/100 DNA Residues	3-meA/7-meG	O^6-meG/7-meG
MNU	+	1.56	0.11	0.10
	−	1.53	0.11	0.09
MNNG	+	1.31	0.14	0.09
	−	1.23	0.13	0.11
NC	+	0.69	0.13	0.11
	−	trace	−	−
C	+	Not detected	−	−
	−	Not detected	−	−

4.9mM Calf Thymus DNA, 10mM agent, 20mM cysteine.

50mM KH_2PO_4, 0.1mM Na_2EDTA, pH 7.4, 5% DMSO, 23°, 34 h.

in vivo.

     N-nitrosocimetidine nitrate was prepared according to Bavin et
al (1980) and incubated with calf thymus DNA in the presence of
cysteine in phosphate buffer at pH 7.4 and 23° for 34 h. N-methyl-N'-
nitro-N-nitrosoguanidine and N-methylnitrosourea were incubated under
similar conditions to serve as positive controls and cimetidine itself
was included. The DNA was rapidly isolated from the reaction mixture,
hydrolysed in mild acid and the mixture analysed by high pressure
liquid chromatography on a Bio Rad Aminex A-9 column, eluting with
500nM ammonium formate, pH 3.0 at 60° and monitoring with a uv detector. These experiments were performed by Dr. David Jensen and a detailed report has been accepted for publication (Jensen & Magee, 1981).
Under the conditions used the DNA was modified to a degree such that
several methylated purines could be detected optically and 7-methyl-
guanine, 3-methyladenine and O^6-methylguanine were quantitated, with
the results shown on Table 6.

     It is clear that nitrosocimetidine reacted with DNA to give a
pattern of methylated purines in identical ratios with those obtained
with the control compounds, N-methyl-N'-nitro-N-nitrosoguanidine and
N-methylnitrosourea, both of which are powerfully carcinogenic and
mutagenic.

     These experiments were then extended to in vivo studies in the
intact rat by Dr. Charles Gombar using methyl labeled [3H]nitroso-

cimetidine, prepared from [N-methyl-^{3}H]cimetidine, supplied by
Amersham Corporation and N-[methyl-^{14}C]Methyl-N'-nitro-N-nitroso-
guanidine (New England Nuclear Corp.) for comparison. Starved female
Wistar rats, approx. 200g body wt., were given the compounds at the
doses indicated in Table 7 by gavage and the specific radioactivities
of the materials were determined by high pressure liquid chromatography and radioassay immediately after administration. The animals
were killed after 12 h and DNA, RNA and proteins were extracted from
various organs. The specific radioactivities of the RNA and protein
were measured and the DNA was hydrolyzed under mild acid conditions and
the purine bases were analysed by high pressure liquid chromatography.
The findings are shown in Table 7.

These results are interesting because of the relatively much
smaller degrees of methylation of DNA found in the organs treated with
nitrosocimetidine than in those of the animals given methylnitro-
nitrosoguanidine. These findings are in contrast with the relative
methylation of DNA in vitro by the same compounds as shown in Table
6 and they illustrate the need for caution in attempts at extrapolation from in vitro studies to the intact animal. They also underline
the need for further work on the pharmacokinetics and metabolism of
nitrosocimetidine and other N-nitroso compounds.

Table 7. Extent of Interaction with Cellular Macromolecules
          by Methylnitronitrosoguanidine (MNNG) and Nitroso-
          cimetidine (NC) in Rat Organs In Vivo

	DNA-Methylation[1]		Total RNA-Incorporation[2]		Total Protein-Incorporation	
Organ	MNNG	NC	MNNG	NC	MNNG	NC
Liver	46	n.d.[3]	99	20	114	17
Stomach	570	n.d.	969	28	103	7
Sm. Intestine	91	11	561	39	124	6
Pancreas	n.d.	n.d.	22	n.d.	50	5
Kidney	n.d.	n.d.	128	12	79	7

$1 - \dfrac{7-meG/10^6 G}{\mu mole/k_g} \times 10^2$ $2 - \dfrac{\mu mole/mg}{\mu mole/k_g} \times 10^8$ 3 - n.d. = none detected

CONCLUSIONS

1) Interaction of N-nitrosoproline with rat liver DNA in vivo could not be detected. This finding, together with the evidence for its rapid elimination from the body and very small degree of metabolism, support the possible use of proline as a safe test compound in studies on endogenous nitrosation in animals and human subjects.

2) There was a very small, but probably real interaction between trichloroethylene and mouse liver DNA in vivo which was very much less than that of the powerful carcinogen, dimethylnitrosamine. These results are compatible with the apparent low carcinogenic potency of trichloroethylene and the doubts expressed concerning its carcinogenicity.

3) N-nitrosocimetidine interacts with calf thymus DNA in vitro, in the presence of sulfhydryl compounds, to yield a pattern of methylated purine bases that is identical with that produced by N-methyl-N'-nitro-N-nitrosoguanidine and other carcinogenic and mutagenic N-nitroso compounds. Preliminary experiments in the intact rat indicate that the extent of methylation of DNA is very much less in organs of animals treated with nitrosocimetidine than in those treated with N-methyl-N'-nitro-N-nitrosoguanidine.

ACKNOWLEDGEMENTS

The authors wish to acknowledge support from grants CA23451 and CA12227 from the National Cancer Institute and from the Samuel S. Fels Fund of Philadelphia.

REFERENCES

Banerjee, S. and Van Duuren, B.L., 1978, Covalent binding of the carcinogen trichloroethylene to hepatic microsomal proteins and to exogenous DNA in vitro, Cancer Res. 38:776 - 780.
Bavin, P.M.G., Durant, G.J., Miles, P.D., Mitchell, R.C. and Pepper, E.S., 1980, Nitrosation of cimetidine [N"-cyano-N-methyl-N'-{2-[(5-methylimidazol-4-yl)methylthio]ethyl}guanidine], J. Chem. Research (S)212-213.
Bonse, G., Henschler, D. and Gehring, P.J., 1976, Chemical reactivity, biotransformation, and toxicity of polychlorinated aliphatic compounds, CRC Critical Reviews in Toxicology, 395-409.
Braunberg, R.C. and Dailey, R.E., 1973, Formation of nitrosoproline in rats, Proc. Soc. Exp. Biol. Med., 142:993-996.
Brookes, P., 1966, Quantitative aspects of the reaction of some carcinogens with nucleic acids and the possible significance of such reactions in the process of carcinogenesis, Cancer Res., 26:1994.

Brookes, P., Druckrey, H., Ehrenberg, L., Lagerlof, B., Litwin, J. and Williams, G.M., The relation of cancer induction and genetic damage, in: "Evaluation of Genetic Risks of Environmental Chemicals", Ramel, C., ed., Ambio Special Report No. 3, pp. 15-16, Royal Swedish Academy of Sciences, Universitets forlaget, Stockholm (1973).

Cardy, R.H., Lijinsky, W. and Hildebrandt, P.K., 1979, Neoplastic and nonneoplastic urinary bladder lesions induced in Fischer 344 rats and B6C3F$_1$ hybrid mice by N-nitrosodiphenylamine, Ecotoxicol. Environ. Safety, 3:29-35.

Dailey, R.E., Braunberg, R.C. and Blaschka, A.M., 1975, The absorption, distribution, and excretion of [^{14}C]nitrosoproline by rats, Toxicology, 3:23-28.

Druckrey, H., Preussmann, R., Ivankovic, S. and Schmähl, D., 1967, Organotrope carcinogene Wirkungen bei 65 verschiedenen N-nitroso-Verbindungen an BD-Ratten, Z. Krebsforsch., 69:103-201.

Elder, J.B., Ganguli, P.C. and Gillespie, I.E., 1979, Cimetidine and gastric cancer, Lancet, 1:1005-1006.

Finkelstein, W., 1978, Cimetidine, N. Engl. J. Med., 299:992-996.

Foster, A.B., Jarman, M. and Manson, D., 1980, Structure and reactivity of nitrosocimetidine, Cancer Letters, 9:47-52.

Henderson, E.E. and Basilio, M., 1980, Evidence for the in vivo damage of DNA by nitrosocimetidine, Proc. Amer. Assoc. Cancer Res., 21:85.

Henschler, D., Eder, E., Neudecker, T. and Metzler, M., 1977, Carcinogenicity of trichloroethylene: fact or artefact? Arch. Toxicol. 37:233-236.

Henschler, D., Hoos, W.R., Fetz, H., Dallmeier, E., and Metzler, M., 1978, Reactions of trichloroethylene epoxide in aqueous systems, Biochem. Pharmacol. 28:543-548.

Hill, M.J., 1979, Gastric cancer in patients who have taken cimetidine, Lancet 1:1235.

Jensen, D.E. and Magee, P.N., 1981, The methylation of DNA by nitrosocimetidine in vitro, Cancer Res., 41:230-236.

Laib, R.J., Stöckle, G., Bolt, H.M. and Kunz, W., 1979, Vinyl chloride and trichloroethylene: comparison of alkylating effects of metabolites and induction of preneoplastic enzyme deficiencies in rat liver, J. Cancer Res. Clin. Oncol. 94:139-147.

Lawley, P.D., 1980, DNA as a target of alkylating carcinogens, Brit. Med. Bull. 36:19-24.

Lutz, W.K., 1979, In vivo covalent binding of organic chemicals to DNA as a quantitative indicator in the process of chemical carcinogenesis, Mutation Res., 65:289-356.

Magee, P.N. and Barnes, J.M., 1967, Carcinogenic nitroso compounds, Advanc. Cancer Res., 10:163-246.

Magee, P.N., Montesano, R. and Preussmann, R., 1976, N-nitroso compounds and related carcinogens, in: "Chemical Carcinogens", A.C.S. Monograph 173, Searle, C.E., ed., pp. 491-625, American Chemical Society, Washington, D.C. (1976).

Miller, E.C., 1978, Some current perspectives on chemical carcinogen-
    esis in humans and experimental animals, Cancer Res., 38:1479-
    1496.
Mirvish, S.S., 1977, N-nitroso compounds: their chemical and in vivo
    formation and possible importance as environmental carcinogens,
    J. Toxicol. Environ. Hlth., 2:1267-1277.
Mirvish, S.S., Bulay, O., Runge, R.G. and Patil, K., 1980, Study of
    the carcinogenicity of large doses of dimethylnitramine, N-
    nitroso-L-proline, and sodium nitrite administered in drinking
    water to rats, J. Nat. Cancer Inst., 64:1435-1440.
Mirvish, S.S., Sams, J., Fan., T.Y. and Tannenbaum, S.R., 1973,
    Kinetics of nitrosation of the amino acids proline, hydroxy-
    proline, and sarcosine, J. Nat. Cancer Inst., 51:1833-1839.
Pegg, A.E., 1977, Formation and metabolism of alkylated nucleosides:
    possible role in carcinogenesis by nitroso compounds and al-
    kylating agents, Advanc. Cancer Res., 25:195-269.
Pool, B.L., Eisenbrand, G. and Schmähl, D., 1979, Biological activity
    of nitrosated cimetidine, Toxicology, 15:69-72.
Roe, F.J.C., 1979, Cimetidine and gastric cancer, Lancet, 1:1039.
Ruddell, W.S.J., 1979, Gastric cancer in patients who have taken
    cimetidine, Lancet, 1:1234.
Sarma, D.S.R., Rajalakshmi, S. and Farber, E., Chemical carcinogenesis:
    interactions of carcinogens with nucleic acids, in: Cancer, Vol.
    1, Becker, F.F., ed., pp. 235-287, Plenum Publishing Corporation,
    New York (1975).
Schwartz, M., Hummel, J. and Eisenbrand, G., 1980, Induction of DNA
    strand breaks by nitrosocimetidine, Cancer Lett., 10:223-228.
Tannenbaum, S.R., Sinskey, A.J., Weisman, M. and Bishop, W., 1974,
    Nitrite in human saliva. Its possible relationship to nitros-
    amine formation, J. Nat. Cancer Inst., 53:79-84.
Uehleke, H., Tabarelli-Poplawski, S., Bonse, G. and Henschler, D.,
    1977, Spectral evidence for 2,2,3-trichloro-oxirane formation
    during microsomal trichloroethylene oxidation, Arch. Toxicol.,
    37:95-105.
Van Duuren, B.L., 1975, On the possible mechanism of carcinogenicity
    of vinyl chloride, Ann. N.Y. Acad. Sci. 246:258-267.
Williams, G.M., 1979, Review of in vitro tests using DNA damage and
    repair for screening of chemical carcinogens, J. Assoc. Off.
    Anal. Chem., 62:857-863.
Williams, G.M., 1980, Classification of genotoxic and epigenetic
    hepatocarcinogens using liver culture assays, Ann. N.Y. Acad.
    Sci., 349:273-282.

CONSTITUTIVE AND CARCINOGEN-DERIVED DNA BINDING

AS A BASIS FOR THE ASSESSMENT OF POTENCY OF CHEMICAL CARCINOGENS

Werner K. Lutz

Institute of Toxicology
ETH and University of Zurich
CH-8603 Schwerzenbach, Switzerland

SUMMARY

1. The hypothesis is presented that a ground-level DNA damage is
    unavoidable and must be regarded constitutive to a cell. The
    genotoxic agents responsible for this type of initiation
    comprise   a number of physiological chemicals which are
    known or suspected to be degraded via chemically reactive
    species, nitrosamines present in the diet or formed from
    amines in the stomach, ubiquitous carcinogenic metals,
    UV- and σ-irradiation, viruses, and other sources in diet and
    environment.

2. It is concluded that what is normally called spontaneous
    tumor incidence is partly due to this constitutive DNA damage.

3. Under this assumption, carcinogens can be divided into two
    classes, the initiating (genotoxic, DNA-damaging) carcinogens
    and the non-genotoxic carcinogens which act by modulating any
    of a number of reactions and side reactions that lead to an
    increase of the constitutive DNA damage or enhance the chance
    for the constitutive DNA damage to proceed to a tumor.

4. The carcinogenic potency of a carcinogen is described as a
    product of persistent DNA damage x mutagenicity x non-geno-
    toxic modulation of the DNA damage, and short-term tests are
    proposed to determine the contributions of the first two
    parameters to the potency of an initiating carcinogen.

5. As an approximation to this theoretical approach, a correla-
    tion is shown of DNA binding in vivo in the form of a

"Covalent Binding Index" to the carcinogenic potency as derived from long-term bioassays. An astonishingly good linear correlation is found with an approximate uncertainty of the estimate of a factor of 10 with a total span of values of 10^6.

6. It is concluded that DNA binding in vivo provides a useful first look at the potency of initiating carcinogens but that additional knowledge is required to assess the organ specificity of initiating carcinogens.

7. The non-genotoxic carcinogens seem to be much less potent than the initiating carcinogens if administered alone. They cannot, however, be spotted on the basis of chemical structure, and there is no short-term test which would allow a good quantit ative approach to their carcinogenic potency as was demonstrated for the initiating carcinogens. One main reason for this lack is the wide variety of different mechanisms of action of non-genotoxic carcinogens.

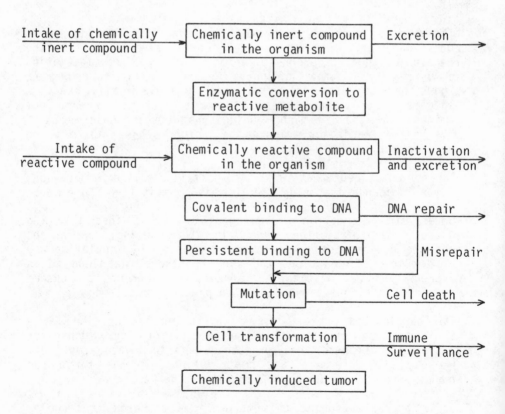

Fig. 1. Sequence of events in the chemical induction of a tumor

1. DEFINITION OF A CHEMICAL CARCINOGEN
AND MECHANISMS IN THE INDUCTION OF A TUMOR

A chemical is called a carcinogen if its administration enhances the incidence of tumors in a long-term bioassay. There is no restriction to this general definition as to the mechanism by which the tumors are induced but the finding that a chemical is carcinogenic can be unique for a specific experimental set-up with its choice of animal species, route of administration, diet, and other conditions.

The current evidence suggests that the chemical induction of a tumor involves the steps depicted in Fig. 1. A central event seems to be the chemical reaction of electrophilic compounds with DNA in a target cell. This so-called "initiation" takes place within hours after uptake of the chemical and is followed by a number of steps which take months to years for completion.

2. OCCURRENCE OF DNA DAMAGE

It is a working hypothesis in this report that not only xenobiotics give rise to DNA damage but that a constitutive or ground-level DNA damage is always present in most cells and is unavoidable. Table 1 lists a number of sources for this type of DNA damage that we cannot escape. We have shown that for instance estrone interacts covalently with liver DNA of rats treated with this hormone (Jaggi et al., 1978). A number of other physiological compounds are activated to chemically reactive intermediates, arachidonic acid to an epoxide generated in the process of prostaglandin synthesis (Eling et al., 1977), cholesterol to its 5,6-oxide (Kelsey and Pienta, 1979). Tryptophane is degraded to aromatic amines which form a well-known class of carcinogens.

None of these compounds have so far been tested for their ability to bind to DNA in vivo and it will only be a matter of the limit of detection that these and many others will be discovered as physiological DNA-damaging agents. In addition, we are sure that we cannot escape minute amounts of nitrosamines always formed

Table 1. Origin of Ubiquitous Initiating Carcinogens
Leading to a Constitutive DNA Damage

- Physiological reactive intermediates
- Nitrosamines formed in vivo from ubiquitous
   dietary amines and nitrite or nitrate
- Irradiation (UV for skin, α-irradiation in general)
- Carcinogenic metals
- Viruses

---

in vivo (Tannenbaum, 1979). Irradiation can produce a number of DNA alterations and carcinogenic metals have been shown to inter- act with DNA in vitro and to affect the fidelity of replication and transcription so that they could also have an initiating share of their carcinogenic activity (Sirover and Loeb, 1976).

We therefore believe that to a certain extent DNA damage is unavoidable and is probably quite remarkable in organs like the liver which takes over most of the degradation of physiological compounds. On the other hand, it is known that the liver is equipped with efficient DNA repair mechanisms so that the relatively low occurrence of primary liver tumors does not invali- date the present hypothesis. It seems rather that those organs with a high activity in the degradation of physiological compounds and xenobiotics have evolved with high DNA repair activity, whereas those cells where DNA damage and mutations must be minimized (for instance germ line cells) almost lack drug-metabo- lizing capacities.

3. CLASSIFICATION AND MODE OF ACTION OF CARCINOGENS

On the basis of the hypothesis presented above, carcinogens can now be divided into two distinct classes as shown in Table 2. Such a separation is not only mechanistical but is important for the conduct and interpretation of long-term bioassays. Two seemingly identical long-term carcinogenicity bioassays with the same test compound and the same animal species and strain can well give different results when the diet with its unavoidable DNA - damaging components is not the same or when the animals are under different stress conditions so that the hormonal status can give rise to a different ground-level DNA damage. Such differences might then be expressed to yield a different tumor incidence.

The above classification will also be useful if it is aimed at assessing the carcinogenic potency from a battery of short-term tests each of which can only deal with a fraction of the reactions leading to a tumor.

Table 2. Classification of Chemical Carcinogens

- Initiating Carcinogens
    (DNA damaging, genotoxic)
- Non-genotoxic Carcinogens
    (Modulate amount or expression
    of the constitutive DNA damage so as to favor
    any of the vertical reactions shown in Fig. 1)

The modulation of the constitutive DNA damage towards a
higher incidence of tumors by the non-genotoxic carcinogens can
occur for any of the steps depicted in Fig. 1. It is also possible
that a test compound is both an initiating and non-genotoxic
carcinogen so that it can modulate its own DNA damage either to
become larger or to be more effectively expressed. Our DNA-binding
experiments with estrone show that such a double function could
be the case for hormones. Some examples for a non-genotoxic mode
of carcinogenic activity are given in Table 3 which is not meant
to be exhaustive.

Depending on the step at which a non-genotoxic carcinogen
exerts its modulatory influence on the process of tumor formation,
this class of carcinogens can be subdivided into different types
as shown on the right of Table 3. It is also possible that a
non-genotoxic carcinogen acts on more than only one level or can
have anticarcinogenic activity in one respect and is a non-genotoxic carcinogen in another (phenobarbital).

Table 3. Mode of Action of Non-Genotoxic Carcinogens

	Type of Modulation

- Formation of more initiator:
  -Induction of activating enzymes
  -Depletion of detoxification pathways
  -Low pH in the stomach,
    speeding up nitrosamine formation
	Pre-Initiation

- Inactivation of DNA repair,
    so that the DNA damage is more persistent
- Higher frequency of mutations:
  -Stimulation of cell division
  -Comutagenic activity at the DNA replication
	Co-Mutagenicity

- Higher rate of transformation of mutated cells - Enhancement of growth of transformed cells - Higher metastatic activity of tumor cells	Post-Initiation*

*The term 'promoter' is often used for this type of mechanism in
a two stage experiment with initiating and non-genotoxic activities of the post-initiation type. We have not used this term in
the present context because a promoter is, as defined by the early
experiments of Berenblum and Shubik (1947), not a carcinogen in
our sense because it did not significantly increase the incidence
of tumors when given alone.

An anticarcinogen can easily be defined in this context as being a compound that modulates constitutive or carcinogen-derived DNA damage in such a way as to favor the horizontal salvage reactions of Fig. 1. As with non-genotoxic carcinogens, this can take place at a pre- or post-initiation level or at the mutational event itself and comprises   activities such as induc ., detoxification pathways, trapping of reactive intermediates, inducing DNA repair, strengthening the immune surveillance, and many others.

4. POTENCY OF A CARCINOGEN AS DERIVED FROM THREE MAIN CONTRIBUTIONS

The potency of an initiating chemical carcinogen is dependent upon three main factors. First, it is correlated with the effectiveness of DNA binding. Second, it is inversely correlated with the speed of error-free DNA repair because a heritable type of DNA damage is produced only if the DNA is replicated before the promutagenic lesion is repaired. Third, the carcinogenic potency is determined by the probability that a given DNA adduct gives rise to mutations.

Most well-studied initiating carcinogens are known to produce more than one DNA adduct, each with its own half life and its own mutagenicity. The carcinogenic potency is therefore based upon the sum of all DNA adducts.

The half life of a DNA adduct is dependent on its chemical stability and the efficiency of the appropriate repair mechanisms, and in a number of cases it has been shown that it is the persistently bound adduct which is important in the induction of tumors.

A combination of these factors now leads to the following equation.

$$\text{Potency} = \sum_{\text{adducts}} \begin{array}{l} \text{Initiation (constitutive +} \\ \quad \text{carcinogen-derived DNA damage)} \\ \text{x Half life of DNA damage} \\ \text{x Mutagenicity of DNA damage} \\ \\ \text{x Non-genotoxic modulation} \end{array}$$

The potency of non-genotoxic carcinogens cannot be so clearly subdivided into different mechanistical steps. This class of carcinogens is therefore included in the above equation merely by multiplying with an additional factor. The real world is obviously more complicated in as much as those non-genotoxic influences are not included that modulate the initiation itself.

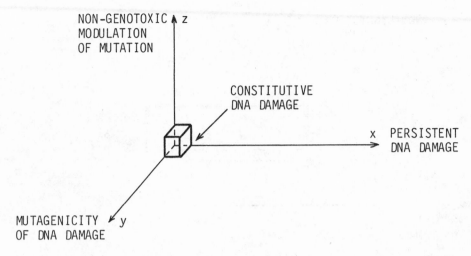

Fig. 2. Carcinogenic potency of the constitutive, ground-level
        DNA damage.

By combining the initial DNA damage with its half-life to
give persistent DNA binding, the carcinogenic potency can be
represented as a product of three factors. This equation can then
be depicted graphically with the persistent DNA binding as the
x-axis, the mutagenicity of the DNA damage as the y-axis, and the
non-genetic influences of the post-initiation type as the z-axis,
and the potency by the volume of the resulting parallelepiped as
shown in Fig. 2.

We have stated in chapter 2 that a ground-level DNA damage
cannot be avoided. The parallelepiped shown in Fig. 2 therefore
represents the potency of this constitutive, ground-level DNA
damage in the induction of what we normally call 'spontaneous'
tumor incidence. Such a hypothesis also means that the non-genetic
contribution on the z-axis cannot be zero. This is easily compatible with the fact that our diet always contains substances with
activities such as those listed in Table 3.

In Fig. 3 it is shown how the potency of two typical repre-
sentatives of the initiating or non-genetic classes of carcinogens, aflatoxin B_1 (A) and saccharin (B), can be illustrated.
Aflatoxin B_1 with its very strong initiating activity bases its
potency primarily on the x- and y-axis whereas saccharin has been
shown not to bind to DNA in vivo on a limit of detection of 10^6
below aflatoxin B_1 (Lutz and Schlatter, 1977) but bases its
carcinogenic activity on the constitutive level of DNA damage and
has an effect primarily along the non-genetic z-axis. The different volumes of A and B are in accordance with the different
carcinogenic potencies.

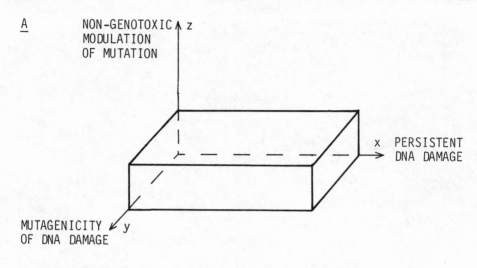

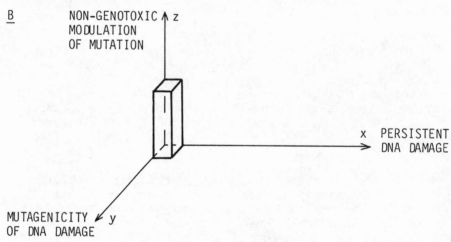

Fig 3.   Graphical representation of the carcinogenic potency of
         an initiating carcinogen (aflatoxin B_1, A) and a non--
         genotoxic carcinogen (saccharin, B).

5. POTENCY OF A CARCINOGEN AS DERIVED FROM SHORT-TERM TESTS

It would, in principle, be possible to calculate the incidence of tumors if the concentrations and kinetics of all
reactions shown in Fig. 1 were known. A battery of short-term
tests, each of which would be good for the determination of one
kinetic parameter, could then serve as a basis to determine the
final tumor incidence from exposure to a carcinogen.

A less detailed subdivision of the carcinogenic process was given in the last section and it is shown here with what type of short-term tests an estimation of the carcinogenic potency could be attempted.

The value for the x-axis can be derived from the measurement of DNA binding of an initiating carcinogen. It is important to determine this activity in an intact mammalian organism because it is well known that the activation/detoxification processes which take place in an organ can never be accurately simulated in an in vitro incubation of a carcinogen with DNA (Ashby and Styles, 1978). Radiolabelled chemical will be required for such a DNA binding assay, and a time lag of a few days between the administration of the compound and the isolation of the DNA will ensure that only the persistent fraction of the DNA adducts is determined.

The y-axis shows how critical a DNA adduct is with respect to the induction of a mutation. It is possible to determine the appropriate value for this in a mutagenicity assay where DNA binding and number of mutations can be scored in the same system. This approach was originally suggested by Sega et al. (1972) with an experiment where they determined ethylations of Drosophila spermatozoan DNA by ethyl methanesulfonate and compared this alkylation with sex-linked recessive lethal frequencies. This approach of a correlation of DNA binding with mutagenicity has in the last years been expanded to a number of other mutagenicity test systems (Manthey et al., 1978; Stark et al., 1979; Newbold et al., 1980; Aaron et al., 1980; references therein).

It is well known that the results from mutagenicity assays are strongly dependent on experimental conditions, especially on

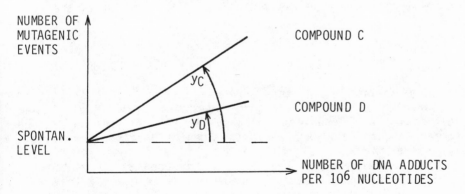

Fig. 4. Mutational events as a function of the number of DNA adducts

dose and toxicity of the compound used as well as on the system chosen for the activation to a reactive metabolite. For a comparison of the mutagenicity of various DNA adducts it will therefore be necessary to find a linear part for the correlation of the number of mutations scored with the amount of DNA adducts and to determine the first deviate as a measure for the mutagenicity of that adduct. This approach is depicted in Fig. 4 and was introduced by Watanabe et al. (this volume) with 1,2-dichloroethane.

For the z-axis, i.e. for the determination of the non-geno-toxic influences of the post-initiation type, we cannot propose any appropriate short-term test. Although a number of in vitro tests have been claimed to recognize this type of non-genotoxic carcinogenicity we believe that a quantification of these processes requires the use of an intact organism because these late events need a complete biology even more than the early stages like the initiation, which involves much more chemistry and enzymology. So new assay systems should be developed in the future.

For the detection of non-genotoxic carcinogens of the pre-initiation type or of the pro-mutagenic type, a somatic mutation assay will be required. Such an assay will show an increase of the 'spontaneous' mutation rate if the test compound has a positive effect on the amount of constitutive DNA damage or on its mutagenicity.

It is obvious from the above discussion that the class of initiating carcinogens is quite well amenable to quantification whereas the non-genotoxic carcinogens are far from being known as well. This difference in knowledge is easily understood if one realizes that initiating carcinogens have one well-defined activity, i.e., covalent binding to cellular macromolecules, whereas non-genotoxic carcinogens can act on such a variety of reactions that a unifying assay is not to be expected in the near future.

6. ASSESSMENT OF CARCINOGENIC POTENCY FROM LONG-TERM BIOASSAYS

For a validation of the above-mentioned use of short-term tests in the assessment of carcinogenic potency, quantitative evaluation of the long-term carcinogenicity data must first be performed. A first approach was introduced by Meselson and Russell (1977) where they defined a $D_{1/2}$ as the daily dose which gives 50% cumulative single-risk incidence of induced cancer after two years of exposure. Because of a number of restrictions, they were able to compile potencies for only 14 chemicals which were listed together with the respective animal species, site of tumor and route of exposure.

Ames and co-workers have announced a comprehensive survey with a calculation of the similar TD_{50} values, i.e., the daily dose needed to reduce the probability for an animal of being tumor-free by one half if treated with the test compound for a standard life time. To our knowledge, this report has not been published yet and only preliminary information on TD_{50} values of 16 chemicals was quoted in an editorial comment without reference to species, site or other experimental conditions (Maugh II, 1978). Six of these were already mentioned by Meselson.

Bartsch et al. (1980) have cited TD_{50} values of 7 alkylating agents in an attempt to correlate carcinogenic potency with mutagenicity in the Ames test. Values were taken from a report quoted as being in preparation by Terracini et al. Again, no indication was given of the experimental conditions.

Finally, Crouch and Wilson (1979) introduced a new definition of carcinogenic potency by defining the probability of getting cancer as $P = \alpha + \beta \cdot d$. α is the probability of getting cancer in the absence of carcinogen, d is the measure of dose of carcinogen (mg/kg·d averaged over a lifetime), and β is the potency of the carcinogen. They compiled the β values from those NCI carcinogenesis bioassays with rats and mice which contained sufficient data so that β values could be computed. Unfortunately, they did not include α values in their report nor did they indicate the site of tumor so that an interested reader must still consult the primary literature. A transformation to TD_{50} values can therefore only be approximated by $TD_{50} = 0.5/\beta$. In addition, a TD_{50} is much more perspicuous than a β value because it is a dose and not its inverse so that the following chapter bases upon TD_{50} values rather than β values.

In summary, although carcinogenic potency derived from long-term bioassays will become more and more important there is a fundamental lack of a comprehensive review of the data, and unpublished figures prevail.

7. APPROXIMATION OF POTENCY OF INITIATING CARCINOGENS ON THE BASIS OF DNA DAMAGE IN VIVO AT THE TIME OF MAXIMUM BINDING ALONE

DNA binding in vivo has already been introduced as a possible way of determining the initiation in vivo (review by Lutz, 1979). Covalent Binding Indices (CBI), i.e., DNA binding per dose, were compiled in this review from the literature and from our own work for more than 80 chemicals. In most of the studies, the DNA damage was measured at the time of maximum binding and only limited information is available on the more critical amount of DNA adducts that persisted. It will be seen in this section that this

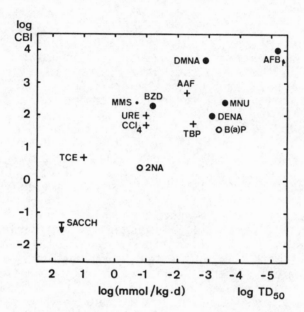

Fig. 5.   Correlation of DNA binding in vivo to carcinogenic potency.

CBI = (μmol chemical bound per mol DNA nucleotides)/(mmol chemical administered per kg animal weight)

Because of the limited experimental information on the long-term data different symbols were used in order to show the varying degree of overlap with respect to animal species, tumor-bearing organ, and route of administration.

The following experimental conditions and references were used for carcinogenicity and DNA binding experiment, respectively (Me = Meselson and Russell, 1977; Ma = Maugh, 1978; Lu = Lutz, 1979).

●: same animal species and organ. AFB_1 = aflatoxin B_1, rat liver (Me/Lu); DMNA = dimethylnitrosamine, mouse liver (Me/Lu); MNU = methylnitrosourea, rat brain (Me/Lu); DENA = diethylnitrosamine, rat liver (Me/Lu); BZD = benzidine, rat liver (Me/Martin and Ekers, 1980).

o: same organ but different animal species. B(a)P = benzo(a)pyrene, hamster/mouse forestomach (Me/Borchert and Wattenberg, 1976); 2NA = 2-naphthylamine, mouse/rat liver (Me/Lu);

+: site for TD_{50} not reported but most probably the same as for CBI. AAF = 2-acetylaminofluorene, rat liver (Ma/Lu); Ure = urethane, mouse liver (Ma/Lu); CCl_4 = carbon tetrachloride, mouse liver (Ma/Lu); TBP = tris-(2,3-dibromopropyl)phosphate, rat kidney (Ma/Dybing et al., this volume); TCE = trichloroethylene, mouse liver (Ma/Magee, this volume)

●: same organ but important difference in the route of administration. MMS = methyl methanesulfonate, mouse (drinking water)/rat (iv) lung (Me/Lu);

▼: limit of detection. SACCH = saccharin, rat bladder (Ma/Lutz and Schlatter, 1977).

---

part information will already provide a useful means for approximating the x-axis of our potency analysis presented in chapter 4.

For a correlation of CBI with TD_{50} values, the limited information on carcinogenic potencies as derived from long-term bioassays discussed in the last chapter was searched in order to find those chemicals for which both TD_{50} values and Covalent Binding Indices to DNA were available. 14 compounds were found and used for the correlation shown in Fig. 5.

A surprisingly good quantitative correlation exists between carcinogenic potency and maximum DNA binding. The uncertainty of an estimation of a carcinogenic potency from a given CBI is about one order of magnitude. The reason for this uncertainty obviously lies in the fact that neither persistence nor mutagenicity of the DNA adducts nor non-genotoxic influences are taken into account. This lack is all the more important the more DNA adduct is formed without mutagenic consequences. With MMS, for instance, most DNA binding results from the methylation at nitrogen 7 of guanine which is probably not a promutagenic lesion. The apparently more dangerous lesions like O^6-guanine- or O^4-thymine-methylation are much less frequent so that too high a total DNA damage (or CBI) results for its carcinogenic potency.

Fortunately, the total range of carcinogenic potencies and CBI cover more than 6 orders of magnitude so that even with an uncertainty of a factor of ten a correlation is found which could be useful for a toxicological evaluation of an initiating carcinogen. To our knowledge, there is no short-term test available which gives a similarly good correlation with TD_{50} values if initiators of all possible chemical classes are taken. A further improvement will be possible with more CBI determined not at the time of maximum binding but after a number of days after the administration of the chemical so that the repair processes are included in the evaluation.

8. IMPORTANCE OF INITIATING AND NON-GENOTOXIC CARCINOGENS IN RISK ASSESSMENT AND ORGAN SPECIFICITY

The last chapter revealed a surprisingly good predictive power of CBI for the carcinogenicity of known carcinogens and the question arises whether the determination of the y- and z-components of Fig. 3 are really needed if the x-axis already gives a reasonable correlation to carcinogenic potency. There are two main reasons why DNA binding is not enough to describe the potency of carcinogens.

(i) DNA binding can only detect initiating carcinogens. It is an important class for the high-risk situation of people who are exposed to large doses of initiating carcinogens due to special dietary habits or to special conditions at the workplace. The vast majority of people are, however, exposed only to low levels of initiating carcinogens but probably to large amounts of non-genotoxic carcinogens. It is therefore of paramount importance to find out how the unavoidable level of DNA damage is dealt with under conditions of an uptake of many modulating factors.

(ii) DNA binding in vivo at the time of maximum binding does not predict the organ specificity of an initiating carcinogen. It has for instance been found that liver DNA is always damaged to an appreciable extent by an initiating carcinogen although it is not always a target organ (Lutz, 1979). If only the persistent fraction of the DNA adducts is measured, an additional organ specific parameter is taken into account but this will still not be enough for a prediction of the susceptibility of a specific cell type. The following criteria must be met by a cell to be susceptible to a carcinogenic stimulus and it is obvious that DNA binding alone can only deal with the first two prerequisites:

- the cell must be able to produce or take up chemically reactive compounds
- the DNA repair mechanism must be inefficient or error-prone
- the cell must divide so that a promutagenic lesion is fixed before the damage is repaired
- the cell must be susceptible to the non-genotoxic influences of the post-initiation type which stimulate the expression of the constitutive or carcinogen-derived DNA damage towards a tumor

It is interesting to note that all strong carcinogens detected so far are of the initiating type and not of the non-genotoxic type. This finding should not, however, be taken to mean that non-genotoxic carcinogens are less important in the induction of tumors. The uptake of non-genotoxic carcinogens is probably much larger than the uptake of known initiating carcinogens. Nevertheless, we believe that it is of prime importance to reduce the exposure of humans to initiating carcinogens because of different dose-effect relationships to be expected for initiating vs. non-genotoxic carcinogens. Whereas true no-effect levels are not to be expected with initiating carcinogens (non-linearities are still possible and clearly of value) such could be the case with non-genotoxic carcinogens. On the other hand, epidemiological evidence suggests that about 50% of human cancers are due to dietary influences (Weisburger, 1979). The initiating carcinogens known so far to contaminate our diet cannot be responsible for this large effect if animal data and estimated doses are extrapolated to the situation of humans. It must therefore be concluded that either hitherto unknown groups of initiating carcinogens are responsible, or that strong synergisms take place, or that the non-genotoxic carcinogens lead to a strong expression of the total DNA damage present in our cells.

All of the above-mentioned reasons probably contribute to the discrepancy between known amounts of initiating carcinogens in our diet and the high incidence of dietary tumors as proposed by epidemiologists. Much weight should therefore be placed upon the elaboration of tests for the quantification of non-genotoxic carcinogens. It will also be of value to determine any possible initiating activity of large constituents in our diet. If, for instance, it can be shown that fat metabolites can bind to DNA even with a minute Covalent Binding Index, this might result in a remarkable total DNA damage if the high dose is accounted for.

If the hypothesis presented above, of a DNA damage constitutive to most cells, can be shown to be valid, this would considerably simplify the current difficulty for a proper definition of the term 'carcinogen'. It would no longer be necessary to propose a different biology for a tumor induction by genotoxic vs. non-genotoxic carcinogens. Nevertheless, it will still be important to distinguish between the two classes because dose-tumor relationships or cell-, tissue-, and species-specificity will clearly not be the same so that an extrapolation of animal data to the human situation will have to take into consideration the different mode of action.

REFERENCES

Aaron, C. S., Van Zeeland, A. A., Mohn, G. R., Natarajan, A. T., Knaap, A. G. A. C., Tates, A. D., and Glickman, B. W., 1980, Molecular dosimetry of the chemical mutagen ethyl methane-sulfonate: quantitative comparison of mutation induction in Escherichia coli, V79 chinese hamster cells and L5178Y mouse lymphoma cells, and some cytological results in vitro and in vivo, Mutat. Res., 69:201.

Ashby, J., and Styles, J. A., 1978, Does carcinogenic potency correlate with mutagenic potency in the Ames assay ? Nature, 271:452.

Bartsch, H., Malaveille, C., Camus, A.-M., Martel-Planche, G., Brun, G., Hautefeuille, A., Sabadie, N., Barbin, A., Kuroki, T., Drevon, C., Piccoli, C., and Montesano, R., 1980, Bacterial and mammalian mutagenicity tests: validation and comparative studies on 180 chemicals, in: Molecular and Cellular Aspects of Carcinogen Screening Tests, R. Montesano, H. Bartsch, and L. Tomatis, eds., IARC Scientific Publications No. 27:179, International Agency for Research on Cancer, Lyon.

Berenblum, I., and Shubik, P., 1947, A new, quantitative, approach to the study of the stages of chemical carcinogenesis in the mouse's skin, Brit. J. Cancer, 1:383.

Borchert, P., and Wattenberg, L. W., 1976, Inhibition of macro-molecular binding of benzo[a]pyrene and inhibition of neo-plasia by disulfiram in the mouse forestomach, J. Natl. Cancer Inst., 57:173.

Crouch, E., and Wilson, R., 1979, Interspecies comparison of carcinogenic potency, J. Toxicol. Envir. Health, 5:1095.

Eling, T. E., Wilson, A. G. E., Chaudhari, A., and Anderson, M. W., 1977, Covalent binding of an intermediate(s) in prostaglandin biosynthesis to guinea pig lung microsomal protein, Life Sciences, 21:245.

Jaggi, W., Lutz, W. K., and Schlatter, C., 1978, Covalent binding of ethinylestradiol and estrone to rat liver DNA in vivo, Chem.-Biol. Interact., 23:13.

Kelsey, M. I., and Pienta, R. J., 1979, Transformation of hamster embryo cells by cholesterol-α-epoxide and lithocholic acid, Cancer Lett., 6:143.

Lutz, W. K., and Schlatter, C., 1977, Saccharin does not bind to DNA of liver or bladder in the rat, Chem.-Biol. Interact., 19:253.

Lutz, W. K., 1979, In vivo covalent binding of organic chemicals
    to DNA as a quantitative indicator in the process of chemical
    carcinogenesis, Mutat. Res., 65:289.
Manthey, B., Lutz, W. K., L'Eplattenier, E., Schlatter, C., and
    Würgler, F., 1978, Binding of the carcinogens benzo(a)pyrene
    and 7,12-dimethylbenz(a)anthracene to Salmonella DNA as
    compared to the corresponding mutagenicity, Experientia,
    34:927.
Martin, C. N., and Ekers, S. F., 1980, Studies on the macro-
    molecular binding of benzidine, Carcinogenesis, 1:101.
Maugh II, T. H., 1978, Chemical carcinogens: how dangerous are
    low doses ? Science, 202:37.
Meselson, M., and Russell, K., 1977, Comparisons of carcinogenic
    and mutagenic potency, in: Origins of Human Cancer, Book
    C:1473, H. H. Hiatt, J. D. Watson, and J. A. Winsten, eds.,
    Cold Spring Harbor Laboratories.
Newbold, R. F., Warren, W., Medcalf, A. S. C., and Amos, J., 1980,
    Mutagenicity of carcinogenic methylating agents is associated
    with a specific DNA modification, Nature, 283:596.
Sega, G. A., Gee, P. A., and Lee, W. R., 1972, Dosimetry of the
    chemical mutagen ethyl methanesulfonate in spermatozoan DNA
    from Drosophila melanogaster, Mutat. Res., 16:203.
Sirover, M. A., and Loeb, L. A., 1976, Infidelity of DNA synthesis
    in vitro: screening for potential metal mutagens or carcino-
    gens, Science, 194:1434.
Stark, A. A., Essigmann, J. M., Demain, A. L., Skopek, T. R., and
    Wogan, G. N., 1979, Aflatoxin B_1 mutagenesis, DNA binding,
    and adduct formation in Salmonella typhimurium, Proc. Natl.
    Acad. Sci. USA, 76:1343.
Tannenbaum, S. R., 1979, Nitrate and nitrite: origin in humans,
    Science, 205:1332.
Weisburger, J. H., 1979, Mechanism of action of diet as a
    carcinogen, Cancer, 43:1987.

RECENT DEVELOPMENTS IN THE MECHANISMS OF DNA REPAIR AND IN THE ROLE

OF DNA REPAIR IN CHEMICAL CARCINOGENESIS AND CANCER CHEMOTHERAPY

J. J. Roberts

Institute of Cancer Research: Royal Cancer Hospital
Pollards Wood Research Station
Chalfont St.Giles, Bucks., HP8 4SP. England

INTRODUCTION

The past decade has seen a continuing accumulation of evidence
to indicate the likely importance of reactions with cellular DNA as
primary events in producing tumours in experimental animals following
administration of a variety of chemical agents. There is also
equally compelling evidence that the cytotoxic effects on cells in
culture or anti tumour effects in vivo by a number of cancer chemotherapeutic agents are similarly the result of reactions with DNA.
Strong support for these views comes from several frequently shared
properties of carcinogenic and cytotoxic agents. Firstly, measurements of the levels of reactions of a number of labelled compounds
with cellular macromolecules have revealed not only the ability of
almost all known carcinogens to react with DNA, but have shown
that, in general, only with DNA is the extent of reaction adequate
to account for some of the biological or biochemical effects
observed. Reactions with RNA were usually too low to inactivate
all types of RNA molecule and with protein were insufficient to inhibit enzyme activity. Moreover those RNA and protein molecules
that are inactivated would be expected to be replaced by subsequent
biosynthesis. Secondly, the initial biochemical effect produced in
dividing cells, after reaction with low doses of carcinogens or
cytotoxic agents, is a selective inhibition in the rate of DNA
synthesis, an effect consequent upon reaction with DNA which is
thereby inactivated as a template for DNA replication rather than
to the inactivation of enzymes involved in DNA replication. Only
after treatment with high doses of certain agents are other biosynthetic pathways inhibited. Thirdly, and as a consequence of
this ability to react with DNA, the majority of both carcinogens
and antitumour cytotoxic agents are powerful mutagens. However

1367

the validity of the above supposition of the importance of interactions with DNA has been particularly supported by studies on the
ability of specific DNA repair processes to modify the cytotoxic,
mutagenic or carcinogenic properties of agents in a variety of cell
and animal systems. It is now evident that lesions on DNA can be
removed by one of a number of excision repair processes. In addition
or alternatively, in some cells, unexcised base damage can be circumvented during DNA replication. This latter process, while not a
DNA repair process in the strict sense that the damage to the DNA
template has been removed, but only circumvented, has, nevertheless,
been referred to, somewhat inappropriately, as post replication
repair. The terms by-pass repair or replication repair have also
been used to describe the process in mammalian cells. For the
purposes of this article I shall use the phrase "replication on a
damaged template" (RDT) to avoid any implication that its mechanism
is currently understood.

EXCISION REPAIR MECHANISM

     Until very recently it could have been presumed that carcinogen-
induced adducts on mammalian cell DNA are removed by one of three
excision repair processes, depending on the nature of the product.
One of these presumed excision repair pathways, the classical nucleotide excision repair is based on what was thought to be the method
by which thymine dimers, induced by ultraviolet light, are removed
from both prokaryotic and eukaryotic cells by four enzymic steps
(fig.1b). The first step was thought to be phosphodiester chain
scission near to the thymine dimer by an endonuclease. This is
followed by exonuclease activity, which removes the damaged piece
of DNA in an oligonucleotide six or seven nucleotides long, and which
acts simultaneously with DNA polymerase activity that resynthesizes
the removed segment of DNA by so-called nonsemiconservative DNA
synthesis or, DNA repair synthesis. The reinserted sections can be
anything from 10 nucleotides to 3000 nucleotides in length, depending
on the organism and the nature of the damage. A 5'-exonuclease
activity is associated with the DNA polymerase activity of E.coli
and is specific for DNA duplexes. Three DNA polymerases have been
isolated from mammalian sources but they are all devoid of exonuclease activity. Finally, a polynucleotide ligase joins the 3'-
hydroxyl group of the reinserted section of DNA to the 5' end of the
pre-existing DNA. While the latter steps of this pathway would

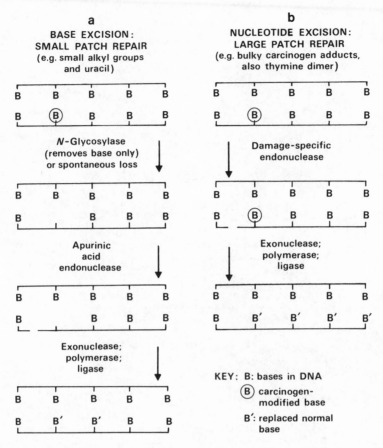

Fig. 1. Mechanisms of excision repair of chemically-damaged DNA.

appear to be correct, recent observations indicate that the first
step in the excision repair of UV-induced thymine dimers is probably
different, at least in some biological system, if not all, from that
originally proposed. It now seems fairly certain that the first
step is similar to that which occurs in another excision repair
process that has been discovered during the past few years to operate
on certain abnormal bases in DNA, such as uracil, hypoxanthine or
3-methyladenine.

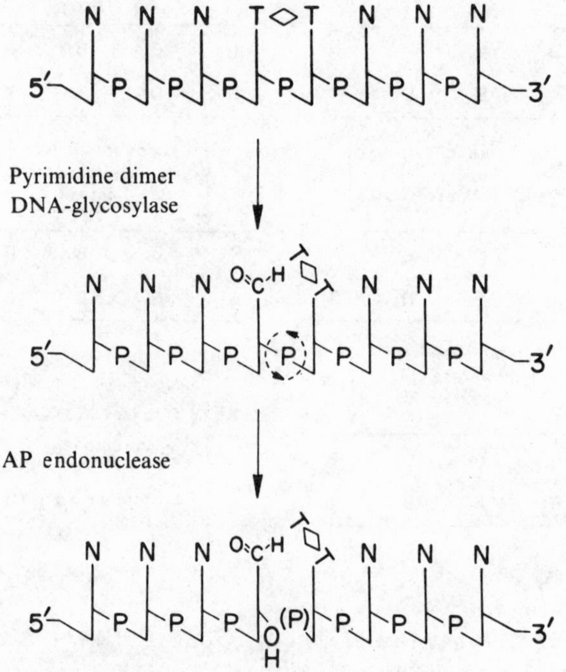

Fig. 2.   A two step model for excision repair of pyrimidine dimers
          (Haseltine et al. 1980).

This process has been called base excision repair and is illustrated in Fig 1a. The first step in this repair pathway is the rupture of the glycosidic bond between the chemically modified base (or abnormal base) and the sugar moiety, a process catalysed by one of a group of enzymes called N-glycosylases. The resultant apurinic (or apyrimidinic) site (AP) is then recognised by an endo-nuclease that cleaves the phosphodiester chain, and this is probably followed by simultaneous exonuclease and polymerase action to restore the original DNA structure as in the above discussed nucleotide repair pathway.

It now appears that the first step in the removal of a thymine dimer is also catalysed by a N-glycosylase that ruptures the bond between one of the thymines involved in the dimer and the sugar residue (Fig.2). The next step is then scission of the phosphodiester bond between the two dimerized pyrimidines. The apyrimidinic site is then repaired, as described above, by exonuclease and polymerase activity. The evidence for this new mechanism of repair of thymine dimers came from an analysis, by gel electrophoresis, of the length of DNA fragments produced when a sequenced region of the E.coli lac operon, containing pyrimidine dimers, was incubated with UV endo-nuclease activities from Micrococcus luteus. It was unexpectedly found by Haseltine et al (1980) that the enzyme-digested fragments were one nucleotide longer than would have been predicted by the existing model and the enzyme had cleaved a phosphodiester bond 5' to a pyrimidine dimer. Subsequently, these authors were able to separate the glycosylase activity from the AP endonuclease activity and show that it was specific for DNA containing pyrimidine dimers, and that the product of the reaction indeed contained AP sites. Finally, after treatment of the UV-DNA with pyrimidine dimer DNA glycosylase, free thymine could be released by photoreversal of the mixed thymine thymidylate dimer with either yeast photolyase of photolysis.

The similarity between the properties of the M. luteus endo-nuclease and the so-called T4 endonuclease V, the product of the V gene of bacterophage T4 soon led to the similar finding of evidence for pyrimidine dimer – DNA glycosylase activity in extracts of T4 V$^+$ – infected Escherichia coli but not in extracts from cells in-fected with the mutant (T4V) (Radany and Friedberg 1980). (Mutations in the V gene of T4 are associated with a marked increase in sensi-tivity to killing by UV irradiation but not to a variety of other forms of base damage to DNA). It remains, however, to be determined whether the V gene codes for an AP endonuclease activity that is physically associated with the pyrimidine dimer – DNA glycosylase activity.

It is now clearly a matter of considerable interest whether modified bases, other than those previously mentioned, and especially those resulting from reaction with bulky carcinogenic agents or DNA

cross linking agents, are similarly removed by glycosylase activity
and particularly in other organisms like E.coli or in mammalian cells.
Thus it may be noted that in E.coli the incision step is controlled
by three genes (uvrA, B and C) and, unlike the enzymes from M.luteus
or T4, the products of these genes recognize other bulky adducts in
DNA as well as the pyrimidine dimer.   Interestingly, it has now been
found that E. coli endonuclease III releases 5,6-hydrated thymines
as free bases and that the UV endonuclease induced by infection of
E.coli with bacteriophage T4, and which is specific for thymine
dimers also acts by the same mechanism as that described above for
the M.luteus UV endonuclease.   In contrast, uninfected E.coli
apparently does not excise pyrimidine dimers via a glycosylase
(Demple & Linn 1980).

        In human cells the excision of thymine dimers is controlled by
at least seven gene products, as indicated by the complementation
groups of the genetic disorder xeroderma pigmentosum (xp) that are
deficient in their ability to incise UV irradiation.   Again it would
seem that the products of at least some of these genes can recognize
some chemically-damaged DNAs as indicated by the inability of XP
cells to excise certain DNA bound adducts and the cross sensitivity
of many of these 'mutant' cells to both UV irradiation and the
chemical agent from which these adducts were derived.

        The question therefore now arises to what extent are there
overlapping pathways for the repair of thymine dimers in DNA and
for the repair of damage induced by a variety of chemical agents?
The original model for DNA excision repair proposed that an endonuclease recognized a common distortion of the DNA induced by these
various lesions.   While this simple model is no longer likely to be
valid there is not a clear consensus of opinion on the extent to
which the mechanisms for the repair of some chemical adducts and
thymine dimers do overlap.   The ability of both rodent and human
cells to excise hydrocarbon- (Dipple and Roberts 1977), or, acetyl-
aminofluorene- (Amacher et al 1977) induced lesions at a similar
rate, in contrast to the apparent decreased ability of rodent cells
compared with human cells to excise thymine dimers, was thought to
indicate the existence of different repair processes for UV irradiation and such chemically-induced damage.   Supportive evidence for
there being different mechanisms for repair of UV irradiation and
chemically-induced damage was initially obtained from a study of one
manifestation of DNA repair after treatment of cells with a combination of the two types of agent.   Thus, additive levels of unscheduled
DNA synthesis were reported to be produced when normal human cells
were treated simultaneously with doses of N-acetoxy-2-acetylamino-
fluorene and ultraviolet irradiation each of which individually
yielded a saturation level of unscheduled DNA synthesis (Ahmed and
Setlow, 1977).   However a similar experimental approach by Brown
et al (1979) gave a conflicting result and it was argued that the
dose of UV irradiation used by Ahmed and Setlow was not saturating,

and it was therefore concluded, on the basis of this and much further
evidence, that there exists overlap between the two repair pathways.

Conflicting reports also exist with regard to the ability of
both UV- and chemically-modified DNAs to act as substrates for a given
enzyme preparation. Thus an enzyme extract isolated from M. leuteus
that incises UV-irradiated DNA failed to incise DNA damaged by either
cis Pt(II)diammine dichloride (Fravel et al. 1978) or 7-bromomethyl-
benz(a)anthracene (7-BMBA) (Roberts & Rawlings unpublished results.
Tomilin et al.1976) even though UV repair-deficient human cells were
cross sensitive to these agents (Fravel et al. 1978; McCaw et al.
1978). Again, Hecht and Thielman (1978) have described the purification of an enzyme from M. leuteus that induces nicks into PM 2
supercoiled DNA that had been modified by treatment with high doses
of the 7-bromomethylbenz(a)anthracene, but not into that which had
been UV-irradiated. Moreover, the properties of this enzyme apparently differed from those of the thymine dimer specific endonuclease
from M.luteus that had been described previously (Tomilin et al.1976.
Kaplan et al. 1969) but what may now be presumed to be the glycosy-
lase-AP endonuclease mixture as discussed above. On the other hand
the extraction and purification of an enzyme from rat liver that
introduces single strand breaks into UV-irradiated DNA and into DNA
containing stably bound benz(a)anthracenylmethyl residues induced
by treatment with 7-BMBA had been reported previously (Maher et al.
1974; Van Lancker & Tomura 1974).

Conflicting reports also exist with regard to the properties of
enzymes or enzyme extracts obtained from E.coli. Earlier it was
reported that hydrocarbon residues in DNA, following its reaction
with the 7-BMBA congener, 7-bromomethyl-12-methylbenz(a)anthracene,
are removed as the substituted base by the action of a N-glycosylase
that was thought to be present in the enzyme preparation from E.coli
designated endonuclease II (Kirtikar et al.1975). On the other hand
Riazuddin and Lindahl (1978) found that similarly modified DNA is not
a substrate for, either the purified or the crude preparation of 3-
methyladenine-N-glycosylase obtained from E.coli. In summary, therefore, it remains to be unequivocably demonstrated, whether any
specific excision repair enzyme acts similarly on UV- and chemically-
damaged DNA to remove a modified base by glycosylase action.

INDUCIBLE EXCISION REPAIR PROCESSES

Prokaryotic Cells-Adaptation and SOS Repair

Recent experiments have demonstrated the existence of another,
as yet not fully defined, mechanism for removing certain specific
lesions in the DNA of E.coli and which is induced by treatment with
low doses of DNA damaging agents. Thus exposure of E.coli to sub-
lethal doses of N-methyl-N'-nitro-N-nitrosoguanidine (MNNG) resulted
in higher survival but in a lower mutation rate when subsequently

exposed to lethal concentrations of the same agent (Samson & Cairns 1977). The process has been called 'adaptation' and is clearly different from 'SOS repair'. The term SOS repair was coined by Radman (1974) to describe a complex group of inducible processes in E.coli which are co-ordinately regulated and which are dependent on recA+ and lexA genes. These responses include inhibition of post-irradiation DNA degradation, induced bacterial mutagenesis, fila-mentation and so-called 'Weigle reactivation phenomena'. The last response describes the process whereby UV-irradiated bacteriophage are able to survive better but yield a higher number of mutant plaques when plated on to previously lightly irradiated bacteria as compared with plating on to unirradiated cells. Much greater concentrations of MNNG were required to elicit the SOS functions than were required to produce the adaptive response. Furthermore while the adaptation is induced under conditions that do not generally alter cell growth or DNA synthesis, SOS functions are induced under conditions that block replication. Again, while UV-radiation and 4-nitroquinoline-1-oxide can induce the SOS function, they do not induce the adaptive response, and MNNG adapted cells are not more resistant to these agents, Jeggo et al (1977). On the other hand cells that have been pretreated with MNNG are more resistant to other alkylating agents like methyl methanesulphonate (MMS), ethyl methanesulphonate (EMS) and N-methyl-N-nitrosourea (MNU). Subsequent studies using mutants deficient in other known pathways of DNA repair for their ability to adapt to MNU indicated that the two parts of the adaptive response must, at least to some extent, be separable and require different gene products (Jeggo et al 1978). Adaptive resistance to the killing effects of MNNG seems to require a functional DNA polymerase I whereas resistance to the mutagenic effects can occur in polymerase I defic-ient strains; similarly, killing adaptation could not be observed in a dam 3 mutant, which was nonetheless able to show mutational adapta-tion. More recent studies now indicate that the adaptive response is a complex set of interacting processes that include a repair pro-cess capable of removing potential sites of mutation, and another process capable of inhibiting MNNG induction of error-prone repair (Schendel et al 1978). In particular it was found that E.coli cells that have been induced for the adaptive response, when subsequently exposed to MNNG, accumulate substantially less O^6-methylguanine in their DNA than control cells and as a consequence of its rapid re-moval from the cell's DNA (Schendel & Robbins 1978). The capacity of the cell for removal of O^6-methylguanine is limited, and ceases to function when too much alkylation has occurred. Thereafter, O^6-methylguanine starts to accumulate and the cells begin to develop mutations at a rate directly proportional to their rate of O^6-methyl-guanine accumulation. These observations lead to the suggestion that the molecules that carry out the initial rapid reaction, leading to the removal of O^6-methylguanine, are used up during their reaction with O^6-methylguanine, functioning only once. Molecular evidence for such a concept came from an in vitro study of the action of cell free extracts from adapted E.coli on DNA containing O^6-methylguanine residues (Karran et al 1979). The O^6-methylguanine was indeed found to disappear from alkylated DNA after incubation with a crude enzyme fraction from adapted cells, although no concomitant release of the methyl group or the alkylated base or nucleotide occurs. It was therefore concluded that this process was due to a DNA repair process, previously unrecognized, involving enzyme-catalysed structural alteration of the alkylated residue. The nature of this modification is still unclear but it was proposed that it could involve either transfer of the methyl group to a biologically less hazardous (non-miscoding) site in DNA, or ring opening of the alkylated residue to give a lesion more readily recognized by other DNA repair enzymes. After such enzymic modification of the alkylated DNA the labelled methyl group appears to be much more acid labile than when present as O^6-methylguanine. As it was not released from the ethanol insoluble precipitate by the action of pronase or RNase it was further concluded that it had not been transferred to either a protein or RNA molecule.

Further indication that the adaptation process involves the induction of new enzyme activity was afforded by studies using E.coli K12 mutants defective in the adaptive response (ada). Extracts from two such mutants were unable to induce the disappearance of detectable amounts of O^6-methylguanine from alkylated DNA even though the bacteria had been treated previously with MNNG.

These various findings on the repair of O^6-methylguanine can be contrasted with the earlier claim that a partially purified preparation of endonuclease II, acting as a glycosylase, could release O^6-methylguanine as the free base from DNA methylated by MNU (Kirtikar and Goldthwait 1974). However, no such activity was found either in extracts of adapted or non adapted E.coli (Karran et al 1979) by a purified 3-methyladenine-N-glycosylase preparation (Riazuddin and Lindahl 1978) or in extracts of M.luteus that were able to remove 3-alkylpurines from alkylated DNA (Shackleton et al.1978).

In Mammalian Cells

Several recent observations now suggest that an inducible process similar to that found in bacteria also operates in mammalian cells to remove O^6-methylguanine residues in alkylated DNA. Thus the removal of O^6-methylguanine from rat liver DNA is more rapid in animals that have had a prolonged exposure to low levels of dimethylnitrosamine (DMN) prior to a large dose of the agent (Montesano et al 1979, 1980). No increased loss of other DNA alkylation products, such as 7-methylguanine or 3-methyladenine, was observed under these conditions. The rate of removal of O^6-methylguanine was dependent on the dose of DMN used for pretreatment, with 2 mg/kg/day giving a maximal response (Montesano et al 1979). It was also dependent on the time of pretreatment, requiring at least 2 weeks of exposure to the same dose of DMN for the maximal effect. O^6-methylguanine was not lost more rapidly from the DNA of kidney or lung under these conditions (Montesano et al 1979). That this increased rate of loss of O^6-methylguanine is likely to be due to the induction or activation of an enzyme comes from the observations that, not only are extracts of rat liver able to cause the loss of O^6-methylguanine from alkylated DNA (Pegg 1978b), but also that the level of this activity increases with prior exposure of rats to DMN (Montesano et al 1980). The inducible process that removes O^6-methylguanine does not release free O^6-methylguanine from alkylated DNA, but instead, converts it into a form in which mild acid treatment releases methanol. A further similarity between the two inducible systems is that both cease to function when the level of alkylation induced by the challenging dose increases above a certain critical level. On the other hand important differences exist between the two systems in the time taken for the induction of this repair process, which is much longer in liver than in bacteria, and in its magnitude, which is much greater in bacteria than it is in rat liver. As found in bacteria, pre-treatment with other methylating (Montesano & Margison 1980) or ethylating (Margison et al. 1979) agents induced the excision process in rat liver. However, unlike in bacteria, the removal of O^6-methylguanine, formed from DMN could also apparently be induced in rat liver by pretreatment with a non alkylating agent, namely, acetylaminofluorene (Buckley et al 1979).

DNA REPAIR AND CARCINOGENESIS

Initial studies of alkylation-induced carcinogenesis attempted, unsuccessfully, to correlate either over-all alkylation of DNA or alkylation of the N-7 position of guanine residues in DNA (which for a number of different methylating agents is the major site of reaction in DNA) with the incidence of tumours in various organs of the rat (see Pegg 1977 and Margison & O'Connor 1979 for references). Subsequently, with the realization that although the major product of reaction with DNA (i.e. N-7 alkylguanine) was the same for a number of closely related alkylating agents, nevertheless the proportions of minor products of alkylation could differ, attention was focussed on these other products as possibly being of greater importance in the aetiology of alkylation-induced carcinogenesis. This approach was stimulated by the findings of Loveless (1969) that reaction with the O^6 position of guanine was associated with mutagenesis in T-even bacteriophage and of Gerchman and Ludlum (1973) that O-alkylguanine adducts could lead to base mispairing with thymine when incorporated into a template for RNA polymerase. It now seems that the reactivity of compounds can be correlated with mutagenicity and carcinogenicity in the sense that those agents that react by an unimolecular substitution reaction (S_N1) are the most potent mutagens and the most potent carcinogens. Moreover S_N1 reactivity has in turn been found to correlate with the ability of compounds to react more extensively with O atom sites in DNA including O^6-guanine O^2- and O^4-thymine and O atoms of phosphate groups. The association between reactions with the O^6 position of guanine and the mutagenic (rather than the cyto-

toxic) action of methylating agents has now been observed in Chinese
hamster cells (Peterson et al 1979; Newbold et al 1980) and in
Drosophila (Vogel & Natarajan). [It should, perhaps, be noted that
in other cell systems in vitro (Roberts et al 1971; Roberts 1978;
Day et al 1980) or animal tissues in vivo (Roberts 1980) reaction
with the O^6-position of guanine or the lack of repair of O^6-methyl-
guanine adducts (see later) has been associated with cytotoxic
effects.]

However, while the relative yields of O^6-alkylguanine could
explain differences in the carcinogenicity of certain closely related
compounds (Frei et al 1978), they did not offer a complete explanation for the tissue specificity exhibited by some alkylating agents,
since the target tissue was not necessarily that in which most O^6-
alkylguanine was produced. A likely explanation for the specificity
of carcinogenic agents was subsequently obtained when the stability
of O^6-alkylguanine residues in the DNA of target and non target
tissues was studied. The first indication of the importance of the
persistence of unrepaired lesions in DNA in inducing cancer came
from the observation of Goth and Rajewsky (1974). They found that
a single injection of N-ethyl-N-nitrosourea (ENU) to new born rats
resulted, specifically, in a high incidence of malignant neuroecto-
dermal tumours. While the respective elimination rates for 7-ethyl-
guanine or 3-ethyladenine were the same for a number of tissues the
rate of loss of O^6-ethylguanine from rat brain DNA was appreciably
slower than from rat liver or from other non target tissues. Other
single-dose studies have similarly shown a correlation between the
persistence of O^6-methylguanine, introduced by either MNU, DMN, 3-
methyl-1-phenyltriazine or 1,2-dimethylhydrazine and the principal
organ for tumour production (see Margison & O'Connor 1979 for references). On the other hand, other related studies which compared
the persistence of O^6-alkylguanine in the DNA of tissues of strains
of animals that differ in their carcinogenic response to a given
agent have not been supportive of this concept (den Engelse 1974;
Bücheler and Kleihues 1977; Cooper et al 1978). Moreover, persistence of O^6-methylguanine in the DNA of the brain of the mongolian
gerbil is not associated with an oncogenic effect in that organ
(Kleihues et al 1980). Again, while a single dose of DMN will induce
liver tumours in the hamster and O^6-methylguanine does in fact per-
sist in that tissue, nevertheless, it appears that lack of repair
of alkylated DNA is not uniquely associated with carcinogenesis in
the Chinese hamster since other tissues also fail to excise O^6-
methylguanine from their DNA (Margison et al 1980) and Chinese hamster cell lines in culture are, in general, unable to excise O^6-
methylguanine from DNA (Warren et al 1979).

As discussed earlier, the stability of O^6-alkylguanine in DNA
has also been examined after chronic exposure to an agent rather than
after a single-dose schedule. Such a protocol has been found to
result in either a persistence, or an induced and rapid removal of

this lesion in DNA, depending on the tissue or the species examined. In some instances, but not in others, the findings have indicated the importance of the persistence of lesions in DNA in inducing cancer. Thus weekly injections of MNU resulted in a greater accumulation of O^6-methylguanine in the DNA from brain, the principal target organ for MNU under these conditions, than in the DNA from kidney or from liver (Margison & Kleihues 1975). Chronic administration of DMN to rats, on the other hand, resulted in an accumulation of O^6-methyl-guanine in the DNA of lung and kidney tissues, which are not the target tissues under this dose schedule, but not in that of the liver in which the tumours arise (Margison et al 1977; Montesano et al 1979, 1980). Clearly, therefore, from all these apparently conflicting findings, factors other than simply the amount of O^6-methylguanine in the DNA of a tissue or its rate of elimination determines the carcinogenic response to a given agent. The physiological state of a tissue, which determines the degree of cell proliferation taking place, will also determine, not only the possible number of base-mis-pairing (mutagenic) events occurring in replicating DNA, but also the cytotoxic effect of these lesions in DNA if they modify, or act as a block to, DNA replication.

DNA EXCISION REPAIR AND CYTOTOXICITY

The role of removal of bulky chemical substituents from the DNA of eukarytic cells in alleviating the toxic effects of either acetyl-aminofluorene (Amacher et al 1977) or 7-bromomethylbenz(a)anthracene (McCaw et al 1978) was evident from studies using repair-deficient human cell lines derived from patients suffering from the sun sensi-tive skin condition, xeroderma pigmentosum. An alternative approach has been to follow the recovery of potentially lethal damage in cells held in the non-dividing, density-inhibited, G_O, state and show that this is accompanied by the loss of lesions on DNA introduced by either UV irradiation (Konze-Thomas et al 1979), cis platinum II diammine dichloride (Fravel & Roberts 1979; Pera & Roberts unpublished results), or N-acetoxy-2-acetylaminofluorene (Heflich et al 1979).

Recent observations on the sensitivity to MNNG of a number of human tumour and normal cell strains, or on their ability to repair MNNG damaged adenovirus 5, have indicated that while normal cells are generally proficient in repair of alkylation damage, the tumour strains may be either proficient or deficient in repair. (Day and Ziolkowski 1979; Day et al 1980a). The repair deficient strains (in contrast to repair proficient ones) were defective in the removal of O^6-methylguanine from DNA (albeit by an undefined mechanism), usually were more sensitive to the cell killing effect of the agent, and to the production of sister chromatid exchanges. They were given the Mer⁻ phenotype, (Day et al 1980b). The MNNG repair deficient tumour cells were also more sensitive to the bifunctional agents, 1-(2-chloroethyl)-1-nitrosoureas, which produce DNA interstrand crosslinks by a reaction sequence of two steps. The first step was proposed to be an addition of a chloroethyl group to a guanine 0^6-position of DNA (Kohn 1977) followed by a slow reaction of the chlorine atom of the newly formed 0^6-chloroethylguanine moiety with a nucleophilic site on the opposing DNA strand (Ewig and Kohn 1978). If excision repair of the initially formed, guanine substituted, mono-adducts occurs before the second, cross linking reaction takes place then fewer crosslinks will be present finally in the cellular DNA. Consistent therefore with their decreased ability to remove O-alkyl groups the drug-sensitive mer⁻ tumour cell strains were found to contain higher interstrand crosslinking levels than the mer⁺ strains (Erickson et al 1980a and b). Clearly therefore the phenotype of the tumour cells, whether mer⁺ or mer⁻, existing in a mer⁺ host will determine their response to this type of chemotherapeutic agent.

Evidence for the repair of alkylating agent-induced DNA inter-strand crosslinks has been available for many years (see Roberts 1978 for early refs.). More recently Ewig & Kohn (1977) have applied the sensitive alkaline elution technique to demonstrate not only loss of DNA crosslinks in mouse L1210 cells treated with nitrogen mustard and nitrosoureas, but also loss of DNA-protein crosslinks induced by these same agents at minimally toxic concentrations. This same technique of alkaline elution has now been used to provide evidence that lack of repair of DNA interstrand crosslinks, introduced by 1-(2-chloroethyl)-3-(4-methylcyclohexyl)-1-nitrosourea, in one human colon tumour cell line, compared with another line in which repair of crosslinks was detected during a 48 hr period, is likely to account for its differential sensitivity to this agent (Erickson et al 1978). However, there was no unequivocal indication that the drug had reacted with both cell lines initially to an equal de-gree; and it should also perhaps be noted that the technique of alkaline elution, while extremely sensitive, can only give an in-direct indication of the nature of the damage to cellular DNA.

REPLICATION OF DNA ON A DAMAGED TEMPLATE (RDT)
(post replication repair)

Despite the multiplicity of mechanisms for excising damage to DNA it has generally been found that the removal of chemical groups from the DNA of mammalian cells can often be a slow process and, under conditions of treatment with minimally toxic concentrations of agents, it can be incomplete before replication of the DNA takes place. Under these conditions it can frequently be observed that the rate of DNA synthesis is slowed down. Measurements of the size of DNA synthesised, both during a short labelling period immediately following treatment and during a subsequent longer chase period, clearly show that this inhibition in rate of DNA synthesis is due to an inhibition of either, replicon initiation (Painter 1977, 1978) or of chain elongation (see Roberts 1978 for refs.), or to both effects, depending on the nature of the agent. But, following treatment of cells with a low dose of an agent such as 7-bromomethyl-

benz(a)anthracene, it can be observed that the nascent DNA in treated
cells can, (given sufficient additional time for synthesis compared
with control cells), reach the size of the template DNA of treated
cells, even though DNA binding data shows this to contain unexcised
damage (Friedlos and Roberts 1978, 1979). Various models have been
proposed to account for the ability of cells to replicate past unexcised lesions in DNA, some of which are based on what is known to
occur in bacteria (fig.3). It has been proposed that the replicating
machinery halts for a time at a lesion, such as a thymine dimer, and
then resumes synthesis at some point beyond it. The resultant hypothetical 'gaps' in nascent DNA are then thought to be 'filled' by
recombinational exchanges (as have indeed been shown to occur in
bacteria) or by de novo DNA synthesis. However not all the facts
are consistent with either of these models. Other evidence suggests
that the replicating machinery stops initially at the lesion and
then synthesizes past it continuously by a process of branch migration
(Fig. 3) but, again subsequent studies did not confirm this model
either. Apparent discontinuities in DNA would result from inhibition
of replicon initiation, in addition to the block to strand elongation.
Currently available techniques do not permit us to distinguish unequivocally between these various possible mechanisms by which cells
replicate past lesions induced by either UV irradiation or a variety
of cytotoxic or carcinogenic chemicals. It should be noted that this
process of RDT is not, strictly speaking, a repair process since
damage in DNA is not being removed. Nevertheless, that it operates
to protect cells from the damaging effect of an agent is evident from
the fact that in rodent cells, but not, in general, in human cells,
it can be adversely influenced by post treatment incubation in the
presence of caffeine. This caffeine-induced inhibition of elongation
of DNA results in a marked increase in cell killing and chromosome
damage (see Roberts 1978 for refs.). Some recent indirect evidence
suggests that caffeine can, in addition to the above effect, also
reverse the agent-induced block to replicon initiation in a way that
results in aberrantly synthesised DNA (Murnane et al 1980).

CONCLUSIONS

During the past decade an increasing number of cytotoxic and
carcinogenic chemicals have been shown to be able to react with
cellular DNA, either directly or following metabolic activation
to reactive electrophiles. Invariably such reactions are accompanied by some manifestations of the currently characterised DNA
repair processes. In some few instances it has been possible to
correlate the extent of repair of a specific lesion in DNA with the
cytotoxic or carcinogenic response. In general, these observations
have reinforced the view that DNA is likely to be the principal
target for both types of agent. It is also apparent that our knowledge of the mechanisms by which cells remove lesions from their
DNA is still fragmentary. Only one example is currently available
of an enzyme, namely 3 methyladenine-N-glycosylase that can speci-

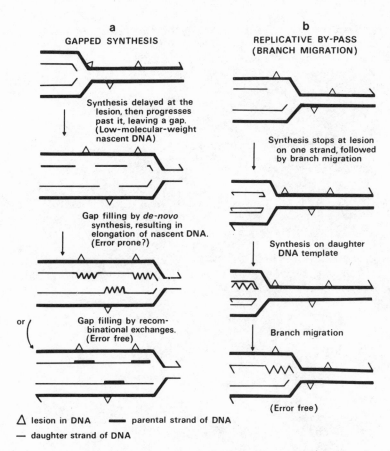

a
GAPPED SYNTHESIS

b
REPLICATIVE BY-PASS (BRANCH MIGRATION)

Synthesis delayed at the lesion, then progresses past it, leaving a gap. (Low-molecular-weight nascent DNA)

Gap filling by *de-novo* synthesis, resulting in elongation of nascent DNA. (Error prone?)

or

Gap filling by recombinational exchanges. (Error free)

Synthesis stops at lesion on one strand, followed by branch migration

Synthesis on daughter DNA template

Branch migration

(Error free)

△ lesion in DNA   ━ parental strand of DNA
─ daughter strand of DNA

Fig. 3. Models for replication of DNA on a chemically-damaged template (RDT)-post replication repair.

fically remove a chemically-modified base from DNA. Moreover, while
it is also evident that replication of DNA can proceed on a damaged
template, it is not known by what mechanism this is accomplished or
to what extent replication errors occur. For some specific types
of base damage it has been possible to demonstrate by the use of
model systems that replication results in base mispairing and hence
the introduction of potentially mutagenic lesions; and such events
have been implicated in the aetiology of nitrosamine-induced carcinogenesis. However it remains to be established whether this is a
generally applicable role for damage to DNA intorduced by other carcinogenic agents. Other data strongly indicate the vital, and possibly additional, role of DNA mediated toxicity which is characteristic
of many carcinogenic agents (Roberts 1980).

REFERENCES

Amacher, D. E., Elliott, J. A. and Lieberman, M. W., 1977, Differences
        in the removal of acetylaminofluorene and pyrimidine dimers
        from the DNA of cultured mammalian cells, Proc. Nat. Acad.
        Sci., 74:1553.
Ahmed, F. E. and Setlow, R. B., 1977, Different rate limiting steps
        in excision repair of ultraviolet- and N-acetoxy-2-acetyl-
        aminofluorene damaged DNA in normal human fibroblasts, Proc.
        Nat. Acad. Sci. (U.S.A), 74:1548.
Brown, A. J., Fickel, T. H., Cleaver, J. E., Lohman, P. H. M., Wade,
        M. H. and Waters,R., 1979, Overlapping pathways for repair
        of damage from ultraviolet light and chemical carcinogens in
        human fibroblasts, Cancer Res., 39:2522.
Bücheler, J. and Kleihues, P., 1977, Excision of O^6-methylguanine
        from DNA of various mouse tissues following a single injec-
        tion of N-methyl-N-nitrosourea, Chem.-Biol. Interactions,
        16:325.
Buckley, J. D., O'Connor, P. J. and Craig, A. W. 1979, Pretreatment
        with acetylaminofluorene enhances the repair of O^6-methyl-
        guanine, Nature, 281:403.
Cooper, H. K., Bücheler, J. and Kleihues, P., 1978, DNA alkylation
        in mice with genetically different susceptibility to 1, 3-
        dimethylhydrazine-induced colon carcinogenesis, Cancer Res.,
        38:3063.
Day III, R. S. and Ziolkowski, C. H. J., 1979, Human brain tumour
        cell strains with deficient host-cell reactivation of N-
        methyl-N'-nitro-N-nitrosoguanidine-damaged adenovirus 5,
        Nature, 279:797.
Day III, R. S., Ziolkowski, C. H. J., Scudiero, D. A., Meyer, S. A.
        and Mattern, M. R., 1980a, Human tumor cell strains defective
        in the repair of alkylation damage, Carcinogenesis 1:21.
Day III, R. S., Ziolkowski, C. H. J., Scudiero, D. A., Meyer, S. A.,
        Lubiniecki, A., Girardi, A., Galloway, S. M. and Bynum, G.
        D., 1980b, Defective repair of alkylated DNA by human tumor
        and SV40 transformed human cell strains, Nature in press.

Demple, B. and Linn, S., 1980, DNA-N-glycosylases and UV repair, Nature, 287:203.

Den Engelse, L. 1974, The formation of methylated bases in DNA by dimethylnitrosamine and its relation to differences in the formation of tumours in the liver of GR and C3Hf mice, Chem.-Biol. Interactions, 8:329.

Dipple, A. and Roberts, J. J., 1977, Excision of 7-bromomethylbenz-(a)anthracene-DNA adducts in replicating mammalian cells, Biochemistry, 16:1499.

Erickson, L. C., Bradley, M. O., Ducore, J. M., Ewig, R. A. G. and Kohn, K. W., 1980a, DNA crosslinking and cytotoxicity in normal and transformed human cells treated with antitumor nitrosoureas, Proc. Nat. Acad. Sci., 77:467.

Erickson, L. C., Laurent, G., Sharkey, N. A. and Kohn, K. W., 1980b, DNA interstrand crosslinking and cytotoxicity due to treatment of human tumor cells with 1(-2 chloroethyl)-1-nitrosourea: dependence on the Mer methylation repair function, Nature in press.

Erickson, L. C., Osieka, R. and Kohn, K. W., 1978, Differential repair of 1(-2 chloroethyl)-3-(4-methylcyclohexyl)-1-nitrosourea-induced DNA damage in two human colon tumor cell lines, Cancer Res., 38:802.

Ewig, R. A. G. and Kohn, K. W., 1977, DNA damage and repair in mouse leukemia L1210 cells treated with nitrogen mustard, 1, 3-bis(2-chloroethyl)-1-nitrosourea, and other nitrosoureas, Cancer Res. 37:2114.

Ewig, R. A. G. and Kohn, K. W., 1978, DNA-protein crosslinking and DNA interstrand crosslinking by haloethyl nitrosoureas in L1210 cells, Cancer Res. 38:3197.

Fraval, H. N. A., Rawlings, C. J. and Roberts, J. J., 1977, Increased sensitivity of UV-repair deficient human cells to DNA bound platinum products which unlike thymine dimers are not recognized by an endonuclease extracted from Micrococcus luteus, Mutation Res., 51:121.

Fraval, H. N. A. and Roberts, J. J., 1979, Excision repair of cis-diamminedichloroplatinum(II)-induced damage to DNA of Chinese hamster cells, Cancer Res., 39:1793.

Frei, J. V., Swenson, D. H., Warren, W. and Lawley, P. D., 1978, Alkylation of deoxyribonucleic-acid in various organs of C57Bl mice by the carcinogens N-methyl-N-nitrosourea and N-ethyl-N-nitrosourea and ethyl methanesulphonate in relation to the induction of thymic lymphoma. Some applications of high pressure liquid chromatography, Biochem, J., 174:1031

Friedlos, F. and Roberts, J. J., 1978a, Caffeine-inhibited repair in 7-bromomethylbenz(a)anthracene-treated Chinese hamster cells: formation of breaks in parental DNA and inhibition of ligation of nascent DNA as a mechanism for enhancement of lethality and chromosome damage, Mutation Res., 50:263.

Friedlos, F. and Roberts, J. J., 1978b, Caffeine inhibits excision of 7-bromomethylbenz(a)anthracene-DNA adducts from exponen-

tially growing but not from stationary phase Chinese hamster
    cells, Nucleic Acids Res., 5:4795.

Gerchman, L. L. and Ludlum, D. B., 1973, The properties of O^6-methyl-
    guanine in templates for RNA polymerase, Biochem. Biophys.
    Acta., 308:310.

Goth, R. and Rajewsky, M. F., 1974, Persistence of O^6-ethylguanine
    in rat brain DNA: correlation with nervous system-specific
    carcinogenesis, Proc. Nat. Acad. Sci., U.S.A., 71:639.

Haseltine, W. A., Gordon, L. K., Lindan, C. P., Grafstrom, R. H.,
    Shaper, N. L. and Grossman, L., 1980, Cleavage of pyrimidine
    dimers in specific DNA sequences by a pyrimidine dimer DNA-
    glycosylase of M. luteus., Nature, 285:634.

Hecht, R. and Thielmann, H. W., 1978, Purification and characteriza-
    tion of an endonuclease from Micrococcus luteus that acts on
    depurinated and carcinogen-modified DNA, Eur. J. Biochem.,
    89:607.

Heflich, R. H., Hazard, R. M., Lommel, L., Scribner, J. D., Maher,
    V. M. and McCormick, J. J., 1980, A comparison of the DNA
    binding, cytotoxicity and repair synthesis induced in human
    fibroblasts by reactive derivatives of aromatic amide car-
    cinogens, Chem.-Biol. Interactions, 29:43.

Jeggo, P., Defais, M., Samson, L. and Schendel, P., 1977, An adaptive
    response of E.coli to low levels of alkylating agents: com-
    parison with previously characterised DNA repair pathways,
    Molec. Gen. Genet., 157:1.

Jeggo, P., Defais, M., Samson, L. and Schendel, P., 1978, The adaptive
    response of E.coli to low levels of alkylating agent: The
    role of pol A in killing adaptation, Molec. Gen. Genet.,
    162:299.

Kaplan, J. C., Kushner, S. R. and Grossman, L., 1969, Enzymatic re-
    pair of DNA. 1. Purification of two enzymes involved in the
    excision of thymine dimers, Proc. Nat. Acad. Sci., U.S.A.
    63:144.

Karran, P., Lindahl,T. and Griffin, B., 1979, Adaptive response to
    alkylating agents involves alteration in situ of O^6-methyl-
    guanine residues in DNA, Nature, 280:76.

Kirtikar, D. M. Dipple, A. and Goldthwait, D. A., 1975, Endonuclease
    II of Escherichia coli: DNA reacted with 7-bromomethyl-12-
    methylbenz(a)anthracene as a substrate, Biochemistry, 14:5548.

Kirtikar, D. M. and Goldthwait, D. A., 1974, The enzymatic release
    of O^6-methylguanine and 3-methyladenine from DNA reacted with
    the carcinogen N-methyl-N-nitrosourea, Proc. Nat. Acad. Sci.,
    71:2022.

Kleihues, P., Bamborschke, S. and Doerjer, G., 1980, Persistence of
    alkylated DNA bases in the mongolian gerbil (Meriones un-
    guiculatus) following a single dose of methylnitrosourea,
    Carcinogenesis 1:111.

Kohn, K. W., 1977, Interstrand crosslinking of DNA by 1,3-bis-(2-
    chloroethyl)-1-nitrosourea and other 1-(2-haloethyl)-1-nitro-
    soureas, Cancer Res., 37:1450.

Konze-Thomas, B., Levinson, J. W., Maher, V. M. and McCormick, J. J., 1979, Correlation among the rates of dimer excision, DNA repair replication, and recovery of human cells from potentially lethal damage induced by ultraviolet radiation, Biophys. J., 28:315.

Loveless, A, 1969, Possible relevance of O-6-alkylation of deoxyguanosine to the mutagenicity and carcinogenicity of nitrosamines and nitrosamides, Nature, 223:206.

Maher, V. M., Douville, D., Tomura, T. and Van Lancker, J. L., 1974, Mutagenicity of reactive derivatives of carcinogenic hydrocarbons, evidence of DNA repair, Mut. Res., 23:113.

McCaw, B. A., Dipple, A., Young, S. and Roberts, J. J., 1978, Excision of hydrocarbon-DNA adducts and consequent cell survival of normal and repair defective cells, Chem.-Biol. Interactions, 22:139.

Margison, G. P., Craig, A. W., Brésil, H., Curtin, N., Snell, K. and Montesano, R., 1979, Chronic administration of N-nitrosamines enhances the removal of O^6-alkylguanines from rat liver DNA in vivo, Br. J. Cancer, 40:815.

Margison, G. P. and Kleihues, P., 1975, Preferential accumulation of O^6-methylguanine in rat brain DNA during repetitive administration of N-methyl-N-nitrosourea, Biochem. J., 148:521.

Margison, G. P., Margison, J. M. and Montesano, R., 1977, Accumulation of O^6-methylguanine in non-target tissue deoxyribonucleic acid during chronic administration of dimethylnitrosamine, Biochem. J., 165:463.

Margison, G. P. and O'Connor, P. J., 1979, Nucleic acid modification by N-nitroso compounds, In Grover P. L. ed. Chemical Carcinogens and DNA Vol.1 pp. 111-159, CRC Press. Florida.

Margison, G. P., Swindell, J. A., Ockey, C. H. and Craig, A. W., 1980, The effects of a single dose of dimethylnitrosamine in the Chinese hamster and the persistence of DNA alkylation products in selected tissues, Carcinogenesis 1:91.

Montesano, R., Brésil, H. and Margison, G. P., 1979, Increased excision of O^6-methylguanine from rat liver DNA after chronic administration of dimethylnitrosamine, Cancer Res., 39:1798.

Montesano, R., Brésil, H., Planche-Martel, G., Margison, G. P. and Pegg, A. E., 1980, Effect of chronic treatment of rats with dimethylnitrosamine on the removal of O^6-methylguanine from DNA, Cancer Res., 40:452.

Montesano, R. and Margison, G. P., 1980, Modulation of repair of DNA damages induced by nitrosamines, In:- Carcinogenesis: Fundamental mechanisms and environmental effects, Eds. Prof. B. Pullman & Prof. P. O. P. TS'O.

Murnane, J. P., Byfield, J. E., Ward, J. F. and Calabro-Jones, P., 1980, Effects of methylated Xanthines on mammalian cells treated with bifunctional alkylating agents, Nature, 285:326.

Newbold, R. F., Warren, W., Medcalf, A. S. C. and Amos, J., 1980, Mutagenicity of carcinogenic methylating agents is associated with a specific DNA modification, Nature (London), 283:596.

Painter, R. B., 1977, Inhibition of initiation of Hela cell replicons by methyl methanesulphonate, Mutation Res. 42:299.

Painter, R. B., 1978, Inhibition of replicon initiation by N-nitroquinoline-1-oxide,adriamycin, and ethyleneimine, Cancer Res., 38:4445.

Pegg, A. E., 1977, Formation and metabolism of alkylated nucleosides: possible role in carcinogenesis by nitroso compounds and alkylating agents, Adv. Cancer Res., 25:195.

Pegg, A. E., 1978, Enzymatic removal of O^6-methylguanine from DNA by mammalian cell extracts, Biochem. Biophys. Res. Commn., 84:166.

Peterson, A. R., Peterson, M. and Heidelberger, C., 1979, Oncogenesis, mutagenesis, DNA damage and cytotoxicity in cultured mammalian cells treated with alkylating agents, Cancer Res., 39:131.

Radany, E. H. and Friedberg, E. C., 1980, A pyrimidine dimer-DNA glycosylase activity associated with the v gene product of bacteriophage T4, Nature, 286:182.

Radman, M., 1974, In molecular and environmental aspects of mutagenesis, eds. L. Prakash, F. Sherman, M. W. Miller, C. W. Lawrence, H. W. Taber, pp.128-142 Springfield Ill.

Riazuddin, S. and Lindahl, T., 1978, Properties of 3-methyladenine-DNA glycosylase from Escherichia coli, Biochemistry, 17:2110.

Roberts, J. J. 1978, Repair of DNA modified by cytotoxic, mutagenic and carcinogenic chemicals. In Advances in Radiation Biology, 7:212.

Roberts, J. J. 1980, Carcinogen-induced DNA damage and its repair, Brit. Med. Bull., 36:25.

Roberts, J. J., Pascoe, J. M., Plant, J. E., Sturrock, J. E. and Crathorn, A. R., 1971, Quantitative aspects of the repair of alkylated DNA in mammalian cells. The effects on Hela and Chinese hamster cell survival of alkylation of cellular macromolecules, Chem.-Biol. Interactions, 3:29.

Samson, L. and Cairns, J., 1977, A new pathway for DNA repair in Escherichia coli, Nature, 267:281.

Schendel, P. F., Defais, M., Jeggo, M., Samson, P. and Cairns, J., 1978, The mechanism of mutagenesis and repair in Escherichia coli exposed to low levels of simple alkylating agents, J. Bacteriol. 135:466.

Schendel, P. F. and Robins, P. E., 1978, Repair of O^6-methylguanine in adapted Escherichia coli, Proc. Nat. Acad. Sci., 75:6017.

Shackleton, J., Warren, W. and Roberts, J. J., 1979, The excision of N-methyl-N-nitrosourea-induced lesions from the DNA of Chinese hamster cells as measured by the loss of sites sensitive to an enzyme extract that excises 3-methylpurines but not O^6-methylguanine, Eur. J. Biochem., 97:425.

Tomilin, V., Paveltchuk, E. B. and Mosevitskaya, T. V., 1976, Substrate specificity of the ultraviolet-endonuclease from Micrococcus luteus: endonucleolytic cleavage of depurinated DNA, Eur. J. Biochem., 69:265.

Van Lancker, J. L. and Tomura, T., 1974, Purification and properties of a mammalian repair endonuclease, Biochim. Biophys. Acta.,

353:99.
Vogel, E. and Nataranjan, A. T., 1979, The relation between reaction
        kinetics and mutagenic action of monofunctional alkylating
        agents in higher eukaryotic systems, Mutation Res., 62:55.
Warren, W., Crathorn, A. R. and Shooter, K. V., 1979, The stability
        of methylated purines and of methylphosphotriesters in the
        DNA of V79 cells after treatment with N-methyl-N-nitrosourea,
        Biochim. Biophys. Acta., 563:82.

THRESHOLD LEVELS IN TOXICOLOGY:

SIGNIFICANCE OF INACTIVATION MECHANISMS

H. Greim, U. Andrae, W. Göggelmann, S. Hesse,
L.R. Schwarz and K.-H. Summer

Department of Toxicology, Gesellschaft für Strahlen-
und Umweltforschung, 8042 Neuherberg, FRG

SUMMARY

Metabolic inactivation of chemicals may prevent toxic effects
of reactive intermediates when present at low levels whereas inactivation may be overcome at high levels changing dose-effect relation. This is demonstrated in various in vitro test systems:
a) Monooxygenase-mediated metabolism causes formation of reactive
oxygen species which induce DNA repair in lymphoblastoid cells.
DNA damage is suppressed in the presence of glutathione (GSH),
catalase or superoxide dismutase. b) Chloroprene is mutagenic in
Salmonella typhimurium but not carcinogenic, possibly due to inactivation by GSH-conjugations. c) Chlorodinitrobenzene is not mutagenic in Salmonella typhimurium in the presence of GSH. However
it is increasingly mutagenic at concentrations exceeding those of
the GSH. d) Suppression of glucuronidation and sulfation in
isolated hepatocytes highly increases irreversible binding of
naphthalene. It is concluded that information on the metabolism
of chemicals is essential for interpretation of toxicity studies
in animals and their relevance to man.

INTRODUCTION

During this meeting there have been several presentations on
metabolic inactivation systems, such as hydrolase activity, glucuronidation, and GSH-conjugation, which are present in cellular or
subcellular systems. Their main function has been reported to be
the metabolic inactivation of reactive intermediates found during
Phase I activation of chemicals. Moreover, catalase and gluta-
thione-peroxidase have been reported to protect the cell from
cytotoxicity of reactive oxygen species. The consequences of such

detoxification effects have been demonstrated mainly in the Ames
test in which mutagenicity of test compounds has been reduced.
The significance of metabolic inactivation processes is further
exemplified by presenting several examples in which information on
these metabolic inactivation systems has helped to clarify the
results of animal studies on toxic or genotoxic effects of chemicals.

METHODS

1. DNA-repair in human lymphoblastoid cells[1]

Cultures of NC37 BaEV cells were incubated in the presence of
1 uM 5-fluorodeoxyuridine and 10 uM 5-bromodeoxyuridine for 1 h.
After addition of 2.5 mM hydroxyurea, the cells were incubated for
3 h with NADPH, liver microsomes (0.5 mg protein/ml) of Clophen A
50-treated rats or phenobarbital-pretreated mice, and the chemical.
Thereafter the cells were labeled with 10 u Ci (^{3}H)-thymidine
(40 Ci/mmol) for 4.5 h, lysed in 0.5% sodium dodecylhydrogensul-
fate, digested with proteinase K (50 ug/ml) and centrifuged in
alkaline CsCl-gradients. The location of the tritium label at
normal density, coincident with the absorbance peak, is indicative
of repair replication.

2. Plate incorporation assay

Nutrient broth culture (8 g nutrient broth and 5 g NaCl per
liter) were inoculated with one of the Salmonella typhimurium
strains TA 1535, TA 100, TA 1538 and TA 98 obtained from frozen
stock cultures. The mutagenicity test was performed according to
Ames et al.[2,3] by using 2.1-2.5 x 10^8 bacteria per plate. The
test chemical dissolved in 10 ul DMSO, 2 ml molten top agar and
0.1 ml bacterial culture were poured immediately onto minimal
glucose agar plates and incubated at 37° for 70-72 h.

3. Determination of glutathione

Samples of the incubates were deproteinized at 0° by perchloric acid at 10% final concentration. After 4 s ultrasonication
and centrifugation the pH of the supernatant was adjusted to 5-6
with 1.65 mM potassium carbonate in 1 M triethanolamine-HCl.
Total glutathione was assayed according to Tietze.[4]

4. Covalent binding in isolated hepatocytes

Hepatocytes were isolated from Sprague-Dawley rats as described recently[5] only using collagenase for 10-15 min. For
details for incubation, extraction and determination of irreversibly bound radioactivity see Hesse et al.[6].

RESULTS

Induced DNA repair due to monooxygenase-dependent H_2O_2 formation

In several presentations during this meeting formation of hydrogen peroxide during monooxygenase functions has been discussed. It is well known that hydrogen peroxide (H_2O_2) effects many kinds of DNA lesions[7-9] and induces chromosome aberrations and unscheduled DNA-synthesis in human fibroblasts.[10]

Recently we demonstrated that reactive oxygen species, which are formed during monooxygenase-mediated metabolism of chemicals, induce DNA repair activity in the human lymphoblastoid cell line NC37 BaEV.[1] This was observed in the course of measuring the induction of DNA repair replication by mutagens requiring metabolic activation. Mutagens such as benzo(a)pyrene and dimethylnitrosamine induce repair activity (Fig. 1). However, in the control

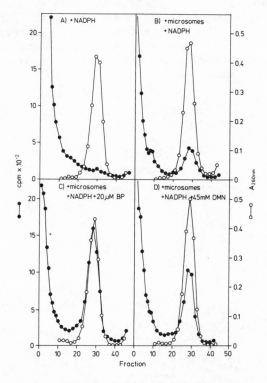

Fig. 1.   Repair replication in NC37 BaEV cells in the presence of
A: NADPH; B: microsomes + NADPH; C: microsomes + NADPH +
20 uM benzo(a)pyrene; D: microsomes + NADPH + 45 mM
dimethylnitrosamine.

experiment without the mutagen a marked incorporation of repair
label was observed. The incubate was comprised of washed microsomes of PB-treated mice, NADPH and hydroxyurea which is used to
suppress semiconservative DNA systhesis. When this inhibitor was
omitted, no repair replication occurred. Hydroxyurea-induced
repair was dependent on the presence of microsomes and NADPH, and
was reduced to 50% by SKF 525-A. This highly indicates that
hydroxyurea-induced repair is a consequence of microsomal monooxygenase-mediated metabolism. Since addition of catalase, glutathione and superoxide dismutase prevents hydroxyurea-induced DNA
repair, H_2O_2 or reactive oxygen species deriving from H_2O_2
are likely to be the genotoxic agents. As H_2O_2 production
generally occurs during monooxygenase functions[11,12] we investigated the effect of ethylmorphine. Accordingly, ethylmorphine
induced DNA repair in the test system but the effect was prevented
in the presence of catalase (Fig. 2). Other in vitro cell systems,
e.g. human A549 lung tumor cells, apparently have a sufficient
H_2O_2-inactivation capability as these systems did not show
enhanced repair activity. The precise mechanisms by which reactive
oxygen species are trapped in vivo remain unknown. GSH peroxidase
is likely to be involved. It is evident, however, that efficient
H_2O_2-inactivation must be present in vivo to prevent reactive
oxygen species from exceeding a no-effect concentration and thus
preventing genotoxic effects.

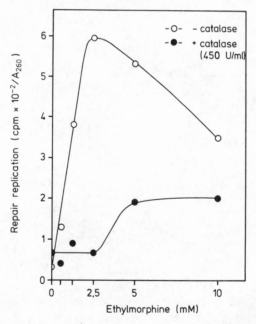

<u>Fig. 2</u>: Repair replication in NC37 BaEV cells in the presence of
microsomes, NADPH and ethylmorphine.

Chloroprene: Mutagenicity but no carcinogenicity due to GSH conjugation

Further evidence for no-effect levels due to metabolic inactivation is given by toxicity studies on chloroprene.

Chloroprene is a reactive chemical which is widely used in the manufacturing of the synthetic rubber neoprene. Chloroprene has been suggested to be responsible for the increased incidence of skin and lung cancer in workers exposed to the chemical. However, the data are questionable and no carcinogenic effects of chloroprene have been noted in animal studies to date.[13,14] Bartsch et al.[15,16] have, however, demonstrated a slight mutagenicity of chloroprene in Salmonella typhimurium strains without metabolic activation. Mutagenicity was increased about 3-fold in the presence of S-9 fractions. These observations were considered to reflect the probable formation of an epoxide intermediate of chloroprene.

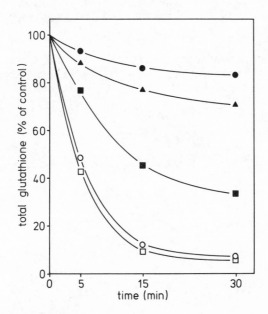

Fig. 3: Glutathione depletion in isolated rat hepatocytes by chloroprene. Data are from one representative experiment out of three with different cell preparations. Control level of cellular GSH amounted to 19.9 ± 2.1 nmol/mg cell protein. Closed symbols: Hepatocytes (8 mg cell protein/ml) of untreated animals incubated with 0.5 mM (●-●), 1.0 mM (▲-▲) and 3.0 mM (■-■) chloroprene. Open symbols: Hepatocytes from either phenobarbital- (○-○) or Clophen A 50- (□-□) pretreated animals incubated with 3 mM chloroprene.

Why was chloroprene shown to be mutagenic but not carcinogenic?

In analogy to vinylchloride and 1'1-dichloroethylene, a subsequent detoxification of the chloroprene by conjugation with glutathione has been suggested.[17,18] Since in vitro mutagenicity
test procedures do not include sufficient glutathione,[19] the
apparent discrepancy of mutagenicity in vitro without producing
carcinogencity in the animal may be explainable. We tested the
possible involvement of glutathione in inactivation in isolated
rat hepatocytes and in the whole animal.[20]

Similar to the rat liver, cellular glutathione decreased in
isolated rat hepatocytes to about 50% of the control values within
15 min in the presence of 3 mM chloroprene (Fig. 3). This depletion was dose-dependent and increased with time. In hepatocytes
of animals pretreated with phenobarbital or Clophen A 50, 3 mM of
chloroprene almost completely depleted glutathione within 30 min.

This strongly indicates that a Phase I reaction, presumably
epoxide formation which is enhanced after Clophen A 50 pretreatment, preceeds glutathione conjugation.

Glutathione-dependent detoxification in the animal was further
verified by determining urinary thioether excretion comprised of
GSH conjugates and mercapturic acids. Chloroprene administration

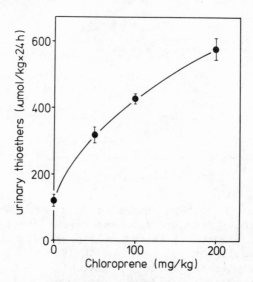

Fig. 4: Chloroprene-dependent excretion of thioethers in the
urine of rats. Chloroprene was administered in olive oil
by stomach tube. Data represent means ± S.D. from 4
animals.

to rats resulted in a dose-dependent increase in the excretion of
urinary thioethers (Fig. 4). This increase was reversible and
completed within 24 hours at all dose levels administered. At dose
levels of 50 and 100 mg chloroprene/kg, the additional excretion
of urinary thioethers was almost 200 and 450 uMol/kg daily, respectively.

It is to be noted that no linear dose-response relationship
was observed in thioether excretion, indicating that at higher
doses of chloroprene the availability of cellular glutathione
becomes rate-limiting. Thus, a dose level at which high concentrations of chloroprene overcome GSH-inactivation is suggested.
Only sufficiently high doses are expected to induce carcinogenicity. They have not been used so far in carcinogenicity studies.

Chlorodinitrobenzene in mutagenicity and GSH conjugation

Chlorodinitrobenzene is a direct mutagen with a dose-dependent increase in mutagenicity in the Salmonella test when added
directly to the bacteria.[21] In the presence of post-mitochondrial
supernatant, the shape of the dose-effect curve was different
(Fig. 5). At low concentrations of chlorodinitrobenzene less mutagenicity was observed than at the same doses without S-9. At
higher doses a disproportionate increase in mutagenicity occurred.

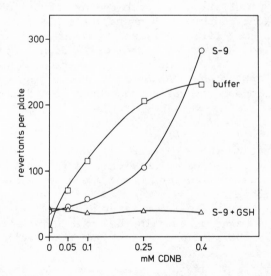

Fig. 5.   Modification of chlorodinitrobenzene (CDNB) mutagen-
            icity in the plate incorporation assay by glutathione
            conjugation. S. typhimurium were incubated with CDNB
            either in the presence of buffer or with postmitochon-
            drial supernatant (S-9) or S-9 and 2.5 mM glutathione.

We have previously shown that the S-9 fraction contains considerable amounts of GSH as well as sufficient GSH S-transferase activities.[19] Addition of GSH to the test system completely abolished
mutagenicity of chlorodinitrobenzene. These data indicate a close
correlation between intracellular glutathione levels and chlorodinitrobenzene genotoxicity. Before mutagenicity of this chemical
was observed it has been used topically to treat alopecia
areata.[22]

Naphthalene: Protein binding and conjugation

     Isolated hepatocytes metabolize naphthalene to water-soluble
compounds.[6] This reaction was linear for 1 hour reaching a
plateau at 2 hours. During biotransformation, naphthalene became
irreversibly bound to cellular protein. Binding almost paralleled
the increase in the formation in metabolites. Formation of water-
soluble compounds and binding was due to metabolism since frozen-
thawed cells in the presence of SKF 525-A showed neither a formation of water-soluble compounds nor binding to cellular macromolecules.

     Naphthol is subjected to glucuronidation and sulfation by the
liver.[23] We inhibited these conjugation mechanisms by interfering
with the synthesis of their respective cofactors. Addition of
D-galactosamine reduces levels of uridine-diphosphoglucuronic acid
by trapping uridine-triphosphate and by inhibiting UDPG dehydrogenase activity.[24] Sulfation is inhibited by incubation of the
cells in sulfate-free medium which decreases synthesis of 3'-phos-
phoadenosin-5-phosphosulfate (PAPS) and, thereby, sulfation.[25]
Incubation of hepatocytes in a sulfate-free medium in the presence
of 3 mMol D-galactosamine did not affect formation of water-soluble
metabolites from naphthalene. However, a drastic increase in
covalently bound metabolites was demonstrated (for details see[6]).

CONCLUSION

     It is evident from these experiments that cellular formation
of reactive and toxic species of chemicals is frequently counter-
acted by metabolic detoxification. As a consequence, two threshold doses can be observed:
a no-effect level where no toxic or genotoxic effects become evident; a disproportionate increase in the toxic effects when high
concentrations of a chemical are present and inactivation can be
overcome.

     For the toxicologist this implies, first of all, that toxico-
logical studies of chemicals in animals require profound information on the metabolic activation and inactivation processes to
which the chemical is subjected. It also questions the relevance
of the extremely high doses frequently used in animal experiments

for studying embryo toxicity, such as teratogenicity, or in other chronic toxicity studies. This has to be considered especially when man is exposed to much lower levels of the chemical in his environment.

REFERENCES

1. U. Andrae and H. Greim, Induction of DNA repair replication by hydroxyurea in human lymphoblastoid cells mediated by liver microsomes and NADPH, Biochem. Biophys. Res. Commun. 87:50-58 (1979).
2. B.N. Ames, W.E. Durston, E. Yamasaki and D.F. Lee, Carcinogens are mutagens: A simple test system combining liver homogenates for activation and bacteria for detection, Proc. Natl. Acad. Sci. (U.S.A.) 70:2281-2285 (1973).
3. B.N. Ames, J. McCann and E. Yamasaki, Methods for detecting carcinogens and mutagens with the Salmonella/ mammalian microsome mutagenicity test, Mutation Res. 31:347-361 (1975).
4. F. Tietze, Enzymic method for quantitative determination of nanogram amounts of total and oxidized glutathione, Anal. Biochem. 27:502-522 (1969).
5. L.R. Schwarz, M. Schwenk, E. Pfaff and H. Greim, Cholestatic steroid hormones inhibit taurocholate uptake into isolated rat hepatocytes, Biochem. Pharmacol. 26:2433-24 (1977).
6. S. Hesse, M. Mezger and L. Schwarz, Formation of reactive metabolites of ^{14}C-naphthalene in isolated rat heptocytes and the effect of decreased glucuronidation and sulfation, This publication (1980).
7. H. Rhaese and E. Freese, Chemical analysis of DNA alterations. I. Base liberation and backbone breakage of DNA and oligodeoxyadenylic acid induced by hydrogen peroxide and hydroxylamine, Biochim. Biophys. Acta 155:476 (1968).
8. H. Rhaese and E. Freese, Chemical analysis of DNA alterations. II. Alteration and liberation of bases of deoxynucleotides and deoxynucleosides induced by hydrogen peroxide and hydroxylamine, Biochim. Biophys. Acta 155:491 (1968).
9. H. Massie, H. Samis and M. Baird, The kinetics of degradation of DNA and RNA by H_2O_2, Biochim. Biophys. Acta 272:539 (1972).
10. M.F. Stich, L. Wei and P. Lam, The need for a mammalian test system for mutagens: Action of some reducing agents, Cancer Let. 5:199-204 (1978).
11. A.G. Hildebrandt and I. Roots, Reduced nicotinadenine dinucleotide phosphate (NADPH)-dependent formation and breakdown of hydrogen peroxide during mixed function oxidation reacations in liver microsomes, Arch. Biochem. Biophys. 171:385-397 (1975).

12. D.P. Jones, H. Thor, B. Andersson and S. Orrenius, Detoxification reactions in isolated hepatocytes, J. Biol. Chem. 253:6031-6037 (1978).

13. L. Fishbein, "Potential Industrial Carcinogens and Mutagens," Elsevier, Amsterdam-Oxford-New York (1979).

14. V. Ponemarkov and L. Tomatis, Long-term testing of vinylidene chloride and chloroprene for carcinogenicity in rats, Oncology 37:136-141 (1980).

15. H. Bartsch, C. Malaveille, R. Montesano, and L. Tomatis, Tissue-mediated mutagenicity of vinylidine chloride and 2-chlorobutadiene in Salmonella typhimurium, Nature 255:641-643 (1975).

16. H. Bartsch, C. Malaveille, A. Barbin and G. Planche, Mutagenic and alkylating metabolites of halo-ethylenes, chlorobutadienes and dichlorobutenes produced by rodent or human liver tissues. Evidence for oxirane formation by P-450 linked microsomal monooxygenase, Arch. Toxicol. 41:249-277 (1979).

17. T.J. Haley, Chloroprene (2-cholor-1,3-butadiene) - What is the evidence for its carcinogenicity? Clin. Toxicol. 13:153-170 (1978).

18. H. Plugge and R.J. Jaeger, Acute inhalation toxicity of 2-chloro-1,3-butadiene (Chloroprene): Effects on liver and lung, Toxicol. Appl. Pharmacol. 50:565-572 (1979).

19. K.H. Summer, W. Göggelmann and H. Greim, Glutathione and glutathione S-transferases in the Salmonella mammalian-microsome mutagenicity test, Mutation Res. 70:269-278 (1980).

20. K.H. Summer and H. Greim, Detoxification of chloroprene (2-chloro-1,3-butadiene) with glutathione in the rat, Biochem. Biophys. Res. Commun. in press (1980).

21. K.H. Summer and W. Göggelmann, 1-Chloro-2,4-dinitrobenzene depletes glutathione in rat skin and is mutagenic in Salmonella typhimurium, Mutation Res. 77:91-93 (1980)

22. R. Happle and K. Echternacht, Induction of hair growth in alopecia areata with dinitrochlorobenzene, Lancet II:1002-1003 (1977).

23. K.W. Bock, G. van Ackeren, F. Lorch and F.W. Bicke, Metabolism of naphthalene to naphthalene dihydrodiol glucuronide in isolated hepatocytes and in liver microsomes. Biochem. Pharmacol. 25:2351-2356 (1976).

24. C. Bauer and W. Reutter, Inhibition of diphosphoglucose dehydrogenase by galactosamine-1-phosphate and UDP-galactosamine, Biochim. Biophys. Acta 293:11-14 (1973).

25. L. Schwarz, Modulation of sulfation and glucuronidation of 1-naphthol in isolated rat liver cells, Arch. Toxicol. 44:137-145 (1980).

TRANSLATION OF PHARMACOKINETIC/BIOCHEMICAL DATA INTO RISK

ASSESSMENT

F.K. Dietz, R.H. Reitz, P.G. Watanabe and P.J. Gehring

Toxicology Research Laboratory
Health and Environmental Sciences, USA
Dow Chemical USA
1803 Building, Midland, Michigan 48640 USA

INTRODUCTION

A basic goal of toxicological research is to provide a rational basis for evaluating the risk to man associated with exposure to potentially harmful chemicals. Chemically induced carcinogenicity is one type of toxic response which has received considerable attention in recent years. Many chemicals, previously shown to be mutagenic and/or carcinogenic in animal studies are consequently considered to be suspect human carcinogens. However, the designation of an agent as a potential human carcinogen does not necessarily imply that concentrations of the chemical present in the environment pose a significant danger to human health. The determination that a potential carcinogen constitutes a public health hazard requires a quantitative risk assessment of the chemical.

Traditional sources have recognized that the chronic animal bioassay is one of the best available data sources from which to make a quantitative risk assessment. Although the administration of high doses in these studies is a useful design for screening chemicals for target organ toxicity in laboratory animals, it often results in frank clinical disease or even death prior to the onset of tumors. Recent studies have demonstrated that a linear extrapolation of the results observed in animal bioassays in order to assess the exposure hazard to man at much lower dose levels is not always sufficient to provide the best estimate of carcinogenic risk. Such an approach fails to consider that many chemicals are capable of saturating various biochemical or physiological transport systems at high dose levels. These processes may involve detoxification or excretion pathways which normally serve under nonsaturating conditions to protect the animal from chemically induced toxicity.

Pharmacokinetics is a discipline which studies the time course or kinetics of all processes including absorption, distribution, biotransformation and excretion which determine the overall biological fate of a chemical within the body. Since most toxic responses, including carcinogenesis, appear to be dependent on both the concentration of toxicant at the target site and the length of time the site is exposed, the pharmacokinetic characteristics of a chemical as a function of administered dose are closely linked to its toxicological effects. However, knowledge of a chemical's pharmacokinetic fate is not always sufficient to explain species and sometimes strain differences with regard to susceptibility to carcinogenesis. Consequently, in using data from chronic animal toxicity studies to ultimately assess the risk of chemical exposure to man, it is also necessary to understand the nature of chemical interactions with intracellular macromolecular sites and appreciate the relationship of these interactions to the toxic response, including carcinogenesis.

The purpose of this presentation is to review some basic pharmacokinetic principles and biochemical mechanisms which must be considered in order to adequately evaluate the risk of exposure to man from potentially carcinogenic chemicals.

PHARMACOKINETIC PRINCIPLES

Pharmacokinetics quantifies the dynamics of many processes, including the absorption of a chemical into the body, distribution to various tissues, metabolism to biologically active or inactive metabolites, reversible or irreversible reactions with macromolecular sites in target tissues and finally the excretion of the chemical from the body. Since the classical introduction of pharmacokinetics by Teorell (1937a and 1937b), pharmacokinetic concepts have been principally applied to the clinical evaluation of therapeutic agents (Wagner, 1968 and 1971; Garrett, 1973; Goldstein et al., 1974 and Levy and Gibaldi, 1975). However, Gehring et al., (1976) have recently shown that pharmacokinetic concepts are readily applicable to assessing the hazard associated with chemical exposure. Indeed, the knowledge of a chemical's pharmacokinetic fate as a function of administered dose is critical for risk assessment.

A classical approach to pharmacokinetic analysis depicts the body as a system of compartments. An individual compartment generally includes all organs, tissues, cells or fluids for which the rate of uptake and clearance of a chemical is sufficiently similar as to preclude further kinetic resolution. Compartments

do not necessarily have direct physiological or anatomical counterparts when analyzed kinetically. Gehring *et al.*(1976, 1978)
provide a more detailed description of pharmacokinetic compartments
and their use in evaluating toxicity data.

Dose-Independent Pharmacokinetics

Over a range of selected dose levels, many chemicals exhibit
first order kinetics which are referred to as being "linear". For
these chemicals, the rates of absorption, distribution, metabolism,
macromolecular interaction and elimination from the body are
proportional to the concentration or amount of the chemical
within the body. The rate constants of all these processes are
thus independent of the administered dose. In a simplified form,
first order elimination kinetics may be expressed by the equation:

$$\text{rate} = -\frac{dC}{dt} = k_e C \tag{1}$$

where C is the concentration of the chemical at time t and k_e is
the overall rate constant for elimination of the chemical from
the body. As long as first order kinetics apply and thus the rate
constants of all processes which dictate a chemical's pharmacokinetic fate are dose-independent, tissue concentration and consequent toxicity will always be proportional to the administered dose
of chemical.

Dose-Dependent Pharmacokinetics

In actuality, most reactions affecting a chemical's pharmacokinetic fate are not independent of administered dose, but are
instead markedly dose-dependent. In this situation, saturable
enzymatic reactions or active transport systems that may play a
key role in preventing or facilitating chemically induced toxicity
are not adequately described by first order kinetics.

As a consequence, high doses of chemicals may saturate or
overwhelm these processes and result in a disproportionate increase
in tissue concentration which may elicit a toxic response.

The rates of saturable processes are best described by
Michaelis-Menten or dose-dependent kinetics in accordance with
the equation

$$\text{rate} = -\frac{dC}{dt} = \frac{\text{Vmax} \cdot C}{\text{Km} + C} . \tag{2}$$

In this equation, dC/dt is the rate of change of the chemical at time t, C is the chemical's concentration at time t, Vmax is the maximum velocity of the process and Km is the Michaelis constant which is equal to the concentration of the chemical at which the rate of the process is equal to one-half the maximum velocity, Vmax.

According to this equation, when the concentration of any chemical C is much less than Km, the rate of the process is approximated by the equation,

$$\text{rate} = -\frac{dC}{dt} \approx k \cdot C \quad (C \ll \text{Km}) \tag{3}$$

where k = Vmax/Km. In this instance, the rate is proportional to the concentration and can be described by apparent first order or "linear" kinetics. However, inspection of Eq. 2 also demonstrates that when the concentration (C) is very large and much greater than Km, Eq. 2 approaches the limiting form of Eq. 4,

$$\text{rate} = -\frac{dC}{dt} \approx \text{Vmax} \quad (C \gg \text{Km}) \tag{4}$$

As a consequence, the rate of the overall process is limited by Vmax, the maximal velocity of the process. In this situation, the rate of this process becomes constant or zero order and will not increase, regardless of the value for C. In this concentration range (C >> Km) the biological processes governed by Michaelis-Menton kinetics are overwhelmed and can be referred to as being saturated or dose-dependent. As a consequence of this saturation, these physiological processes no longer function as efficiently as they might at lower doses.

THEORETICAL MODEL DESCRIBING THE PHARMACOKINETICS OF CHEMICAL INTERACTIONS WITH MACROMOLECULES

The irreversible binding of a chemical or its metabolic products to cellular macromolecules such as DNA, RNA and protein is thought to be responsible for many manifestations of chemically induced toxicity, including carcinogenesis. A model recently proposed by Gehring and Blau (1977) incorporates the concept that chemical interaction with macromolecules is a function of several processes

which are influenced by the administration of high doses. This
model offers advantages in that it realistically depicts certain
enzymatic and active transport processes as being dose-dependent.

In this model (Figure 1), all rate processes which are thought
to be saturable or dose-dependent are designated by a name rather
than a given rate constant (k). Excretion (e) of a chemical (C)
as an unchanged compound (Ce), bioactivation (a) of the chemical
to a reactive metabolite (RM) and detoxification (d) of the reactive
metabolite to an inactive metabolite (IM) are all processes frequently mediated by enzymatic reactions and are thus conceived to
be saturable processes.

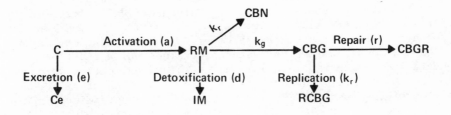

C   = Chemical
RM   = Reactive Metabolite
Ce   = Excreted Chemical
IM   = Inactive Metabolite
CBN   = Covalent Binding, Nongenetic
CBG   = Covalent Binding, Genetic
CBGR   = Repaired Covalently Bound Genetic Material
RCBG   = Retained Genetic Program, Critical & Noncritical

Fig. 1. Hypothetical model illustrating the processes determining
         the fate of a chemical (C) within the body. Rate con-
         stants (k) are assigned to first order processes. All
         other rate processes thought to be saturable or dose-
         dependent are designated by name.

In addition to inactivation, reactive metabolites may covalently bind to intracellular macromolecules such as DNA, RNA, lipids
or protein. These reactions are likely to be non-enzymatic and only
a function of the chemical reactivity of the ultimate toxicant. The
consequences of these reactions generally fall into two categories;
those which cause only structural or functional alterations of the
cell (CBN) and those which may lead to heritable genetic transformations (CBG). The latter reactions will occur with both critical
and non-critical receptor sites on DNA and possibly RNA.

After interaction or binding with genetic material, cellular
repair processes may correct the chemically induced alterations
in nucleic acids by eliminating the altered bases from DNA.
According to the model of Figure 1, repair (r) is represented by a
saturable process since it is enzymatically mediated and likely to
be saturated when large numbers of defects occur in DNA. Genetic
material which has undergone an error free repair process is
indicated by CBGR in the model. Products that escape cellular
repair mechanisms are fixed into the cell's chromosomes by
replication and are represented by RCBG. Although post-replicative
repair of altered genetic material occurs, it is not considered by
this model.

Gehring and Blau (1977) have quantitatively assessed the
significance of this model by assigning plausible values to the
kinetic constants describing Figure 1 and numerically integrating
a series of differential equations which describe the model by
using a computer. The processes of excretion, activation, detoxification and repair were given saturable values while reactions with
genetic and nongenetic material and cellular replication were considered to be first order processes. As a function of different
dose levels, values for CBN, CBGR and RCBG were determined. These
values were all normalized by dividing each quantity by the initial
dose level (Co). Figure 2 depicts the results of these integrations.
If each of the indicated parameters of Figure 2 were independent of
the administered dose and thus non-saturable, each of the respective
lines of Figure 2 would be parallel to the abscissa. However,
inspection of this simulation indicates that at dose levels above
10^{-3} moles/kg, there is a disproportional increase in the quantity
of genetic material covalently bound to reactive metabolites which
undergoes replication (RCBG/Co). There is also a disproportional
decrease in the amount of altered genetic material that undergoes
cellular repair (CBGR/Co). It is in this concentration range that
the biochemical processes responsible for the repair of chemically
altered genetic material have become saturated, resulting in an
increase in altered genetic material available to undergo replication. It is clear from this model, that reactions of chemicals
with genetic material are not related in a linear fashion across
dose levels and are thus dose-dependent. Consequently, for

chemicals acting through this mechanism under these conditions, toxicity would also not be expected to be linear throughout a broad dose range.

This discussion of a theoretical model describing the irre-versible interactions between chemicals and macromolecular target sites stresses the importance of a knowledge of the dose-related fate of a chemical (pharmacokinetics) and its interaction with macromolecular target sites (kinetic-macromolecular reactions).

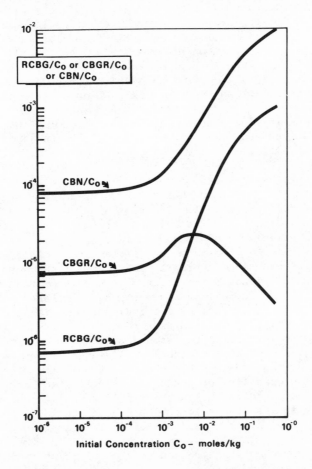

Fig. 2. Effects of increasing dose (Co) on the fraction of a chemical covalently bound to: 1) non-genetic macro-molecules (CBN/Co); 2) genetic macromolecules followed by subsequent repair (CBGR/Co) and 3) genetic macromole-cules in which the chemically induced alteration escapes repair and is fixed by replication (RCBG/Co). The amounts of chemical covalently bound to each material have been normalized for the administered dose (Co).

MECHANISMS OF TUMORIGENICITY

Recent evidence suggests that there are at least two basic mechanisms that are responsible for the observation of increased tumor frequencies in the typical animal bioassay. The genetic or genotoxic mechanism is simply described as a direct chemical interaction with genetic materials by the test substance (or active metabolites produced from the test substance), in accordance with the somatic mutation theory of chemical carcinogenesis. This interaction is exemplified by an electrophilic attack by many potent carcinogens or their metabolites on nucleophilic sites of DNA. Once transformed, those DNA molecules which escape repair may undergo replication and result in transformed cells which ultimately may lead to cancer. While normal DNA repair mechanisms tend to inhibit this toxicological progression of biochemical events, exposure to chemicals which may elicit a carcinogenic response via this mechanism should be carefully monitored as the initial genetic alterations cannot always be clinically observed and toxicity does not become readily apparent until tumors result.

The second mechanism responsible for tumorigenesis differs from the genetic mechanism in that it produces tumors through a cytotoxic or nongenetic effect. In this instance, tissue injury occurs to such a degree that cellular death results, leading to an increase in cellular division to replace damaged or dead tissue. It is important to note that this mechanism differs from that mediated by genetic effects in that the chemical does not primarily interact with genetic material but rather produces cytotoxicity. Since there is a small but nevertheless finite chance for error in each replication cycle of genetic material, an increase in cellular division as a response to tissue injury can contribute to increasing the spontaneous mutation rate. Another consequence of an increase in cellular division is that the relative rates of DNA synthesis and/or repair are altered, increasing the probability of transformed cells which may ultimately lead to tumor formation. This mechanism of tumor induction is important in risk assessment in that tissue damage associated with cytotoxic mechanisms should be clinically detectable and reversible, once recognized. Thus, exposure that does not cause any tissue injury should not result in the induction of any tumors via this mechanism.

Several authors have recently reported the various methodologies utilized to experimentally assess a chemical carcinogen's mechanism of tumorigenesis (Reitz et al., 1980b; Weisburger and Williams, 1980). Briefly, these include DNA alkylation and DNA repair as indicators of a genetic mechanism while DNA synthesis and histopathology are used to confirm nongenetic or cytotoxic mechanisms.

DNA alkylation, measured by incorporation of radioactivity from a test material into highly purified DNA preparations is a primary indicator of genetic mechanisms. If a chemical has a potential to cause genotoxicity by irreversible covalent binding to DNA, an irreversible incorporation of radioactivity into DNA is observed. DNA repair is an additional indicator of preceding genotoxicity. The repair of DNA is measured by the incorporation of tritiated thymidine into DNA following exposure to the test material in the presence of hydroxyurea to block normal endogenous DNA synthesis. While this procedure is less sensitive than the DNA alkylating method, it does not require the use of a radiolabeled chemical for testing.

According to the conditions responsible for an nongenetic mechanism of tumorigenesis, DNA damage or DNA repair are not radically affected, but the rate of normal DNA synthesis is increased above the normal endogenous level. This is due to the increase in cellular division in response to a chemical's cytotoxic effects. This increase in DNA synthesis can be measured directly by the administration of tritiated thymidine to animals at a time when cellular regeneration in response to chemical exposure is occurring. Since DNA synthesis may be affected by several factors, including age, time of day, etc., the use of this procedure requires a matched control group to adequately define treatment-related effects. The microscopic examination of tissue sections from treated animals is also a valuable technique to determine if a carcinogen produces cytotoxic effects. Histopathology provides information on the relative sensitivity of target organs, the persistence of toxic effects and a qualitative index of the severity of cytotoxic lesions.

USE OF PHARMACOKINETIC/BIOCHEMICAL CONCEPTS IN RISK ASSESSMENT

Utilizing concepts of pharmacokinetics to study chemical interactions with intracellular macromolecules is an important aspect in interpreting results of chronic animal bioassays for quantitative risk assessment. Since chemically induced toxicity is frequently mediated by reactions with molecular target sites in sensitive tissue, the pharmacokinetics of a chemical at the molecular level frequently provides insight into differential species and strain sensitivities to toxicity, including carcinogenicity. Studies on vinyl chloride and chloroform illustrate how pharmacokinetics and biochemical data on mechanisms of tumorigenesis can be used to understand the differential carcinogenic response between species with particular emphasis on risk assessment.

Vinyl Chloride

An example of dose-dependent pharmacokinetics that directly

relates to carcinogenic risk estimation is that of inhaled vinyl chloride (VC). These concepts will be demonstrated by carrying out a risk analysis for this industrial chemical.

Animal Carcinogenicity Data. This extrapolation begins with the results of a long-term animal bioassay reported by Maltoni and Lefemine (1975). In this chronic investigation, rats were exposed to various concentrations of VC 4 hr/day, 5 days/week for up to 12 months and then held for observation until death. The incidence of angiosarcomas found in animals of this study are given in Table 1, column 3.

Pharmacokinetics/Metabolism. Numerous studies have suggested that VC requires metabolic activation in rats in order to produce tumors (Barbin et al., 1975; Bartsch et al., 1975; Bolt et al., 1975; Malaveille et al., 1975; Kappus et al., 1976 and Watanabe et al., 1978). The covalent binding of ^{14}C to hepatic macromolecules in rats exposed to ^{14}C-VC has been shown to be dependent on VC bioactivation (Watanabe et al., 1978). It is presently felt that this covalent interaction of an electrophilic metabolite from VC with nucleophilic sites on hepatic DNA is responsible for VC induced angiosarcoma.

Consistent with the results of previous studies (Bolt, et al., 1976 and Watanabe et al., 1976a-c), Gehring et al., (1978) have shown that the biotransformation of VC in rats is a saturable process (Figure 3) that occurs in accordance with Michaelis-Menton kinetics and is described by:

$$v = \frac{Vmax \cdot S}{Km + S} \qquad (5)$$

In this equation v and Vmax are the velocity and maximum velocity, respectively, for the biotransformation of VC expressed as microgram equivalents of VC metabolized daily. S and Km are the inhaled concentration of VC and the Michaelis constant expressed as micrograms of VC per liter of air, respectively.

As a consequence of the dose-dependent metabolism of VC, the toxicity or carcinogenicity resulting from VC exposure can not be directly related to the exposure concentration of VC. This results from a saturation of pathways responsible for VC activation at high exposure concentrations which serves to limit the amount of toxic metabolite produced.

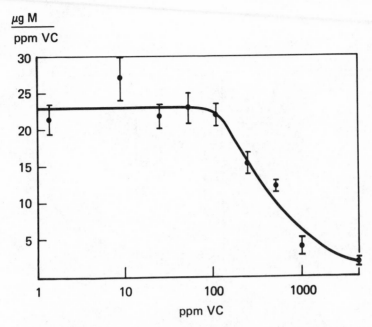

Fig. 3. The ratio between the amount of vinyl chloride (VC) metabolized (μg M) in rats and six hour exposure concentration of vinyl chloride (ppm VC). Note that at exposure concentrations above 100 ppm, VC metabolism proceeds in a non-linear or dose-dependent fashion, indicating a saturable process. Data from Gehring *et al.* (1978).

Indeed, if the data of Maltoni and Lefemine are plotted as a function of the exposure concentration versus tumor incidence (Figure 4B), an essentially flat dose-response curve at exposure levels above 1000 ppm is readily apparent. However, if the incidence of angiosarcoma in rats is plotted as a function of the amount of vinyl chloride metabolized (v) using Eq. 5, the results are clearly more interpretable (Figure 4A). These observations further support the contention that VC induced tumor incidence in rats is not related to VC exposure concentration in a linear fashion, but rather to the body burden of metabolized VC resulting from VC exposure.

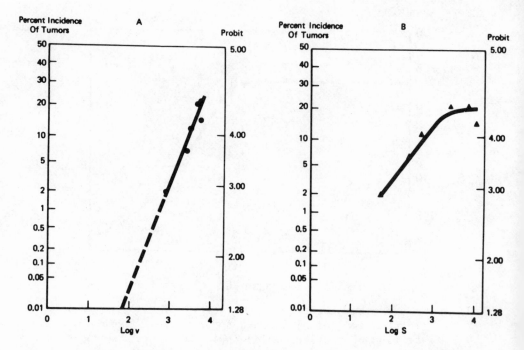

Fig. 4.   (A) Metabolism of vinyl chloride expressed as log v (μgm
          of vinyl chloride metabolized per 4 hour exposure) versus
          percentage incidence of hepatic angiosarcoma (probability
          scale).   (B) Exposure concentration expressed as log S
          (ppm) versus the percentage incidence of hepatic angio-
          sarcoma. The solid line is the best fit for experimentally
          observed responses while the dashed line represents an
          extrapolation below those doses producing an observable
          response, assuming no threshold exists.

Integration of Vinyl Chloride Pharmacokinetic/Metabolism Data
for Risk Assessment. As a consequence of the saturable metabolism
of VC, any model used for risk assessment must be initially fit to
the data which relates the amount of VC metabolized (v) as a
function of exposure concentration versus the resulting tumor
incidence. Numerous models are available to extrapolate a carcinogenic response observed at high exposure concentrations to the
expected response at much lower exposure levels. Four models
which are commonly used in risk assessment have been evaluated by

Gehring *et al.* (1979) for predicting the incidence of carcinogenesis
due to VC exposure. These models include:

    A) Probit % = a+b (log v)
              = -1.625+1.543 (log v).
    B) Linear % = a+b (v)
              = $-1.48+0.389 \times 10^{-2}$(v).
    C) Linear Forced Through Origin
          % = bv
            = 0.3565×10^{-2}(v)
    D) One-hit Model (Cancer Assessment Group, National Cancer
        Institute) $\% = 1-\exp^{-\beta(v)}$
              $= \{1-\exp(-0.38 \times 10^{-4}xv)\}100$.

The constants given for the preceding models were determined by
linear regression analysis and represent the best fit of the data
for the incidence of angiosarcoma versus the amount of VC metabolized on a daily basis in rats as a function of exposure to
various concentrations of VC (Gehring *et al.*, 1978).

The data used to derive the constants are given in Table 1.
Using the equations derived for each of the above models, the
predicted incidence of angiosarcoma for groups of rats exposed to
different concentrations of VC was determined. These results are
also shown in Table 1. Inspection of these data reveals that all
four models overestimated the incidence of hepatic angiosarcoma
in rats exposed to 10,000 ppm VC. This result was expected as this
exposure level resulted in excessive mortality which was unrelated
to the development of angiosarcoma.

With respect to the predicted versus observed incidence of
hepatic angiosarcoma found in groups of rats exposed to the lower
concentrations of VC, no one of the four models is most reliable.
As a consequence, any one of the four models adequately predicted
the experimental data observed in rats. Since no model appeared
to be unequivocally best, all four will be used to predict the
incidence of hepatic angiosarcoma experienced or to be experienced
in workers exposed to VC.

In order to calculate the equivalent dose of VC metabolite
in exposed workers, the following assumptions were made:

    1)   The metabolism of VC in man is a dose-dependent process as
           in experimental animals.
    2)   A 1 year exposure period in rats is equivalent to a 35
           year exposure period in man (a time period approximately
           equal to 50% of each species' lifetime).
    3)   The rate of VC metabolism in each species is low relative

to the rate of VC equilibration between lungs and plasma. This means that the inhalation of equivalent vapor concentrations of VC in each species would result in equal plasma concentrations of VC at steady state.

4) The amount of VC transformed to a toxic metabolite by man on an equivalent basis to rats during an 8 hour work day is given by:

$$v(\mu g/8 \text{ hr}) = \frac{1675 \ \mu g/hr \cdot S \ \mu g/l}{860 \ \mu g/l + S \ \mu g/l} \tag{6}$$

This equation assumes that there are no significant differences in the Km between the two species while the Vmax between rat and man differs as a function of body surface area (Gehring *et al.*, 1978).

Using Eq. 6 to determine the amount of VC metabolite produced in workers on a daily basis (v), it is possible to use each of the four models for risk assessment to predict the incidence of angiosarcoma in workers exposed to any concentration of VC for a given fraction of their working life (35 years). Since a comprehensive epidemiological survey of VC workers has recently been

Table 1. Predicted Incidence Of Angiosarcoma In Rats Exposed To Vinyl Chloride (VC) Using Various Models Versus Dose (v, μg VC Metabolized Per Day) Compared To Experimental Results

Exposure[a]	Dose[b] (v, μg/day)	Experimental[a] (%)	Predicted percentage (Models)[c]			
			A	B	C	D
10,000	5521	14.8 (9/61)	19.8	20.0	19.7	18.9
6,000	5403	21.7 (13/60)	19.3	20.5	19.3	18.6
2,500	5030	22.0 (13/59)	17.9	18.1	17.9	17.4
500	3413	11.9 (7/50)	12.1	11.8	12.2	12.2
250	2435	6.8 (4/59)	8.1	8.0	8.7	8.8
50	739	1.7 (1/59)	1.4	1.4	2.6	2.8

[a] Data from Maltoni and Lefemine (1975).

[b] Calculated from $v = \{5706(\mu g \ VC/4 \ hr) \cdot S(\mu g/liter)\}/\{860(\mu g/liter)+ S(\mu g/liter)\}$, where $S = 2.56$ (μg/ppm/liter) x exposure (ppm). Data from Gehring et al. (1978).

[c] Model A, Probit % = -1.625 + 1.543 Log (dose); Model B, % = -1.48 + 0.389 x 10^{-2} (dose); Model C, % = 0.3565 x 10^{-2} (dose); Model D, % = $\{1-e^{-\beta(dose)}\}100$, where $\beta = 0.38$ x 10^{-4}.

published (Equitable Environmental Health, Inc., 1978), Gehring
et al., (1979) were able to assess the reliability of
each model to predict the incidence of VC induced angiosarcoma in
man. The occurence of angiosarcoma in subgroups of workmen exposed
to VC as a function of their duration of exposure is given in
Table 2.

In order to carry out a risk estimation using this data
base, an estimation of human exposure was necessary. Unfortunately,
the atmospheric concentrations of VC to which these workers had
been exposed were not quantitated, but was subjectively divided
into high, medium and low categories. However, since the time-
weighted-average (TWA) exposure recommended by the American
Conference of Governmental Industrial Hygienists prior to 1972
was 500 ppm and subsequently 200 ppm until adoption of the Occupational Safety and Health Act Standard of less than 1 ppm in 1974, it
was assumed that even those exposures subjectively listed as being
low in this survey were in actuality, high. It is even reasonable
to expect that the TWA exposures of 200 ppm and greater on a daily
basis were common rather than exceptional.

Table 2. Incidence Of Angiosarcoma In 9677 Workmen Exposed To
Vinyl Chloride[a]

Population	Exposure duration (years)	Mean exposure duration (years)[b]	Fraction of working life[c]	Observed Angiosarcomas
4384(0.31)[d]	<4	2	0.06	1 (18 yrs)[e]
2339(0.27)	5-9	7	0.20	0
946(0.58)	10-14	12	0.34	2 (15 & 23 yrs)
1007(1.0)	15-19	17	0.49	2 (18 & 19 yrs)
677(1.0)	20-24	22	0.63	0
324(1.0)	25+	27	0.77	0

Total = 9677

[a]From Equitable Environmental Health, Inc. (1978)

[b]Assumed mean duration of exposure.

[c]Mean duration of exposure divided by 35 years.

[d]Fraction of population incurring first exposure 15 or more years prior to study.

[e]Duration between first exposure and diagnosis.

Table 3 depicts the average theoretical rat equivalent daily doses of metabolic products of VC received by workers exposed to 200 or 500 ppm for 8 hr daily, 5 days/week after adjustment for the fraction of their working life during which exposure occurred. The doses (v) are expressed as micrograms of VC metabolized per 8 hours and were calculated by using Eq. 6 multiplied by the fraction of an assumed 35-year working life during which exposure occurred.

Table 3. Average Theoretical Rat Equivalent Daily Doses (v) Of Biotransformation Products Received By A 70-Kg Human Exposed To 200 or 500 ppm Vinyl Chloride (VC) For Various Fractions Of An Assumed 35-Year Working Life

Mean Exposure duration (years)	Fraction of working life	v, Micrograms per 8 hr averaged over 35 years[a]	
		500[b]	200[b]
2	0.06	60	38
7	0.20	200	125
12	0.34	341	213
17	0.49	491	306
22	0.63	637	394
27	0.77	771	481

[a] $v = \{1675\ \mu g/8\ hr \cdot S\ \mu g/liter\}/\{860\ \mu g/liter + S\ \mu g/liter\} \times$ (fraction of working life). Data from Gehring et al.(1979).

[b] Parts per million (ppm) VC. 1 ppm of VC = 2.56 µg of VC/liter of air.

With these estimates for the daily amount of VC metabolized in man it is possible to compare the predicted incidence of angiosarcoma in workers exposed to 200 or 500 ppm VC by using each of the four models versus the actual number of tumors observed. These results are shown in Table 4. Inspection of the data in this table reveals that models C and D overestimated the actual incidence of hepatic angiosarcoma experienced by these workers while model B underestimates the incidences.

Table 4. Predicted Number Of Angiosarcomas In Workers Exposed To 500 or 200 ppm Vinyl Chloride 8 Hr/Day, 5 Days/Week For Various Fractions Of A 35-Year Working Life Versus The Numbers Derived Using Biotransformation And Angiosarcoma Incidence Data From Rats And Four Models For Extrapolation[a]

| Population | Observed Angiosarcomas[b] | Angiosarcomas predicted (Models)[c] | | | | | | | |
| | | A | | B | | C | | D | |
		500[d]	200	500	200	500	200	500	200
4384	1	0.2	0	0	0	9.4	5.9	10.0	6.3
2339	0	2.4	0.8	0	0	16.7	10.4	17.7	11.1
946	2	3.0	1.1	0	0	11.5	7.2	12.1	7.6
1007	2	6.8	2.7	4.0	0	17.5	11.0	18.6	11.7
677	0	7.3	3.0	6.6	0.3	15.2	9.5	16.0	10.1
324	0	4.9	2.1	4.9	1.3	8.9	5.6	9.4	5.9
Total 9677	5	25	10	16	2	79	50	84	53

[a] Data from Gehring et al. (1979).

[b] Observed number of angiosarcomas from Equitable Environmental Health, Inc. (1978).

[c] Model A, Probit % = $-1.625 + 1.543$ log (dose); Model B, % = $-1.48 + 0.389 \times 10^{-2}$ (dose); Model C, % = 0.3565×10^{-2} (dose); Model D, % = $\{1-e^{-\beta(dose)}\}100$, where $\beta = 0.38 \times 10^{-4}$. The constants used in these models were calculated using biotransformation and angiosarcoma incidence data for rats exposed to vinyl chloride (Gehring et al., 1978).

[d] Parts per million.

However, when a TWA of 200 ppm is utilized to represent the atmospheric concentrations to which these workers were exposed, the predicted incidence of 10 cases of argiosarcoma using Model A compares favorably with the actual experienced number of 5 cases. For a significant number of workers, the time elapsed since initial exposure has yet to reach 15 years, which is the shortest time needed for the development of angiosarcoma in any of the afflicted workers (Table 2) thus, it may be anticipated that a few additional cases of angiosarcoma may develop in this population, however, it is reassuring to note that an epidemic of cancer cases is not anticipated.

In summary, the foregoing analysis has demonstrated that data collected in experiments on rats can be used to predict the incidence of hepatic angiosarcoma in humans with reasonable accuracy by using a probit percentage incidence model (Model A). The concepts evolved from this analysis demonstrate why the dose-dependent or saturable pharmacokinetic fate of VC must be considered for predicting the risk associated with VC exposure. Since VC induced angiosarcoma is felt to be mediated via a reactive metabolite, this analysis predicts that at concentrations of VC less than those which saturate VC bioactivation small animal species with their greater rate of metabolic activity would be more sensitive to the tumorigenic effects of VC than would larger species such as man.

Partial support for this concept is found in the work of Buchter et al., (1978) and Filser and Bolt (1979) in which mice were found to metabolize VC approximately twice as fast as rats (Table 5). These authors also found that these two species metabolized VC at a rate which was 5-12 times that for humans. These results would predict that man, by nature of his slower rate of metabolic activation of VC, would be less sensitive to the tumorigenic effects of VC than these two laboratory species. It is of interest to note, that using a model which appeared to be the most reliable for quantitative risk assessment (model A, probit %), Gehring et al., 1979 predicted that the exposure of workers to 1 ppm of VC for 35 years would only be expected to result in 0.015 tumors per million exposed individuals, which is clearly a rather minor risk.

Chloroform

Chloroform is produced in small amounts during the chlorination of drinking water. Since studies at the National Cancer Institute (NCI) have demonstrated that high doses of chloroform induce liver and kidney tumors in mice and rats (National Cancer Institute, 1976), the United States Environmental Protection Agency (EPA) has established a Maximum Contamination Level of 100 ppb (ng/ml) of chloro-

Table 5.   First Order Metabolic Clearance Rates For Vinyl Chloride
In Man Versus Other Species[a]

Species	Clearance (ℓ/hr/kg body wt.)
Man	2.0
Rabbit	2.7
Rat	11.0
Gerbil	12.5
Mouse	25.6

[a]Data from Buchter *et al.* (1978) and Filser and Bolt (1979).

form in drinking water. The risk estimation procedure cited by the
EPA for deriving this limit was based on the assumptions that man is
more sensitive to the toxic effects of chloroform than experimental
animals and that a genetic mechanism is responsible for the tumorigenic response to chloroform observed in chronic bioassays.

The validity of these assumptions has been recently addressed
by Reitz *et al.* (1978 and 1980a). In predicting species sensitivity to chloroform exposure, the original risk estimation assumed
that the toxic and carcinogenic potential of chloroform was due to
chloroform *per se*. This assumption predicts that man would be
more sensitive than the mouse to chloroform toxicity, including
carcinogenicity, because of his less effective metabolic and
excretory processes which frequently function as routes for
xenobiotic detoxification and elimination from the body (Weiss
et al., 1977). However, experimental evidence is inconsistent with
this assumption. Chloroform appears to be metabolized *in vivo* to
a reactive intermediate which is capable of covalently binding to
tissue macromolecules (Ilett *et al.*, 1973 and Pohl *et al.*, 1977).
The extent of this activation and subsequent binding is well
correlated with the toxic actions of chloroform in the liver and
kidneys of treated animals. For example, inducing the enzymes
responsible for chloroform metabolism greatly enhances chloroform
hepatotoxicity while inhibiting these enzymes reduces chloroform
toxicity (Ilett *et al.*, 1973).

Since bioactivation appears to be responsible for the toxic
and carcinogenic potential of chloroform, any interspecies variations in chloroform metabolism should markedly alter the toxicity
and carcinogenicity of this chemical. In studies of chloroform
metabolism in various species, Brown *et al.* (1974) have found
that mice metabolize chloroform more readily than rats, while
monkeys metabolize chloroform less readily than either species of

1418 F. K. DIETZ ET AL.

rodents. Results from the studies of Fry *et al.*, (1972) indicate
that man metabolizes chloroform more like the monkey than the mouse.
Collectively, these data predict that the relative species sensitivity to the carcinogenic potential of chloroform will be greater in
mice, less in rats and further diminished in monkeys and man. Indeed, the chronic bioassay performed for the NCI indicated that mice
developed more tumors than rats after exposure to equivalent doses
of chloroform.

Using B6C3F1 mice (a species and strain shown by the NCI study
to be greatly susceptible to the tumorigenic effects of chloroform
at high doses), Reitz et al., (1980a) determined DNA alkylation after
the administration of ^{14}C-chloroform as an indicator of the potential for chloroform to cause genetic alterations. Chloroform caused
little alkylation in DNA isolated from the liver or kidneys of male
mice, despite the administration of a single tumorigenic dose of 240 mg
chloroform/kg (Table 6). In contrast, high levels of DNA alkylation
were observed in target organs of various species after exposure to
single carcinogenic doses of dimethylnitrosamine, dimethylhydrazine
or N-methyl-N-nitrosourea. Figure 5 shows the ratio of DNA repair in
treated groups administered various doses of dimethylnitrosamine
(DMN) and chloroform versus controls. In contrast to the dose-
related increase in DNA repair observed after DMN, chloroform was

Table 6. Alkylation Of DNA Isolated From Target Organs After
Treatment With A Carcinogenic Dose Of Various Chemicals[a]

Chemical (dose)	Species/Organ	Alkylation (Mole % x 10^4)
Chloroform (240 mg/kg, po)	Mouse/Liver	3
Chloroform (240 mg/kg, po)	Mouse/Kidney	1
Dimethylnitrosamine (10 mg/kg, ip)	Rat/Liver	933
Dimethylhydrazine (15 mg/kg, sc)	Mouse/Colon	260[b]
N-methyl-N-nitrosourea (80 mg/kg, iv)	Rat/Brain	1460[b]

[a] Data from Reitz *et al.* (1980a).
[b] Based on % alkylation of guanine.

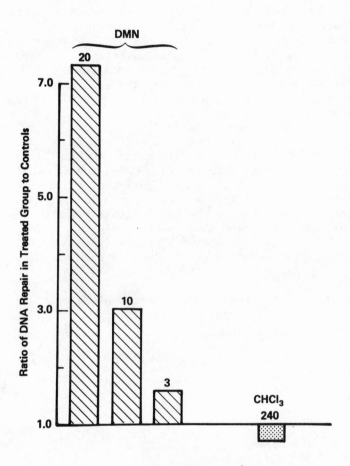

Fig. 5. Hepatic DNA repair in mice following the intraperitoneal administration of dimethylnitrosamine (DMN) or oral administration of chloroform. Repair is calculated as the hydroxyurea resistant uptake of ^{3}H-thymidine relative to controls. Data from Reitz et al. (1980a).

inactive in this system. Collectively, these data suggest that the
genetic alterations to DNA after chloroform are appreciably less
than those produced by carcinogens such as DMN.

In an investigation of the possible nongenetic or cytotoxic
effects of chloroform, Reitz et al. (1980a) found that chloroform
produced severe tissue damage after the administration of doses
which were carcinogenic in chronic bioassays. Histopathological
lesions were found in the same organs which are sensitive to
tumorigenesis. In addition, cytotoxicity as indicated by an
increase in DNA synthesis in both the liver and kidney of chloro-
form-treated mice increased in a dose-related fashion from 60 to
240 mg/kg but was not elevated in animals administered a non-
tumorigenic dose of 15 mg/kg (Table 7).

Collectively, these data indicate that the toxicity and
carcinogenicity of chloroform is mediated by the production of a
reactive metabolite in vivo. Consequently, a consideration of
metabolic rates alone, suggests that man should be less sensitive
than experimental rodents to the carcinogenic effects of chloroform.
In addition, the assumption that chloroform produces tumorigenicity
via a direct chemical interaction with DNA (genotoxicity) is not
supported by in vivo studies of chemical interaction with DNA or

Table 7.   Cellular Regeneration In Tissues Of Male B6C3F1 Mice
After A Single Dose Of Chloroform[a]

Tissue	Dose	DPM/mg DNA ± SD	Ratio[b]
Liver	Control	3.6±1.5	–
	15 mg/kg	3.4±2.2	0.94
	60 mg/kg	8.0±3.3	2.2
	240 mg/kg	50.6±13	14.0
Kidney	Control	2.4±0.21	–
	15 mg/kg	1.9±0.32	0.79
	60 mg/kg	19.6±16.2	8.2
	240 mg/kg	59.4±10.6	24.8

[a]Estimated by determination of the relative incorporation of ^{3}H-
thymidine into DNA. Data from Reitz et al. (1980a).
[b]Treated/Control

in vitro studies demonstrating a lack of mutagenic activity for chloroform (Simon *et al.*, 1977 and Uehleke *et al.*, 1977). It appears that the primary mechanism responsible for chloroform carcinogenicity in animal bioassays is mediated by recurrent cytotoxicity accompanied by chronic tissue regeneration.

An understanding of the species dependent metabolic activation of chloroform and biochemical nature of chloroform induced tumorigenicity has led Reitz *et al.*, (1980a) to conclude that the original risk estimation was unjustified and overestimated the risk of chloroform exposure.

CONCLUSION

The concepts discussed in this presentation stress the importance of integrating the results of chronic animal bioassays with pharmacokinetics and macromolecular events in assessing the risk to chemical exposure. The examples demonstrate how an understanding of the relationships of these processes to dose-response data is necessary to ultimately control chemical exposure in a rational fashion. While it is impossible to prove that any chemical, either man-made or naturally occuring, is non-carcinogenic in man, our precision of estimating the relative degree of risk associated with exposure to these agents can be increased as our knowledge of pharmacokinetics and macromolecular interactions progresses.

REFERENCES

Barbin, A., Bresil, H., Croisy, A., Jacquignon, P., Malaveille, C., Montesano, R. and Bartsch, H., 1975, Liver-microsome-mediated formation of alkylating agents from vinyl bromide and vinyl chloride, Biochem. Biophys. Res. Commun., 67:596.

Bartsch, H., Malaveille, C. and Montesano, R., 1975. Human, rat and mouse liver mediated mutagenicity of vinyl chloride in salmonella typhimurium strains, Int. J. Cancer., 15:429.

Bolt, H. M., Kappus, H., Buchter, A. and Bolt, W., 1976, Disposition of (1,2,-^{14}C) vinyl chloride in the rat, Arch. Toxicol., 35:153.

Bolt, H. M., Kappus, H., Kaufmann, R., Appel, K. E., Buchter, A. and Bolt, W., 1975, Metabolism of ^{14}C-vinyl chloride in vitro and in vivo, Inserm., 52:151.

Brown, D. M., Langley, P. F., Smith, D. and Taylor, D. C., 1974, Metabolism of chloroform I. The metabolism of (^{14}C) chloroform by different species, Xenobiotica, 4:151.

Buchter, A., Bolt, H. M., Filser, J. G., Goergens, H. W., Laib, R. J.
     and Bolt, W., 1978, Pharmacokinetik und karzinogenese von
     vinylchlorid. Arbeitsmedizinische risikobeurteilung,
     Verh. Dtsch. Ges. Arbeitsmed., 18:111.
Equitable Environmental Health Inc., 1978, Epidemiological study of
     vinyl chloride workers: Final report, Prepared for Manu-
     facturing Chemists Association, Washington, D. C., USA.
Filser, G. and Bolt, H. M., 1979, Pharmacokinetics of halogenated
     ethylenes in rats, Arch. Toxicol., 42:123.
Fry, B. J., Taylor, T. and Hathaway, D. E., 1972, Pulmonary
     elimination of chloroform and its metabolite in man,
     Arch. Int. Pharmacodyn. Ther., 196:98.
Garrett, E. R., 1973, Classical pharmacokinetics to the frontier,
     J. Pharmacokin. and Biopharm., 1:341.
Gehring, P. J., 1978, Chemobiokinetics and metabolism, in:
     "Principles and Methods for Evaluating the Toxicity of
     Chemicals Part I.," World Health Organization, Geneva.
Gehring, P. J. and Blau, G. E., 1977, Mechanism of Carcinogenesis:
     dose-response, J. Environm. Path. Toxicol., 1:163.
Gehring, P. J., Watanabe, P. G. and Blau, G. E., 1976, Pharmaco-
     kinetic studies in evaluation of the toxicological and
     environmental hazard of chemicals, in: "New Concepts in
     Safety Evaluation", M. A. Mehlman, R. E. Sharpiro and
     H. Blumental, eds., Hemisphere Publ. Corp., Washington, D.C.
     1(1):195.
Gehring, P. J., Watanabe, P. G. and Park, C. N., 1978, Resolution of
     dose-response toxicity data for chemicals requiring metabolic
     activation: Example-vinyl chloride, Toxicol. Appl.
     Pharmacol., 44:581.
Gehring, P. J., Watanabe, P. G. and Park, C. N., 1979, Risk of
     angiosarcoma in workers exposed to vinyl chloride as pre-
     dicted from studies in rats, Toxicol. Appl. Pharmacol.,
     49:15.
Goldstein, A., Aronow, L. and Kalman, S. M., 1974, "Principles
     of Drug Action: The Basis of Pharmacology," John Wiley and
     Sons, New York, N.Y.
Ilett, K. F., Reid, W. D., Sipes, I. G. and Krishna, G., 1973,
     Chloroform toxicity in mice: Correlation of renal and
     hepatic necrosis with covalent binding of metabolites to
     tissue macromolecules, Exptl. Molec. Path., 19:215.
Kappus, H., Bolt, H. M., Buchter, A. and Bolt, W., 1976, Liver
     microsomal uptake of (^{14}C) vinyl chloride and transformation
     to protein alkylating metabolites in vitro, Toxicol. Appl.
     Pharmacol., 37:461.
Levy, G. and Gibaldi, M., 1975, Pharmacokinetics, in: "Handbook of
     Experimental Pharmacology New Series," O. Eichler, A. Farah,
     H. Herken and A. D. Welch, eds., Springer-Verlog, New York
     28(3):1.

Malaveille, C., Bartsch, H., Barbin, A., Camus, A. M. and Montesano,
    R., 1975, Mutagenicity of vinyl chloride, chloroethylene-
    oxide, chloroacetaldehyde and chloroethanol, Biochem.
    Biophys. Res. Commun., 63:363.
Maltoni, C. and Lefemine, G., 1975, Carcinogenicity assays of
    vinyl chloride: Current results, Ann. N.Y. Acad. Sci.,
    246:195.
National Cancer Institute, 1976, "Carcinogenesis Bioassay of
    Chloroform", Natl. Tech. Info. Service No. PB264018/AS,
    March 1.
Pohl, L. R., Bhooshan, B. and Krishna, G., 1977, Phosgene: A
    potential metabolite of chloroform, Pharmacologist, 19:192.
Reitz, R. H., Gehring, P. J. and Park, C. N., 1978, Carcinogenic
    risk estimation for chloroform: An alternative to EPA's
    procedures, Food Cosmet. Toxicol., 16:511.
Reitz, R. H., Quast, J. F., Stott, W. T., Watanabe, P. G. and
    Gehring, P. J., 1980a, Pharmacokinetics and macromolecular
    effects of chloroform in rats and mice: Implications for
    carcinogenic risk estimation, in: "Water Chlorination:
    Environmental Impact and Health Effects, Vol. III"
    R. L. Jolley, W. A. Brungs, and R. B. Cummings, eds.,
    Ann Arbor Press, Ann Arbor, Michigan (In Press).
Reitz, R. H., Watanabe, P. G., McKenna, M. J., Quast, J. G., and
    Gehring, P. J., 1980b, Effects of vinylidene chloride on DNA
    synthesis and DNA repair in the rat and mouse: A compara-
    tive study with dimethylnitrosamine, Toxicol. Appl. Pharma-
    col., 52:356.
Simon, V. F., Kauhanen, K. and Tardiff, R. G., 1977, Mutagenic
    activity of chemicals identified in drinking water, 2nd
    International Meeting of the Environmental Mutagen Society,
    Edinburgh, Scotland.
Teorell, T., 1937a, Kinetics of distribution of substances adminis-
    tered to the body. I. The extravascular modes of adminis-
    tration., Arch. Intern. Pharmacodynamic., 57:205.
Teorell, T., 1937b, Kinetics of distribution of substances
    administered to the body. II. The intravascular modes of
    administration., Arch. Intern. Pharmacodynamic., 57:227.
Uehleke, H., Werner, T., Greim, H. and Kramer, M., 1977, Metabolic
    activation of haloalkanes and tests in vitro for mutageni-
    city, Xenobiotica, 7:393.
Wagner, J. G., 1968, Pharmacokinetics, Ann. Rev. Pharmacol., 8:67.
Wagner, J. G., 1971, "Biopharmaceutics and Relevant Pharmacoki-
    netics," 1st ed., Drug Intelligence Publications, Hamilton,
    II.
Watanabe, P. G., Hefner, R. E., Jr. and Gehring, P. J., 1976a,
    Vinyl chloride induced depression of hepatic non-protein
    sulfhydryl content and effects on bromosulphthalein (BSP)
    clearance in rats, Toxicology, 6:1.

Watanabe, P. G., McGowan, G. R., Madrid, E. O. and Gehring, P. J., 1976b, Fate of (^{14}C) vinyl chloride after single oral administration in rats, Toxicol. Appl. Pharmacol., 36:339.

Watanabe, P. G., McGowan, G. R., Madrid, E. O. and Gehring, P. J., 1976c, Fate of (^{14}C) vinyl chloride following inhalation exposure in rats, Toxicol. Appl. Pharmacol., 37:49.

Watanabe, P. G., Zempel, J. H., Pegg, D. G. and Gehring, P. J., 1978, Hepatic macromolecular binding following exposure to vinyl chloride, Toxicol. Appl. Pharmacol., 44:571.

Weisburger, J. H. and Williams, G. M., 1980, Chemical carcinogens, in: "Toxicology: The Basic Science of Poisons," J. Doull, C. D. Klaassen and M. O. Amdur, eds., Macmillan Publ. Co., New York, New York.

Weiss, M., Sziegoleit, W. and Forster, W., 1977, Dependence of pharmacokinetic parameters on the body weight, Int. J. Clin. Pharmacol., 15:572.

FUTURE DIRECTIONS IN THE STUDY OF BIOLOGICAL REACTIVE INTERMEDIATES

James A. Miller

McArdle Laboratory for Cancer Research
University of Wisconsin Medical Center
Madison, Wisconsin 53706

I am pleased to note that the speakers and discussants in this symposium have made my task in this closing presentation much easier, for future directions have been repeatedly voiced throughout this comprehensive and very stimulating meeting. So I would like to briefly sketch just a few paths in the future that appear attractive to me in basic and applied research on biological reactive intermediates.

A large variety of reactive species was discussed in this symposium. Most of these species were derived from xenobiotics or compounds apparently foreign to living cells. Most workers seem to accept this concept of "foreignness", but I have been slowly adopting a less definite view as more and more metabolites of cells resemble many synthetic chemicals. So I wonder if it is really true that cells have never before experienced structures or atomic groupings such as those found in many synthetic chemicals. The long evolutionary history of cells in adjusting to their changing environments, which included other cells and their products, the decomposition products of cells, and the pyrolysis products of organic matter, suggests that man has not produced much that is really new to cells in the living world. Thus, living cells appear to be the most experienced and clever organic and inorganic chemists of all. I have therefore come to view the precursors of reactive intermediates as a continuum of structures with varying degrees of similarity to cell metabolites and with varying probabilities of being found naturally in some living cells. Clearly, however, many exogenous compounds may be essentially foreign to many living systems and the levels of exposure may be measures of "foreignness."

Early in this symposium a fundamental group of reactive intermediates was discussed at length in several very illuminating, if
not entirely concordant, talks. This is the group comprised of
various endogenous activated forms of the oxygen atom and the oxygen
molecule or dioxygen. Several of these reactive intermediates
would be classified as being electron-deficient or electrophilic
in a general sense. The oxygen atom in the terminal cytochrome
P-450-oxygen-substrate complex appears to be electrophilic (Ullrich,
1979; White and Coon, 1980) and to react with many nucleophilic
substrates. Some of the products of these reactions are themselves
electrophilic (e.g., some epoxides). Furthermore, some of the
nucleophilic hydroxylated products can yield reactive electrophiles
by conversion to products with good negatively charged leaving
groups or functions that can potentially yield such leaving groups.

A wide variety of reactive intermediates that involve carbon,
nitrogen, and sulfur atoms as reactive electrophilic centers was
discussed at these meetings. One wonders how many other elements
could serve as electrophilic centers in reactive forms that might
be generated in the lipid and aqueous phases of cells. For example,
could oxygen or phosphorous atoms serve as reactive centers in
certain transient metabolites? This kind of question comes to
mind in the case of an interesting selenium compound. Selenium
sulfide has been used for some time for the treatment of seborrheic
dermatitis, dandruff, and some fungal infections of the skin.
Many shampoos which contain low concentrations of selenium sulfide
in detergent solutions are sold over the counter. Selenium sulfide
is not a well-defined compound and appears to be a mixture of
selenium mono- and disulfides in solid solution with elemental
sulfur and selenium (Windholz, 1976). In any event, recent tests
of this so-called selenium sulfide by the National Cancer Institute
in the United States indicated that this substance, given by
gavage at a level of 15 mg/kg body weight/day for about 100 weeks,
was hepatocarcinogenic in female and male rats (Carcinogenesis
Testing Program, 1980a). An increased incidence of tumors was
also noted in the livers and lungs of female mice treated in a
similar manner. These effects were not noted in male mice, but
these animals did not survive as long as the female mice. In other
tests, prolonged dermal application of selenium sulfide in mice did
not produce any tumors (Carcinogenesis Testing Program, 1980b).
Selenium compounds have long been known to produce toxic effects
in mammals, but selenium is also now recognized as an essential
nutritional trace element (Venugopal and Luckey, 1978; Jansson,
1980). This element is a component of several enzymes, such as
glutathione peroxidase in mammals and formate dehydrogenase and
glycine reductase in bacteria (Stadtman, 1980). The nutritional
level of selenium that is required is at least 100-fold lower than
that required for toxic effects. Selenium has had a varied history
in its relation to carcinogenesis in the liver. In an early

test KNH₄Se appeared to be a weak hepatocarcinogen in the rat.
However, later tests did not support the hepatocarcinogenicity
of selenium and other studies indicated that selenium had anti-
carcinogenic effects (Jansson, 1980). Thus, identification of the
active component of selenium sulfide in its hepatocarcinogenicity
in rats and studies of its presumed reactive forms in vivo and in
vitro would be of great interest, especially in view of the
apparent existence of a practical threshold somewhere between the
low nutritional and putative anti-carcinogenic levels of selenium
and the high toxic and possibly carcinogenic levels in mammals.
Of course, one should not lose sight of the possibility that a
sulfur atom in what is called selenium sulfide might be the electrophilic site in any reactive form in vivo. Electrophilic sulfur
atoms have been noted in the microsomal metabolism of carbon disulfide and other compounds (Dalvi et al., 1974, 1975; Neal, this
volume). These data on selenium sulfide exemplify the complexity
that may be encountered as our man-made and natural chemical
environments are examined ever more closely.

Most of the reactive intermediates reported at this meeting
appear to be species containing carbon or nitrogen atoms that
are deficient or potentially deficient in electron-pairs. These
species enter into reactions that fall into the broad SN1-SN2
spectrum of nucleophilic substitutions. New synthetic and naturally occurring compounds that are metabolizable to these types of
reactive intermediates will continue to be found for the indefinite
future.

Free radical reactive intermediates of halogenated hydrocarbons and other agents were also discussed to a considerable extent
at these meetings. These species are electrophilic in a general
sense and certainly react with a variety of nucleophiles as well
as with other free radicals. It seems likely that there will
be much more work on radicals as reactive intermediates of many
classes of compounds. The propensity of radicals to take part
in hit-and-run reactions will make it difficult to determine
the chemical changes induced by radicals in cellular informational
macromolecules such as nucleic acids and proteins in addition
to those caused by formation of adducts with these biopolymers.

In addition to electrophiles such as two-electron deficient
species and those with unpaired electrons, other electron-deficient
reactive species are known to organic chemists. Carbenes, for
example, appear to be reactive intermediates for some halogenated
and other compounds (Reiner and Uehleke, 1971; Mansuy et al., 1974;
Ullrich, 1980; Bridges, this volume). Carbenes are highly reactive
neutral species in which the carbon has a valence of two and can
exist in singlet and in triplet biradical states. These species
undergo a variety of interesting reactions involving additions to

unsaturated centers such as those in alkenes, insertions at carbon-
hydrogen bonds, and rearrangements to alkenes. These reactions may
occur to a far greater extent in the metabolism of carbon compounds
than is now appreciated. Similarly, the metabolism of nitrogen-
containing compounds to the corresponding electron-deficient
nitrenes may occur much more frequently than is now recognized.

Indeed, I would go farther and state my conviction that most,
if not all, of the various types of reactive intermediates discovered
by organic chemists - even those found under reaction conditions
that seem far removed from intracellular conditions - are likely
to occur in the metabolism of some organic compounds in the various
aqueous and hydrophobic regions of cells. Whether or not these
intermediates can be detected in vivo and in vitro will depend on
how well investigators can match the great subtlety exhibited by
enzymes and transient reactive intermediates. So it seems to me
that the full range of different metabolically formed reactive
intermediates is not yet fully appreciated.

Many other aspects of studies of reactive intermediates are
likewise in their early stages of development. One topic is the
formation of reactive intermediates in different cellular sites.
Cells clearly consist of many cooperative metabolic compartments,
and different sites of formation of a given reactive intermediate
could lead to very different toxicological outcomes. The occurrence
of oxidations in the nuclear envelope versus the occurrence of
similar oxidations in the endoplasmic reticulum is a case in point
(Kasper, 1976).

Another topic of interest is the intracellular transport
of reactive intermediates by processes other than by diffusion
through lipid bilayer membranes. This possibility has been proposed and is likely to receive considerably more attention.
Especially interesting is the concept of transport of reactive
intermediates through aqueous phases while being protected by
hydrophobic regions in proteins and lipoproteins (Tanford, 1980)
as suggested by recent studies (Mainigi and Sorof, 1977; Ketterer,
1980).

It is evident that various gross phenotypic expressions of
reactions in vivo of reactive intermediates have been described.
Carcinogenic, mutagenic, teratogenic, necrogenic, and immunogenic
effects have been prominent among these expressions. Still other
acute and long-term consequences of the effects of reactive intermediates in vivo seem likely to be recognized as research in this
field progresses, especially with the advancement of new techniques
which will permit old and new concepts to be tested.

The pharmacokinetics of reactive intermediates will certainly
also receive more attention in the future in the prediction and

explanation of various toxic effects of chemicals. This was exemplified at this meeting in an impressive manner in Dr. Jollow's
use of Dr. Gillette's kinetic model (Gillette et al., 1978).
As important details in the formation and disposition of reactive
intermediates become known, these kinetic approaches will be
necessary. This is surely the case, for example, in studies
on the metabolism of 2-acetylaminofluorene (Razzouk et al., 1980;
Roberfroid, this volume) and in the interrelated metabolisms of
phenacetin and acetaminophen (Gillette, this volume), where there
are many detailed data now available.

Covalent binding of reactive intermediates has received
much attention at this meeting. While I remain persuaded that
covalent binding is required for the carcinogenic action of most,
if not all, of the well-studied chemical carcinogens, I also
remain on the look-out for carcinogens that may interact with
informational macromolecules by non-covalent linkages as noted
for many pharmacological substances (Martin, 1978). These bonds
may be easily disrupted in the usual searches for adducts. And,
of course, free radicals may initiate chemical changes, in DNA
for example, without adduct formation. These possibilities apply
equally to toxic processes other than carcinogenesis.

The detection and study of reactive intermediates has been
accomplished by many chemical and biological assays, and further
developments in increased sensitivity and specificity can be
expected. For example, in addition to the detection of covalently
bound adducts of intermediates with components of nucleic acids
and proteins by use of radioactive precursors, antibodies of
high specificity for some of these adducts provide an additional
and more sensitive tool (Poirier et al., 1977, 1979, 1980; Muller
and Rajewsky, 1978; DeMurcia et al., 1979). An ultrasensitive
enzymatic radioimmunoassay for the measurement of DNA adducts
of 2-acetylaminofluorene has been reported (Hsu et al., 1980).

Increases in sensitivity of assays for biological activities
such as mutagenicity can also be expected. Indeed, very recently,
a bioluminescence assay for mutagenicity was published (Ulitzur
et al., 1980) which has the potential of extending the sensitivity
of the Ames assays by a factor of 100 or more. However, I must
admit that, in this case, I fear the public misunderstanding
this may bring. The health hazards of the mutagens now detectable
in all manner of biological material, including foods, are certainly
not clear at present. Assays that detect smaller and smaller
amounts of mutagenic substances or very weak mutagens will surely
compound our present problems in the assessment of the importance
of these agents in our environment.

Lastly, I want to mention a practical aspect of research
on reactive intermediates that I believe will demand more and

more attention. This is the detection and estimation of metabolites and especially of adducts of reactive intermediates in
the available tissues, body fluids, and excreta of humans exposed
to various exogenous toxic agents. The extremely sensitive assays
that I just mentioned will be of great value here, for the levels
of metabolites and adducts would be a measure of the degree of
exposure to toxic agents. Cancer patients receiving chemotherapy
with known amounts of alkylating agents would constitute useful
subjects for basic studies of this sort. The data derived from
such studies should assist in the difficult task of extrapolating
data with toxic agents in experimental animals to humans. The
same approach would apply in the study of toxic agents in human
cell and organ cultures.

Now I wish to make a personal comment. My wife and I have
had the good fortune of participating in the early studies on
biological reactive intermediates and in seeing this field grow
in importance. We now find ourselves overwhelmed with the literature in this field and, even worse, we find the sophistication
of some of this literature, especially in biophysics and quantum
chemistry, is beyond our full comprehension. Of course, this
is the usual progression of research and we are really delighted
to see that this field has come so far, that it is so vigorous
and so full of promise, and that it is in such good hands.

I would like to close by expressing on behalf of all of us
who journeyed to this delightful campus our very great thanks to
Professor Parke, to Dr. Gibson and his colleagues, to Dr. Snyder
and the Organizing Committee, and last, but not least, to the
several academic and corporate sponsors in this country and the
United States for this outstanding and highly stimulating meeting.

REFERENCES

Carcinogenesis Testing Program and National Toxicology Program,
     1980a, Bioassay of selenium sulfide for possible carcinogen-
     icity (gavage study), National Cancer Institute, DHHS Public
     No. (NIH) 80-1750, U. S. Department of Health and Human
     Services, Washington, D. C.
Carcinogenesis Testing Program and National Toxicology Program,
     1980b, Bioassay of selenium sulfide for possible carcinogen-
     icity (dermal study), National Cancer Institute, DHHS Public
     No. (NIH) 80-1753, U. S. Department of Health and Human
     Services, Washington, D. C.
Dalvi, R., Hunter, A., and Neal, R., 1974, Studies of the metabo-
     lism of carbon disulfide by rat liver microsomes, Life Sci.,
     14:1785.

Dalvi, R., Hunter, A., and Neal, R., 1975, Toxicological implica-
     tions of the mixed-function oxidase catalyzed metabolism of
     carbon disulfide, Chem.-Biol. Interact., 10:349.
De Murcia, G., Lang, M.-C. E., Freund, A.-M., Fuchs, R. P. P.,
     Daune, M. P., Sage, E., Leng, M., 1979, Electron microscopic
     visualization of N-acetoxy-2-acetylaminofluorene binding sites
     in ColEl DNA by means of specific antibodies, Proc. Natl.
     Acad. Sci. USA, 76:6076.
Gillette, J. R., Hinson, J. A., and Andrews, L. S., 1978, Pharma-
     cokinetic aspects of the formation and inactivation of
     chemically reactive metabolites, in: "Polycyclic Hydrocarbons
     and Cancer: Chemistry, Molecular Biology and Environment,"
     vol. 1, p. 375, H. V. Gelboin and P. O. P. Ts'o, eds.,
     Academic Press, New York.
Hsu, I. C., Poirier, M. C., Yuspa, S. H., Yolken, R. H., and
     Harris, C. C., 1980, Ultrasensitive enzymatic radioimmuno-
     assay (USERIA) detects femtomoles of acetylaminofluorene-
     DNA adducts, Carcinogenesis, 1:455.
Jansson, B., 1980, The role of selenium as a cancer-protecting
     trace element, in: "Metal Ions in Biological Systems," H.
     Sigel, ed., p. 281, Marcel Dekker, New York.
Kasper, C. B., 1976, Chemical and enzymic composition of the
     nuclear envelope, in: "Cell Biology, I," P. L. Altman and
     D. D. Katz, eds., p. 395, Fed Amer. Soc. Exper. Biol.,
     Bethesda, Maryland.
Ketterer, B., 1980, Interactions between carcinogens and proteins,
     Brit. Med. Bull., 36:71.
Mainigi, K. D., and Sorof, S., 1977, Evidence for a receptor
     protein of activated carcinogen, Proc. Natl. Acad. Sci.,
     USA, 74:2293.
Mansuy, D., Nastainczyk, and Ullrich, V., 1974, The mechanism
     of halothane binding to microsomal cytochrome P-450, Naunyn-
     Schmiedebergs Arch. Pharmakol., 285:315.
Martin, Y. C., 1978, "Quantitative Drug Design," Chapter 3:
     Noncovalent interactions of importance in biological systems,
     Marcel Dekker, New York.
Muller, R., and Rajewsky, M. F., 1978, Sensitive radioimmunoassay
     for detection of O^6-ethyldeoxyguanosine in DNA exposed to
     the carcinogen ethylnitrosourea, Z. Naturforsch., 33c:897.
Poirier, M. C., Dubin, M. A., and Yuspa, S. H., 1979, Formation
     and removal of specific acetylaminofluorene-DNA adducts
     in mouse and human cells measured by radioimmunoassay, Cancer
     Res., 39:1377.
Poirier, M. C., Santella, R., Weinstein, I. B., Grunberger, D.,
     and Yuspa, S. H., 1980, Quantitation of benzo(a)pyrene-deoxy-
     guanosine adducts by radioimmunoassay, Cancer Res., 40:412.
Poirier, M. C., Yuspa, S. H., Weinstein, I. B., and Blobstein,
     S., 1977, Detection of carcinogen-DNA adducts by radioimmuno-
     assay, Nature, 270:186.

Razzouk, C., Mercier, M., and Roberfroid, M., 1980, Characteriza-
    tion of the guinea pig liver microsomal 2-fluorenylamine and
    N-2-fluorenylacetamide N-hydroxylase, Cancer Letters, 9:123.
Reiner, O., and Uehleke, H., 1971, Bindung von Tetrachlorkohlen-
    stoff an reduziertes mikrosomales Cytochrome P-450 und an
    Häm, Hoppe-Seyler's Z. Physiol. Chem., 352:1048.
Stadtman, T. C., 1980, Selenium-dependent enzymes, Ann. Rev.
    Biochem., 49:93.
Tanford, C., 1980, "The Hydrophobic Effect: Formation of Micelles
    and Biological Membranes,", 2nd edition, John Wiley and Sons,
    New York.
Ulitzur, S., Weiser, I., and Yannai, S., 1980, A new, sensitive
    and simple bioluminescence test for mutagenic compounds,
    Mutation Res., 74:113.
Ullrich, V., 1979, Cytochrome P-450 and biological hydroxylation
    reactions, in: "Topics in Current Chemistry," vol. 83,
    p. 67, F. L. Boschke, ed., Springer-Verlag, New York.
Venugopal, B., and Luckey, T. D., 1978, "Metal Toxicity in
    Mammals," vol. 2, Plenum Press, New York.
White, R. C., and Coon, M. J., 1980, Oxygen activation by Cyto-
    chrome P-450, Ann. Rev. Biochem., 49:315.
Windholz, M., ed., 1976, "The Merck Index," 9th edition, entry
    No. 8187, Merck & Co., Rahway, N.J.

DISCUSSION

Ivanetich opened the discussion with a question for De Matteis concerning the evidence available to indicate that the modified porphyrin diffuses off of cytochrome P-450, thus allowing P-450 to bind heme from the heme pool. De Matteis responded that when cytochrome P-450 is destroyed by 2-allyl-2-isopropylacetamide (AIA) both *in vivo* and *in vitro*, one can demonstrate the distribution of radioactivity from heme into the supernatant fraction of the liver homogenate and this is formed entirely from the modified heme product. De Matteis further stated that there is evidence both *in vitro* from his laboratory and *in vivo* from that of the San Francisco group that exogenous heme can be utilized for the production of this particular compound. Others have obtained evidence that along with the ability of partially destroyed cytochrome P-450 to produce heme pigments there is also a partial reconstitution of enzymatic activity suggesting that the heme which one injects can be taken up by the apoprotein to regenerate the active cytochrome P-450; in doing so it can be acted on in a suicidal manner and give rise to the modified heme. In addition, it has been shown that the partially depleted apoprotein of cytochrome P-450 resulting from treatment with this unsaturated drug can be partially reconstituted by exogenous heme. Sugimura commented that AIA suppresses the amount but not the activity of liver catalase when given to rats or mice and that it was once understood that AIA inhibited the biosynthesis of the apoprotein of liver catalase *in vivo*. De Matteis responded that the most attractive explanation is that AIA destroys heme in the liver by a reactive process such that the cytochrome P-450 pool of heme is depleted as well as the precursor pool of heme which is responsible for the regulation of AIA synthesis; it is this pool which provides heme for the biosynthesis of hemoglobin. Consequently catalase biosynthesis may also be limited by the supply of heme. There are, in fact, some experiments reported in the Japanese literature which indicate that the apoprotein of catalase which has been isolated is partially heme deficient. De Matteis feels that this is a most attractive unifying hypothesis. De Matteis then responded affirmatively to a question from Balazs who asked if the cholestatic effects of ethynylestradiol or norethynodrel may possibly be attributable to their porphyrin-inducing properties.

1433

Following the presentation by Lutz on covalent binding to DNA as a quantitative indicator of carcinogenesis, Magee asked whether the covalent binding index (CBI) was capable of reflecting marked differences in species susceptibility to a carcinogen as, for example, in the case of aflatoxin binding in the rat vs the mouse. Lutz stated that he had studied that point and had found that the CBI did accurately reflect such differences in the covalent binding. For example, binding to mouse liver DNA was forty times lower than the binding to rat liver DNA. However, as to differences in sensitivity to carcinogenicity, the species differences are more qualitative than quantitative. Kalf then commented on Lutz's presentation by stating that we tend to equate covalent binding to DNA with binding to nuclear DNA. He pointed out that there is a second genetic system in the cell, located in the mitochondrion and that recent work from several laboratories has shown that many of these carcinogenic compounds are capable of binding covalently to mitochondrial DNA to a far greater extent than they bind to nuclear DNA. In fact, recent work by Avadhani has demonstrated that aflatoxin not only covalently binds to DNA but that it can be activated by a mixed function oxidase system in the organelle. Furthermore, covalent binding of both aflatoxin and benzo(a)pyrene has been shown to interfere with both mitochondrial DNA and RNA synthesis. Since the mitochondrion is a very lipophilic organelle, these compounds, which tend to be lipophilic, are concentrated within the organelle where they find a mitochondrial genome which is "naked" in the sense that it is not protected by being organized into a highly compact structure like the nucleosome. This should be kept in mind when we study covalent binding of toxic compounds and we should not just assume that all of the covalent binding is to nuclear DNA and reflects the only detrimental effect.

Lutz asked Magee how much DNA was present in the scintillation vial in the experiment without proline in order that he might calculate the limit of detection of a CBI for nitrosoproline. Magee responded that the vial contained 5 mg of DNA whereupon Lutz calculated a limit of detection for the CBI of about 0.5 indicating no cytotoxicity for nitrosoproline. Lutz also questioned whether the figure of 7 could have resulted from the incorporation of ^{14}C into the bases of the small intestine in the case of nitrosocimetidine. Magee stated that the label was 7-methylguanine. Archer pointed out that nitrosoproline has 3 hydrogen atoms alpha to the N-nitroso group and hence should be capable of being metabolized to an alkylating agent. Archer has shown that it can be chemically oxidized to yield an intermediate that alkylates p-nitrobenzyl pyridine. On the basis of this evidence, Archer felt that the lack of carcinogenic activity by nitrosoproline is probably related to its rapid excretion from the intact animal as indeed Magee's results might also suggest. Very little of the administered compound probably gets to sites inside possible target cells where it can be activated. Magee responded that small amounts of radioactivity were found in the various fractions of the homogenate.

Snyder asked Roberts after his talk on advances in DNA repair whether there is any indication that one of the things that reactive metabolites might do is to damage the DNA repair mechanism. Roberts answered that this has been proposed, but stated that if one gives a high dose, repair will seem to be inhibited; however, one doesn't know if repair is being inhibited or the repair mechanism is being saturated. Magee made the observation that some time ago Dr. A.E. Pegg reported that labeled methanol could be recovered from an incubation medium which contained DNA that had been labeled with a radioactive carcinogenic methylating agent and inquired as to the position regarding this at the present time. Roberts responded that there is currently still no real evidence for the removal of O^6 methylguanine by these enzymes.

After Greim's presentation on threshold levels in toxicology, Suetter commented that dinitrochlorobenzene (DNCB) is clinically useful because of its ability to provide almost immediate contact sensitization, thus mobilizing the immunological system in a more or less nonspecific reaction. In Suetter's view this gives reason to doubt the value of any mutagenicity test since DNCB, or products formed from it, will be trapped by the immunological system in such a way that they cannot reach the physiological system. Greim stated that he is aware of the effect of DNCB on the immunological system and feels that topical application of such a highly reactive and mutagenic compound, which will decrease glutathione levels in the skin, is rather dangerous. Vainio pointed out that there are some disputable epidemiological data on the carcinogenicity of chloroprene as well as some positive short-term studies on the potential genotoxicity of the compound. He then inquired on what basis Greim had suggested that while high dose studies were of doubtful value, yet his (Greim's) results suggested just the opposite – that the high dose study is of interest because it shows toxic potential; and it then requires interpretation to decide whether toxicity could possibly be achieved and whether hazards are involved. McLean also stressed that we should not imply that compounds that require high doses to produce cancer pose no great risk to the public. Greim was of the opinion that McLean was just looking at the problem from the other side and pointed out that there are two objectives in running these toxic studies. One is to demonstrate the toxic effects which can be obtained at extremely high doses and the other is to discuss the relevance of such high dose studies to assessing cancer risk. In discussing relevance one has to be aware of the inactivating processes and the fact that they can be overwhelmed.

In a discussion of risk assessment, Boobis asked Watanabe, who had presented the paper of Dietz et al., whether one of the consequences of his results with chloroform is that other cytotoxic compounds such as acetaminophen will also enhance tumor formation. For example, if a low dose of a carcinogen such as dimethylnitrosamine (DMN) is given after chloroform, will this lead to a higher tumor incidence from DMN? Watanabe suspected that might be the case and predicted that any stimulus would probably increase tumorgenicity in an organ that is also assaulted with a genetically active chemical. Watanabe further stated that it would not be surprising to him to see an enhancing effect but he does not use the term promoting effect; promotion or promoting effect has other connotations and there are particular aspects that are peculiar or restricted to skin. People tend to generalize from skin to other body organs Watanabe stated that for these reasons he tends to use the term "enhancing" rather than "promoting" for these situations and that such activities in association with initiators could be expected to enhance that effect.

LIST OF PARTICIPANTS

Ahmed, A.E., University of Texas Med. Branch, Galveston, TX, USA
Ahr, H.J., Universitat D. Saarlandes, Homburg, Fed. Rep. Germany
Albano, E.,Brunel University, Uxbridge, Middx. UK
Albrecht, J.,Polish Academy of Sciences, Warsaw, Poland
Amos, H.E., ICI Pharmaceuticals, Macclesfield, Cheshire, UK
Anders, M.W.,University of Minnesota, Minneapolis, MN, USA
Archer, M.C., Ontario Cancer Institute, Toronto, Canada
Aune, T.,National Inst. of Public Health, Oslo, Norway
Awouters, F., Dept. of Pharmacology, Janssen Pharmaceutica, Belgium
Baarke, M., University of Turku, Finland
Baillie, T., University of California, San Francisco, CA, USA
Baines, P.J., Beecham Pharmaceuticals Research, Epsom, Surrey, UK
Balazs, T.,Food and Drug Admin., Washington, DC, USA
Bartsch, H., Int. Agency for Research on Cancer, Lyon, France
Becher, G., Central Inst. for Industrial Research, Oslo, Norway
Bend, J.R., N.I.E.H.S., Research Triangle Park, NC, USA
Benedetto, C., Brunel University, Uxbridge, Middx., UK
Bentley, P., Ciba-Geigy AG, Basel, Switzerland
Bock, K.W., University of Gottingen, Fed. Rep. Germany
Bolt, H.M., University Mainz, Fed. Rep. Germany
Bonanomi, L., Zambon Farmaceutici S.p.A., Bresso, Italy
Boobis, A.R., Royal Postgraduate Medical School, London, UK
Bories, G., INRA, Toulouse, France
Boyd, M.R., National Cancer Institute, Bethesda, MD, USA
Brabec, M.J. University of Michigan, Ann Arbor, MI, USA
Bradshaw, T.K., Shell Research Centre, Sittingbourne, Kent, UK
Bratt, H., ICI Ltd., Alderley Park, Cheshire, UK
Brooks, G.T., University of Sussex, Brighton, Sussex, UK
Bridges, J.W., University of Surrey, Guildford, Surrey, UK
Comporti, M., Sienna University, Italy
Capdevila, J. University of Texas Health Science Center, Dallas,
  TX., USA
Charlesworth, J.D., ICI, Macclesfield, Cheshire, UK
Chasseaud, L.F., Huntingdon Research Centre, Huntingdon, UK
Cheeseman, Mr., Brunel University, Middx., UK
Chipman, J.K., St. Mary's Hospital Medical School, London, UK
Corcoran, G.B., III, Baylor College of Medicine, Houston, TX., USA

Cottrell, R.C., B.I.B.R.A., Carshalton, Surrey, UK
Damani, L.A., University of Manchester, Manchester, UK
Dansette, P., Ecole Normale Superieure, Labo de Chimie, Paris,
  France
Davies, D.S., Royal Postgraduate Med. School, London, UK
Davis, A.F., Beecham Products, Leatherhead, Surrey, UK
Dent, J.G., C.I.I.T.,Research Triangle Park, NC, USA
De Matteis, F., M.R.C. Toxicology Labs., Carshalton, Surrey, UK
Demeunynck, M., Universite des Sciences et Techniques de Lille,
  Lille, France
De Pierre, J.W., Arrhenius Laboratoriet, Stockholm University,
  Stockholm, Sweden
Devalia, J.L., Rayne Institute, University College, London, UK
Dock, L., Karolinska Institute, Stockholm, Sweden
Dow, J., Centre De Recherche Delalande, Rueil Malmaison, France
Drew, A., Information Retrieval Ltd., London, UK
Dybing, E., National Institute of Public Health, Oslo, Norway
Eggers, R., Medizinische Universitatsklinik, Essen, Fed. Rep.
  Germany
Elbers, R., Pharmakologisches Institut, Munchen, Fed. Rep. Germany
Elcombe, C.R., Central Toxicology Lab., ICI Ltd., Macclesfield,
  Cheshire, UK
Elliott, B.M., ICI Ltd., Macclesfield, Cheshire, UK
Elsom, L.F., Huntingdon Research Centre, Huntingdon, Cambs., UK
Estabrook, R.W., Texas Southwestern Medical School, Dallas, TX, USA
Eyer, P., University Munchen, Munchen, Fed. Rep. Germany
Faigle, J.W., Ciba-Geigy Ltd., Basle, Switzerland
Farmer, P.B., M.R.C. Toxicology Unit, Carshalton, Surrey, UK
Fawell, J.K., Water Research Centre, Marlow, Bucks, UK
Ferrari, V. Zambon S.p.A., Bresso, Milan, Italy
Finch, R., Rayne Institute, London, UK
Ford-Hutchinson, A.W., Kings College Hospital Med. School, London, UK
Forster, R., Water Research Centre, Marlow, Bucks, UK
Foster, P., B.I.B.R.A., Carshalton Beeches, Surrey, UK
Frank, H. Institute f. Toxikologie, Tubingen, Fed. Rep. Germany
Franklin, M.R., University of Utah, Salt Lake City, UT, USA
Foster, A.B., Chester Beatty Research Inst., Fulham Road, London, UK
Freedman, R.B., University of Kent, Canterbury, Kent, UK
Furukawa, K., Fermentation Research Institute, Japan
Gehring, P.H., Dow Chemical Co., Midland, MI, USA
Gelbke, P., BASF Ludwigshafen, Fed., Rep. Germany
Gibson, G.G., University of Surrey, Guildford, Surrey, UK
Gibson, J.E., C.I.I.T., Research Triangle Park, NC, USA
Gielen, J., University Liege, Liege, Belgium
Gillette, J.R., National Heart, Lung and Blood Inst., Bethesda,
  MD, USA
Golberg, L., C.I.I.T., Research Triangle Park, NC, USA
Goldstein, B.D., Rutgers Medical School, Piscataway, NJ, USA
Gooderham, N.J., Chelsea College, University of London, UK
Gorrod, J.W., Chelsea College, University of London, UK

Green, M.H.L., University of Sussex, Brighton, Sussex. UK

Greim, H.A., Dept. of Toxicology, Neuherberg, Fed. Rep. Germany

Gribble, A.D., Smith Kline & French Ltd., Welwyn Garden City, Herts, UK

Gry, J, National Food Institute, Denmark

Guengerich, P., Vanderbilt University, Nashville, TN, USA

Guenthner, T.M., University of Mainz, Fed. Rep. Germany

Gumbrecht, J., University of Utah, Salt Lake City, UT, USA

Gwynn, J., Smith Kline & French Ltd., Welwyn Garden City, Herts, UK

Hamnett, A.F., University of Sussex, Brighton, Sussex, UK

Hansen, L.G., University of Illinois, Urbana, IL, USA

Hanzlik, R.P., University of Kansas, Kansas City, KS, USA

Hawkins, D.R., Huntingdon Research Centre, Huntingdon, Cambs., UK

Hemminki, K., Inst. of Occupational Health, Helsinki, Finland

Henschler, D., University of Wurtzburg, Fed. Rep. Germany

Herzberg, S., Shell Internationale Research, The Hague, Netherlands

Hesse, S., Ges. F. Strahlen u Umweltforschung, Munchen, Fed. Rep. Germany

Hibberd, A.R., Northwick Park Hospital, Harrow, Middx., UK

Hildebrandt, A., Free University of Berlin, Fed. Rep. Germany

Hirom, P.C., St. Mary's Hospital Med. School, London, UK

Hjordis, T., Karolinska Institute, Stockholm, Sweden

Hobbs, D.C., Pfizer Inc., Groton, CT, USA

Hodgson, E., North Carolina State University, NC, USA

Holden, K.G., Smith Kline & French Labs., Philadelphia, PA, USA

Hook, J.B., I.C.I. Ltd., Alderley Park, Cheshire, UK

Housley, J.R., Boots Co. Ltd., Nottingham, Notts., UK

Hsieh, D.P.H., University of California, Davis, CA, USA

Huckle, K.R., St. Mary's Hospital Medical School, London, UK

Irons, R.D., C.I.I.T., Research Triangle Park, NC, USA

Ivanetich, K.M., Medical School, S. Africa

Jansson, I., University of Connecticut Health Center, Farmington, CT, USA

Jauch, R., Hoffman La Roche, Basle, Switzerland

Jensen, A.A., Copenhagen, Denmark

Jerina, D.M., National Institute of Health, Bethesda, MD, USA

Johnson, P.M., Fisons Ltd., Leics, UK

Jollow, D.J., Medical University of South Carolina, Charleston, SC., USA

Jones, R.B., I.C.I. Ltd., Cheshire, UK

Juchau, M.R., University of Washington, Seattle, WA, USA

Kaderbhai, M.A., University of Kent, Canterbury, Kent, UK

Kalf, G.F., Thomas Jefferson University, Philadelphia, PA, USA

Kampffmeyer, University Munchen, Munchen, Fed. Rep. Germany

Kappus, H., University Dusseldorf, Fed. Rep. Germany

Katerberg, G.J., Philips-Duphar B.V., Netherlands

Kato, R., Keio University, Tokyo, Japan

Ketterer, B., Courtauld Inst. of Biochemistry, Middlesex Hospital Med. School, London, UK

Kitta, D., German Cancer Research Centre, Heidelberg, Fed. Rep. Germany

Klaassen, C.D., University of Kansas Medical Center, Kansas City, KS, USA

Kocsis, J.J., Thimas Jefferson University, Philadelphia, PA, USA

Kolar, G.R., German Cancer Research Center, Heidelberg, Fed. Rep. Germany

Laaksonen, M., University of Kuopio, Finland

Lake, B.G., B.I.B.R.A., Carshalton, Surrey, UK

Larsen, J.C., National Food Institute, Denmark

Leibman, K.C., University of Florida, Gainesville, FL, USA

Lener, M., Rome, Italy

Lenk, W., Pharmakol. Inst., University Munchen, Munchen, Fed. Rep. Germany

Lewis, J., Harrow College of Further Education, Harrow, Middx., UK

LHomme, J., University Lille, Lille, France

Lington, A.W., Exxon Corporation, USA

Lock, E.A., I.C.I. Ltd., Macclesfield, Cheshire, UK

Longacre, S.L., Thomas Jefferson University, Philadelphia, PA, USA

Lowery, C., I.C.I. Pharmaceuticals, Macclesfield, Cheshire, UK

Lutz, W.K., University of Zurich, Switzerland

Luu-The, V., University of Louvain, Brussels, Belgium

Magee, P.N., Temple University, Philadelphia, PA, USA

Malik, J.K., Punjab Agricul. University, Ludhiana, India

Malvoisin, E., University of Louvain, Brussels, Belgium

Manson, M., MRC Toxicology Unit, Carshalton, Surrey, UK

Marletta, M., Massachusetts Institute of Technology, Cambridge, MA, USA

McIntosh, P., University of Kent, Canterbury, Kent, UK

McLean, A.E.M., University College School of Medicine, London, UK

Meijer, J., University of Stockholm, Sweden

Menzel, D.B., Duke University, Durham, NC, USA

Menzer, R.E., University of Maryland, MD, USA

Mercier, M., University of Louvain, Brussels, Belgium

Metcalfe, S., M.R.C. Labs., Carshalton, Surrey, UK

Meyer, D.J., Courtauld Inst. of Biochemistry, Middlesex Hospital Med. School, London, UK

Metzler, M., University of Wurzburg, Fed. Rep. Germany

Midgley, I., Huntingdon Research Centre, Huntingdon, Cambs., UK

Midskov, C., A/S Dumex, Copenhagen, Denmark

Millburn, P., St. Mary's Hospital Medical School, Univ. of London, UK

Miller, E.C., University of Wisconsin, Madison, WI, USA

Miller, J.A., University of Wisconsin, Madison, WI, USA

Mitchell, J.R., Baylor College of Medicine, Houston, TX, USA

Mitchell, S.C., University of Birmingham, UK

Moldeus, P., Karolinska Institute, Stockholm, Sweden

Nakano, R., Japan

Nastainczyk, W., University des Saarlands, Fed. Rep. Germany

Neal, G.E., M.R.C. Labs, Carshalton, Surrey, UK

Neal, R.A., Vanderbilt University, Nashville, TN, USA

Nelson, S., University of Washington, Seattle, WA, USA

Norling, A., Oslo, Finland

Notarianni, L.J., University of Bath, Bath, Avon, UK
Oesch, F., University of Mainz, Fed. Rep. Germany
Orrenius, S., Karolinska Institute, Stockholm, Sweden
Orton, T., I.C.I. Ltd., Alderley Park, Cheshire, UK
Ottenwalder, H., University Dusseldorf, Fed. Rep. Germany
Pai, V., Chelsea College, University of London, UK
Parke, B.K., Liverpool University, Liverpool, UK
Parke, D.V., University of Surrey, Guildford, Surrey, UK
Parker, Dr., I.C.I. Ltd., Alderley Park, Cheshire, UK
Parkinson, A., University of Guelph, Ontario, Canada
Phillips, J.C., B.I.B.R.A., Carshalton, Surrey, UK
Piloth, A., Stockholm University, Sweden
Powis, G., Mayo Clinic, Rochester, MN, USA
Pratt, G.E., University of Sussex, Brighton, Sussex, UK
Prough, R.A., University of Texas, Dallas, TX, USA
Prout, M.S., I.C.I. Ltd., Alderley Park, Cheshire, UK
Pue, M. A., St. Mary's Hospital Medical School, University of
    London, UK
Quattrucci, E., University of Surrey, Guildford, Surrey, UK
Rahimtula, A., Karolinska Institute, Stockholm, Sweden
Rasmussen, F., University of Oslo, Oslo, Norway
Razzouk, C., University de Louvain, Brussels, Belgium
Recknagel, R.O., Case Western University, Cleveland, OH, USA
Reed, D.J., Oregon State University, Corvallis, OR, USA
Remmer, H., University of Tubingen, Fed. Rep. Germany
Renwick, A.G., University of Southampton, Southampton, Hants, UK
Reynolds, E.S., University of Texas Medical Branch, Dallas, TX, USA
Rhodes, C., I.C.I. Ltd., Macclesfield, Cheshire, UK
Roberfroid, M.B., University de Louvain, Brussels, Belgium
Roberts, J.J., Institute for Cancer Research, Pollards Wood Research
    Station, Chalfont St. Giles, Bucks, UK
Robertson, J.A., Inveresk Research International Ltd., Edinburgh, UK
Rose, M.S., I.C.I. Ltd., Macclesfield, Cheshire, UK
Rowe, J.S., School of Pharmacy, University of London, London, UK
Salmon, A.G., I.C.I. Ltd., Macclesfield, Cheshire, UK
Saxholm, Dr., University of Oslo, Norway
Schelin, C., Biochemistry Chemical Centre, Lund, Sweden
Schenkman, J.B., University of Connecticut Medical School,
    Farmington, CT, USA
Scheulen, M.E., West German Tumor Center, Essen, Fed. Rep. Germany
Schleyer, H., University of Pennsylvania, Philadelphia, PA, USA
Schmidt, U., Bayer AG, Wuppertal, Fed. Rep. Germany
Schrenk, D., German Cancer Research Center, Heidelberg, Fed. Rep.
    Germany
Schwarz, L.R., Department of Toxicology, GSF, Neuherberg, Fed. Rep.
    Germany
Schwarz, M., German Cancer Research Center, Heidelberg, Fed. Rep.
    Germany
Seago, A., Chelsea College, London University, UK
Seidegard, J., Stockholm University, Stockholm, Sweden

Seutter, E., University of Nijmogen, Netherlands
Seutter-Berlage, F., University of Nijmogen, Netherlands
Sims, P., Chester Beatty Research, London, UK
Singh, J., Institute fur Toxicologie und Biochemie, Neuherberg,
   Fed. Rep. Germany
Sipes, I.G., University of Arizona, Tucson, AZ, USA
Slater, T.F., Brunel University, Uxbridge, Middx., USA
Smith, C.V., Baylor College of Medicine, Houston, TX, USA
Smith, M., University College, London, UK
Smith, M.T., Biochemistry Department, Med. College of St. Bartholo-
   mew's Hospital, London, UK
Smuckler, E., University of California,San Francisco, CA, USA
Snyder, R., Thomas Jefferson University, Philadelphia, PA, USA
Soderlund, E., National Institute of Public Health, Oslo, Norway
Srai, K.S., Courtauld Inst. of Biochemistry, Middlesex Hospital
   Med. School, London, UK
Stripp, B., National Heart, Lung & Blood Inst., MD, USA
Sugimura, T., National Cancer Research Institute, Tokyo, Japan
Svensson, S-A, Karolinska Institute, Stockholm, Sweden
Sweatman, T.W., University of Southampton, Hants, UK
Tennekes, H., German Cancer Research Center, Heidelberg, Fed. Rep.
   Germany
Thorgeirsson, S.S., National Cancer Institute, Bethesda, MD, USA
Timbrell, J.A., Royal Postgraduate Medical School, University of
   London, UK
Tomasi, A., Brunel University, Uxbridge, Middx., UK
Tunek, A., University Lund, Sweden
Uehleke, H., Fed. German Health Office, Berlin, Fed. Rep. Germany
Ullrich, V., University of Saarland, Fed. Rep. Germany
Upadhyaya, B.K., Wellcome Research Labs., Beckenham, Kent, UK
Vadi, H., Oregon State University, Corvallis, OR, USA
Vainio, H., Institute of Occupational Health, Helsinki, Finland
Van Bladeren, P.J., University of Leiden, Netherlands
Van Cantfort, J., University Liege, Belgium
Van Dyke, R.A., Mayo Clinic, Rochester, MN, USA
Vodicnik, M.J., Medical College of Wisconsin, Madison, WI, USA
Volp, R.F., University Gottingen, Fed. Rep. Germany
Walker, C.H., Reading University, Berks, UK
Walklin, C.M., University of Kent, Kent, UK
Walsh, C.,M.I.T., Cambridge, MA, USA
Walsh, S.T., I.C.I. Ltd., Alderley Park, Cheshire, UK
Wareing, M., Chelsea College, University of London, UK
Waring, R.H., University of Birmingham, UK
Watanabe, P.G., Dow Chemical, Midland, MI, USA
Watson, W.P., Shell Research, Sittingbourne, Kent, UK
Weaver, R.J., University of Sussex, Brighton, Sussex, UK
Weinmann, I., Schaper & Brummer, Salzgitter, Fed. Rep. Germany
Wells, P.G., Vanderbilt University, Nashville, TN, USA
Weston, J.B., Wellcome Foundation, Berkhamsted, Herts, UK
White, I., M.R.C. Labs, Carshalton, Surrey, UK

Wiebkin, P., University of Texas Health Science Center, Dallas,
   TX, USA
Witmer, C.M., Thomas Jefferson University, Philadelphia, PA, USA
Witts, D., University College, London, UK
Wolff, T., Ges. F. Strahlen u Umweltforschung, Abtlg. Toxikologie,
   München, Fed. Rep. Germany
Wright, A.F., I.C.I. Ltd., Alderley Park, Macclesfield, Cheshire, UK

NOTE: Part A consists of pages 1 to 846, Part B consists of pages
      847 to 1476.

INDEX